Topics in
Current Physics

HAROLD MAR

Topics in Current Physics Founded by Helmut K. V. Lotsch

Volume 39 **Nonequilibrium Vibrational Kinetics**
Editor: M. Capitelli

Volume 40 **Microscopic Methods in Metals**
Editor: U. Gonser

Volume 41 **Structure and Dynamics of Surfaces I**
Editors: W. Schommers and P. v. Blanckenhagen

Volumes 1–38 are listed on the back inside cover

Monte Carlo Methods

in Statistical Physics

Edited by K. Binder

With Contributions by
K. Binder D. M. Ceperley J.-P. Hansen
M. H. Kalos D. P. Landau D. Levesque
H. Müller-Krumbhaar D. Stauffer J.-J. Weis

Second Edition

With 97 Figures

Springer-Verlag
Berlin Heidelberg New York Tokyo

Professor Dr. Kurt Binder

Fachbereich Physik, Johannes-Gutenberg-Universität, Postfach 39 80,
D-6500 Mainz 1, Fed. Rep. of Germany

ISBN 3-540-16514-2 2. Auflage Springer-Verlag Berlin Heidelberg New York Tokyo
ISBN 0-387-16514-2 2nd edition Springer-Verlag New York Heidelberg Berlin Tokyo

ISBN 3-540-09018-5 1. Auflage Springer-Verlag Berlin Heidelberg New York
ISBN 0-387-09018-5 1st edition Springer-Verlag New York Heidelberg Berlin

Offset printing and bookbinding: Konrad Triltsch, Graphischer Betrieb, 8700 Würzburg.
2153/3150-543210

Preface to the Second Edition

In the seven years since this volume first appeared, there has been an enormous expansion of the range of problems to which Monte Carlo computer simulation methods have been applied. This fact has already led to the addition of a companion volume ("Applications of the Monte Carlo Method in Statistical Physics", Topics in Current Physics, Vol. 36), edited in 1984, to this book. But the field continues to develop further; rapid progress is being made with respect to the implementation of Monte Carlo algorithms, the construction of special-purpose computers dedicated to execute Monte Carlo programs, and new methods to analyze the "data" generated by these programs.

Brief descriptions of these and other developments, together with numerous additional references, are included in a new chapter, "Recent Trends in Monte Carlo Simulations", which has been written for this second edition. Typographical corrections have been made and fuller references given where appropriate, but otherwise the layout and contents of the other chapters are left unchanged. Thus this book, together with its companion volume mentioned above, gives a fairly complete and up-to-date review of the field. It is hoped that the reduced price of this paperback edition will make it accessible to a wide range of scientists and students in the fields to which it is relevant: theoretical physics and physical chemistry, condensed-matter physics and materials science, computational physics and applied mathematics, etc.

The editor is grateful to his colleagues for their valuable contributions to this book and he thanks them, as well as many other colleagues, for stimulating comments and advice.

Mainz, February 1986 *Kurt Binder*

Preface to the First Edition

The "Monte Carlo method" is a method of computer simulation of a system with many degrees of freedom, and thus has widespread applications in science. It has its name from the use of "random numbers" to simulate statistical fluctuations in order to numerically generate probability distributions (which otherwise may not be known explicitly since the considered systems are so complex). While the method would work in principle also with random numbers generated at a roulette table, an effective and economic use of this method requires the use of high-speed digital computers. Thus the first successful application of this method to a problem of statistical thermodynamics dates back only to 1953, when Metropolis and co-workers studied a "fluid" consisting of hard disks. Since then this technique has experienced an impetuous development which is likely to even speed up in the future, since better computers now available allow many fascinating applications.

What are then the specific advantages of Monte Carlo "computer experiments"? To answer that question, one first notes that Monte Carlo methods yield information on "model systems" (where specific assumption about the effective forces between the atoms have been made) which in principle is numerically exact, i.e., the results are accurate apart from statistical erros which can be made as small as desired if only enough computing time is invested. Thus the purpose of the Monte Carlo method is twofold: comparing the results with data from experiments on real systems, one checks the extent to which a model system approximates a real system; and comparing the results to analytic theories starting with the same model, one checks the validity of various approximations made in the analytic treatment. In direct comparisons of theory and experiment it is often hard to separate the influence of errors due to inappropriate models and errors due to inappropriate approximations. It is one of the main purposes of the present book to clearly bring out this interplay between computer simulation and experiment on the one side and theory on the other side, and the progress which thereby is obtained. One more advantage then is that one obtains a microscopic information on the system both in space and time; this information can be much more detailed than what is available from experiments on real systems, and hence it gives insight into some problems which cannot be obtained else. This insight may stimulate new theoretical descriptions as well as new experiments. For all these reasons, Monte Carlo computer experiments are a valuable tool for many branches of statistical physics, and since their application is relatively simple they will become a standard method of scientific research in the near future.

In the present book, first the theoretical background and the efficient practical implementation of this method are described, and then an up-to-date review of widespread applications is given: calculation of thermal properties and scattering functions for dense gases, fluids and plasmas; short- and long-range order properties of metallic alloys and various magnetic systems including the behavior near "critical points"; ground state properties of quantum fluids; thermal properties and structure of microscopically small liquid droplets and solid "clusters" of molecules; relaxation phenomena and diffusion in solids; kinetics of crystal growth and of other phase transformations; structural and magnetic properties of surfaces and adsorbed layers, and the kinetics of adsorption on surfaces; structural and thermal properties of disordered systems like glasses and amorphous magnets, etc. So far, these applications have been scattered throughout the literature, and it was necessary to bring together from several countries a team of leading experts in this field to accomplish such a review for the first time: D. Ceperley and M.H. Kalos from New York University, D.P. Landau from the University of Georgia, D. Levesque, J.P. Hansen, and J.J. Weis from Orsay, H. Müller-Krumbhaar from Jülich and D. Stauffer from Cologne. It is a pleasure to thank them for their fruitful collaboration.

Jülich, September 1978 *Kurt Binder*

Contents

List of Contributors

BINDER, KURT

Fachbereich Physik, Johannes-Gutenberg-Universität, Postfach 3980
D-6500 Mainz 1, Fed. Rep. of Germany

CEPERLEY, DAVID M.

c/o. Prof. G. Jacucci, Dipartimento di Fisica, Libera Universita di Trento
I-38050 Povo, Italy and National Resource for Computation in Chemistry,
Lawrence Livermore Laboratory, Berkeley, CA 94720, USA (permanent address)

HANSEN, JEAN-PIERRE

Laboratoire de Physique Théorique des Liquides, Université Pierre et Marie Curie,
75230 Paris Cédex 05, France

KALOS, MALVIN H.

Courant Institute of Mathematical Sciences, New York University,
251 Mercer Street, New York, N.Y. 10012, USA

LANDAU, DAVID P.

Department of Physics and Astronomy, The University of Georgia,
Athens, GA 30602, USA

LEVESQUE, DOMINIQUE

Laboratoire de Physique Théorique et Hautes Energies,
Université de Paris-Sud, 91405 Orsay, France

MÜLLER-KRUMBHAAR, HEINER

Institut für Festkörperforschung der Kernforschungsanlage Jülich GmbH
Postfach 1913, D-5170 Jülich 1, Fed. Rep. of Germany

STAUFFER, DIETRICH

Institut für Theoretische Physik, Universität Köln
D-5000 Köln 41, Fed. Rep. of Germany

WEIS, JEAN-JAQUES

Laboratoire de Physique Théorique et Hautes Energies,
Université de Paris-Sud, 94105 Orsay, France

1. Introduction: Theory and "Technical" Aspects of Monte Carlo Simulations

K. Binder

With 5 Figures

An outline is given of the physical problems which can be treated by Monte Carlo sampling and which are described in the later chapters of this book. Then the theoretical background is described for the application of this technique to calculate statistical ensemble averages of classical interacting many-body systems. The practical realization of the method is discussed, as well as its limitations due to finite time averaging, finite size and boundary effects, etc. It is shown how to extract meaningful information from the "raw data" of such a "computer experiment". The stochastic simulation of kinetic processes is also treated, with particular emphasis on the interpretation of the results near phase transitions in the system. Finally some approximative variants of the technique are discussed which might become useful to simulate critical phenomena.

1.1 Purpose of the Monte Carlo Method

The Monte Carlo method in statistical physics studies models of equilibrium and nonequilibrium thermodynamic systems by stochastic computer simulation [1.1]. Starting from a description of the desired physical system in terms of a model hamiltonian, one uses (pseudo-)random numbers to construct the appropriate probability with which the various generated states of the system have to be weighted. By appropriate probability one means for equilibrium systems the thermodynamic probability defined according to either the microcanonic, canonic, or grand-canonic ensemble. It has been realized [1.2-4] that even then, in its actual realization, the Monte Carlo method performs a "time" averaging of a model with (often artificial) stochastic kinetics rather than an ensemble averaging (i.e., time plays the role of a label characterizing the sequential order of the states, and need not be related to physical times). While time averages and ensemble averages agree for ergodic systems, this dynamic aspect nevertheless presents the theoretical basis for another purpose a Monte Carlo simulation may have, namely the simulation of relaxation and transport phenomena.

It should be obvious that the Monte Carlo method in statistical physics as defined above is a special case of more general Monte Carlo simulation in science: rather than thermodynamic averaging, one may wish to simulate any other averaging with a probability distribution one encounters in the study of a scientific problem. Also kinetic phenomena can be simulated, which are not related with the relaxation towards thermal equilibrium. Well-known examples are the simulation of the multiscattering processes of energetic elementary particles in matter, or the transmission of cold neutrons through total-reflecting beam tubes, etc. Such types of Monte Carlo simulations are outside consideration here, since several reviews exist [1.5-7]. Also, we do not consider applications where the random numbers are used to model some physical randomness in a system which is then treated by other numerical techniques (for instance, one may simulate the disorder in the exchange constants of a ferromagnetic random alloy and then obtain the spin wave spectrum by diagonalization of the resulting random matrix [1.8]).

The Monte Carlo method in statistical physics is less straightforward and thus requires a more detailed theoretical analysis, which will be attempted in this book. There are several sources of difficulties: (I) Usually one wants to study a macroscopic system, with $N \approx 10^{22}$ atoms, starting from a model hamiltonian describing interactions on molecular scales. Since systems with $N \leq 10^3$-10^6 (the precise value depends on the type of problem) are accessible to the simulation, a nontrivial extrapolation to much larger systems has to be performed. (II) Averages over the probability distribution describing thermal equilibrium are replaced by sums over a finite subset of points which approximate the answers to a degree that can only be gauged during "the course of a Monte Carlo simulation". The judgement required to do this is not at all trivial. (III) It is an intrinsic feature of many of the kinetic models simulated that they contain relaxational modes with very low frequencies. Such slow relaxation occurs in all models which exhibit conservation laws ("hydrodynamic slowing down" [1.9]), further at all phase transitions of second order ("critical slowing down" [1.10]) or first order ("metastable states" [1.11]), and even more generally at temperatures low in comparison to some energy barriers in the system, such that freeze-in phenomena occur (as in glasses, quenched alloys, etc.). Then it is very hard to reach thermal equilibrium.

The following section of this introductory chapter will deal with these difficulties in more detail. We shall discuss the question of how large the size of the system to be simulated must be chosen for the problems considered, and in which way an extrapolation to the thermodynamic limit can be performed. We shall discuss the question of how one estimates the statistical accuracy of the averages obtained from the simulation, making use of the dynamic interpretation of the Monte Carlo process. We shall discuss various choices of boundary conditions and point out in which way they correspond to situations of physical interest. We shall discuss various choices of transition probabilities and point out in which way they correspond to dynamic models of physical systems. Finally, we shall discuss variants of the Monte Carlo method which have recently been suggested for the treatment of critical phenomena. Our treatment will refer to classical statistical mechanics throughout; the particular features of Monte Carlo applications to quantum statistics can be found in Chap.4. It is hoped, however, that the present introduction provides sufficient background on the technical aspects of the Monte Carlo simulations described in the remaining chapters of this book. In these chapters, a representative survey will be given on applications of the Monte Carlo method in wide-ranging areas of statistical physics. Those aspects of a problem particularly suitable for Monte Carlo study, and how progress in our physical understanding is reached by an interplay of computer simulation, experiment, and theory will be emphasized. The applications discussed concentrate on liquids (Chap.2), calculations of phase diagrams (Chap.3), kinetics of phase changes and other relaxation phenomena (Chaps.6,7), and disordered solids, particularly magnets (Chap.8). Furthermore, the applications in which some of the restrictions of the Monte Carlo method mentioned above are turned into an advantage are stressed: studies of finite size effects for small particles, "clusters", etc. (Chap.5) and studies of surface phenomena (Chap.9). We are well aware, of course, that the selection of the applications reviewed in this book is fairly subjective, and some interesting topics (such as the study of excluded volume problems for polymer chains [1.12], or the simulation of solid crystals and their thermal and elastic properties [1.13], etc.) have been omitted. It is hoped, however, that the references contained in the chapters are fairly complete for the topics treated, so that this book should provide a useful guideline to the recent relevant literature.

1.2 Description of the Monte Carlo Technique in Classical Statistical Mechanics

1.2.1 Computation of Static Averages in the Canonic Ensemble

We treat a system of N particles in a volume V at specified temperature T. Additional thermodynamic coordinates like electric field $\underline{E}$ or magnetic field $\underline{H}$, etc., may have to be included if the particles possess an electric or magnetic moment.[1] We emphasize the point that we have at least three independent thermodynamic variables (N,V,T) rather than only two (v = V/N, T), since we are dealing with a finite system [1.14]. Only if we treat lattice models where the particles are restricted to be on

---

[1] In some cases it may be advantageous to include "fields" which have no simple realization in the laboratory, like the staggered field H_s which is useful for the study of antiferromagnets, or a surface field H_1 which is useful for surface magnetism, etc.

the sites of a (typically regular) lattice (as in the case of the Ising- Heisenberg model of magnetism), the variable V can be omitted.

We describe now each particle (labelled by index i) by a set of dynamical variables $\{\alpha_i\}$. In a monatomic liquid, $\{\alpha_i\}$ will be the position vector $\underline{r}_i$ of a molecule. In a liquid containing rigid diatomic molecules, $\{\alpha_i\}$ will contain both $\underline{r}_i$ and the angle Ω_i describing the molecular orientation. In the rigid lattice model of a solid alloy consisting of ν components $\{K\}$, the set $\{\alpha_i\}$ becomes the set of local concentration variables $\{C_i^k = 0,1\}$. In the Heisenberg model mentioned above, $\{\alpha_i\}$ is the unit vector $\underline{S}_i$ in the direction of the magnetic moment. The dynamic variables are thus chosen ad hoc in terms of the model description one wishes to use for the system desired. The set $\{\{\alpha_1\},\{\alpha_2\}, \ldots,\{\alpha_N\}\}$ then describes a configuration or phase space point $\underline{x}$ of the model system. The statistical mechanics of the system is nontrivial due to the presence of interactions between the particles, which are described in terms of a suitable hamiltonian $\mathcal{H}_N(\underline{x})$, omitting kinetic energy terms. Thermodynamic averages of any observables $A(\underline{x})$ are then given by

$$\langle A\rangle = \frac{\int_\Omega d\underline{x} A(\underline{x}) \exp[-\mathcal{H}_N(\underline{x})/k_B T]}{\int_\Omega d\underline{x} \exp[-\mathcal{H}_N(\underline{x})/k_B T]} \quad . \tag{1.1}$$

In the case where the α_i can take on discrete values only, for instance for the above rigid alloy model, the integrals over the phase-space volume Ω in (1.1) stand symbolically for the appropriate sums.

The basic idea of the Monte Carlo method is now to calculate the phase space integrals in (1.1) numerically. Standard numerical integration routines could be used, in principle, where an integral $\int f(\underline{x})d\underline{x}$ is approximated by a sum with a finite number of terms $\sum_\nu f(\underline{x}_\nu)\Delta\underline{x}_\nu$. It is obvious, however, that the usual scheme where the set of points $\{\underline{x}_\nu\}$ forms a regular lattice cannot be used if the integration space Ω is high dimensional. Suppose the integration space is an N'-dimensional hypercube $\Omega = L^{N'}$ and $\Delta\underline{x}_\nu = (\Delta L)^{N'}$; then we would have a set of $M = n^{N'}$ points with $n \equiv L/\Delta L > 2$, where $(n-2)^{N'}$ points are in the interior of the integration hypercube. Since N' is of the order of N in (1.1), the number of points M needed would be prohibitively large for all cases of interest. Thus one has to resort to the well-known fact that one has to use a set of points $\{\underline{x}_\nu\}$ chosen randomly in Ω, if one wants to do numerical integrations in a high-dimensional space. Such a calculation with simple random sampling of the points is already a Monte Carlo method in a more general sense [1.5-7].

Unfortunately, this simple random sampling method is not of great use for problems in statistical mechanics. The reason is that the integrand $\exp[-\mathcal{H}_N(\underline{x})/k_B T]$ will have a variation over many orders of magnitude at most of the interesting temperatures, for which $\langle\mathcal{H}_N\rangle \gtrapprox Nk_BT$. The Monte Carlo method introduced by METROPOLIS et al. [1.1] is a sampling algorithm based on the idea of "importance sampling" [1.5-7]. Here the M phase space points $\{\underline{x}_\nu\}$ are not chosen completely at random but are selected according to a probability $P(\underline{x}_\nu)$. Then (1.1) is approximated by [1.15]

$$\langle A\rangle \approx \bar{A} = \frac{\sum_{\nu=1} A(\underline{x}_\nu)P^{-1}(\underline{x}_\nu)\exp[-\mathcal{H}_N(\underline{x}_\nu)/k_BT]}{\sum_{\nu=1}^{M} P^{-1}(\underline{x}_\nu)\exp[-\mathcal{H}_N(\underline{x}_\nu)/k_BT]} \quad . \tag{1.2}$$

The simplest and most natural possibility is choosing $P(\underline{x}_\gamma) = P_{eq}(\underline{x}_\gamma) \propto \exp[-\mathcal{H}_N(\underline{x}_\gamma)/k_BT]$, for which (1.2) reduces to a simple arithmetic average

$$\bar{A} = (1/M)\sum_{\nu=1}^{M} A(\underline{x}_\nu) \quad . \tag{1.3}$$

Since $P_{eq}(\underline{x}_\nu)$ is not known explicitly in the cases of interest here, the realization of (1.3) is not completely straightforward, of course. It is possible to construct a random walk of points $\{\underline{x}_\gamma\}$ through the space via a Markov process, such that $P(\{\underline{x}_\nu\})$ tends towards $P_{eq}(\{\underline{x}_\nu\})$ as $M \to \infty$. This Markov process is defined by specifying a transition probability $W(\underline{x}_\nu \to \underline{x}_{\nu'})$ from one phase space point $\underline{x}_\nu$ to another point $\underline{x}_{\nu'}$. In order that the Markov process has the desired property that $P(\{\underline{x}_\nu\})$ converges towards $P_{eq}(\{x_\gamma\})$, it is sufficient (but not necessary [1.15-17])[2] to impose the detailed balance condition

$$P_{eq}(\underline{x}_\nu)W(\underline{x}_\nu \to \underline{x}_{\nu'}) = P_{eq}(\underline{x}_{\nu'})W(\underline{x}_{\nu'} \to \underline{x}_\nu) \quad , \tag{1.4}$$

which means that the ratio of transition probabilities depends on the change in energy $\delta\mathcal{H} = \mathcal{H}_N(\underline{x}_{\nu'}) - \mathcal{H}_N(\underline{x}_\nu)$ only

$$\frac{W(\underline{x}_\nu \to \underline{x}_{\nu'})}{W(\underline{x}_{\nu'} \to \underline{x}_\nu)} = \exp[-\delta\mathcal{H}/k_BT] \quad . \tag{1.5}$$

Equation (1.5) does obviously not specify $W(\underline{x}_\nu \to \underline{x}_{\nu'})$ uniquely. The choices of W commonly used are [1.2]

---

[2]More general conditions on the transition probabilities are formulated in [1.4, 15-17]. But choices of W which do not satisfy (1.4,5) have been of no practical importance so far.

$$W(\underline{x}_\nu \to \underline{x}_{\nu'}) = \frac{1}{2\tau_s}\,[1-\tanh(\delta\mathcal{H}/2k_BT)] = \frac{1}{\tau_s}\,\frac{\exp(-\delta\mathcal{H}/k_BT)}{1+\exp(-\delta\mathcal{H}/k_BT)} \tag{1.6a}$$

or

$$W(\underline{x}_\nu \to \underline{x}_{\nu'}) = \begin{cases} \dfrac{1}{\tau_s}\exp(-\delta\mathcal{H}/k_BT) & \text{if } \delta\mathcal{H} > 0 \\[2ex] \dfrac{1}{\tau_s} & \text{otherwise,} \end{cases} \tag{1.6b}$$

where τ_s is some arbitrary factor which does not affect the detailed balance, and hence usually one takes $\tau_s = 1$. The fact that (1.4-6) lead to the desired convergence property can be proven formally [1.5-8,6] making use of the central limit theorem of probability theory [1.18].

We prefer to reproduce here a less rigorous but physically more intuitive argument [1.1], where one considers a large number of Markov processes together. Suppose that N_r systems are in state $\underline{x}_r$ and N_s systems in state $\underline{x}_s$ at the ν'th step of the chains, with $\mathcal{H}_N(\underline{x}_r) < \mathcal{H}_N(\underline{x}_s)$. Using random numbers, one may construct moves $\underline{x}_r \to \underline{x}_s$, the "a priori probability" of which is $W_{rs} = W_{sr}$ [i.e., the probability without condition (1.4)]. But transition probabilities which are in accord with (1.4,5) can now easily be constructed by taking

$$W(\underline{x}_r \to \underline{x}_s) = W_{rs}\exp(-\delta\mathcal{H}/k_BT) = W_{rs}\exp\{-[\mathcal{H}_N(\underline{x}_s) - \mathcal{H}_N(\underline{x}_r)]/k_BT\} \tag{1.7a}$$

and

$$W(\underline{x}_s \to \underline{x}_r) = W_{sr} = W_{rs} \quad . \tag{1.7b}$$

Thus we find for the total number $N_{r\to s}$ of transitions from $\underline{x}_r$ to $\underline{x}_s$,

$$N_{r\to s} = N_r W(\underline{x}_r \to \underline{x}_s) = N_r W_{rs}\exp\{-[\mathcal{H}_N(\underline{x}_r) - \mathcal{H}_N(\underline{x}_s)]/k_BT\} \quad , \tag{1.8a}$$

and for the reverse process

$$N_{s\to r} = N_s W(\underline{x}_s \to \underline{x}_r) = W_{rs}N_s \tag{1.8b}$$

Then the *net* number of transitions $\Delta N_{r\to s} = N_{r\to s} - N_{s\to r}$ is

$$\Delta N_{r\to s} = N_r W_{rs}\left\{\frac{\exp[-\mathcal{H}_N(\underline{x}_s)/k_BT]}{\exp[-\mathcal{H}_N(\underline{x}_r)/k_BT]} - \frac{N_s}{N_r}\right\} \quad . \tag{1.9}$$

As long as N_s/N_r is smaller than the canonic value we have $\Delta N_{r\to s} > 0$ and hence N_s/N_r increases, while $\Delta N_{r,s} < 0$ and hence N_s/N_r decreases if N_s/N_r is too large. Thus asymptotically for $\nu\to\infty$ a steady state must be reached where N_s/N_r has precisely the canonic value. Instead of considering many Markov chains together, we may also consider the "pieces" of one very long chain.

The above method of Monte Carlo sampling in classical statistical mechanics straightforwardly applies to the quantum-mechanical case as well, if eigenstates $|\underline{x}\rangle$ and eigenvalues $\mathcal{H}_N(\underline{x})$ of the hamiltonian $\hat{\mathcal{H}}_N(\underline{x})$ are known $\{\hat{\mathcal{H}}_N(\underline{x})|\underline{x}\rangle = \mathcal{H}_N(\underline{x})|\underline{x}\rangle\}$. An important example is the Ising model of magnetism [1.19]

$$\mathcal{H}_N = -\sum_{i\neq j} J_{ij}\hat{S}_i^z\,\hat{S}_j^z - g\mu_BH\sum_{i=1}\hat{S}_i^z \quad , \tag{1.10}$$

where only the z components of spin operators $\hat{\underline{S}}_i$, $\hat{\underline{S}}_j$ are coupled together by exchange constants J_{ij}, and also the magnetic field $\underline{H}$ which couples to the magnetic moment $g\mu_B\underline{S}_i$ of the spin is applied in the z direction.

1.2.2 Estimation of the Free Energy. Practical Realization. Other Ensembles

In the steps from (1.1) to (1.3) we can still obtain meaningful estimates of any quantities which are thermal averages of observables A, for example, the internal energy $U_N = \langle\mathcal{H}_N\rangle$ itself, but we have lost the information on the denominator in (1.1). Thus one does not have knowledge on the partition function Z

$$Z_N = \left(\frac{\hbar^2}{2\pi m k_BT}\right)^{-3N}\frac{1}{N!}\int_\Omega d\underline{x}\,\exp[-\mathcal{H}_N(\underline{x})/k_BT] \equiv \left(\frac{\hbar^2}{2\pi m k_BT}\right)^{-3N}\frac{Z_N'}{N!} \quad , \tag{1.11}$$

where we have assumed a free energy contribution appropriate for N (undistinguishable) particles of mass m ($\hbar$ is Planck's constant). Thus one does not know the free energy F_N and entropy S_N also

$$F_N = -k_BT\ln(Z_N), \quad S_N = (U_N - F_N)/T \quad . \tag{1.12}$$

The problem of estimating the free energy by the METROPOLIS et al. [1.1] Monte Carlo method has received considerable attention in the literature [1.20-24]. One reason for this interest comes from systems which may occur in several phases under different thermodynamic conditions. A method to locate where a phase transition occurs would be to compare the free energies of the various phases.

From (1.1,11) it is recognized that also Z can be written as an ensemble average, namely

$$Z_N = \left(\frac{\hbar^2}{2\pi m k_B T}\right)^{-3N} \frac{1}{N!} \frac{\Omega}{\langle \exp[\mathscr{H}_N(\underline{x})/k_B T] \rangle} \quad . \tag{1.13}$$

For a monatomic liquid, for instance, we have $\int_\Omega d\underline{x} \ldots = \int_V d\underline{r}_1 \int_V d\underline{r}_2 \ldots \int_V d\underline{r}_N \ldots$ and hence $\Omega = V^N$. It has been suggested [1.20] to compute Z_N and hence F_N, S_N from (1.13)by estimating $\langle \exp[\mathscr{H}(\underline{x}_\nu)/k_B T] \rangle \approx \frac{1}{M}\sum_{\nu=1} \exp[\mathscr{H}(\underline{x}_\nu)/k_B T]$, i.e., applying (1.3) and the usual importance sampling of METROPOLIS et al. [1.1]. This method is impractical, however, for the same reason that the simple random sampling does not work well:the integrand $\exp[\mathscr{H}(\underline{x})/k_B T]$ has a variation over too many orders of magnitude. Note also that the variance $\langle A^2 \rangle - \langle A \rangle^2$ of $A(\underline{x}) \equiv \exp[\mathscr{H}_N(\underline{x})/k_B T]$ diverges for a continuous system with repulsive forces; the usual central limit theorem of statistics hence does not hold, and the convergence is very slow. The method which is, hence, most commonly applied determines the free energy integrating suitable derivatives, for instance [1.25],

$$\frac{1}{k_B T} F_N - \lim_{T\to\infty} \frac{F_N}{T} = \int_0^{1/k_B T} U_N(T') d \frac{1}{k_B T'} \quad . \tag{1.14}$$

This method is also somewhat impractical, since one has to determine U_N for a broad range of temperatures T' between the considered T and $T = \infty$. Since one often wants to study a model over a broad range of temperatures anyhow, this disadvantage does not matter too much: this type of method was applied successfully in fact both in the case of Ising magnets [1.25], where reliable analytic results compared favorably to the Monte Carlo estimates of F_N, and for various liquid models [1.22,26].

A more sophisticated method was recently investigated by VALLEAU and co-workers [1.22-24] by using the full information on the distribution $p(\mathscr{H})d\mathscr{H} \equiv \exp[-\mathscr{H}(\underline{x})/k_B T] d\underline{x}/Z'_N$ rather than only its first moment $U_N \equiv \int_0^\infty \mathscr{H} p(\mathscr{H}) d\mathscr{H}$ as done in (1.14). These authors suggested studying the function $\tilde{p}(\mathscr{H}) \equiv Z'_N p(\mathscr{H}) \exp[\mathscr{H}(\underline{x})/k_B T] = d\underline{x}/d\mathscr{H}$. Obviously one then finds Z'_N[(1.11)] and hence Z_N from the relation $Z'_N = \int d\mathscr{H} \tilde{p}(\mathscr{H}) \exp(-\mathscr{H}/k_B T]$, while $\int d\mathscr{H} \tilde{p}(\mathscr{H}) = \Omega$. The METROPOLIS et al. [1.1] importance sampling provides us with samples drawn from the *normalized* distribution $\tilde{p}_n(\mathscr{H}) \equiv \tilde{p}(\mathscr{H})/\int \tilde{p}(\mathscr{H}) d\mathscr{H}$ only, however [from the normalized distribution $p_n(\mathscr{H}) \equiv p(\mathscr{H})/\int p(\mathscr{H}) d\mathscr{H}$ which is immediately recorded from the sampling, we find $\tilde{p}_n(\mathscr{H}) = p_n(\mathscr{H}) \exp(\mathscr{H}/k_B T/\int p_N(\mathscr{H}) \exp(\mathscr{H}/k_B T) d\mathscr{H}]$. In order to find the factor of proportionality between $\tilde{p}(\mathscr{H})$ and

$\tilde{p}_n(\mathscr{H})$, one has to generate M phase-space points $\underline{x}_\nu$ by simple random sampling, which yields $\tilde{p}(\mathscr{H}) = \Omega n(\mathscr{H})/M$, $n(\mathscr{H})$ being the number of observations in the interval $[\mathscr{H}, \mathscr{H} + \Delta\mathscr{H}]$ with $\sum n(\mathscr{H}) = M$. Of course, by this run, $\tilde{p}(\mathscr{H})$ is found with sufficient accuracy only in a restricted interval of values for $\mathscr{H}$, which is not the same interval over which $\tilde{p}_n(\mathscr{H})$ is known accurately. But it may happen that these intervals have a regime of sufficient overlap. Then one may estimate there the constant of proportionality and know $\tilde{p}(\mathscr{H})$ also in that regime, where one cannot get enough statistics with simple random sampling. In case there is not sufficient overlap between $p_n(\mathscr{H})$ and $p(\mathscr{H})$, one may use (1.2) to generate distributions "intermediate" between the simple random sampling $[P(\underline{x}_\nu) = \text{const}]$ and the METROPOLIS et al. sampling $\{P(\underline{x}_\nu) \propto \exp[-\mathscr{H}_N(\underline{x}_\nu)/k_BT]\}$, which have overlap with $p_n(\mathscr{H})$ and $p(\mathscr{H})$. Since it is well known that $p(\mathscr{H})$ becomes extremely sharply peaked as $N\to\infty$, the interval over which $p_n(\mathscr{H})$, and hence $\tilde{p}_n(\mathscr{H})$, is known accurately becomes narrower as N is larger. Then it may happen that several intermediate distributions are required. Here we do not discuss the best possible choices of the weighting functions $P(\underline{x}_\nu)$ of these intermediate distributions in this so-called "multistage" [1.22] or the related "umbrella" [1.23] sampling, however. A quantitative accuracy analysis of this method is obviously rather difficult, and remains to be given. Successful applications are available for extremely small N only (typically $N \approx 32$ [1.22,23]). For a useful application of the free energy estimates to locate a first-order phase transition, a rather high accuracy is required, however, often the changes in slope which one has to look for are quite small. The ordinary scheme [1.1], on the other hand, yields directly rather precise estimates for the derivatives of the free energy. At first-order phase transitions these derivatives have a jump. Due to hysteresis effects it may again be difficult to precisely locate where the jump occurs, however (see Sect.1.3.3). Practical

Next we consider the *question of the way in which the METROPOLIS et al. [1.1] scheme is practically realized*. Application One starts by specifying an initial condition for the set $\{\alpha_i\}$ of dynamic variables. While in principle this choice is completely arbitrary, accuracy considerations may well favor one choice over another (Sect. 1.2.4). Then one selects one particle (i) for which one (or more) dynamical variable(s) will be changed randomly from α_i to $\alpha_{i'}$. The index i is selected by going through the array of particle labels either regularly or randomly. The move $\alpha_i \to \alpha_{i'}$ is the transition denoted formally by $\underline{x}_\gamma \to \underline{x}_{\gamma'}$ in (1.4-8). Next one computes the change in energy $\delta\mathscr{H}$ produced by the trial move, and computes the transition probability W (1.6). Then one selects a random number z in the interval $0 < z < 1$. If $\tau_s W < z$, the move is rejected, and the state with the old configuration (α_i) is counted once more as a "new" state in the averaging, (1.3). If $\tau_s W > z$, the move is accepted, i.e., the state with the new coordinate $(\alpha_{i'})$ but otherwise unchanged is taken as a new configuration in the averaging, (1.3). This procedure is repeated very often, and thus a Markov chain of M events is generated.

Since subsequent points $\underline{x}_\nu$ in the chain may differ by one coordinate only, they are highly correlated. Rather than using (1.3) for estimates $\bar{A}$ where the calculation of $A(\underline{x})$ is time consuming, one thus uses

$$\bar{A} = \frac{1}{m} \sum_{\mu=\mu_0}^{m=\mu_0} A(\underline{x}_\nu), \quad \nu = \mu n, \qquad n = M/(m + \mu_0) \quad , \tag{1.15}$$

where m is chosen such that n is an (integer) number at least as large as N. The number μ_0 is chosen such that the $A(\underline{x}_\nu)$ for $\nu > \mu_0 n$ differ from $\bar{A}$ by no more than expected statistical deviations. Since these "expected statistical deviations" depend on averages like $\langle A^2 \rangle - \langle A \rangle^2$ which are not known beforehand but have to be estimated from the same calculation, the choice of μ_0 is of course somewhat ad hoc and it is often recommended to choose it somewhat larger than necessary, to eliminate the influence of the initial condition as much as possible (see also Sect.1.2.4). – One also wants to estimate the "statistical error" δA of the estimate $\bar{A}$ in (1.15). The standard answer is [1.16]

$$(\delta A)^2 = \frac{1}{m(m-1)} \sum_{\mu=\mu_0}^{m+\mu_0} [A(\underline{x}_\nu) - \bar{A}]^2 \quad , \qquad m \gg 1 \quad , \tag{1.16}$$

i.e., the error becomes smaller the larger m is chosen. On the other hand, (1.16) *applies only* if the subsequent $A(\underline{x}_\nu)$ used in (1.15,16) are *not correlated*. Since each configuration of the Markov chain is generated from the previous one by only a small change ($\alpha_i \to \alpha_{i'}$), such correlations between subsequent $A(\underline{x}_\nu)$ are expected to occur except, however, if one chooses n large enough, and then the largest possible value for m follows from M via (1.15). The nature of these correlations, and the corresponding knowledge about the choice of n, will be discussed in Section 1.2.3 and 1.2.4. Here we only mention that an empirical way to check the appropriate choice of n is to compute the quantity $(\delta A)^2 m$ for various choices of n: if the $A(\underline{x}_\nu)$ are already uncorrelated for these choices, then $(\delta A)^2 m$ should be independent of m and hence n, while in the presence of correlations there is some systematic trend.

It often happens that for many transitions $\underline{x}_\nu \to \underline{x}_{\nu'}$ one would have energy changes $\delta\mathcal{H} \gg k_B T$. Then the probability $w \equiv \tau_s W$ that the transition is accepted would be very small, (1.6), and hence the convergence would be very slow, since the system moves through phase space very slowly. For classical systems where the variables $\{\alpha_i\}$ have a continuous variation, one can remove this difficulty by introducing a parameter Δ which limits the allowed changes: $|\alpha_i - \alpha_{i'}| < \Delta$. For instance, α_i may be a spatial coordinate x_i: then one may choose $\alpha_{i'} = \alpha_i + z\Delta$, with z a random number in the interval $-1 < z < +1$.) The magnitude of Δ is chosen such that the average acceptance rate $\langle w \rangle$ is close to 1/2.

No such option exists for systems with discrete variables (like Ising models). But there it is often convenient to use a somewhat different scheme [1.30,31]. Suppose that a particle (i) is in state ν and may undergo a discrete number (ℓ) of transitions to (other) states ν' with rates $w_{\nu\nu'}(i)$, its total transition rate being $w_t = \sum_{\nu(\neq\nu')} w_{\nu\nu'}(i)$. Then only a finite set of possible numbers $\{w_t^{(k)}\}$, $k = 1, \ldots, n$ exists, and we may group all particles into one of the n classes $\{w_t^{(k)}\}$. If we now select a particle (i) not at random but according to the probability $w_t^{(k)}$, then correct weighting is obtained if this particle is moved to some new state (ν'). The appropriate new state is selected with probability $w_{\nu\nu'}^{(i)}/w_t^{(k)}$. This can be done by computing the set $\{p_\mu, \mu = 1, \ldots\}$ where $p_\mu = \sum_{\nu'=1} w_{\nu\nu'}^{(i)}/w_t^{(k)}$ and choosing a random number z with $0 < z < 1$, ν' ($\neq \nu$) is then determined from the condition $p_{\nu'} < z < p_{\nu'+1}$. At each move the class division must be updated, of course; thus the programming of this scheme is rather difficult and has not been applied very often. More practical variants of this scheme will be discussed in connection with time-dependent properties in Sect.1.3.1.

We conclude this subsection by a discussion of how one realizes statistical ensembles other than the canonic one in a Monte Carlo calculation. In the microcanonic ensemble, the energy $\mathscr{H}$ is held fixed and thus the internal energy U_N rather than the temperature T is an independent variable (and hence an input of the calculation). It is clear that this ensemble can be realized by considering special changes $\underline{x}_\nu \to \underline{x}_{\nu'}$ which leave the energy unchanged. Since, then, $\delta\mathscr{H} = 0$ by definition, detailed balance (1.4) is obeyed for any symmetric transition probability $w(\underline{x}_\nu \to \underline{x}_{\nu'}) = w(\underline{x}_{\nu'} \to \underline{x}_\nu)$. While such a calculation may be interesting for obtaining dynamic properties of a system with conserved energy [1.27], it is clearly disadvantageous for obtaining static properties: I) While the probability of occurence of states $\underline{x}_\nu$ in the Markov chain is proportional to the thermodynamic probability $\rho(\underline{x}_\nu)$, the constant of proportionality is again unknown. Thus it is not possible to calculate the entropy $S_N = -k_B \int \rho \ln[\rho(2\pi\hbar)^{3N}]d\underline{x}$ and hence the temperature $1/T = (\partial S_N/\partial U_N)v$ from runs at closely neighboring values of U_N. II) The "conservation law" for the energy leads to an approach to thermal equilibrium which is much slower than in the canonic case (see Sect.1.3.2).

More interesting are calculations where neither energy nor density $\rho = N/V$ are held fixed, namely the grand canonical ensemble (μVT), where the chemical potential rather than particle number N is kept fixed, or the isothermal-isobaric ensemble (NpT), where the pressure p rather than the volume V is kept fixed. The grand-canonical ensemble is trivially applicable to lattice gases [1.28]: a lattice gas with lattice cells of volume v being either empty ($\rho_i = 0$) or full is exactly equivalent to an Ising magnet with $N = V/v$ spins $\sigma_i = +1$ (if $\rho_i = 0$) or $\sigma_i = -1$ (if $\rho_i = 1$) in a magnetic field $\mu_B H \equiv \mu$, as is well known [1.24]. A grand-canonical calculation of a lattice gas is thus a reformulation of a canonic calculation of an Ising magnet only. (For a discussion of nontrivial μVT Monte Carlo calculations see Chap.2.) We now proceed to the case of pVT ensembles. The basic relation is

$$\langle A\rangle = \frac{\int_0^\infty dV \exp[-(pV/k_BT)] Z_N(V,T) A}{\int_0^\infty dV \exp[-(pV/k_BT)] Z_N(V,T)} \tag{1.17a}$$

or, equivalently

$$\langle A\rangle = \frac{\int_0^\infty dV \int d\underline{x} \exp\left\{-[pV + \mathscr{H}_N(\underline{x})/k_BT]\right\} A}{\int_0^\infty dV \int d\underline{x} \exp\left\{-[pV + \mathscr{H}_N(\underline{x})/k_BT]\right\}} \tag{1.17b}$$

One readily notes that the average in (1.17b) has a structure similar to that in (1.1), the only difference being that (1.17b) contains an integration space with an additional variable (V) and a modified weighting factor. It is clear that a Monte Carlo estimate for (1.17b) is obtained in close analogy to that for (1.1), i.e., (1.14) remains valid if we understand that $\underline{x}_\nu$ comprises the extra variable V and (1.5) has to be replaced by

$$\frac{W(\underline{x}_\nu \rightarrow \underline{x}_{\nu'})}{W(\underline{x}_{\nu'} \rightarrow \underline{x}_\nu)} = \exp[-(\delta\mathscr{H} + p\delta V)/k_BT] \quad . \tag{1.18}$$

In (1.18), δV denotes the change of volume involved in the transition $\underline{x}_\nu \rightarrow \underline{x}_{\nu'}$. The practical realization of NpT calculations was described by WOOD [1.16] at great length and thus need not be discussed here.

1.2.3 Dynamic Interpretation of the Monte Carlo Process

It was already mentioned in the last subsection that one expects *correlations* between the configurations generated subsequently one after the other in the Markov chain. These correlations strongly influence the accuracy which can be reached with a given number of total steps, of course. The correlation with the initial state determines the number of steps $M_0 (= \mu_0 n)$ which must be excluded from the averaging [cf. (1.15)]. By a dynamic interpretation of the Monte Carlo process [1.2-4], these correlations can be understood in terms of dynamic correlation functions of a well-defined model with stochastic kinetics. Apart from its use for the discussion of accuracy questions (Sect.1.2.4), this interpretation also provides the theoretical background for the simulation of kinetic processes (Sect.1.3).

This dynamic interpretation is in fact very simple. Obviously it is natural to associate a scale of "time" t with the scale ν of subsequent configurations. We may normalize the time such that $N\tau_s^{-1}$ single particle transitions are performed within *unit time* [τ_s^{-1} being the prefactor of the transition probability W, (1.6)]. If we write $P(\underline{x}_\nu) \equiv P(\underline{x},t)$, the dynamic evolution of $P(\underline{x},t)$ is governed by the *master equation* [1.2-4,32]

$$\frac{dP(\underline{x},t)}{dt} = -\sum_{\underline{x}'} W(\underline{x}\rightarrow\underline{x}')P(\underline{x},t) + \sum_{\underline{x}} W(\underline{x}'\rightarrow\underline{x})P(\underline{x}',t) \quad , \tag{1.19}$$

where $W(\underline{x} \rightarrow \underline{x}')$ [as given by (1.6), for instance] has now the meaning of a transition probability per unit time. Equation (1.19) describes the balance which was already considered in (1.7,8) in terms of a rate equation: the first sum describes the rate of all processes where one jumps out of the considered state (and hence decreases its probability), while the second sum describes the rate of all processes where one jumps into the considered state (and hence increases its probability). In thermal equilibrium $\{P(\underline{x},t) = P_{eq}(\underline{x})\}$ these two sums exactly cancel due to the detailed balance condition (1.4): $P_{eq}(\underline{x})$ is the steady-state distribution of the above master equation. From (1.4) it also follows that $\lim P(\underline{x},t) = P_{eq}(\underline{x})$, i.e., the kinetic model described by (1.19) is ergodic.

In (1.19) we have replaced $\Delta P(\underline{x},t)/\Delta t$ by $dP(\underline{x},t)/dt$. Since the system described by (1.19) relaxes on a time scale τ_s while $\Delta t \equiv \tau_s/N$ (for single particle transitions), this approximation becomes exact in the thermodynamic limit and produces no serious errors for the values of N which are of interest. Time-dependent averages $\langle A(t)\rangle$ are then defined by

$$\langle A(t)\rangle = \sum_{\{\underline{x}\}} P(\underline{x},t)A(\underline{x}) \quad , \tag{1.20a}$$

where we assume that A depends on time only implicitly (i.e., through the time dependence of the probability of the states). It can be shown that (1.20), where the observable $A(\underline{x})$ is considered fixed but the probability develops with time, is precisely equivalent to an averaging over the initial state $P(\underline{x},t_0)$ while the state $\underline{x}(t)$ and hence the observable A develops according to the Markov process described [1.33]

$$\langle A(t)\rangle = \sum_{\{\underline{x}\}} P(\underline{x},t_0)A[\underline{x}(t)] \quad . \tag{1.20b}$$

Equation (1.20b) is readily generalized to time-dependent correlation functions

$$\langle A(t)B(t_0)\rangle = \sum_{\{\underline{x}\}} P(\underline{x},t_0)A[\underline{x}(t)]B[\underline{x}(t_0)] \quad . \tag{1.21}$$

A well-known example of stochastic models of this type which has found widespread attention is the kinetic Ising model [1.33,34] defined by (1.6a,10,19). The Ising hamiltonian, (1.10), by itself does not lead to any (quantum mechanical) time evolution of any function $f(\{S_k^z\})$ depending on the set of spin variables only: $\hbar(d/dt)f(\{S_k^z\}) = i[f(\{S_k^z\}),\mathscr{H}_N^{\text{Ising}}] \equiv 0$. The kinetics provided by (1.6a,19) can be interpreted in terms of a (weak) coupling of the spins to a heat bath (the thermal vibrations of an underlying crystal lattice, for instance), which induces random spin flips in the system.

It must be emphasized that most other systems (classical Heisenberg magnets, fluids, etc.) do have a time evolution in terms of deterministic kinetic equations for their variables, e.g., Newton's law describes the motion of molecules in a fluid. Then the "time" evolution of the system according to the Monte Carlo process is not consistent with its actual physical evolution, even if one starts with the same initial state. Although both the (artificial) stochastic dynamics of (1.6,19) and the physical dynamics lead (for $N\to\infty$) to the same thermal equilibrium distribution, there is in general little similarity between this stochastic dynamics and the actual dynamics. Consider, for example, a Heisenberg ferromagnet [1.29]

$$\mathscr{H}_N^{\text{Heisenberg}} = -\sum_{i\neq j} J_{ij}\underline{S}_i\underline{S}_j - g\mu_B H S\sum_i S_i^z \quad , \tag{1.22}$$

where J_{ij} is the exchange constant between spins at sites i,j and $\underline{S}_i$ is a unit vector in the direction of the magnetic moment in the classical limit $(S\to\infty)$. There the equation of motion

$$\hbar\frac{d}{dt}S_k^z(t) = i[S_k^z(t),\mathscr{H}_N^{\text{Heisenberg}}] = -2\sum_{j(\neq k)} J_{kj}(S_k^yS_j^x - S_k^xS_j^y) \tag{1.23}$$

leads to the well-known phenomena of spin waves $(T < T_c)$ and spin diffusion $(T > T_c)$ [1.35]. The kinetics provided by (1.6,8) does not yield either of these modes, but rather yields purely relaxational behavior.

While in the Ising case, simulation of kinetic behavior was attempted [1.36,37] even before a proper formulation of the theoretical background was available, this kinetic aspect has consistently been neglected in the simulations of other systems, particularly of fluids. Often it is said (e.g., [1.16]) that the Monte Carlo method yields configuration space averages in contrast to the molecular dynamics method

[1.38] which yields time averages. We would like to draw attention to the misunderstanding that this statement can engender: The usual METROPOLIS et al. [1.1.] technique which samples by means of a Markov process inevitably yields a sequence of highly correlated states and thus "time" averages analogous to those obtained in molecular dynamics. A computationally practical method for sampling statistically independent states - with no such time correlations - of a system with many degrees of freedom has yet to be demonstrated. Of course, the kinetics used in molecular dynamics is that which follows from the hamiltonian, while the kinetics of the Monte Carlo method is an (often artificial) stochastic one. Nevertheless, this kinetics is of intrinsic interest: it will become apparent in Chap.6 that kinetic Ising models similar to that mentioned above constitute reasonable approximations to a surprising variety of physical systems. The point is that computational efficiency often requires the reduction of the physical complexity by using a simplified model, which contains only a part of the degrees of freedom of the system. The other degrees of freedom then act like a heat bath, inducing random transitions in the considered variables. In fact one can conceive simulations intermediate between the Monte Carlo technique (where all the kinetics are random transitions) and the molecular dynamics (where all the kinetics is deterministically derived from a hamiltonian) by using a modified molecular dynamics method where part of the kinetics is deterministic and part of it stochastic due to random forces [1.39].

We conclude this subsection by briefly defining the relaxation functions which are conveniently used. Suppose that the initial state of a simulation can be discribed as thermal equilibrium with some thermodynamic coordinate $\ell = \ell_0$, which takes on the value $\ell = \ell_0 + \delta\ell$ in the state under consideration. Then a nonlinear relaxation function $\phi_A^{\delta\ell}(t)$ is defined by [1.40]

$$\phi_A^{\delta\ell}(t) = \frac{\langle A(t)\rangle_{\ell 0} - \langle A(\infty)\rangle_{\ell 0}}{\langle A(t_0)\rangle_{\ell 0} - \langle A(\infty)\rangle_{\ell 0}} \quad , \tag{1.24}$$

where $\langle A(t)\rangle_{\ell 0}$ means an average, (1.20), where $P(\underline{x},t_0)$ is a thermal equilibrium distribution (with $\ell = \ell_0$). As an example, if one studies a ferromagnet starting with all spins aligned parallel, this would mean that the temperature is zero, and hence $\ell_0 = T_0 = 0$, $\delta\ell = \Delta T = T - T_0 = T$. If one starts with a random spin configuration, this would correspond to $T_0 \to \infty$. In thermal equilibrium, the relaxation of fluctuations is described by a linear relaxation function [1.41]

$$\phi_{AB}(t) = \frac{\langle A(0)B(t)\rangle - \langle A\rangle\langle B\rangle}{\langle AB\rangle - \langle A\rangle\langle B\rangle} \tag{1.25}$$

for any dynamic observables A,B. Furthermore it can be shown that [1.40]

$$\lim_{\delta\ell\to 0} \phi_A^{\delta\ell}(t) = \phi_{AC}(t) \quad , \tag{1.26}$$

where C is the variable conjugate to ℓ (if ℓ is the temperature, C is the internal energy; if ℓ is a magnetic field, C is the magnetization, etc.).

The practical realization of the averages (1.24,25) is easy. The average $\langle ... \rangle$ is performed by repeating the run with different initial conditions (which belong to the same equilibrium distribution with $\ell = \ell_0$) and otherwise identical conditions. Since fluctuations of A are of order 1/N for large N, often a single run (i.e., where $\langle ... \rangle_{\ell 0}$ is omitted) yields a resonable estimate for $\phi^{\delta\ell}(t)$ [1.36,42]. In the case of (1.25), one averages $A(t_0)\, B(t_0^A + t)$ over the time t_0 along the chain [1.2,32,42].

1.2.4 Accuracy Considerations: Pseudorandom Numbers, Finite Time Averaging, Initial Condition, etc.

With the theoretical knowledge described so far, we are now able to meaningfully discuss more practical questions of accuracy. In this subsection we are only concerned with the extent to which a Monte Carlo estimate $\bar{A}$ approximates the canonic average $\langle A \rangle$ for the same N and same choice of boundary condition. The appropriate choice of boundary conditions and the N dependence of the results are then discussed in the following subsections.

Monte Carlo averaging, as we have seen, rests on the use of random numbers. Digital computers produce pseudorandom numbers only; we thus have to discuss to what extent some nonrandomness of these numbers may affect the results.

For that purpose it is helpful to briefly recall the pseudorandom number generation in more detail. It usually rests on the fact that the machine system allows integer numbers I only up to some maximal number I_{max} which takes all possible digits (in binary code), for instance $I_{max} = 2^{m-1}-1$. If we now multiply two integer numbers I_ν, J, the result $I_{\nu+1} = J \times I_\nu$ will be truncated if $I_{\nu+1}$ exceeds I_{max}: one keeps the m-1 last bits of this product. It turns out that the first digits of the numbers $\{I_\nu\}$ produced by subsequent application of this recursion formula are approximately random if J is chosen suitably [1.43]. In order to restrict the range of numbers generated between I and $I_{max}-1$, one has to replace $I_{\nu+1}$ by $I_{\nu+1} + I_{max} - 1$ in case $I_{\nu+1}$ would otherwise be negative. Pseudorandom numbers ξ with $0 < \zeta < 1$ are then obtained from $\xi_\nu = I_\nu / I_{max}$. It should be noted that the "quality" (i.e., the randomness of the ξ_ν) of such multiplicative congruential generators depends very much on the number of bits m kept after the multiplication (while 32 bits are hardly satisfactory, 48 or 64 bits often are sufficient, cf. KNUTH [1.43]). While this algorithm is clearly simple and its realization on the computer is clearly very quick, and hence a scheme of this type is usually provided as part of the software in recent computers, it is a clearly deterministic procedure and even periodic. While this periodicity is large enough in comparison with actual Markov chain lengths and thus does not provide a very serious practical limitation, short-range correlations between pseudorandom numbers have to be taken seriously. For instance, if one uses subsequent ξ_ν to choose the coordinates (h,k,l) of an $L \times L \times L$ lattice at random ($h = \xi_\nu L + 1$, $k = \xi_{\nu+1}L + 1$, $l = \xi_{\nu+2}L + 1$), then it often happens that some sites will never be reached, while others are reached with a probability which is by far too high. Such obvious shortcomings can be removed by suitable mixing together random pieces of the same chain $\{I_\nu\}$. In any case it is necessary to check that supposedly random procedures (like the random choice of particles, etc.) are actually close to random by performing extended statistical tests.

Clearly the above discussion is not entirely satisfactory without a clear-cut estimate of the inaccuracy of final results due to this nonrandomness. This question is most simply checked by making calculations for systems for which the statistical mechanics can be worked out exactly. FRIEDBERG and CAMERON [1.44] made runs of 10^6-10^7 single steps for a 4×4 periodic Ising lattice; BINDER [1.25] made runs of similar length for a $3 \times 3 \times 6$ Ising lattice with free surfaces; STOLL et al. [1.42], and LANDAU [1.45] studied periodic $N \times N$ Ising lattices for various N up to N = 110. Statistical analyses did not reveal any systematic deviations from the exact results. For other hamiltonians, unfortunately rather few limiting cases are exactly soluble and hence useful for a test: one-dimensional classical Heisenberg chains, Heisenberg magnets at $(k_BT)^{-1} = 0$ but H/k_BT finite (one may then check that the magnetization actually follows the Langevin function [1.46]), ideal gases, and harmonic crystal lattices. But often systematic series expansions with some exactly known terms are available, like the virial expansion for nonideal gases, high-temperature expansions for Ising and Heisenberg magnets, low-temperature expansions for Ising magnets, and spin-wave theory for Heisenberg magnets, etc. Extensive comparisons have been made with these various expansions in their regimes of applicability [1.2-4,46-48]. These comparisons showed that systematic deviations which would have to be attributed to bad pseudorandom numbers never occurred. The actual relative accuracy $(\bar{A} - \langle A\rangle)/\langle A\rangle$ which is typically reached with 10^3-10^4 Monte Carlo steps/particle typically is of the order of 10^{-2}. We do not go into these comparisons in more detail since recent reviews are available [1.3,4]. For simulations of kinetic properties, a comparison was made for the one-dimensional kinetic Ising model [1.42] with the exact solution of GLAUBER [1.34], and again good agreement was found by STOLL et al. [1.42].

We next discuss the influence of the initial condition on the accuracy of Monte Carlo averages. If we define the relaxation time which is characteristic for the approach to equilibrium by $\tau_A^{\delta\ell}$,

Condition for eliminating influence of initial condition.

$$\tau_A^{\delta\ell} = \int_0^\infty dt\, \phi_A^{\delta\ell}(t) \ , \tag{1.27}$$

a formal condition for eliminating the influence of the initial condition is $\tau_A^{\delta\ell} \ll M_0/N$ where M_0 is the number of single steps omitted from the averaging (note that our time unit is a Monte Carlo step/particle). Of course, this condition is useful only if some a priori knowledge on $\tau_A^{\delta\ell}$ is available, as for instance in the case of kinetic Ising models. If such knowledge is not available, it may be helpful to consider the question if in some initial state there are some energy barriers $\Delta U \gg k_BT$. If so, one may estimate $\tau_A^{\delta\ell}$ from the theory of thermally activated processes [1.49] as $\tau_A^{\delta\ell} = \tau_0 \exp(\Delta U/k_BT)$, where τ_0 is a time of the order of the relaxation time of fluctuations within the metastable state which one gets into. Such metastable states are familiar from many kinds of first-order phase transitions [1.11], where usually the energy barrier is that for "nucleation" of the new phase. But such barriers occur in other problems, too, for instance in "spin glasses" (see Chap.9), and hence the Monte Carlo method then is hampered by very slow relaxation [1.50]. Another reason for slow relaxation towards thermal equilibrium is that the final state (T) is close to a critical point (T_c) where a second-order phase transition occurs. In that case $\tau_A^{\delta\ell}$ of an infinite system would diverge as

$$\tau_A^{\delta T} \propto |T - T_c|^{-\Delta_A^{\delta T}} \ , \tag{1.28}$$

where $\Delta_A^{\delta T}$ is the associated "critical exponent" [1.42]. For a finite system this divergence is rounded off, of course (see Sect.1.2.6). It turns out, however, that (1.28) is not the most serious limitation in this case, since the linear relaxation time τ_{AB}

$$\tau_{AB} = \int_0^\infty dt\, \phi_{AB}(t) \tag{1.29}$$

diverges with an exponent Δ_{AB} [1.41]

$$\tau_{AB} \propto |T - T_c|^{-\Delta_{AB}} \ , \tag{1.30}$$

which satisfies the inequality $\Delta_{AB} > \Delta_A^{\delta T}$ [1.51].

Another case where extremely slow relaxation is expected is that the "susceptibility" $\chi_{AA} = (N/k_BT)[\langle A^2\rangle - \langle A\rangle^2]$ is very large or even divergent in the thermodynamic limit $N \to \infty$. Applying the van Hove "conventional theory" of slowing down [1.10], one may relate τ_{AA} and χ_{AA} as follows

$$\tau_{AA}^{conv} = \chi_{AA}/L_{AA} \quad , \tag{1.31}$$

where L_{AA} is a kinetic coefficient which can be approximated as constant. Thus, if χ_{AA} is very large, τ_{AA}^{conv} (which turns out to be a lower limit to the true τ_{AA}) is also very large, and also $\tau_A^{\delta\ell}$ is expected to be large. Such a behavior occurs in cooperative systems which "nearly" undergo a second-order phase transition (like the Heisenberg model of magnetism in two dimensions which [1.52] has no spontaneous magnetization M but whose magnetic susceptibility becomes so large that it even is consistent with a critical divergence [1.53]) or which undergo a phase transition without long-range order but divergent susceptibility (like the XY model in two dimensions [1.54]). From this point of view, the apparent breakdown of convergence of the Monte Carlo method reported in studies [1.55] of the two-dimensional Heisenberg model is not at all surprising. The same is true for a recent study of the two-dimensional XY model [1.56]

$$\mathcal{H} = -J \sum_{\langle i,j\rangle} (S_i^x S_j^x + S_i^y S_j^y) \quad , \tag{1.32}$$

where the behavior clearly depended on the initial condition (Fig.1.1). Starting with the spins aligned in the x axis, one ends up with spin configurations with no vortices; starting with the spins lying randomly in the xy plane, one obtains states with some vortices; starting with the spins aligned in the z axis, one obtains states with many vortices at low temperatures. The direct observation of the time evolution of the spin configurations reveals that the relaxation times are so large that the configurations are nearly frozen-in (note the similarity of states at t = 500 and t = 1000 MCS/spin in Fig.1.1).

This example shows how one can combine the knowledge on "pathological" behavior of a susceptibility with the conventional theory of slowing down to interpret the relaxation found in the Monte Carlo simulations. In many interesting cases, however, lack of precise knowledge of static properties makes the interpretation of very slow relaxation ambiguous, for instance in spin glasses (Chap.9).

Another case of slow relaxation concerns Monte Carlo simulations where one observable A is kept constant (in simulations of binary alloys, for instance, it makes sense to keep the concentration c constant). If, in addition, the changes $\underline{x} \to \underline{x}'$ involve only *local* changes (in the case of the alloy this means the exchange of neighboring atoms rather than the exchange of arbitrary atoms in the

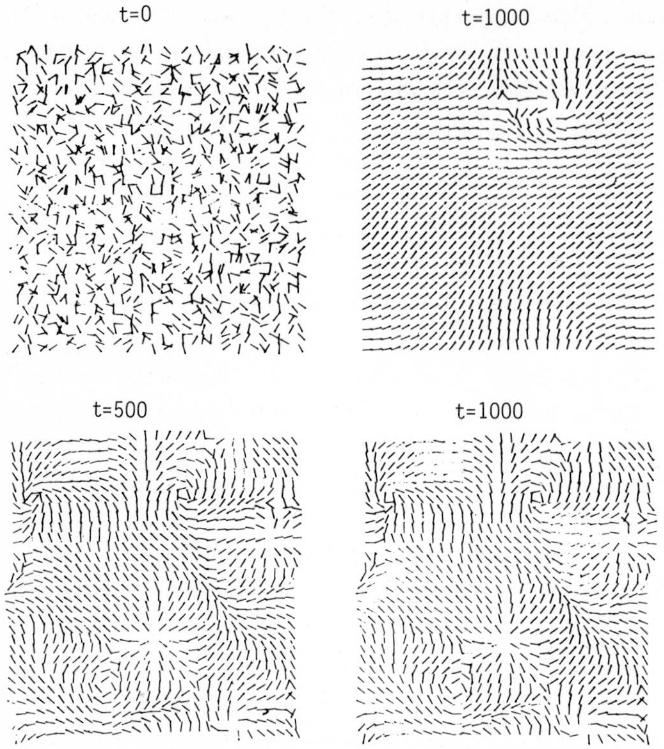

Fig.1.1 Spin configurations of a periodic 30 × 30 classical XY model at $k_BT/J = 0.01$ starting with a random configuration in the XY plane (upper part) and starting with a configuration aligned in the +z direction (lower part). Time at which these "snapshots" have been made is shown in Monte Carlo steps/spin [1.56]

system), then it can be shown that Fourier components $A(\underline{k})$ exhibit the "hydrodynamic slowing down" [1.9]

$$\tau_{AA}(\underline{k}) = (D_{AA}k^2)^{-1} \quad , \tag{1.33}$$

where D_{AA} is some transport coefficient. The nonlinear relaxation time is equal to $\tau_{AA}(\underline{k})$ for small $\underline{k}$ [1.57]. In the example of an alloy, D_{AA} would be the concentration diffusivity. The relaxation time (1.33) becomes very large for small wave vectors $\underline{k}$. In the example of an alloy, the structure factor $S(\underline{k})$, which is the Fourier transform of the correlation function of the local concentration variables c_j,

$$S(\underline{k}) = \sum_j \exp[i\underline{k}(\underline{r}_j - \underline{r}_0)][\langle c_j c_0\rangle - \langle c_j\rangle\langle c_0\rangle] \quad ,$$

thus hardly reaches its equilibrium value for small k during the simulation, if its initial value distinctly differs from the equilibrium value. This example illustrates a general point: one has to be very careful in the judgement if thermal equilibrium is achieved in a simulation run, since the answer may be quite different for different quantities. In the above example, $S(\underline{k})$ for $\underline{k}$ large reaches equilibrium quite quickly, and so does the internal energy. Often it is said that the latter quantity is used as an indicator of thermal equilibrium of the whole system and this may be quite misleading.

Next we discuss the limitation of accuracy due to finite-time averaging. As it was necessary to request

$$\tau_A^{\delta\ell} \ll M_0/N \tag{1.34a}$$

to eliminate the influence of the initial condition, it is now necessary to request (M is the total number of configurations generated, while M_0 are omitted from the averaging)

$$\tau_{AA} \ll (M - M_0)/N \quad , \tag{1.34b}$$

in order that averages over equilibrium states are taken over a time which is much larger than the time over which the observables are correlated. If either one of the conditions of (1.34a) or (1.34b) is violated, one cannot obtain any reliable results from the simulation for the observable A. The same reasons for slow relaxation apply for τ_{AA} as did for $\tau_A^{\delta\ell}$:high energy barriers ΔU, critical slowing down, large (or divergent) χ_{AA}, and hydrodynamic slowing down. If, on the other hand, one is able to satisfy the conditions of (1.34a,b), then one may apply the analysis of "statistical errors" as described earlier in (1.16). Moreover, one may then make quantitative predictions on the expected error $\delta\bar{A}$ of an average $\bar{A}$, namely [1.2-4][3]

$$(\delta\bar{A})^2 = \frac{N}{M - M_0}[\langle A^2\rangle - \langle A\rangle^2](1 + 2\tau_{AA}) \quad . \tag{1.35}$$

Apart from the factor $1 + 2\tau_{AA}$, (1.35) would be the same as (1.16) if one assumes that observations taken at subsequent Monte Carlo steps *per particle* are uncorrelated. The fact that the actual errors are much larger than according to that assumption has been observed by various authors [1.44,45] who called this the "statistical inefficiency" of the Monte Carlo method. We feel that this term is better avoided since (1.35) is an obvious consequence of the dynamic nature of

---

[3] A factor N has to be removed from the denominator in [1.3], (16).

the Monte Carlo process, (1.19). Approximating τ_{AA} by τ_{AA}^{conv} and putting $L_{AA} = 1$ if time is measured in Monte Carlo steps per particle, one may use the fluctuation relation $N[\langle A^2\rangle - \langle A\rangle^2] = k_BT\,\chi_{AA}$ to simplify (1.35) to $(\delta\bar{A})^2 \approx k_BT\,\chi_{AA}(1 + 2\,k_BT\,\chi_{AA})/(M - M_0)$.

Estimating both A, χ_{AA} and $\delta\bar{A}$ [using (1.16)] from the simulation, one may compare the "experimental" and "theoretical" error estimates $\delta\bar{A}$. This was done in the case of the magnetization in simulations of the classical Heisenberg model [1.58], and good agreement between experimental and theoretical error estimates was obtained. Unfortunately, it becomes much more cumbersome to estimate the error of quantities like dynamic correlation functions $\langle A(0)A(t)\rangle - \langle A\rangle^2$. However, the same difficulty applies to the molecular dynamics methods, of course [1.59]. In most practical cases one has to resort to experimental error estimates [e.g., using (1.16) and checking that subsequent $A(\underline{x}_\nu)$ are uncorrelated]. Equation (1.35) clearly shows that the accuracy of the Monte Carlo method breaks down near the critical point T_c of a second-order phase transition, where one expects $\chi_{AA} \propto |1 - T/T_c|^{-\gamma_{AA}}$ and hence $(\delta\bar{A})^2 \propto |1 - T/T_c|^{-(\gamma_{AA}+\Delta_{AA})}/(M - M_0)$. On the other hand, critical singularities are rounded off in finite systems when $\xi \propto |1 - T/T_c|^{-\nu}$ exceeds the linear dimension $L \propto N^{1/d}$ of the d-dimensional model system, and therefore the error stays finite

$$(\delta\bar{A})^2_{T_c} \propto N^{(\gamma_{AA}+\Delta_{AA})/d\nu}/(M - M_0)\ , \qquad M - M_0 \gg N^{1+\Delta_{AA}/d\nu}\ , \tag{1.36}$$

but clearly it is very difficult to obtain meaningful estimates close to T_c. In the case of the two-dimensional Ising model, for instance, $d\nu = 2$, $\Delta_{AA} \approx 2$, $\gamma = 1.75$, the number of steps per spin has to be much larger than N, e.g., $M - M_0 = 10N$, and nevertheless the error of the magnetization is still rather large, $(\delta\bar{m})^2_{T_c} \approx 10^{-1}\,N^{-1/8}$.

A further consequence of the statistical inaccuracy is the possible occurrence of a bias in some estimates; such a bias may occur, for instance, if a quantity is defined as $\langle c\rangle = \langle A\rangle/\langle B\rangle$, since then the errors of A and B are correlated if these estimates are drawn from the same Monte Carlo run. Such effects make it particularly difficult to assess the accuracy of estimates for dynamic correlation functions [1.42] [cf. (1.25)].

1.2.5 Appropriate Choice of Boundary Conditions

In the Monte Carlo method, one studies the properties of a finite system of N particles (typically $N \approx 10^3$-10^4) while one is usually interested in the bulk properties of an infinite system ($N \to \infty$). In order to be able to perform a meaningful extrapolation to the thermodynamic limit (see Sect.1.2.6), the question of boundary conditions must be given some consideration.

The physically simplest boundary condition is that of "free surfaces", simulating a small particle. Interactions across the surface of the particle are then put equal to zero. Such a boundary condition is meaningful if one wants to study the magnetic properties of superparamagnetic particles, for instance (see Chap.5). The calculation is quite straightforward since the shape of the particle is held fixed during the simulation (rectangular hypercubes, or approximately spherical particles, etc.). Less straightforward is the study of particles whose shape is not fixed but may fluctuate like liquid droplets in a gas, or grains precipitated from a supersaturated solid solution, or polymers in aqueous solution, etc. Then it may be convenient to treat a (larger) system which contains the desired system as a subsystem. For instance, one may put the liquid droplet in a box; if the box has suitable volume an equilibrium between the droplet and surrounding (supersaturated) gas will be reached. In this way also the properties of "clusters" in the lattice gas model have been studied. For details of how these and related simulations are performed, see Chap.5.

If the number of degrees of freedom of particles in the box is large in comparison to that of the subsystem considered, most of the computing time is spent on the (less interesting) properties of particles in the box rather than of particles in the desired subsystem. This fact is particularly cumbersome in the study of polymers in aqueous solution. Therefore the idea originated [1.60] to take only a small part of water molecules which are closely attached to the considered polymer explicitly into account, and to represent the remaining water molecules in the box by a suitably constructed "stochastic boundary condition" [1.61]. At the surface of the part of the system which is treated explicitly, one allows action of random forces which are constructed such that their correlation function is the same as that of actual forces acting through a thought-dividing surface of a bulk system of water molecules. While such techniques are far from being well controlled, they are potentially interesting since they might open up new areas of applications to the Monte Carlo method.

The conventional boundary condition of a finite system used to simulate infinite systems is the periodic or (toroidal) boundary condition. This applies to systems whose shape is that of a hypercubic box with linear dimensions $L_1, L_2, \dots, L_d$ (d being the dimensionality). Defining vectors $\underline{L}_1, \underline{L}_2, \dots, \underline{L}_d$ by $\underline{L}_1 = (L_1, 0, \dots, 0)$, $\underline{L}_2 = (0, L_2, 0, \dots, 0)$, $\dots$, $\underline{L}_d = (0, 0, \dots, 0, L_d)$, the boundary condition is then specified mathematically by requesting that for arbitrary observables A and points $\underline{x}$

$$A(\underline{x}) = A(\underline{x} \pm \underline{L}_\nu) \quad , \quad \nu = 1, \dots, d \quad . \tag{1.37}$$

Clearly, (1.37) is inconvenient if the potential function $U(\underline{x}_1, \dots, \underline{x}_N)$ is given by a sum of pair potentials $u(r) \neq 0$ for any finite r, (1.37) requires then that

$[\underline{L}_{\underline{m}} = m_1\underline{L}_1 + m_2\underline{L}_2 + \dots m_d\underline{L}_d,\ \underline{m} = (m_1, m_2, \dots, m_d),\ \{m_\nu\}$ integers]

$$U(\underline{x}_1, \dots, \underline{x}_N) = \frac{1}{2}\sum_{i\neq j} u(|\underline{x}_i - \underline{x}_j|) + \frac{1}{2}\sum_{\underline{m}}{}' \sum_{i\neq j} u(|\underline{x}_i - \underline{x}_j + \underline{L}_{\underline{m}}|) \tag{1.38}$$

In order to avoid the infinite summation over $\underline{m}$, one either has to truncate the potential [1.16] or to use Ewald summation techniques [1.62] (see also Chap.2). Furthermore, it must be stressed that choosing hypercubic boxes is particularly dangerous if one wants to study transitions from one lattice structure to another or liquid-solid transitions, etc. Clearly a particular choice of values L_1, L_2, ..., L_d and N may favor one lattice structure over others, and in order to get meaningful results it may be necessary to study a variety of choices.

An important consequence of the periodic boundary condition is that Fourier transforms $A(\underline{k}) = \sum \exp(i\underline{k}\cdot\underline{x})A(\underline{x})$ are meaningful for discrete points $\underline{k}_\nu$ in $\underline{k}$ space only, $\underline{k}_\nu = (\nu_1, \nu_2, \dots, \nu_d)\pi/L$, with $\nu_1, \dots \nu_d$ integers. This fact hampers, for instance, studies of the structure factor of alloys in the small k region [1.63]. Also the correlation function $\langle A(\underline{x})B(\underline{x}')\rangle$ approximates that of an infinite system only for distances $|\underline{x} - \underline{x}'| < L/2$ and only if the correlation length $\xi(T) \ll L/2$ (see [1.56], Fig.4 for an example of correlation functions with $\xi(T) > L/2$).

Although the periodicity is a well-defined boundary condition for any finite system, it is not necessarily the best approximation for an infinite system. This consideration gave rise to the introduction of the "selfconsistent effective field" - boundary condition [1.3,4,58,64-66]. Since this method so far is restricted to (ferromagnetic) spin systems and has been reviewed recently [1.4,65], we only summarize the main points briefly. One uses free boundary conditions together with an "effective field" which acts on the free surfaces of the system only. The magnitude of the field is determined in an iteration procedure by the condition that the gradient of the order parameter (the magnetization m) at the surface be zero (Fig.1.2)

$$\left.\frac{\partial m(z)}{\partial z}\right|_{\text{surface}} = 0 \quad . \tag{1.39}$$

While such a condition eliminates boundary effects completely within the Ginzburg-Landau theory of phase transitions [cf. 1.67], this is only approximately true in the general case. In fact, this method leads to a sharp critical point but with mean field exponents, as one would expect in interpreting this method as a high-order generalization of the BETHE [1.68] approximation. Nevertheless this method is advantageous: using free surfaces alone, or periodic boundary conditions, one would have a rounding of the critical anomalies close to T_c. Here a crossover from

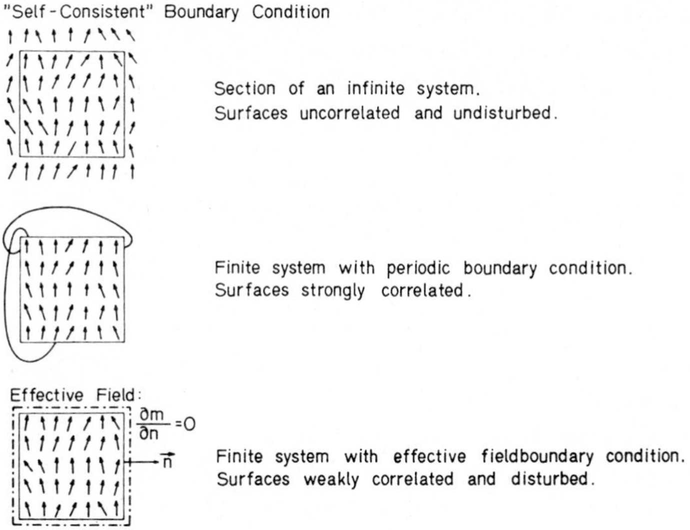

Fig.1.2 Schematic spin projections of one (arbitrary) plane of a simple cubic Heisenberg system, for the explanation of the "self-consistent" boundary condition [1.65]

the true critical behavior to the mean field critical behavior occurs rather than the rounding. It turns out, however, that the mean field region around T_c is somewhat narrower than the region where rounding effects (cf. Sect.1.2.6) are important. Thus the convergence to the thermodynamic limit is somewhat quicker than in the case of periodic boundary conditions, and reliable estimates can be drawn from rather small systems [1.58,64,65].

A further sophistication was introduced by BOLTON and JOHNSON [1.66], who replaced the *constant* effective field by a randomly fluctuating effective field with the same mean value. It must be emphasized, however, that all these techniques are probably rather difficult if several order parameters have to be considered, as occurs in systems with several ordered phases. Then one would need an effective boundary "field" for each of these ordered phases. These fields must be determined in a joint iteration process. Such a procedure has not yet been attempted since one has to expect convergence difficulties.

Finally we mention the antiperiodic boundary condition where, for the degrees of freedom $A(\underline{x})$, (1.37) is replaced by

$$A(\underline{x}) = -A(\underline{x} \pm \underline{L}_\mu) \tag{1.40}$$

for one direction μ. This boundary condition is sometimes applied if one wishes to introduce an interface into the system [for instance, if $A(\underline{x})$ is a spin variable

$\underline{S}(\underline{x})$, one generates a Bloch wall], which then is not fixed in position, however. This boundary condition is hence useful for the Monte Carlo study of two-phase equilibria and also kinetic problems like crystal growth from the vapor, etc. (cf. Chap.7).

1.2.6 Finite Size Problems. The Extrapolation to the Thermodynamic Limit

Here we consider the question of how one extracts estimates for observables $\langle A\rangle_T$ of an infinite system from Monte Carlo averages $\bar{A}_{T,N}$ on finite systems with linear dimension $L(N = L^d$ in d dimensions), using one of the boundary conditions discussed in the previous subsection (free, periodic, self-consistent field, etc.). In some cases there may be good reasons to believe that $|\bar{A}_{T,\infty}-\bar{A}_{T,N}|$ does not exceed the "statistical" error of $\bar{A}_{T,N}$, and then no particular analysis is necessary (this is often claimed to be true in studies of Lennard-Jones liquids [1.16]). In many cases, however, this is not obvious, or even not true, and then a *finite size analysis* of the data has to be made, making runs for a variety of choices for L and extrapolating to $L\to\infty$. This analysis involves two questions: (I) the appropriate analytic L dependence on which this extrapolation is based. (II) the appropriate range of L values such that one is already in the asymptotic regime where the assumed L dependence holds.

In order to illustrate these questions, we show in Fig.1.3 some recent data [1.47] on the magnetization of nearest neighbor square Ising lattices with either periodic or free boundary conditions. The data are plotted vs L^{-1}, which allows linear extrapolation if

$$\bar{A}_{T,N} = \bar{A}_{T,\infty} + \bar{A}_{T,s}/L, \quad L\to\infty \quad , \tag{1.41}$$

where $\bar{A}_{T,s}$ is some finite size correction. In the case of free surfaces, $\bar{A}_{T,s}$ is interpreted as a surface correction. Equation (1.41) applies in the paramagnetic temperature regime even in the periodic case, since $\sqrt{\langle M^2\rangle} = 1/\sqrt{N}\,\sqrt{k_BT\chi}$, where χ is the magnetic susceptibility, and $1/\sqrt{N} = 1/L$ in two dimensions. For the paramagnetic temperature regime of three-dimensional systems, this argument shows that the appropriate abscissa variable for linear extrapolation is $L^{-3/2}$ rather than L^{-1}. Below the critical temperature, one expects a nonzero spontaneous magnetization $\langle M\rangle_N$ to be metastable [1.69] in finite systems, and hence the above fluctuation relation is modified to

$$\sqrt{\langle M^2\rangle} = \sqrt{\langle M\rangle^2 + k_BT\chi/N} = \langle M\rangle_N + k_BT\chi/2\langle M\rangle_N N \quad .$$

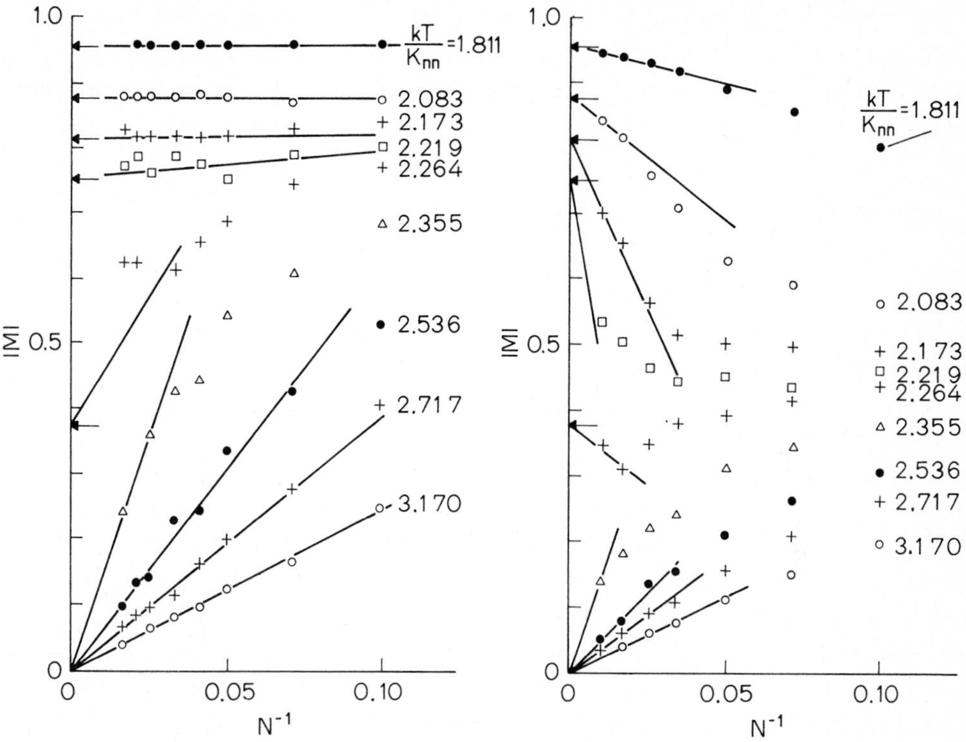

Fig.1.3 Magnetization of nearest neighbor Ising square lattice with periodic boundary conditions (left part) and free boundary conditions (right part) plotted vs the inverse linear dimension of the lattices at various temperatures. Arrows show the rigorous results for infinite lattice size [1.45]

In the periodic case (where surface effects are eliminated) one expects the leading finite size correction to come from the second term, and hence $N^{-1} = L^{-d}$ rather than L^{-1} would be the appropriate abscissa variable for linear extrapolation. No such correction is present in the internal energy U, however, where one expects [1.70] corrections which are exponentially small,

$$\langle \mathcal{H} \rangle_{T,N} - \langle \mathcal{H} \rangle_{T,\infty} \propto \exp[-L/\lambda(T)] \quad , \qquad (1.42)$$

$\lambda(T)$ being proportional to the correlation length $\xi(T)$ of magnetization fluctuations. In that case $\langle \mathcal{H} \rangle_{T,\infty}$ could be estimated by requesting that log $|\mathcal{H}_{T,N} - \langle \mathcal{H} \rangle_{T,\infty}|$ vs L should give a linear variation with L. Hence, we note that a linear extrapolation in a $A_{T,N}$ vs L^{-1} plot is appropriate in the case of free boundary conditions, while in the periodic case the behavior is more complicated, and each quantity requires separate consideration. Furthermore, Fig.1.3 demonstrates that for many temperatures rather large L are required to really see the asymptotic L dependence. In fact, for $T = T_c$, all the above formulae break down as [1.29]: $\langle M\rangle_\infty \propto (1 - T/T_c)^\beta \to 0$, $\chi \propto |T/T_c - 1|^{-\gamma} \to \infty$, $\langle M\rangle_s \propto (1 - T/T_c)^{-\beta_s} \to \infty$, $\xi(T) \propto (1 - T/T_c)^{-\nu} \to \infty$. These divergencies mean that at T_c (1.41,42) are replaced by a much slower variation with L, namely [1.64]

$$\sqrt{\langle M^2\rangle_{N,T_c}} \propto N^{-1/(\delta+1)} \; , \quad \delta = 1 + \beta/\nu \tag{1.43a}$$

and

$$\langle \mathcal{H}\rangle_{T_c,N} - \langle \mathcal{H}\rangle_{T_c,\infty} \propto L^{-(1-\alpha)/\nu} \; , \quad \alpha = 2 - \gamma - 2\beta \tag{1.43b}$$

Equation (1.43) will be shown below to result from the finite size scaling theory [1.71]. For the two-dimensional Ising model the critical exponents are [1.29] $\beta = 1/8$, $\gamma = 7/4$, $\delta = 15$, $\alpha = 0$, $\nu = 1$; hence the decay of the magnetization with N^{-1} is extremly slow. Figure 1.3 and the above equations illustrate the fact that a finite system cannot have a sharp phase transition; thus the location of a critical temperature from Monte Carlo calculations is a subtle question (see also Sect.1.3.3), and requires careful finite size analysis in any case.

The finite size scaling theory [1.71] referred to above assumes the singular part of the free energy of a magnet to be a generalized homogeneous function of the variables L, $T - T_c$, H, i.e.,

$$F_s(T,H,L) = L^{-(2-\alpha)/\nu}\, \tilde{F}[(T - T_c)L^{1/\nu} \; , \; HL^{\beta\delta/\nu}] \; . \tag{1.44}$$

For $T \neq T_c$, H fixed, the expansion $\tilde{F}(x,y) \xrightarrow[x,y\to\infty]{} x^{2-\alpha}\tilde{f}(x^{-\beta\delta}y) + O(1/L)$ yields ordinary static scaling [1.29] for the free energy of an infinite system. In order to derive (1.43), we differentiate $F_s(T,H,L)$ with respect to magnetic field (or temperature) to obtain the singular part of the magnetization (or energy)

$$\langle M\rangle = L^{-(2-\alpha-\beta\delta)/\nu} \frac{\partial \tilde{F}(x,y)}{\partial y} \propto L^{-\beta/\nu} \propto N^{-\beta/d\nu} \; , \tag{1.45a}$$

$$\langle \mathcal{H}\rangle_s = L^{-(1-\alpha)/\nu} \frac{\partial \tilde{F}(x,y)}{\partial x} \; . \tag{1.45b}$$

Equation (1.43) result from the fact that $\partial\tilde{F}/\partial y|_{x,y=0}$ and $\partial\tilde{F}/\partial x|_{x,y=0}$ stay finite [and note the scaling law [1.29] $d\nu = \beta(\delta + 1)$].

Equation (1.45) suggests also that $\langle M \rangle L^{\beta/\nu}$ should (for H = 0) depend on the scaled variable $(T - T_c)L^{1/\nu}$ only, rather than on the variables T, L separately. Figure 1.4 shows, replotting the data of Fig.1.3, that in this way a whole family of curves now actually fit two curves only (one for $T > T_c$ and one for $T < T_c$). The scaling representation has now been verified by many examples (the first study [1.72] referred to three-dimensional 55 × 55 × L Ising films). It is the most reliable method to extract estimates of both the critical exponents and of T_c from Monte Carlo calculations.

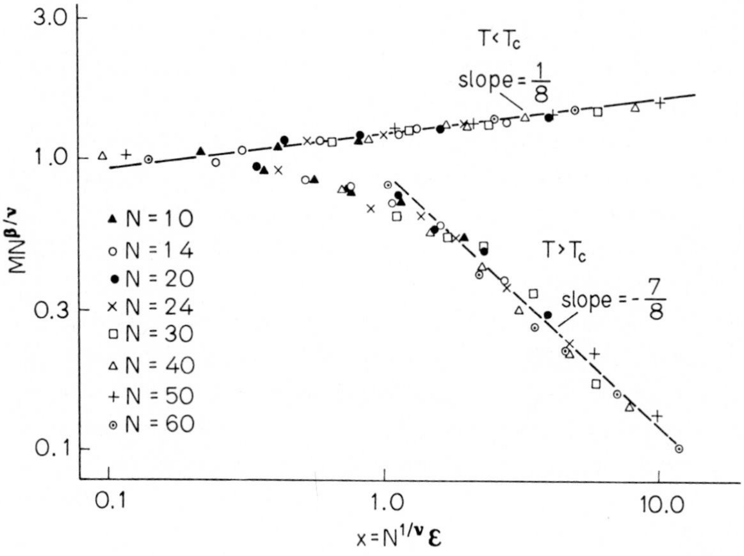

Fig.1.4 Same data as Fig.1.3 replotted in scaled form [1.45]

It should be emphasized that (1.42,44) have an intuitive physical interpretation: basically the free energy and its derivatives are functions of the ratio of linear dimension and correlation length $L/\xi(T)$. The asymptotic regime where (1.41,42) are valid is reached for $L/\xi(T) \gg 1$.

From this interpretation, one already expects an anomalous L dependence in cases where a correlation length diverges away from T_c. This occurs, for instance, in the case of the Heisenberg magnet below T_c and H = 0. In the case of free boundaries, one finds from the decay law $m(z) \propto 1/z$ [1.73] for the local magnetization in distance z from a surface that

$$\langle M\rangle_s = \sum_{z=1}^{L/2} [m(\infty)-m(z)] \propto \log(L) \text{ and hence}$$

$$\langle M\rangle_{T,N} - \langle M\rangle_{T,\infty} \propto \log L/L \quad . \tag{1.46}$$

In the two-dimensional XY model below T_c, the correlation function $\langle \underline{S}_i \underline{S}_j \rangle$ behaves as $\langle \underline{S}_i \underline{S}_j \rangle \propto |\underline{r}_i - \underline{r}_j|^{-\eta(T)}$ [1.54], and hence one finds that $\sqrt{\langle M^2 \rangle} = \sqrt{\sum_{ij} \langle \underline{S}_i \underline{S}_j \rangle / N} \propto \sqrt{\int_0^L r^{-\eta(T)} r dr / L^2} \propto L^{-\eta(T)/2}$ [1.46]. Since $\eta(T) = k_B T/\pi J$ for low T [1.54], it is clear that for such a model extremely large systems would be required to faithfully simulate the behavior of an infinite system [1.56].

We have chosen examples of magnetic systems to illustrate the problems of estimating reliably the properties of an infinite system with simulations of small systems. But it is clear that a similar analysis applies to other systems as well. Figure 1.3 also vividly illustrates that a broad range of values of L is necessary to definitely assess the magnitude of these finite size effects. Note that for $N \lesssim 500$ the size dependence is usually much weaker than for large N - hence the use of too small systems may be quite misleading.

For systems like alloys, liquid-gas, or liquid-solid systems, etc., it may occur that for some range of parameter values the thermodynamic equilibrium does not consist of a single phase but rather of two coexisting phases (e.g., two alloy phases with different concentration, or gas and liquid with densities according to the coexistence curve, etc.). Then the equilibrium state of the system necessarily contains one (or more) interface(s), and one gets 1/L corrections due to these interfaces even though periodic boundary conditions are used.

1.3 Aspects of Simulations of Kinetic Processes

1.3.1 Various Monte Carlo Realizations of the Same Master Equation

Here we focus interest on the Monte Carlo realization of the master equation, (1.19), for the simulation of kinetic processes. Such master equations may approximately describe the dynamics of many physical systems (see Chap.6). Since some of the rate constants involved may be very small and the relaxation hence be very slow, it may be advantageous to use algorithms different from those described in Sect.1.2.2. We reproduce here a survey recently given by KALOS [1.31].

To be specific, it is assumed that the transition $\underline{x} \to \underline{x}'$ in (1.19) consists of a change in the state of one "object" or particle (ν) ($\nu = 1,2,\ldots, N$) from its previous state into one (m) of M (ν) states, possibly including its original state (n). The transition probability per unit time (or evolution rate, respectively) be $w_{nm}^{(\nu)}$. In this description, the evolutions from a fixed initial state n to final states m are also considered to occur independently (but nonetheless competitively) of each other, as are the evolutions of the various objects. The total rate of evolution for object ν is

$$\hat{r}_\nu = \sum_m w_{nm}^{(\nu)} \quad , \tag{1.47a}$$

while the total rate for object ν to *change* is

$$r_\nu = \sum_{m \neq n} w_{nm}^{(\nu)} \quad . \tag{1.47b}$$

The rates for the entire system to evolve (or to change, respectively) are then

$$\hat{R} = \sum_\nu \hat{r}_\nu \ , \quad R = \sum_\nu r_\nu \quad . \tag{1.47c}$$

Before considering Monte Carlo algorithms to follow this kind of time evolution, we give some necessary mathematical statements: Considered by itself, the ν th object has probability $P_\nu(t) = \exp(-r_\nu t)$ that it is unchanged at time t. The probability that all objects are unchanged hence is

$$P(t) = \prod_\nu P_\nu(t) = \prod_\nu e^{-r_\nu t} = e^{-\sum r_\nu t} = e^{-Rt} \quad . \tag{1.48a}$$

The probability that the earliest change of state occurs between t and t + dt hence is

$$P(t) = P(t)\ R\ dt = Re^{-Rt}\ dt, \tag{1.48b}$$

and the probability that the first such change is that of object ν (which is in state n) to state m in dt is

$$P_{nm}(t)dt = \frac{w_{nm}^{(\nu)}}{R}\ Re^{-Rt}\ dt \quad . \tag{1.48c}$$

We now list and discuss several algorithms.

Algorithm 1

(I) Consider t as a random variable, which is generated according to the distribution R exp(-Rt).

(II) Select ν with probability r_ν/R. Since each object ν must be in a specified state, n is then also fixed.

(III) Select m (conditional on n) with probability $w_{nm}^{(\nu)}/r_\nu$. An alternative is to combine steps (II) and (III) and pick ν, m together with probability $w_{nm}^{(\nu)}/R$.

While this algorithm is straightforward, it requires that all r_ν (or $w_{nm}^{(\nu)}$) be known in advance and R be known explicitly. Such knowledge is possible to obtain in each state of the system, but it requires computation and bookkeeping that can be avoided by other algorithms. The time t so generated is understood to be the simulated time at which the next change in state occurs. Note that R (and the $\{r_\nu\}$) may change when the system changes.

Algorithm 2

(I) Let t_j be chosen from $r_j \exp(-r_j t_j)$, $j = 1,2, \ldots, N$.

(II) Set $t = \mathrm{Min}\,\{t_j\}$.

(III) Now ν is found from the condition $t = t_\nu$.

(IV) Select m with probability $w_{nm}^{(\nu)}/r_\nu$.

For proving the correctness of this algorithm, one uses the fact that the probability that the event $n \to m$ for the ν-th object is the probability that t_ν is in $[t,t + dt]$ and all other t_j exceed t_ν

$$P_\nu(t)dt = r_\nu e^{-r_\nu t} \prod_{j \neq \nu} e^{-r_j t}\, dt = r_\nu e^{-Rt}\, dt \,, \tag{1.49a}$$

and

$$P_{nm}(t) = \frac{w_{nm}^{(\nu)}}{r_\nu} P_\nu(t) = w_{nm}^{(\nu)}\, e^{-Rt} \,, \tag{1.49b}$$

in agreement with (1.48c). This algorithm has the advantage that R need not be known, and the choice of ν is very simple. The necessity of selecting N times from an exponential distribution, however, is a serious disadvantage. In the case where all or many of the r_ν are equal, both algorithms become much faster.

Before going on to the next class of algorithms, we discuss an *Auxiliary Algorithm*

(I) Let $R' \geq R$ and $t = 0$.

(II) Select t' from $R' \exp(-R't)$.

(III) Replace t by $t' + t$ with probability R/R' and use this as elapsed time to next event.

(IV) Repeat steps (II) and (III).

Let $P(T)$ be the probability that no event before T occurs. In dT, $P(T)$ decreases by the probability that a successful event is found, namely $P(T)R'dT\ R/R'$. Hence

$$P(T + dT) = P(T) - P(T)R\ dt \ , \quad \frac{dP}{dT} = -P(T)RdT \quad , \quad P(T) = e^{-RT} \quad . \tag{1.50}$$

Thus those events accepted in step (III) have the correct distribution for time intervals. This makes it possible to use artificially large rates if convenient. It is, in fact, very convenient since one does not have to know all the r_ν or $w_{nm}^{(\nu)}$ before sampling an event. Thus one can use

Algorithm 3. Let $r \geq \mathrm{Max}\ \{r_\nu\}$.

(I) Set $t = 0$.

(II) Sample t' from $Nre^{-Nrt'}$; replace t by $t + t'$.

(III) Sample ν at random, $1 \leq \nu \leq N$.

(IV) With probability r_ν/r, accept the event at elapsed time t.

(V) Else repeat steps (II) through (IV).

(VI) Select m with probability $w_{nm}^{(\nu)}/r_\nu$.

Here r_ν is not computed until the event $n \rightarrow m$ is selected. This avoids much bookkeeping and usually considerable redundant computation. On the other hand, if $r \gg r_\nu$ for most ν, much computing time is wasted to determine r_ν for moves that are rejected. One part of the computation which can be economized is that of t. To sample t' from $R' \exp(-R't)$, set $t' = -\ln(\xi)/R'$, where ξ' is a uniform random number with $0 < \xi < 1$. Then while R' is constant, $t = \sum_i t_i' = \ln(\Pi\xi_i)/R'$. This is the algorithm most commonly used [1.63].

In order to avoid much wasted time associated with rejected events, one can use different upper bounds for different classes of events. Thus, considering the kinetics of metallic alloys in terms of models where interstitials and vacancies can migrate, it would be inefficient to use the same bound for interstitials and for vacancies, since they have such different rates of migration. In the last algorithm we allow different bounds, now on the rates $w_{nm}^{(\nu)}$.

Algorithm 4. Let $r'_\ell \geq w_{nm}^{(\nu)}$ for $\sum_{k=1}^{\ell-1} N_k < \nu < \sum_{k=1}^{\ell} N_k$, $\ell = 1,2, \ldots, L$. That is, we suppose that there are L classes of states, and that ν is ordered so that the classes are segregated together, and that each class has a separate bound r'_ℓ, and N_ℓ members. Then the algorithm has the following steps:

(I) Set $t = 0$.

(II) Set $R' = \sum_\ell r'_\ell N_\ell$, sample t' from $R' \exp(-R't')$, and replace t by $t + t'$.

(III) Choose a class ℓ with probability $r'_\ell N_\ell / R'$ for $1 < \ell < L$.

(IV) Pick any member μ of class ℓ uniformly and at random, $\mu < N_\ell$.

(V) Select m uniformly and at random among all alternatives available, $m < M(\nu)$.

(VI) Compute $w_{nm}^{(\mu)}$; with probability $w_{nm}^{(\mu)}/r_\ell$, accept the event with elapsed time t.

(VII) Else repeat steps (II) to (VI).

This division into classes is, of course, arbitrary. With suitable definition of the classes it should be possible to find a convenient reasonably efficient algorithm. It must be emphasized, in summary, that in spite of their apparent differences, all these algorithms yield simulations of systems that have, on the average, exactly the same kinetics and hence describe equivalent physical systems. The alternative of using one or another rests upon convenience, economy in programming, and efficient use of computer time. Although apparently there has been little use of these methods so far (Algorithm 1 was used by BORTZ et al. [1.30] and PUNDARIKA et al. [1.74]), we have discussed these methods in detail since we feel that they will become increasingly important in the future.

1.3.2 Computations with Conservation Laws: "Hydrodynamic" Slowing Down

If the kinetic evolution defined by the master equation, (1.19), is such that an extensive variable $A(t) \equiv 1/V \int d^d\underline{r}a(\underline{r},t)$ exists for which

$$\frac{dA(t)}{dt} \equiv 0 \quad , \tag{1.51}$$

we call (1.51) a conservation law. In canonic ensemble calculations of a fluid, for instance, the particle number N and hence the density $\rho = N/V$ is constant; in this case $a(\underline{r},t) = \rho(\underline{r},t)$, the local density. For a binary alloy of constant chemical composition, $a(\underline{r},t)$ would be the local concentration. In a microcanonic calculation also the energy would be conserved. It is possible, of course, to construct transition probabilities $W(\underline{x} \to \underline{x}')$ in (1.19) which keep arbitrary other extensive variables fixed ("conserved"). One just has to calculate the change in these variables produced by a transition $\underline{x} \to \underline{x}'$; if the change is nonzero $w(\underline{x} \to \underline{x}') = 0$. For computational reasons it is convenient, of course, to construct prescriptions for "moves" $\underline{x} \to \underline{x}'$ which take the conservation laws automatically into account. In an Ising magnet with conserved magnetization and changes where two (opposite) spins are flipped at the same time, the conservation law is automatically obeyed if one exchanges the two spins under consideration [1.33].

The conservation law (1.50) has important consequences for the kinetics of fluctuations [1.9]. If we introduce a current density $\underline{j}_A(\underline{r},t)$ for the quantity A, one can write down instead of (1.50) a continuity equation for the density of A,

$$\frac{\partial a(\underline{r},t)}{\partial t} + \nabla \underline{j}_A(\underline{r},t) = 0 \tag{1.52}$$

Suppose now the transitions $\underline{x} \to \underline{x}'$ involve only "local" changes of $a(\underline{r},t)$, i.e., $a(\underline{r},t)$ changes only within some volume v around $\underline{r}$, with v staying finite as $V \to \infty$. This assumption holds, for instance, if in a model of alloy kinetics the transitions $\underline{x} \to \underline{x}'$ mean the interchange of *neighboring* atoms in the system; it would not hold if arbitrary atoms in the system would be interchanged. For fluids this assumption holds if $\underline{x} \to \underline{x}'$ means a change $\underline{r} \to \underline{r} + \underline{\xi}\nabla$, where ∇ is a constant and $\underline{\xi} = (\xi_1, \ldots, \xi_d)$ an array of random numbers with $0 < \xi_i < 1$. This is a type of evolution commonly used, cf. [1.16] and Chap.2). Then we may approximate the current as

$$\bar{\underline{j}}_A(\underline{r},t) = -D_{AA}\,\nabla a(\underline{r},t) \quad , \tag{1.53}$$

if the deviations of $a(\underline{r},t)$ from thermal equilibrium are small. The minus sign in (1.53) means that the system tends to smooth out local fluctuations and evolve towards a translationally invariant homogeneous thermal equilibrium state. From (1.52,53) we immediately find a diffusion equation

$$\frac{\partial a(\underline{r},t)}{\partial t} = D_{AA}\,\nabla^2 a(\underline{r},t) \quad . \tag{1.54}$$

Considering Fourier components $A(\underline{k},t) = V^{-1} \int d^d\underline{r}\, \exp(i\underline{k}\underline{r}) a(\underline{r},t), a(\underline{r},t) = \int d^d\underline{k}$ $\exp(-i\underline{k}\underline{r}) A(\underline{k},t)$ we find

$$\frac{\partial A(\underline{k},t)}{\partial t} = -D_{AA}\, k^2 A(\underline{k},t) \quad , \tag{1.55a}$$

which is integrated as

$$A(\underline{k},t) = A(\underline{k},\infty) + [A(\underline{k},0) - A(\underline{k},\infty)]\, e^{-D_{AA}k^2 t} \quad . \tag{1.55b}$$

Equation (1.55b) exhibits the "hydrodynamic slowing down", $\tau_{AA}(\underline{k}) = (D_{AA}k^2)^{-1}$, as mentioned earlier [cf. (1.33)]. For a more rigorous justification of (1.54,55) see KAWASAKI [1.33]. We emphasize that (1.55) has important consequences for the rate at which equilibrium is approached in the simulations, as discussed already in Sect.1.2.4. In simulations of alloys or fluids, the structure factor will approach thermal equilibrium very slowly if k is small, and hence the concentration correlation $\langle c(\underline{r})c(\underline{r}')\rangle$ or density correlation $\langle \rho(\underline{r})\rho(\underline{r}')\rangle$ will approach equilibrium very slowly if $|\underline{r} - \underline{r}'|$ is large. For these quantities, one hence expects a quicker approach to equilibrium if the grand-canonic rather than the canonic ensemble is used.

1.3.3 Slowing Down at Phase Transitions. How to Estimate the Order of a Transition

It has already been mentioned in Sect.1.2.4 that the relaxation times of the stochastic model (1.19) diverge at a critical point. Apart from this slowing down at second-order phase transitions, there is also a slowing down associated with first-order phase transitions. Therefore the interpretation of Monte Carlo calculations becomes difficult if the system is in the vicinity of any phase transitions. On the other hand, there is intrinsic interest in estimating phase diagrams from Monte Carlo simulations (cf. Chap.2-4,8). Thus we discuss this slowing down here in some detail, and use the case of an anisotropic antiferromagnet in a magnetic field as a simple example (Fig.1.5) [1.48]. If the field $H_{\parallel}$ in the direction of the easy axis is varied, the system may undergo transitions of first-order (a,b) or second-order (c). In the first-order case, the magnetization $M_{\parallel}$ has a jump, while in the second-order case it only has a point with infinite slope (infinite susceptibility). Similar divergencies may (b) or may not (a) occur in the first-order case. The last possibility (a) is expected to involve the existence of metastable states (it is this case that one expects to apply to transitions in alloys, adsorbed layers, fluid-solid-gas systems, etc.). The behavior of the respective relaxation times is sketched also in Fig.1.5, middle part. Since the relaxation time τ_M diverges in all three cases (a)-(c), it will exceed the computing time for some range of values of the field $H_{\parallel}$. Thus the simulation will not reproduce the upper part of Fig.1.5 correctly, but rather one will observe *hysteresis* in all three cases (a)-(c); the result for the magnetization will depend on the initial state (one will get different results starting from values characteristic for the high-field side and the low-field side of the transition). Such a hysteresis hence indicates that the system undergoes a phase transition, although one cannot definitely distinguish between the behavior where a true transition occurs ($\tau_M \to \infty$) and where the relaxation time only becomes very large but stays finite. Since this hysteresis occurs also in the second-order case, it is difficult to estimate the *order* of the phase transition. The only possibility to distinguish

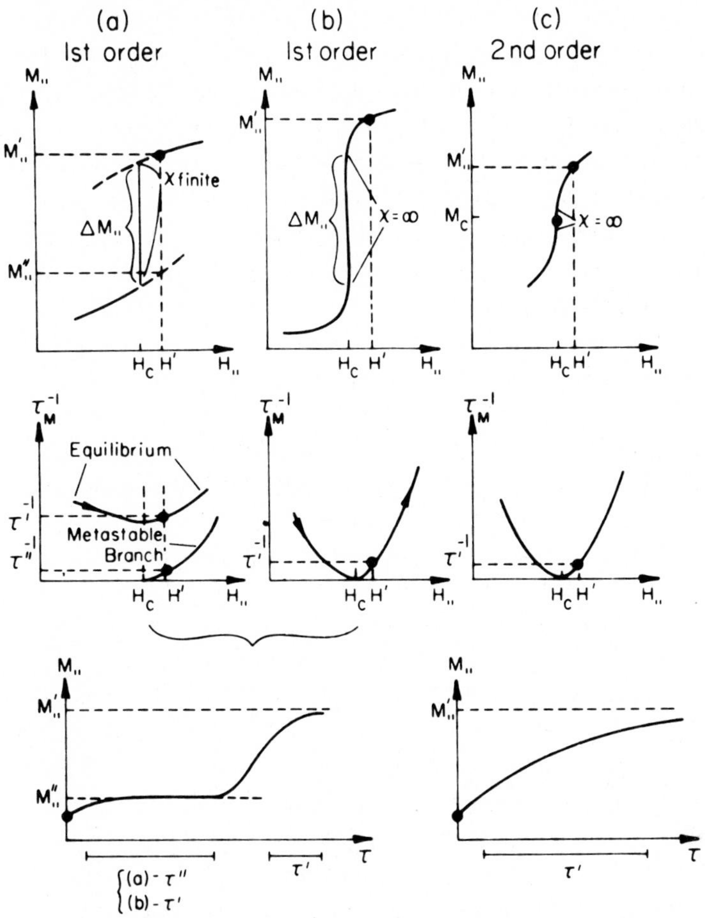

Fig.1.5a-c Relaxation effects near phase boundaries in an anisotropic antiferromagnet: (a) first-order phase change with a finite susceptibility as $H \to H_C$. The relaxation time of the metastable state is τ'' and of the equilibrium state τ'; (b) first-order phase change with $\chi \to \infty$ as $H \to H_C$. The relaxation times are the same (τ') near the ends of the magnetization discontinuity; (c) second-order phase change with a single relaxation time (τ') to equilibrium [1.48]

the two cases comes from a kinetic analysis (lower part of Fig.1.5). Since the magnetization relaxes quicker in the "forbidden region" of the jump than close to the "metastable" value $M''_{\|}$, one observes [1.48,75] distinct "two-step" relaxation, while the rate of relaxation is monotonic in the second-order case.

In principle, nothing can be said concerning equilibrium thermodynamics in the region where this hysteresis occurs. In practice, one often tries to estimate the magnitude of the jump from a Maxwell equal-area rule [1.48] (cf. Chap.3). We emphasize that such a prescription implicitly assumes that the relaxation times at the two sides of the transition are approximately symmetric with respect to it, since the hysteresis in our case is of purely kinetic origin. While such a symmetry assumption is reasonable for magnetic transitions, it is probably unreasonable for transitions like the liquid-solid transition. In principle, the correct way to locate the transition and hence estimate the magnitude of the jump would be to compare the free energies of the two phases. But as we have seen (Sect.1.2.2), the estimation of the free energy from the Monte Carlo method is difficult.

The situation is even more difficult if a first derivative of the appropriate thermodynamic potential is used as an independent variable (in a magnet this would mean doing calculations at constant magnetization rather than at constant field; in a fluid-gas system it means doing calculations at constant density; in an alloy it means doing calculations at constant concentration rather than constant chemical potential difference, etc.). In the region of the first-order transition where this variable has a jump, thermal equilibrium then consists of a coexistence of two phases. If one uses a one-phase state as an initial condition in that region, the approach towards thermal equilibrium is so slow [1.63] that in many cases it is impossible to derive accurate estimates for the equilibrium properties.

1.4 Variants of the Monte Carlo Method

1.4.1 The Approximation of Alexandrowicz

While in the methods described so far, one starts with a quite arbitrary configuration of the system and lets it then dynamically "relax" to thermal equilibrium states, ALEXANDROWICZ et al. [1.76-79] suggested an approximate but immediate construction procedure of an equilibrium configuration, based on adding particles gradually to an empty volume. Suppose n such configurations have been constructed. If the method were exact, their respective probabilities should be given by $P_i = Z^{-1} \exp(-E_i/k_BT)$, where E_i is the energy of the i th configuration. In practice, their probability will be given by some different distribution P_i which depends on the details of the construction procedure. Since the minimum principle for the free energy functional states that

$$F(\underline{P}') \equiv \sum_{\text{all } i} P_i'(E_i + k_BT \log P_i') > F(\underline{P}) \equiv \sum_{\text{all } i} P_i(E_i + k_BT \log P_i) \quad , \qquad (1.56)$$

it is appealing to use $\underline{P}'$ which depends on a set of parameters $\underline{x}$, and minimize $F(\underline{P})$ with respect to $\underline{x}$. Thus one obtains the "best" approximate distribution consistent with the constraints imposed by the construction procedure. We explain the latter for the case of the two-dimensional Ising model. There one starts choosing the spins of the first row of the lattice completely at random. The orientations of the spins in the second (and all subsequent) rows is specified by choosing them according to some transition probability $P(\underline{x})$, which depends on the configuration of the preceding row and on the set of parameters $\underline{x}$ mentioned above. These parameters describe the magnitude of the effect of each spin in the last row on the spin considered, and are chosen such that they decrease with increasing distance. Clearly, this is an approximation, since the probability for spin σ_k depends on its four neighbors on the lattice, only two of which are already known. The sign of the two unknown spins is "guessed" in some way in terms of the signs of the other spins in the last row. For the second configuration of the lattice, one uses the last row of the first lattice as neighboring row to the first row and thus generates a nonrandom configuration, and so on. Then $F(\underline{P}')$ in (1.56) is approximated as

$$F_n(\underline{x}) = n^{-1} \sum_{i=1}^{n} [E_i + k_B T \log P_i(\underline{x})] \quad , \tag{1.57}$$

where $P_i(\underline{x})$ is the product of the N transition probabilities according to which the spins have been chosen. Typically one chooses $n \approx 10^3$. In contrast to the states generated in the usual Monte Carlo method, there is only very little correlation between the subsequently generated lattice configurations. Thus it suffices to exclude the first configuration (which is somewhat influenced by the initial random row) from the averaging. At each temperature one has then to seek the optimum set $\underline{x}^*$ which minimizes (1.57). For this set $\underline{x}^*$, one then calculates average energy $E_n(\underline{x}^*)$, magnetization $M_n(\underline{x}^*)$, entropy $S_n(\underline{x}^*)$; specific heat and susceptibility are obtained from standard fluctuation relations. Clearly, an advantage of this scheme is that it yields entropy and free energy of the system directly.

This approximation so far has been applied to both the two- and three-dimensional Ising model [1.76,79], to polymers with excluded volume interaction [1.77], and to a fluid model consisting of hard cubic molecules with an attractive potential [1.78]. In the Ising case the method compares favorably with both exact results [1.70,80] and conventional Monte Carlo calculations. It seems that the method yields useful results even right at T_c, where the conventional method can hardly be used for large lattices [see (1.36) and associated discussion]. On the other hand, the method has some inherent deficiencies: spin correlations parallel to the rows and perpendicular to them are not precisely identical [1.76], etc. In addition, there is no general rule according to which one can find a parametric representation for a suitable transition probability for a system considered, and it is unclear how the accuracy depends on the numbers of parameters $\underline{x}$ which are optimized in the general case. We feel that in order to demonstrate the general usefulness of an approximation it is necessary but not sufficient to show that it works well for the two-dimensional Ising model. For instance, there are many "real space renormalization group" -

approximations [1.81] which work surprisingly well for nearest neighbor Ising lattices, but are less useful for other systems [1.82].

1.4.2 Renormalization Group Treatments Utilizing Monte Carlo Methods

In view of the great recent interest in critical properties at second-order phase transitions (critical exponents, amplitudes, scaling functions, etc.) [1.10,29,83], there have been various attempts to estimate these properties with Monte Carlo methods (see [1.3,4] and Chap.3 for a review). Typically, both the conventional method and the ALEXANDROWICZ approximations [1.79] yield estimates which are accurate to about 10%. This accuracy is not sufficient to answer most questions of interest. While the renormalization group approach [1.84] yields very precise estimates for systems with dimensionality close to 4, or with order parameter dimensionality close to infinity, estimates for physical systems at two and three dimensions are mostly based on extrapolations whose accuracy is not known precisely. Thus these techniques have been complemented by "real space" renormalization group methods [1.81] which can be applied directly for two- and three-dimensional lattice systems, but involve complicated numerical calculations. It is at that point that Monte Carlo techniques can be useful [1.85,86].

The renormalization transformation is effected by (I) grouping neighboring lattice sites into cells; (II) associating with each cell a new variable (cell spin) which reflects a collective property of the initial variables (site spins) inside the cell, (III) finding a hamiltonian for the cell spins in such a way that the partition function will be conserved. For an Ising model, this means

$$\exp[-\mathcal{H}'(\{\sigma'_\alpha\})/k_BT] = \sum_{\{\sigma_i\}} \prod_\alpha \frac{1}{2} [1 + \sigma'_\alpha f_\alpha(\sigma_{\alpha,1}, \ldots, \sigma_{\alpha,L})] \exp[-\mathcal{H}(\{\sigma_i\})/k_BT] \quad , \tag{1.58}$$

where $\sigma_{\alpha,1}, \ldots, \sigma_{\alpha,L}$ are the L site spins of cell α, σ'_α is the cell spin, $\mathcal{H}$ and $\mathcal{H}'$ are the original and renormalized hamiltonians, and $f_\alpha = \mathrm{sign}(\sum_{i=1}^{L} \sigma_{\alpha,L})$ [1.81]. One then assumes that $\mathcal{H}'/k_BT$ can be approximated by $\mathcal{H}'/k_BT = -K' \sum_{<\alpha,\beta>} \sigma'_\alpha \sigma'_\beta$, if $\mathcal{H}/k_BT = -K \sum_{<i,j>} \sigma_i\sigma_j$, and the sums are on nearest neighbors only. FRIEDMAN and FELSTEINER [1.86] stated that (1.58) is then equivalent to the condition

$$<f_\alpha f_\beta>_{\mathcal{H}} = <\sigma'_\alpha \sigma'_\beta>_{\mathcal{H}'} = \tanh(2dK') \tag{1.59}$$

in d dimensions. The correlation function $<f_\alpha f_\beta>_{\mathcal{H}}$ requires thermal averaging in a system of $L = \ell^d$ spins, and this is done by ordinary Monte Carlo sampling. One thus obtains $<f_\alpha f_\beta>$ as a function of K. The point $K = K^*$ where this function intersects the function $\langle\sigma'_\alpha\sigma'_\beta\rangle_{K'}$ is then the fixed point of the renormalization group transformation, and one finds the "thermal eigenvalue" [1.81] and associated exponent ν from the relation between K and K' in the vicinity of the fixed point, $\lambda_T = \partial K'/\partial K|_{K^*}$, $\lambda_T \propto \ell^{1/\nu}$ as $\ell \to \infty$. Unfortunately, it turns out that the convergence of the estimates for ν converge very slowly with ℓ (probably logarithmically), but perhaps better results could be obtained with other choices of the somewhat arbitrary function f_α.

The earlier work of MA [1.85] attempted to determine the parameters of $\mathcal{H}'(\{\sigma'_\alpha\})$ in (1.58) directly, by determining flip rates of cell spins back and forth and using the detailed balance condition, (1.5). This approach has the immediate advantage that it allows one to estimate dynamic critical properties as well. Furthermore, a next-nearest neighbor interaction was taken into account. This fact is important since the transformation (1.58) generates higher neighbor interactions in $\mathcal{H}'$ even if one starts out with purely nearest neighbor interaction in $\mathcal{H}$. The results obtained for the two-dimensional Ising model are quite encouraging, but both a systematic investigation of this method and applications to other systems remain to be given.

1.5 Conclusions

Here we first summarize the main results and main problems discussed in this introductory chapter, and then give a brief list of check points which should enable the nonspecialist to identify "good" Monte Carlo work.

(I) The "conventional" Metropolis Monte Carlo method for the computation of static averages in the canonic ensemble is by now fairly well investigated. In cases where exact answers are available, the method yields satisfactory results with a fair amount of computing time. Finite size problems occurring particularly close to second-order phase transitions can be systematically analyzed in terms of finite size scaling theory. Statistical errors are interpreted in terms of the dynamic properties of the associated stochastic model. A few cases where the method does not give satisfactory answers are explained in terms of nearly divergent susceptibilities and relaxation times.

(II) Recent developments concern variants of the scheme which allow better estimation of the free energy (Sect.1.2.2), or a quicker realization of the master equation (Sect.1.3.1), or approximate methods which are more suitable for investigating second-order phase transitions (Sect.1.4). While we feel that most of these developments require further investigation in order to convincingly prove their usefulness, these approaches are potentially very interesting, and further studies along these lines will be valuable.

(III) Although application of the METROPOLIS Monte Carlo method is by now a standard procedure in statistical physics, nevertheless, quite a lot of work of poor quality appears even in the recent literature. This may be due to the fact that many people are not fully aware of many of the difficulties which occur (cf. Sects.1.2.4-1.2.6), and hence underestimate the amount of computing time which is necessary for good results. Hence, following KALOS [1.87], we mention a few points which in good Monte Carlo work should be given some consideration:

1) *Error estimates* must be included, and the procedure by which they have been obtained should be mentioned.

2) *Bias* of Monte Carlo estimates may occur, and if so, it should be estimated, or bounded.

3) *Long runs* are necessary such that the averaging time is large compared to the relaxation time. *Very long runs* in a few cases are useful to show that no surprises appear, and that the asymptotic regime of validity of the central limit theorem has actually been reached, where meaningful error estimates and confidence limits can be established.

4) *Existence (and even smallness) of the variance* is necessary in order to obtain a meaningful estimate for a quantity. If the variance of a quantity is too large, it is practically not possible to satisfy condition 3).

5) *The size dependence* must be considered. A few runs at fairly larger system sizes (and also at smaller ones) help to give a feeling for the magnitude of the effects occurring. And where possible, one should check estimates of asymptotic size dependence of the important results.

6) *The insensitivity of the results to details of the method* should be checked. For instance, it should not matter if one uses a different transition probability, or different move parameter for generating a new configuration, or different initialization of random numbers, etc.

Good Monte Carlo work obviously is not cheap. From the point of view of an economical use of computer time, it is clear that not all of these points need be considered in every study. But one has to watch out for cases where a "small" change in a parameter produces dramatic changes in system behavior and hence quality of the Monte Carlo work.

These criteria may also help the nonspecialist to assess the credibility of published Monte Carlo results.

Acknowledgements. The author is indebted to M.H. KALOS for unpublished notes on which Sect.1.3.1 is based, and for pertinent and very helpful comments on the manuscript. He also thanks D.P. LANDAU and H. MÜLLER-KRUMBHAAR for detailed information on their research over many years. Thanks are also due to D. LEVESQUE and H. MÜLLER-KRUMBHAAR for useful comments on the manuscript.

References

1.1 N. Metropolis, A.W. Rosenbluth, M.N. Rosenbluth, A.H. Teller, E. Teller: J. Chem. Phys. *21*, 1087 (1953)
1.2 H. Müller-Krumbhaar, K. Binder: J. Statist. Phys. *8*, *1* (1973)
1.3 K. Binder: Advan. Phys. *23*, 917 (1974)
1.4 K. Binder: In *Phase Transitions and Critical Phenomena*, ed. by C. Domb, M.S. Green, Vol. 5b (Academic Press, New York 1976) p. 1
1.5 J.M. Hammersley, D.C. Handscomb: *Monte Carlo Methods* (Methuen and Co., London 1964)
Yu. A. Shreider: *The Monte Carlo Method* (Oxford University Press, Oxford 1966)
1.6 L.L. Carter, E.D. Cashwell: *Particle Transport Simulation with the Monte Carlo Method*, available as TID-26607, National Technical Information Service
1.7 S.M. Ermakow: *Die Monte Carlo Methode und verwandte Fragen* (R. Oldenburg Verlag, München 1975)
1.8 R. Alben, S. Kirkpatrick, D. Beeman: Phys. Rev. B *15*, 346 (1977)
1.9 L.P. Kadanoff, P.C. Martin: Ann. Phys. (NY) *24*, 419 (1963)
1.10 L. van Hove: Phys. Rev. *93*, 1374 (1954). For an up-to-date review in this rapidly developing field see
B.I. Halperin, P.C. Hohenberg: Rev. Mod. Phys. *49*, 435 (1977)
1.11 For a recent introduction to relaxation kinetics associated with metastable states see, e.g.,
K. Binder: In *Fluctuations, Instabilities and Phase Transitions*, ed. by T. Riste (Plenum Press, New York 1975) p. 53
1.12 See, e.g., S. Windwer: *Markov Chains and Monte Carlo Calculations in Polymer Science* (Marcel Dekker, New York 1970)
1.13 See, e.g., N.S. Gillis, N.R. Werthamer, T.R. Koehler: Phys. Rev. *165*, 951 (1968)
1.14 T.L. Hill: *Thermodynamics of Small Systems* (Benjamin, New York 1963)
1.15 L.D. Fosdick: Methods Comp. Phys. *1*, 245 (1963)
1.16 W.W. Wood: In *Physics of Simple Liquids*, ed. by H.N.V. Temperley, J.S. Rowlinson, G.S. Rushbrooke (North-Holland, Amsterdam 1968) p. 115
1.17 I.Z. Fisher: Sov. Phys. Uspekhi *2*, 783 (1959)
1.18 J.L. Doob: *Stochastic Processes* (Wiley and Sons, New York 1953)
1.19 E. Ising: Z. Physik *31*, 253 (1925)
1.20 Z.W. Salsburg, J.D. Jacobsen, W. Fickett, W.W. Wood: J. Chem. Phys. *30*, 65 (1959)
1.21 I.R. McDonald, K. Singer: J. Chem. Phys. *47*, 4766 (1967); *50*, 2308 (1969)
1.22 J.P. Valleau, D.N. Card: J. Chem. Phys. *57*, 5457 (1972)
1.23 G.M. Torrie, J.P. Valleau: Chem. Phys. Lett. *28*, 578 (1974); J. Comp. Phys. *23*, 187 (1977)
1.24 C.H. Bennett: J. Comp. Phys. *22*, 245 (1976)
1.25 K. Binder: Physica *62*, 508 (1972)
1.26 D. Levesque, L. Verlet: Phys. Rev. *182*, 307 (1969)
J.P. Hansen, L. Verlet: Phys. Rev. *184*, 151 (1969)
J.P. Hansen: Phys. Rev. A *2*, 221 (1970)
1.27 A. Sadiq: Phys. Rev. B *9*, 2299 (1974)
1.28 D.A. Chesnut, Z.W. Salsburg: J. Chem. Phys. *38*, 2861 (1963)
1.29 See, e.g., M.E. Fisher: Rept. Progr. Phys. *30*, 615 (1967)
1.30 A.B. Bortz, M.H. Kalos, J.L. Lebowitz: J. Comp. Phys. *17*, 10 (1975)
1.31 M.H. Kalos: To be published
1.32 The first calculation of dynamic correlation functions on this basis was given in
H. Müller-Krumbhaar, K. Binder: Intern. J. Magn. *3*, 113 (1972)
1.33 K. Kawasaki: In *Phase Transitions and Critical Phenomena*, ed. by C. Domb, M.S. Green (Academic Press, New York 1972), Vol. 2, p. 443
1.34 Yu.Ya. Gotlib: Sov. Phys.-Solid State *3*, 1574 (1962)
R.J. Glauber: J. Math. Phys. *4*, 294 (1963)
1.35 See, e.g., W. Marshall, R.D. Lowde: Rept. Progr. Phys. *31*, 705 (1968)

1.36 N. Ogita, A. Ueda, T. Matsubara, H. Matsuda, F. Yonezawa: J. Phys. Soc. Jpn. *26*, S-145 (1969)
1.37 P.A. Flinn, G.M. McManus: Phys. Rev. *124*, 54 (1961)
J.R. Beeler, Jr., J.A. Delaney: Phys. Rev. *130*, 962 (1963)
1.38 For a review, see B. Alder: In *Physics of Simple Liquids*, ed. by H.N.V. Temperley, J.S. Rowlinson, G.S. Rushbrooke (North-Holland, Amsterdam 1968) Chap.4
1.39 D.L. Ermak: J. Chem. Phys. *62*, 4189 (1975)
P. Turq, F. Lantelme, H.L. Freedman: J. Chem. Phys. *66*, 3039 (1977)
1.40 K. Binder: Phys. Rev. B *8*, 3419 (1973)
M. Suzuki: Intern. J. Magn. *1*, 123 (1971)
1.41 H. Yahata, M. Suzuki: J. Phys. Soc. Jpn. *27*, 1421 (1969)
1.42 E. Stoll, K. Binder, T. Schneider: Phys. Rev. B *8*, 3266 (1973)
H. Müller-Krumbhaar, D.P. Landau: Phys. Rev. B *14*, 2014 (1976)
1.43 P.A.W. Lewis, A.S. Goodman, J.M. Miller: IBM Syst. J. *2*, 145 (1969)
See also D.E. Knuth: *The Art of Computer Programming* (Addison-Wesley, Reading Mass. 1969)
B. Jansen: *Random Number Generators* (Petersson, Stockholm 1976)
1.44 R. Friedberg, J.E. Cameron: J. Chem. Phys. *52*, 6049 (1970)
1.45 D.P. Landau: Phys. Rev. B *13*, 2997 (1976)
1.46 This was done for instance by K. Binder, D.P. Landau: Phys. Rev. B *13*, 1140 (1976)
1.47 D.P. Landau: Phys. Rev. B *14*, 255 (1976)
1.48 D.P. Landau, K. Binder: Phys. Rev.B *17*, 2328 (1978)
1.49 R. Landauer, J.A. Swanson: Phys. Rev. *127*, 1668 (1961), and references contained therein
1.50 K. Binder, K. Schröder: Phys. Rev. B *14*, 2142 (1976)
1.51 Z. Racz: Phys. Rev. B *13*, 263 (1976)
1.52 N.D. Mermin, H. Wagner: Phys. Rev. Lett. *17*, 1133 (1966)
1.53 G.S. Rushbrooke, P.J. Wood: Mol. Phys. *1*, 257 (1958)
H.E. Stanley, T.A. Kaplan: Phys. Rev. Lett. *17*, 913 (1966)
1.54 V.L. Berezinski: Zh. Eksp. J. Teor. Fiz. *59*, 907 (1970; *61*, 1144 (1971)
J.M. Kosterlitz, D. Thouless: J. Phys. C *5*, L 124 (1972); C *6*, 1181 (1973)
J. Zittartz: Z. Physik B *23*, 55, 63 (1976)
1.55 R.E. Watson, M. Blume, G.H. Vineyard: Phys. Rev. B *2*, 684 (1970)
1.56 C. Kawabata, K. Binder: Solid State Commun. *22*, 705 (1977)
1.57 K. Binder: Phys. Rev. B *15*, 4425 (1977)
1.58 K. Binder, H. Müller-Krumbhaar: Phys. Rev. B *7*, 3297 (1973)
1.59 R. Zwanzig, N.K. Ailawadi: Phys. Rev. *182*, 280 (1969)
1.60 See the Proc. CECAM Workshop on Protein Dynamics, Université de Paris XI, Orsay (1976)
1.61 G. Ciccotti, A. Tenenbaum: J. Stat. Phys. *23*, 767 (1980)
1.62 S.G. Brush, H.L. Sahlin, E. Teller: J. Chem. Phys. *45*, 2102 (1966)
1.63 A.B. Bortz, M.H. Kalos, J.L. Lebowitz, M.A. Zendejas: Phys. Rev. B *10*, 535 (1974)
J. Marro, M.H. Kalos, J.L. Lebowitz: Phys. Rev. B *12*, 2000 (1975)
1.64 H. Müller-Krumbhaar, K. Binder: Z. Physik *254*, 269 (1972)
1.65 H. Müller-Krumbhaar, K. Binder: In *Proc. 3rd Internat. Colloquium on Advanced Computing Methods in Theoretical Physics*, ed. by A. Visconti (Centre de Phys. Theor. CNRS, Marseille 1974), p. B IX-1
1.66 H.C. Bolton, C.H. Johnson: Phys. Rev. B *13*, 3025 (1976)
1.67 K. Binder, P.C. Hohenberg: Phys. Rev. B *6*, 3461 (1972)
1.68 H. Bethe: Proc. Roy. Soc. (London) A *150*, 552 (1935)
1.69 For a finite system there does not exist any spontaneous magnetization which is thermodynamically stable, $\langle M \rangle_{NT} \equiv 0$ in thermal equilibrium [1.27]
1.70 A.E. Ferdinand, M.E. Fisher: Phys. Rev. Lett. *19*, 169 (1967)
1.71 M.E. Fisher: In *Critical Phenomena*, ed. by M.S. Green (Academic Press, New York 1971)
M.E. Fisher, M.N. Barber: Phys. Rev. Lett. *28*, 1516 (1972)
1.72 K. Binder: Thin Solid Films *20*, 367 (1974)
1.73 K. Binder, P.C. Hohenberg: Phys. Rev. B *9*, 2194 (1974)

1.74 E.S. Pundarika, K. Hanson, J.W. Morris, Jr.: Bull. Am. Phys. Soc. *22*, 302 (1977)
1.75 K. Binder, H. Müller-Krumbhaar: Phys. Rev. B *9*, 2328 (1974)
1.76 Z. Alexandrowicz: J. Chem. Phys. *55*, 2765 (1971); J. Statist. Phys. *5*, 19 (1972)
1.77 Z. Alexandrowicz: J. Chem. Phys. *55*, 2780 (1971)
1.78 Z. Alexandrowicz, M. Morlow: J. Chem. Phys. *56*, 1274 (1972)
1.79 H. Meirovitch, Z. Alexandrowicz: J. Statist. Phys. *16*, 121 (1977)
1.80 L. Onsager: Phys. Rev. *65*, 117 (1944)
C.N. Yang: Phys. Rev. *85*, 809 (1952)
1.81 Th. Niemeyer, M.J.J. van Leeuwen: In *Phase Transitions and Critical Phenomena*, ed. by C. Domb, M.S. Green, Vol. 6 (Academic Press, New York 1976)
1.82 The results are much less satisfactory already in the presence of a next-nearest neighbor interaction. For a discussion see
K. Binder: In Proceedings of the Int. Symp. on Order and Disorder in Solids, Paris, July 1977; J. Physique *38-C7*, 396 (1977)
1.83 H.E. Stanley: *An Introduction to Phase Transitions and Critical Phenomena* (Oxford University Press, Oxford 1971)
C. Domb, M.S. Green (eds.): *Phase Transitions and Critical Phenomena*, Vols. 1-6 (Academic Press, New York 1972-1977)
1.84 For review, see Ref. [1.83], Vol. 6
K.G. Wilson, J. Kogut: Phys. Rep. C *12*, 75 (1974)
1.85 S.K. Ma: Phys. Rev. Lett. *37*, 461 (1976)
1.86 Z. Friedman, J. Felsteiner: Phys. Rev. B *15*, 5317 (1977)
1.87 M.H. Kalos: Unpublished notes

Addendum

The use of Monte Carlo methods for the renormalization group approach has been developed further recently [1.88-90]. It has been used to evaluate the cumulant expansion of the "blockspin"-interaction for three dimensional Ising models [1.88] and has been applied to the percolation problem [1.89]. A particular relevant progress was obtained by KINZEL [1.90] who succeeded in treating Ising, Heisenberg and XY models in $d = 2$, $d = 3$ and $d = 4$ dimensions by a new combination of Monte-Carlo evaluation of blockspin properties together with a meanfieldtreatment of blockspin-interactions. While this method seems very encouraging for ordinary phase transitions, it fails for the $d = 2$ XY and Heisenberg models. These deficiencies probably are not due to the use of Monte Carlo but to the meanfield-blockspin-interaction.

1.88 Z. Racz, P. Rujan: Z. Phys. B *28*, 287 (1977)
1.89 P.J. Reynolds, H.E. Stanley, W. Klein: J. Phys. A *11*, L199 (1978)
1.90 W. Kinzel: Phys. Rev. B *19*, 4584 (1979)

2. Simulation of Classical Fluids

D. Levesque, J. J. Weis, and J. P. Hansen

A concise review is given of Monte Carlo calculations for liquid systems, starting from hard-sphere models and Lennard-Jones fluids both in three and in two dimensions, and describing then ionic systems in detail; both fully ionized matter (i.e., dense plasmas) and electrolyte solutions, molten salts, etc. are treated. The third part of this review considers molecular fluids, while in the last section the gas-liquid interface is treated. Attention is paid both to technical problems, like the appropriate treatment of long-range interactions by the Monte Carlo method, and to the physical consequences of the results obtained. A detailed comparison of the numerical results with several analytic theories of the respective fluid systems is performed.

2.1 Overview

Liquids and highly compressed neutral or charged fluids present none of the simplifying features of the other states of matter; they lack the symmetry and periodicity of crystalline solids as well as the low density of gases. Consequently the statistical mechanics of liquids and compressed fluids cannot be based on a systematic treatment starting from either the perfect crystal or the ideal gas. For that reason, our understanding of the disordered high-density states of matter is fairly recent, and it is fair to say that the significant progress in the field over the last twenty years owes much to computer "experiments", by which we mean both Monte Carlo (MC) calculations of thermodynamic averages and static distribution functions, and "molecular dynamics" (MD) simulations of the time evolution of small samples of a few hundred or thousand interacting particles.

Monte Carlo computations of canonic ensemble (NVT) averages are practically all based on the fundamental algorithm of METROPOLIS et al. [2.1] and its extensions to isothermal-isobaric (NPT) and grand-canonic ensembles. This algorithm is introduced and discussed in Chap.1. The present chapter reviews essentially the *results* obtained for a wide variety of liquids and compressed fluids, using this algorithm.

We shall not be concerned with methodology, except for some special topics which deal with specific problems, like absolute entropy calculations or the treatment of long-range forces between particles. We only recall briefly that the Metropolis algorithm samples configurations, for systems of a few hundred or thousand particles interacting generally through given pairwise additive forces, according to a probability density which is proportional to a Boltzmann factor. In order to minimize surface effects, periodic boundary conditions are used almost exclusively. Averages are taken over the total number of configurations. The statistical quality of a MC run depends on the total number of configurations generated, and statistical errors can be estimated from the scatter of subaverages. There are, of course, large variations in the statistical quality of published results. The dependence of these results on the number N of particles in the sample is also an important feature which is not always sufficiently analyzed. It is clear that, for these two reasons, the published MC results cannot all be accepted with the same degree of confidence.

In view of the very extensive literature dealing with applications of the Monte Carlo method to various liquids and fluids (several hundred papers have already appeared on the subject!) we shall stress *recent* results, in particular on ionic and molecular liquids obtained over the last ten years. Results on more simple liquids (like hard-sphere or Lennard-Jones systems) will be dealt with more briefly, since they are adequately reviewed in other articles [2.2-6].

The MD method, which was first introduced by ALDER and WAINWRIGHT [2.7] yields averages over a microcanonic (constant energy) ensemble and its applications are in many respects similar to those of the MC method. It appears, hence, somewhat artificial to distinguish between both methods when they are applied to liquids and compressed fluids; for that reason we refer to MD results whenever they are complementary to those obtained by MC calculations; this excludes of course dynamic properties, which are solely accessible by the MD method.

The present chapter is restricted to *classical* systems, since quantum liquids will be treated separately in Chap.4. The review is organized in a rather arbitrary order, starting, roughly speaking, from the simpler systems and going to more and more complicated systems. In Sect.2.2 we review computer "experiments" on spherical, hard-core and discontinuous potentials, both in two and in three dimensions. Section 2.3 deals with spherical, continuous potentials, essentially with the MC results obtained for inverse power potentials and Lennard-Jones systems. Section 2.4 is devoted to ionic systems: strongly coupled plasmas, electrolyte solutions, molten salts, and liquid metals. In Sect.2.5 we consider systems of particles interacting through anisotropic force laws, essentially polar and molecular liquids. Finally Sect.2.6 briefly reviews the MC work on liquid surfaces.

2.2 Hard Core and Discontinuous Potentials

This section will deal with the computer simulation of systems of particles interacting by a potential with a "hard core" repulsive part. For these systems, simulations have been done in three and two dimensions. There are also some simulations for one-dimensional systems, but these are essentially devoted to the study of dynamic properties [2.2] and outside the scope of this review.

We shall consider both three- and two-dimensional systems, and in each case both homogeneous systems and mixtures will be discussed.

2.2.1 Hard-Sphere System

a) The Pure Hard-Sphere System

The pure hard-sphere system has been well studied by both MC and MD methods. Insofar as equilibrium properties are concerned, the MC method is certainly easier to program and is probably more efficient for the computation of the distribution functions. This is because the most efficient version of the MD method generates configurations only where two hard spheres are in contact. Also the MC method permits the use of unbiased or biased sampling of the configurations of the hard-sphere system, thus allowing the excess chemical potential or the entropy, quantities which are not so directly accessible by the MD method, to be calculated. Although in the following sections we shall briefly describe some work already well reviewed [2.3-6], our main emphasis will be put on simulations done since 1972.

The Equation of State

The first computation of the equation of state of a hard-sphere system was done by M. ROSENBLUTH and A. ROSENBLUTH in 1954 [2.8], by MC simulation. In 1957, the MD simulations of ALDER and WAINWRIGHT [2.7] revealed the existence of a fluid-solid phase transition in the hard-sphere system. This result was in apparent conflict with the calculations of [2.8]. This difficulty was solved by WOOD and JACOBSON [2.9], who obtained good agreement between their MC computations and the results in [2.7], and explained the initial disagreement between the two methods by the small number of configurations generated in the MC simulation of [2.8].

In their publication [2.10], ALDER and WAINWRIGHT gave detailed results for the equation of state for systems ranging in size from 4 to 500 hard spheres and did a careful study in the vicinity of the phase transition. However, they could not estimate the free energy from their MD computations and hence did not determine precise values for the transition densities. ROTENBERG [2.11] compared the equation of state obtained with 32 and 864 hard spheres. He found that the discrepancies occurring in the phase transition region are greater than the estimated errors. As did ALDER and WAINWRIGHT, ROTENBERG computed the pressure of the solid and fluid phases at the same density, but he did not observe a jump from one phase to the other which could have permitted the determination of the free energy of the solid phase.

REE and HOOVER [2.12,13] used the MC method for calculating the fifth, sixth, and seventh virial coefficients of the hard spheres. With these virial coefficients, they constructed the different Padê approximants of the equation of state. They found that the (3,3) Padê approximant including only six virial coefficients gives the best agreement with the simulation data in the fluid phase.

The accurate location of the transition plateau was obtained by HOOVER and REE [2.14,15]. They noted that the free energy of the hard-sphere system in the solid phase is identical to the free energy of a hard-sphere system where each particle is confined to a Wigner-Seitz cell of a face-centered-cubic lattice. So, computing the equation of state of a system where at all densities each hard sphere is artificially constrained to stay inside a Wigner-Seitz cell, HOOVER and REE determined the free energy by integrating the equation of state from zero to solid density. The MC simulations were done for 32, 108, 256 and 500 hard spheres, which made possible an extrapolation to the infinite system.

Their final results are that the hard-sphere solid and fluid phases are in thermodynamic equilibrium between the densities $0.667 \pm 0.003\ \rho_0$ and $0.736 \pm 0.003\ \rho_0$ at the pressure $(8.27 \pm 0.13)\rho_0 k_B T$, where ρ_0 is the density of close packing, and that the variation in entropy is $\Delta S/(Nk_B) = 0.312 \pm 0.015$.

A more direct way to locate the transition plateau is that proposed by TORRIE et al. [2.16]. This method is based on the observation that the transition pressure is that at which the difference in the chemical potentials between the system constrained to single occupancy of the Wigner-Seitz cells and the unconstrained system is zero. If Z_c and Z are, respectively, the configurational integrals of an N particle system with constraint and without constraint then the quantity

$$-\log(Z_c/Z) = \Delta S/k_B$$

is the difference in entropy between the two systems. The difference in the chemical potential is easily obtained from $\Delta S/k_B$. The essential difficulty is to obtain the value of Z_c/Z, which is the fraction of the unconstrained configurations which show single occupancy with respect to a lattice of Wigner-Seitz cells, superimposed on the cube containing the system of N hard spheres. Since a direct computation of Z_c/Z is impossible, TORRIE et al. used a biased multistage sampling method.

A stage consists of generating configurations with at least J singly occupied cells and counting the fraction of configurations with at least M singly occupied cells for all $M > J$. If we call $\Gamma(J)$ the volume of phase space corresponding to the configurations with J singly occupied cells, it is clear that

$$Z = \sum_{J=0}^{N} \Gamma(J) \quad \text{and} \quad Z_c = \Gamma(N) \quad .$$

Z_c/Z can be written

$$\frac{Z_c}{Z} = \frac{\Gamma(N)}{\sum\limits_{J=\ell_k}^{N} \Gamma(J)} \times \frac{\sum\limits_{J=\ell_k}^{N} \Gamma(J)}{\sum\limits_{J=\ell_{k-1}}^{N} \Gamma(J)} \times \cdots\cdots \times \frac{\sum\limits_{J=\ell_2}^{N} \Gamma(J)}{\sum\limits_{J=\ell_1}^{N} \Gamma(J)} \quad ,$$

where $\ell_1 = 0$, and the ℓ_i are a set of increasing positive integers $\leqq$ N and where each ratio corresponds to a stage.

For the ith stage, it is possible to introduce the fraction of configurations with exactly J singly occupied cells $f_i(J)$

$$\frac{\sum\limits_{J=\ell_i}^{N} \Gamma(J)}{\sum\limits_{J=\ell_{i-1}}^{N} \Gamma(J)} = \sum_{J=\ell_i}^{N} f_i(J) = F(i) \quad .$$

F(i) is the fraction of configurations with at least ℓ_i singly occupied cells. In order to increase the efficiency of the sampling, TORRIE et al. introduced a weight W(J) for the configurations with J singly occupied cells; W(J) increases with J. For the ith stage, they have

$$\frac{\sum\limits_{J=\ell_i}^{N} \Gamma(J)}{\sum\limits_{J=\ell_{i-1}}^{N} \Gamma(J)} = \frac{\sum\limits_{J=\ell_i}^{N} \frac{1}{W(J)} W(J)\Gamma(J)}{\sum\limits_{J=\ell_{i-1}}^{N} \frac{1}{W(J)} W(J)\Gamma(J)} \times \frac{\sum\limits_{J=\ell_{i-1}}^{N} W(J)\Gamma(J)}{\sum\limits_{J=\ell_{i-1}}^{N} W(J)\Gamma(J)}$$

and with the definition for $f_i'(J)$,

$$\frac{\sum_{J=\ell_i}^{N} \frac{1}{W(J)} W(J)\Gamma(J)}{\sum_{J=\ell_{i-1}}^{N} W(J)\Gamma(J)} = \sum_{J=\ell_i}^{N} \frac{f_i'(J)}{W(J)} ,$$

they finally obtained

$$F(i) = \frac{\sum_{J=\ell_i}^{N} f_i'(J)/W(J)}{\sum_{J=\ell_{i-1}}^{N} f_i'(J)/W(J)} .$$

The computation of $f_i'(J)$ is done by a "standard" Metropolis method. A configuration of N hard spheres, with k singly occupied cells ($k \geqq \ell_{i-1}$) is generated; then one sphere is moved at random in a small volume surrounding its initial position. If this displacement corresponds to a configuration with overlapping hard spheres or a number of singly occupied cells less than ℓ_{i-1}, the configuration is rejected; if this displacement corresponds to a configuration with no overlapping spheres and a number of singly occupied cells $k'>k$, the configuration is accepted with a probability 1, and if $k'<k$, it is accepted with a probability $W(k')/W(k)$. The next move is then performed. $f_i'(J)$ is the number of configurations with J singly occupied cells divided by the total number of configurations generated in the Markov chain described above.

TORRIE et al. studied systems of 32 and 108 hard spheres at $\rho = 0.60\ \rho_0$. With 32 hard spheres only a single stage is necessary to obtain Z_c/Z.

$$\frac{Z_c}{Z} = \frac{(32)}{\sum_{J=0}^{N} \Gamma(J)} = (1.8 \pm 0.3) \times 10^{-4}$$

and $\Delta S/(Nk_B) = 0.27 \pm 0.006$.

For 108 hard spheres, four stages are necessary with $\ell_1 = 0$, $\ell_2 = 94$, $\ell_3 = 102$, $\ell_4 = 104$ and one obtains

$$\frac{Z_c}{Z} \simeq 4.3\times10^{-14} \quad \text{and} \quad \Delta S/(Nk_B) = 0.285 \pm 0.007 .$$

ρ equal to 0.60 ρ_0 is not the density corresponding to the transition pressure, so some extrapolation must be done by using the equation of state to determine the pressure at which the difference in the chemical potentials between the constrained and unconstrained systems is zero.

At the transition TORRIE et al. obtained

$$\Delta S/(Nk_B) = 0.291 \pm 0.008 \quad .$$

Their result is in good agreement with that of HOOVER and REE [2.14,15]. Similar agreement was obtained for the pressure and for the densities of the coexisting phases.

We have described this computation [2.16] in detail because it illustrates very well the possibilities of the multistage sampling method [2.4,5] and the advantage of introducing "importance sampling" when the usual MC method is impracticable.

The work of TORRIE et al. gives the difference in entropy between the solid and fluid phases. However, the entropy of the fluid phase must still be determined by integrating the equation of state. ADAMS [2.17] showed that it is possible to compute directly the chemical potential of the hard-sphere system in the density domain $0.0707\ \rho_0 \leqq \rho \leqq 0.565\ \rho_0$. He used the following formula for the excess chemical potential μ':

$$\mu'/(k_BT) = -\log[P(\sigma)]$$

where σ is the hard-sphere diameter and $P(\sigma)$ is the probability that if an arbitrary point is chosen in the hard-sphere system, it would be possible to add another sphere at that point without a hard-sphere overlap occurring.

$P(\sigma)$ is also equal to

$$P(\sigma) = \int_\sigma^\infty 4\pi m(r) r^2 dr \quad ,$$

where $4\pi m(r)r^2dr$ is the probability that the nearest particle to a randomly chosen point is between r and $r + dr$. The compressibility factor z can be expressed as a function of $m(\sigma)$ and $P(\sigma)$. One obtains

$$z = 1 + 2\pi\, \sigma^3 m(\sigma)/[3P(\sigma)] \quad .$$

Thus if $m(r)$ is known, z and μ' are easily calculated. ADAMS computed $m(r)$ in the canonic ensemble for the systems of 32, 256, and 864 hard spheres. For 32 particles, after each step in the usual Markov-Metropolis process one point is chosen at random in the system and the distance of the nearest sphere is determined. For larger systems, the cube containing the N hard spheres is subdivided into 8 subcells. Every 250 MC steps, for the 256 particle system, 250 random points are chosen within the subcells (for the 864 particle system 250 is replaced by 500). From the computed distances of the nearest spheres, m(r) is easily obtained.

The results of ADAMS for μ' are in fairly good agreement with the μ' values calculated by integration of the equation of state. ADAMS completed his work with a determination of the equation of state using MC sampling in the grand-canonic ensemble, in place of the canonic ensemble. The MC procedure for sampling configurations in the grand-canonic ensemble was described by ADAMS and also very clearly in [2.6]. Thus we shall not give more details here. The equation of state computed by this method is also in agreement with the preceding simulations. However, the attempt to compute the compressibility is not successful at the densities larger than 0.50 ρ_0.

Recently WOODCOCK [2.18] did more MD simulations for the hard-sphere system, especially for the densities larger than the maximum stable density of the fluid phase. From these simulations, WOODCOCK attempted to prove "that the supercooled metastable hard-sphere system undergoes a glass transition, i.e., a higher order discontinuity in its free energy derivatives at a density intermediate between the equilibrium fluid freezing density and the density of amorphous close packing". The results of [2.18] are not convincing, since no real proof of a discontinuity in any of the free energy derivatives is reported.

From all this work which has been done since 1954, the equation of state of the hard-sphere system is very well known. It is interesting to observe that for the fluid phase, the results of the simulations can be reproduced within an error of a few percent, by

$$z = (1 + \eta + \eta^2 - \eta^3)/(1 - \eta)^3 \quad ,$$

where z is the compressibility factor and $\eta = \pi\rho\sigma^3/6$.

The simple formula, obtained by CARNAHAN and STARLING [2.19] was followed by more complicated expressions quoted in [2.20-24]. ADAMS [2.17] analysed carefully the respective merits of the Carnahan and Starling formula and of the Lefevre formula [2.20]; he concluded that the Carnahan and Starling formula is better for z, but when it is integrated, the Lefevre equation is better for μ'.

The Correlation Functions of the Hard-Sphere System

The two-body correlation function g(r) was computed in [2.8] and by ALDER et al. [2.25] for two values of the density $\rho = 0.2\ \rho_0$ and $\rho = 0.724\ \rho_0$. REE et al. [2.26] studied the expansion in ρ of g(r) and used the MC method to compute the term proportional to ρ^3. ALDER and HECHT [2.27] gave a tabulation of g(r) for the densities 0.333 ρ_0, 0.50 ρ_0, 0.588 ρ_0, 0.625 ρ_0 for $\sigma \leq r \leq 2.50\ \sigma_0$.

REE et al. [2.28] tabulated g(r) for ρ equal to 0.298 ρ_0 and 0.372 ρ_0. These rather dispersed computations of g(r) are completed by the systematic work of BARKER and HENDERSON [2.29] who published eight g(r) functions for the densities from 0.1414 ρ_0 to 0.6364 ρ_0 with an interval of 0.0707 ρ_0 and for $\sigma \leqq r \leqq 2.30\ \sigma$.

Recently PATEY and TORRIE [2.30] have done a MC calculation of the function

$$y(r) = \exp[\beta v(r)]g(r) \quad ,$$

where v(r) is the pair potential. This function can be written

$$y(r_{12}) = \frac{V^2}{Z} \int_V \exp\{-\beta[U - v(r_{12})]\}d\underline{r}_3 \ldots d\underline{r}_N \quad ,$$

where U is the internal energy. Obviously, for the case of the hard-sphere system, y(r) has a simple interpretation; y(r) is proportional to the distribution function of a pair of "cavities", which interact by a hard-core potential with the other hard spheres, but may overlap each other. The value of y(r) for r = 0 is related to μ' by the formula

$$\log[y(0)] = \mu' \quad .$$

For small r and high density, y(r) is steep. Thus its evaluation requires the use of an importance sampling method. PATEY and TORRIE defined

$$\frac{P(r_{12})}{4\pi\, r_{12}^2} = \frac{2\int \exp\{-\beta[U - v(r_{12})]\}d\underline{r}_3 \ldots d\underline{r}_N}{\int_C \exp\{-\beta[U - v(r_{12})]\}d\underline{r}_1 \ldots d\underline{r}_N}$$

$$= \frac{2y(r_{12})Z}{V^2\, Z_c} \quad .$$

P(r) is the distribution function of the two cavities in a system of N-2 hard spheres, where the center of mass of the cavities is fixed and their distance r varies within some predetermined interval. In principle, P(r) could be calculated in this constrained ensemble by the Metropolis method and y(r) recovered by multiplication of P(r) by the constant $V^2 Z_c/(2Z)$. This constant is easily determined because, for $r \geqq \sigma$, y(r) is equal to g(r) which is very well known from the above-mentioned work. However, the rapid variation of P(r) especially as $r \Rightarrow 0$ requires that biased sampling be used.

Then $P(r)$ is written

$$\frac{P(r_{12})}{4\pi\, r_{12}^2} = \frac{\int w(r_{12}) \exp\{-\beta[U - v(r_{12})]\}d\underline{r}_3 \ldots d\underline{r}_N}{w(r_{12}) \int_c \frac{1}{w(r_{12})} w(r_{12}) \exp\{-\beta[U - v(r_{12})]\}d\underline{r}_1 \ldots d\underline{r}_N}$$

$$= \frac{P_w(r_{12})}{4\,\pi\, r_{12}^2\, w(r_{12}) \left\langle \frac{1}{w} \right\rangle_w}$$

The index w means that, in the Metropolis sampling of the configurations, a configuration is obtained with a probability proportional to $w(r) \exp\{-\beta[U - v(r)]\}$. PATEY and TORRIE chose $r^2 w(r)$ equal to an approximate formula [2.31] for $y(r)^{-1}$. If this formula were exact, $P_w(r_{12})$ would be constant for all r_{12}. This would lead to a uniform and consequently very efficient sampling distribution for $0 \leqq r \leqq \sigma$.

Here we have again given the details on this work as an example of the advantage of the importance sampling method when the usual MC method is impracticable.

At $\rho = 0.5657\ \rho_0$, PATEY and TORRIE gave results for a system of two cavities plus 106 or 862 hard spheres. The agreement of the computed $y(r)$ with the formula of GRUNDKE and HENDERSON [2.31] is remarkable. This means that a similar agreement is probable for the hard-sphere system at all fluid densities.

Finally a very simple analytical representation of $g(r)$ was proposed by VERLET and WEIS [2.32]

$$g(r/\sigma,\rho\sigma^3) = 0 \qquad r < \sigma$$

$$g(r/\sigma,\rho\sigma^3) = g_{PY}(r/\sigma',\rho\sigma'^3) + A \exp[-\mu_0(r-\sigma)]\cos[\mu_0(r-\sigma)]/r \qquad r \geqq \sigma \quad, \tag{2.1}$$

where $g_{PY}(r/\sigma',\rho\sigma'^3)$ is the known solution of the PERCUS and YEVICK equation (PY) [2.33] for hard spheres of diameter σ' and density ρ. VERLET and WEIS found the choice

$$\left(\frac{\sigma'}{\sigma}\right)^3 = 1 - \pi\rho\sigma^3/96$$

to be satisfactory. A and μ_0 are determined by using the equation of state of [2.19] in the relations

$$g(\sigma) = 3(z-1)/(2\pi\rho\sigma^3) \quad \text{and} \quad \frac{k_B T \partial\rho}{\partial P} = 1 + \rho\int[g(r) - 1]\, d\underline{r} \quad .$$

Formula (2.1) reproduces the simulation results with a precision $\leq$ 1% for all values of r. The statistical errors on the simulation results are, sometimes, smaller than 1%, but due to the effects of the finite size or of the periodic boundary conditions it is dubious that the MC or MD results are valid for an infinite system to within an error smaller than 1%.

For the higher order correlation functions, the only work was done by ALDER [2.34] who computed the three-body correlation function and examined the validity of the superposition approximation

$$g_3(r_{12},r_{13},r_{23}) = g(r_{13})g(r_{12})g(r_{23})$$

for this function.

ALDER concluded that the superposition approximation is rather good, especially for the case where the three arguments of g_3 are equal. For the other cases, the ratio of $g_3(r_{12},r_{13},r_{23})$ to the product $g(r_{12})g(r_{13})g(r_{23})$ varies between 0.80 and 1.40, but its value is generally 1.0 ± 0.1.

To end this section on the pure hard-sphere system, it is worth noting the high degree of precision of all the simulation work which has been done. The MD and MC computations have been done with millions of collisions or configurations, and the statistical errors and finite system size effects have been carefully analyzed.

b) Hard-Sphere Mixture

Mixtures of hard spheres are characterized by the choice of the hard-core diameters of the various components, and by the choice of the kind of interaction between these components. Generally, if the diameters of the ith and j^{th} components are σ_i and σ_j, the ith and j^{th} components interact by a hard-core potential of diameter σ_{ij} equal to $(\sigma_i + \sigma_j)/2$. However, it is possible that this interaction does not give typical phenomena of fluid mixtures, such as demixing or partial mixing of the components. Systems where σ_{ij} has the form

$$\sigma_{ij} = (\sigma_i + \sigma_j)(1 + \Delta)/2$$

with $\Delta > 0$ or < 0 have also been studied. In the first case ($\Delta = 0$) the diameters of the hard spheres are said to be additive; in the second case ($\Delta \neq 0$), nonadditive.

Mixtures of Hard Spheres with Additive Diameter

To date, the MC or MD simulations have concerned only systems with two components. ALDER [2.35] considered an equimolar mixture of hard spheres with σ_1/σ_2 equal to 3. η the fraction of the total volume V occupied by the spheres is given by

$$\eta = \pi(N_1\sigma_1^3 + N_2\sigma_2^3)/(6V) \quad ,$$

where N_1 and N_2 are the number of particles of each component. The equation of state is calculated in the range $0.24 \leqq \eta \leqq 0.52$. A comparison with the PY equation extended to mixtures of hard spheres [2.36], shows that this approximation is valid for $\eta \leqq 0.30$. The changes in the thermodynamic function upon mixing are computed. They are found, at constant pressure, to be very small. No fluid-fluid phase transition is observed, and convincing arguments are developed for an identical result for the other values of σ_1/σ_2.

By MC simulations, SMITH and LEA [2.37], and ROTENBERG [2.38] calculated the equation of state of binary mixtures, with $\sigma_1/\sigma_2 = 1.667$ and $\sigma_1/\sigma_2 = 1.1$. In [2.37], the simulations were performed for a system of 32 molecules in the domain $0.157 \leqq \eta \leqq 0.445$; for most densities the systems are equimolar. For $\eta = 0.26$ and $\eta = 0.366$ the compressibility factor is computed as a function of the composition. ROTENBERG considered only equimolar systems of 500 or 512 hard spheres for $0.11 \leqq \eta \leqq 0.64$. In this density domain, 27 values were considered and the corresponding partial pressures were tabulated. The fluid-solid transition was estimated to be at $\eta \approx 0.48$. This transition was not located by free energy measurements. The precision of these results [2.37,38] is poor (5%), but they confirm the conclusions of ALDER [2.35] concerning the changes in the thermodynamic functions upon mixing and the absence of a fluid-fluid phase transition.

The Carnahan and Starling formula has been generalized to the case of the mixtures of hard spheres with additive diameter [2.39,40]. The expression of the compressibility factor is for n components:

$$Z = (1+\eta+\eta^2)/(1-\eta)^3 - \frac{\pi}{2NV}\left[\sum_{i<j=1}^{n} N_iN_j(\sigma_i-\sigma_j)^2(\sigma_i+\sigma_j+\sigma_i\sigma_jX)\right]/(1-\eta)^3$$

$$- \frac{6}{\pi\rho}\,\eta X^3/(1-\eta)^3$$

where $N = \sum_i N_i$, $X = \pi(\sum_i N_i\sigma_i^2)/(6V)$ and $\eta = \pi(\sum_i N_i\sigma_i^3)/(6V)$.

The Verlet and Weis representation of the hard-sphere two-body correlation function has also been extended to binary mixtures [2.41,42]. A comparison with the results of MC simulations was done for the 1-2, 1-1, 2-2 correlation functions in

[2.41]. At $\sigma_1/\sigma_2 = 0.90$, $\eta = 0.49$ and 0,47, and $\sigma_1/\sigma_2 = 0.3$, $\eta = 0.49$, the agreement is found to be satisfactory.

Mixtures of Hard Spheres with Nonadditive Diameter

From very simple arguments, a fluid-fluid phase transition is expected in these systems. This is because for $\Delta > 0$, if the pressure increases, the components separate in order to minimize the volume excluded by the hard-sphere interactions. From the same argument, for $\Delta < 0$, no phase transition exists in the fluid phase. The case $\Delta > 0$ was investigated by MELNYK and SAWFORD [2.43] and the case $\Delta < 0$ by ADAMS and McDONALD [2.44].

MELNYK and SAWFORD used the MD method to compute the equation of state for 256 and 500 hard spheres of equal diameter with $\Delta = 0.2$. They considered equimolar mixtures in the range $0.15 < \eta < 0.35$, and at $\eta = 0.30$ several mixtures where the concentration of one component was allowed to increase to a maximum of 0.1 were studied. From the first set of computations they deduced an estimation of the critical point associated with demixing to be

$$\eta_c = 0.22 \pm 0.02 \quad \text{and} \quad z_c = 0.85 \pm 0.10 \quad .$$

From the second set, they obtained the values of the concentrations at the fluid-fluid phase transition. The concentration is 0.03 for one of the components and obviously 0.97 for the other, and z is equal to 4.22 ± 0.05.

The methods for finding the critical point and the location of the phase transition remain qualitative because the free energy is not computed. For the large values of η, the MD calculations give incorrect results due to the finite size of the system. Namely, when the two phases coexist, the number of particles on the surface of separation is almost equal to the number of particles in the bulk of each phase or of one of the phases.

A comparison with the thermodynamic properties calculated from various theories tends to prove that the best values are given by the theory developed by LEONARD et al. [2.45-47] and called MIX1 by MELNYK and SAWFORD. Certainly more computations must be done in order to obtain a precise description of the phase transition.

ADAMS and McDONALD considered binary mixtures of 64 hard spheres with Δ equal to -0.1, -0.2, -0.3, -0.4, and -0.5. The density range which was investigated is $0.1 < \eta < 0.8$. There was no phase transition and comparison showed that the analytical theories [2.45-47] are not very successful for $\Delta < -0.2$ and $\eta > 0.4$. The agreement is better for $\Delta \simeq -0.1$, but the errors are 15% at $\eta \simeq 0.6$. For the two-body correlation functions the theories are not more satisfactory. The functions calculated in the simulations are not tabulated.

2.2.2 Hard Spheres with Discontinuous Short-Range Potential

In this section, we consider systems of hard spheres plus a short range potential. Computations have been done for the square well potential and the triangular well potential. They are very crude models of the real fluids, but they have two phase transitions: a fluid-solid transition and a gas-liquid transition.

a) Hard Spheres with a Square Well

The first MC simulations were made by ROTENBERG [2.48] on a system of 256 spheres with an attractive square well of range 1.5σ. The depth of the well ε, in unit of k_BT, varies from 0 to 2.0. The main and surprising result of these simulations is the existence of two van der Waals loops in the equation of state for ε larger than 1. One loop at $\eta \approx 0.18$ is easily interpreted as a gas-liquid transition. The other loop, at high density $\eta \approx 0.55$, was shown by LADO and WOOD [2.49] to be an artifact of the periodic boundary conditions and of the smallness of the system. They showed that, in increasing the size of the system from 256 to 864 particles, at $\eta = 0.625$, z varied from -2.82 to 1.09.

ALDER et al. [2.50] calculated by MD simulations, the equation of state of a square well fluid. The range of the attractive part was 1.5σ. The results cover the density domain $0.10 < \eta < 0.50$ and the temperature range 0.2 to ∞ (the temperature is in units of k_B/ε). These results are used to locate the critical point and, also, to analyze carefully the convergence of the free energy expansion in reciprocal temperature [2.51]. The terms of this expansion up to the fifth are directly evaluated by MD calculations on the hard-sphere system which is considered as the square well fluid at infinite temperature. The terms of higher order were estimated by extrapolation of the results for finite temperature.

The critical constants deduced from the MD simulations are estimated for 500 particles

$$\eta_c = 0.174 \pm 0.009 \qquad k_BT_c/\varepsilon = 0.26 \pm 0.005 \qquad P_cV_c/k_BT_c = 0.287 \quad .$$

The critical indices [2.52] are calculated from the MD data. Their values are compatible with the values obtained from the van der Waals equation of state with the exception of the critical indices α and α' related to the specific heat at constant volume. As long as α and α' are not considered, the results agree perfectly with the fact that the free energy of a finite system is an analytic function of ρ and T. Then, as was proved by FISHER [2.52], this fact implies that the critical indices are equal to the critical indices of the van der Waals equation of state. Also the values of α and α' contradict the scaling laws [2.53]. Finally, the unexpected values of α and α' are probably an artifact of the periodic boundary conditions and of the small system size.

The conclusions about the convergence of the free energy expansion are discussed in detail in [2.6]. Briefly, an excellent value of the free energy is obtained only when the first three terms of the expansion are used. This is not very useful, because the last of the three terms involves the computation of the four-body correlation function.

For a square well of range 1.5σ, ten simulations were done by HENDERSON et al. [2.54] for the densities and temperatures $0.25 < \eta < 0.4$, and $0.7 < k_BT/\varepsilon < 2$. The essential purpose is to establish the domain of validity of the "hypernetted chain" (HNC) approximation [2.55]. In [2.54], tabulations of the two-body correlation functions are given for $\sigma < r < 1.85\sigma$.

ROSENFELD and THIEBERGER [2.56] compared the results of approximate theories [2.51,57] with their MC simulations for square wells of range $1.2042\ \sigma$ and $1.3675\ \sigma$, in the temperature domain 1 to ∞ and the density domain $0.1 < \eta < 0.38$. They reported fairly good agreements.

b) Hard Spheres with a Triangular Well

CARD and WALKLEY [2.58] studied by MC simulations a system of hard spheres with a triangular well, in order to check the perturbation theories proposed by BARKER and HENDERSON [2.59,60] and WEEKS et al. [2.57]. They published the thermodynamic properties for temperatures and densities ranging from 0.8 to 5 and from 0.093 to 0.449. These extensive results permitted CARD and WALKLEY to find the domain of temperature and density where each perturbation theory is the most successful. The computed two-body correlation functions were plotted for representative values of the density and temperature, and they were compared to the pure hard-sphere results and to an approximation proposed by SMITH et al. [2.61].

c) The Hard Spheres with a Repulsive Square Well

Some computations were done by LEVESQUE and WEIS [2.62] in order to calculate the equilibrium structure factor of a system of hard spheres with a repulsive square well of range 2.1σ. The aim was to show that, for some precise temperature and density, the equilibrium structure factor has a shoulder on the right side of the main peak. For other temperatures and densities, the equilibrium structure factors are similar to those of a hard-sphere system. In the same article, the equilibrium structure factor for an attractive square well of range 2.1σ was shown to have a smaller shoulder on the left sides of the secondary peaks.

d) Mixtures of Hard Spheres and Hard Spheres Plus Square Well

A system of hard spheres mixed with square well molecules was studied by MD simulations by ALDER et al. [2.63]. The equation of state, the energy, and the entropy were obtained for a system of 500 particles with η increasing from 0.1 to 0.49. The temperature varied between 1.4 and 2.47 and the concentration between 0.25 and 0.80. The validity of the free energy expansion [2.51] in the inverse T was again tested. For the equation of state, good results were obtained with the first two terms of the expansion, but for the computation of the other excess thermodynamic properties more terms need to be included.

e) Conclusion

Most of the simulations on square wells or similar potentials were made to check the approximate estimations of the free energy of systems where the potential is a continuous function of the distance of a pair of molecules. In these theories the potential is divided in two parts: an attractive part and a repulsive part. The free energy F_0 of the system with an interaction identical to the repulsive part of the potential is the leading term of a perturbation expansion. The infinite series of the other terms takes into account the attractive part of the potential. In the theories [2.57,59], F_0 is calculated from the free energy of a hard-sphere system and the choice of the hard-sphere diameter is made in order to optimize the convergence of the series.

2.2.3 Two-Dimensional Systems

Hard disks and hard parallel squares are two-dimensional systems with a hard-core potential, which have been investigated by the MD and MC methods. The main aim of these computations is to study the fluid-solid phase transition.

a) The Hard-Disk System

Some results concerning the equation of state of the hard disk system are given in the well-known article of METROPOLIS et al. [2.1], which is the origin of the application of the MC method in statistical mechanics. If A_0 is the area of the system at close packing, the range of the reduced area A/A_0, which is explored in this article, goes from 1.04 to 7.

Values of the equation of state which are much more precise were reported in the work of ALDER and WAINWRIGHT [2.64], who by using an 870 hard-disk system, were able to locate the fluid-solid transition by the observation of a van der Waals loop at $1.25 < A/A_0 < 1.32$ and $PA_0/Nk_BT \simeq 7.8$. The observation of a van der Waals loop is possible because in this system the fluid and crystalline regions can coexist. This coexistence is never obtained for the hard-sphere systems studied by simulations.

In [2.12,13], REE and HOOVER computed the virial coefficients of hard disks up to the seventh. ROTENBERG [2.11] did extensive MC calculations for a 224 hard-disk system. He could not reproduce the van der Waals loop found in [2.64]. HOOVER and ALDER [2.65] compared the pressure obtained by the MD and MC methods. They obtained excellent agreement, if the MD results were corrected by a factor $N/(N - 1)$. This factor took into account the effect of the fixed velocity of the center of mass, which is zero in the MD simulations.

The dependence on N of the entropy change $\Delta S/Nk_B$ in the fluid-solid transition was also discussed in detail. A $\log(N)/N$ variation was reported. A precise value of $\Delta S/Nk_B$ was given in [2.14,15] by HOOVER and REE. This value is obtained in the same manner as that used for hard spheres. HOOVER and REE found $\Delta S/Nk_B = (0.05 \pm 0.01)$, and a transition located between 1.31 A_0 and 1.25 A_0 at pressure $PA_0/Nk_BT = 8.08$.

Careful computations on hard disks were done by WOOD [2.66-68] using the MC method. In [2.67,68] WOOD did calculations in the isobaric-isothermal ensemble (NPT ensemble). The first article [2.67] is devoted to the description of MC sampling in the NPT ensemble. WOOD showed that in this ensemble, the hard-disk partition function is equivalent to an N-particle system of fixed volume where the particles interact by a soft N-body pseudopotential. In the second article [2.68] the results of the computation are given. They agree with those of ALDER and WAINWRIGHT [2.64], with the (3,3) Padê approximant for the equation of state proposed by HOOVER and REE [2.13], and also with the MC calculation of CHAE et al. [2.69].

In [2.69] CHAE et al. made an evaluation of the relative merits of the various approximate theories [2.33,55] which have been applied to hard disks. The PY equation [2.33] proves to be the best approximation. The correlation functions are tabulated for A/A_0 equal to 2.5, 2.0 and 1.666 and for $\sigma < r < 4.0\ \sigma$.

Recently COLDWELL [2.70] has proposed MC techniques to evaluate the partition function Z of a hard-disk system. The technique changes with the density of the system. Here we give a description of the method used for the case of low densities and we refer to the original article for the other cases. COLDWELL considers a system of N hard disks of radius d. A first point is chosen at random in the area of the system and is considered as the center of a disk of radius 2d. A second point is chosen at random in the area which is not excluded by the disk and is considered also as the center of a disk of radius 2d. Then a third point is chosen at random in the area which is free, etc. . Obviously the large disks may overlap. When N successive points are chosen without overlap, the probability w to obtain this configuration is easily calculated. But, if an overlap occurs, the procedure restarts at the beginning and w is taken infinite. If n(w) is the number of configurations corresponding to a finite value of w, wn(w)dw is the probability of finding w in the range dw. Let us call P(w) this probability; then

$$\int_0^\infty n(w)\, dw = \int_0^\infty \frac{P(w)}{w}\, dw = \left\langle 1/w \right\rangle$$

is proportional to the number of N point configurations acceptable as the centers of N hard disks (in the integral w stays in the domain $0 < w < \infty$) and so is proportional to Z. Now if n configurations are generated, we have obviously

$$\langle 1/w \rangle = \lim_{n \to \infty} \left(\sum_{i=1}^{n} 1/w_i\right)/n$$

because the algorithm generates acceptable configurations with probability P(w) and because for the case of unacceptable configurations $1/w_i$ is zero. This method seems to work well and Coldwell reports results in agreement with those of ALDER and WAINWRIGHT [2.64].

An empirical equation of state for the hard disks has been derived by HENDERSON [2.71]. Its form is

$$PA/Nk_BT = (1 + y^2/8)/(1 - y)^2$$

when $y = \pi N\sigma^2/(2A)$.

b) The Mixtures of Hard Disks

Mixtures of hard disks are investigated by BELLEMANS et al. [2.72] and by DICKINSON [2.73]. The aim of BELLEMANS et al. is to evaluate the first terms of a perturbation expansion of the Gibbs excess free energy G^E. The parameter a of the expansion is the size difference in diameter of the two components of the mixture. If the terms of order larger than a^3 are neglected, G^E can be evaluated if the equation of state and some values of the three-body correlation function g_3 for the pure hard-disk system are known. BELLEMANS et al. do MC simulations of a system of 100 hard disks. They calculate the pressure and g_3 for the case of three disks where two disks are in contact with the third. The conclusions are that for $a < 0.1$, there is no tendency for the system to separate into two phases and that the superposition approximation for the calculated values of g_3 is poor.

The MD simulations of DICKINSON [2.73] concern binary mixtures where the hard-core diameters of the interactions satisfy or do not satisfy the additive diameter rule. The equation of state of the additive hard-disk mixtures is computed for σ_1/σ_2 = 0.667, 0.551 and 0.459. In the case of equimolar mixtures the first two values of σ_1/σ_2 are considered. The excess pressure is reported in the density range $0.5 < \lambda < 0.65$. λ is defined by

$$\lambda = \pi N(x_1\sigma_1^2 + x_2\sigma_2^2)/A$$

where A is the area of the system and x_1 and x_2 are the mole fractions. For $\lambda = 0.553$, $\sigma_1/\sigma_2 = 0.667$, and $\lambda = 0.568$, $\sigma_1/\sigma_2 = 0.551$ the excess compressibility factor is obtained for three values of x_1. The simulations on the nonadditive hard-disk mixture were done for $x_1 = 0.5$, $\sigma_1/\sigma_2 = 1$, $\Delta = 0.2$ and $0.3 < \lambda < 0.6$, and also for $x_1 = 0.5$, $\sigma_1/\sigma_2 = 0.667$, $\Delta = 0.2$ and $\lambda \approx 0.4$. For $\sigma_1/\sigma_2 = 1$ the critical density is estimated to be $\lambda_c = 0.45$. The two-body correlation functions are plotted for three sets of values of λ, x_1, and σ_1/σ_2, with $\Delta = 0$. The conclusions of the comparison with approximate theories [2.45-47] are similar to the conclusions in the hard-sphere case. The precision of the results in [2.73] looks poor, because the averages are taken only over $80\text{-}100 \times 10^3$ collisions for 100 particle systems.

c) System with a Repulsive "Step" Potential

YOUNG and ALDER [2.74] present results of MD simulations for a hard-disk system with a repulsive square well in the range $\sigma < r < c\sigma$. The melting curve of this system has maxima and minima similar to those existing for caesium and cerium. The simulations are done for $c = 1.2$ and 1.5. For $k_BT/\varepsilon > 0.5$ the perturbation theory [2.51] accurately represents the equation of state, but very extensive computer runs were required in the temperature range $0.1 < k_BT/\varepsilon < 0.5$ where the perturbation theory fails.

d) The Hard Parallel Square System

The hard parallel square system is obviously very different from real fluids, but it is considered to be a useful model against which to test theoretical hypotheses and to study phase transitions.

The first results for the equation of state were found by HOOVER and ALDER [2.65] who give the equation of state of four parallel squares with both rigid and periodic boundaries. In the latter case a first-order phase transition is reported at $A/A_0 = 2.25$.

Two very extensive calculations of the equation of state have been done by CARLIER and FRISCH [2.75] and F. REE and T. REE [2.76]. The first authors find a fluid-solid phase transition at $A/A_0 \approx 1.3$ with a very small change in density. However, in [2.76] no phase transition was found. The difference between the two equations of [2.75] and of [2.76] is larger than the quoted errors, but it is smaller than 2%. To our knowledge, no definitive answer about the existence of this phase transition has been published. MEHTA [2.77] discusses the problem for the case of the hard squares on lattices, but he cannot give a definitive answer.

A special version of the Widom-Rowlinson model has also been investigated by FRISCH and CARLIER [2.78]. In [2.78], the system is formed of two free gases, but the particles which do not belong to the same gas, interact by a hard parallel square interaction. The demixing of the two gases seems to occur at $A/A_0 \approx 1.8$ for an equimolar mixture.

2.3 Soft Short-Range Potentials

2.3.1 Inverse Power Potentials

Repulsive inverse power pair potentials of the general form

$$v(r) = \varepsilon\left(\frac{\sigma}{r}\right)^n \tag{2.2}$$

are of theoretical interest because of their great simplicity; in particular the homogeneity of these potentials implies that the excess thermodynamic properties and the scaled pair distribution functions $g(r/\sigma)$ of systems of particles interacting through (2.2) depend on a *single* parameter

$$x = \rho\sigma^3(\varepsilon/k_BT)^{3/n} \quad . \tag{2.3}$$

This scale invariance means that the equation of state calculated along a single isotherm suffices to determine the thermodynamic properties in the whole (ρ,T) plane.

The hard-sphere system is a special case of (2.2) with $n = \infty$. Another extreme case is the Coulomb potential ($n = 1$), which will be considered in Sect.2.4, where the parameter $\Gamma = (4\pi/3)^{1/3}x^{1/3}$, rather than x, will be used.

a) The Inverse-12 Potential

Apart from the cases $n = \infty$ and $n = 1$, the most thoroughly studied inverse power potential is the inverse-12 potential; its importance stems from its obvious relation to the widely used Lennard-Jones potential, which will be discussed in a subsequent section. The equation of state has been determined as a function of x independently by HOOVER et al. [2.79] and by HANSEN [2.80], both in the fluid and solid phases through a series of MC computations. HOOVER et al. used systems of 32 and 500 particles, whereas Hansen's computations are for 864 particles. Both sets of data agree to better than 1% for all values of x. HANSEN proposed a simple equation of state which fits the MC data very well over the whole fluid range:

$$\frac{\beta P}{\rho} - 1 = B_1x + B_2x^2 + B_3x^3 + B_4x^4 + B_{10}x^{10}$$

with $B_1 = 3.629$ (exact second virial coefficient), $B_2 = 7.2641$, $B_3 = 10.4924$, $B_4 = 11.459$, $B_{10} = 2.1762$.

In both papers, the single-occupancy constraint of HOOVER and REE [2.14] has been used to stabilize the solid phase artificially down to zero density, thus allowing the computation of the solid phase free energy by integration of the equation of state. The Maxwell double - tangent construction then allows the determination of the values of x for the coexisting solid (S) and fluid (F) phases. The results are the following (the numbers of HOOVER et al. between parentheses): $x_S = 0.844 \pm 0.002$ (0.844), $x_F = 0.814 \pm 0.002$ (0.813) and the melting pressure is given by

$$P_m = 16 \pm 0.1 \quad T^{*5/4}$$

$$(= 15.95 \quad T^{*5/4})$$

where $T^* = k_B T/\varepsilon$ and $P^* = P\sigma^3/(\sqrt{2}\varepsilon)$.

HANSEN used thermodynamic perturbation theory to relate the equation of state of the Lennard-Jones (LJ) fluid to that of the inverse - 12 potential fluid. By comparing to MC data for the full LJ potential, he finds that first-order perturbation theory gives excellent results for $T^* \geqq 3$.

The pair distribution functions of the inverse - 12 fluid are tabulated by HANSEN and WEIS [2.81] for 10 values of x, and compared to the predictions of a perturbation theory which is based on a hard-sphere reference system: the agreement is not very satisfactory.

b) The Inverse-9, -6 and -4 Potentials

HOOVER et al. [2.82] have extended the MC computations to inverse-9, -6 and -4 potentials. For each of these systems they have determined the fluid-solid phase transition. Combining these results with the earlier data on the inverse -12 potential, these authors were able to exhibit a rather universal behavior for all inverse-power potentials. In particular they proposed a universal equation of state by scaling the "thermal" energy of the fluid (i.e., the excess internal energy minus the static lattice energy at the same density) with the solid-like Einstein frequency. They also showed that a certain number of melting criteria are roughly independent of the exponent n: the Lindemann ratio at melting (i.e., the ratio of the root mean square displacement of an atom around its equilibrium position over the nearest neighbor distance) and the thermal part of the excess Helmholtz free energy which is roughly equal to 6 Nk_BT along the melting line [2.83]. This work was complemented by HANSEN and SCHIFF [2.84] who computed the structure factor of the fluid at crystallization for the various inverse power potentials, and compared the results to the known structure factors of the hard-sphere fluid and the one component plasma (cf. Sect.2.4.2a). They concluded that the structure factor is very insensitive to n; in particular the amplitude of the first peak is nearly constant, and equal to its hard-sphere value ($\simeq 2.85$).

2.3.2 The Lennard-Jones (LJ) Potential

Systems of particles interacting by a LJ potential have been well studied by both MD and MC simulations. All this work constitutes an extensive literature. Here we emphasize the most recent works and describe briefly the others.

The LJ potential used in the simulations has the form

$$v(r) = 4\varepsilon[(\sigma/r)^{12} - (\sigma/r)^{6}] \quad .$$

For the comparisons with the experimental data for argon ε and σ are chosen: $\varepsilon/k_B = 119.8$ K and $\sigma = 3.405$ A. Also for the comparisons with the perturbation theories we divide the LJ potential in two parts $v_0(r)$ and $\lambda w(r)$. We have: $v(r) = v_0(r) + \lambda w(r)$. For instance WEEKS et al. [2.57] chose:

$$v_0(r) = v(r) + \varepsilon \text{ for } r \leq (2)^{1/6}\sigma \text{ and } 0 \text{ for } r > (2)^{1/6}\sigma$$

$$w(r) = -\varepsilon \qquad \text{for } r \leq (2)^{1/6}\sigma \text{ and } v(r) \text{ for } r > (2)^{1/6}\sigma \quad .$$

a) The Pure Lennard-Jones (LJ) System

The Thermodynamic Properties

The MC method was first applied to the LJ system by WOOD and PARKER [2.85]. Their article contains a very illuminating description of the use of the MC method in statistical mechanics, and also some results for the equation of state and for the internal energy at the temperature $k_BT/\varepsilon = 2.74$. RAHMAN [2.86] has done pioneer work in the application of the MD method to continuous potentials. He published essentially results for dynamical quantities. VERLET and LEVESQUE [2.87] gave some results limited to the isotherm $k_BT/\varepsilon = 1.35$.

A systematic investigation of the equation of state is done in the publications of McDONALD and SINGER [2.88-90] and VERLET [2.91]. In [2.88] McDONALD and SINGER propose a method which permits the data of MC simulation at given temperature and volume to be used to find the thermodynamic properties of a system at the same volume but with a slightly different temperature. The method consists of accumulating a histogram $f_T(U)$ of the number of configurations with a given energy U, in the sequence of configurations weighted by $\exp(-U/k_BT)$. It is obvious that for the temperature T', $f_{T'}(U) = f_T(U) \exp[-U(\beta' - \beta)]$ with $\beta = 1/k_BT$ if the volume stays the same. So the internal energy, or the specific heat at constant volume, is easily obtained by average on this new histogram. For the pressure, the average values of the virial of the configuration with energy U are kept during the simulation at temperature T and reweighted at temperature T' with $f_{T'}(U)$. This last procedure is not exact, but the approximation seems good. Another method is proposed for calculating the thermodynamic properties of a LJ system with parameters ε' and σ' from the MC data for a LJ system with parameters ε and σ, if ε' and σ' are not very different from ε and σ. A third method proposed for computing directly the partition function seems very difficult to use in practice.

In the references [2.89,90] the thermodynamic properties of about 80 thermodynamic states are reported for systems of 32 or 108 particles. In [2.88,89] six isotherms are considered for the densities $0.6 < \rho\sigma^3 < 0.9$ and four isochores for $\rho\sigma^3$ equal to 0.6798, 0.5099, 0.34 and 0.17 in the temperature range $1.44 < k_BT/\varepsilon < 3.54$. In [2.90] new results are given for the liquid phase at temperature $k_BT/\varepsilon < 1.3$; a comparison with the properties of argon yields fair agreement. An estimation of the contribution of the three-body interactions to the pressure and to the internal energy is given.

VERLET [2.91] has done an extensive study of the liquid phase by MD simulations, using a 864 particle system. The equation of state of five isochores $\rho\sigma^3 = 0.85$, 0.75, 0.65, 0.5426 and 0.4 are calculated for a large temperature domain. This article contains many methodological details about the MD method.

The work of HANSEN and VERLET [2.92] completes the preceding publications by determining the phase coexistence lines in the domain $0 < \rho\sigma^3 < 1.2$ and the temperature range $0.7 < k_BT/\varepsilon < 3$. In order to locate the fluid-solid phase transition, they use the same method as HOOVER and REE [2.14] and an equivalent method for the gas-liquid transition. The volume of the system is divided in ν cubic cells of equal size. In these cells the fluctuation in the number of particles is kept between imposed limits. This constraint prevents the large density fluctuation in the two-phase region. The integration of the equation of state for the constrained system into a liquid density where the density fluctuation stays within the imposed limits, permits the free energy to be calculated. The liquid-gas coexistence line is closed in the critical region by using the values of the critical constants given in [2.87] on the basis of approximate theories. HANSEN and VERLET note that the finite size of the systems used in the MC simulations prevents a precise determination of the critical constants of a LJ system.

In [2.93] LEVESQUE and VERLET give some simulation results for a LJ system for the purpose of comparing with the perturbation theory of BARKER and HENDERSON [2.59]. This point is well discussed in [2.6].

TORRIE and VALLEAU [2.94,95] apply the multistage sampling method and the importance sampling method to a direct calculation of the free energy. As in the hard-sphere case, the aim is to by-pass the tedious work of computing the equation of state for a large domain of densities. Details on the method are given in [2.95]. The results of TORRIE and VALLEAU agree very well with the other computations.

In their controversial article, STREETT et al. [2.96] make very precise simulations in the vicinity of the fluid-solid transition. Results for 140 temperature-density points are given for $0.74 < k_BT/\varepsilon < 1.17$ and $0.80 < \rho\sigma^3 < 0.99$. The agreement with the preceding results on the equation of state is excellent. However, despite of this fact, STREETT et al. concluded that the location of the fluid-solid transition found by HANSEN and VERLET is incorrect [2.92]. This surprising conclusion is based on a new determination of the free energy of the solid phase, which is calculated by means of a van der Waals loop between the liquid and the solid. But this van der Waals loop is qualitative, because in the simulations the system does not jump from one phase to the other and back, and also, because no coexistence between liquid and crystalline regions is observed. Very convincing arguments unfavorable to this new location of the fluid-solid transition are given by HANSEN and POLLOCK [2.97]. More comments on the fluid-solid transition were made by RAVECHE and MOUNTAIN [2.98], but this work concerns essentially the solid phase.

ROWLEY et al. [2.99] applied the MC method in the grand-canonic ensemble to the LJ system. They proposed a new way to sample the configurations which differs by many details from ADAMS' method [2.17]. The reference [2.6] contains a detailed description of this new method. In [2.99] the results disagree with those of HANSEN and VERLET [2.92], but this discrepancy is not real. It is due to the fact that the long-range correction must be properly taken into account in the calculation of the energy difference between successive configurations in the sampling process (private communication quoted in [2.99]).

Other computations for the LJ systems in the grand-canonic ensemble have been done by NORMAN and FILINOV [2.100] and in the NPT ensemble by VORONTSOV-VEL'YAMINOV et al. [2.101].

A new calculation of the gas-liquid coexistence line at low temperature has been made by ADAMS [2.102]. He used a combination of canonic and grand-canonic ensemble MC simulations. Precise results were limited to the domain $k_BT/\varepsilon < 1.1$. These new data are in agreement with those of HANSEN and VERLET [2.92] and also with the empirical equation of state of McDONALD and SINGER [2.103]. In this publication McDONALD and SINGER reported new MC simulations for the LJ system in the domain $0.551 < k_BT/\varepsilon < 1.237$ and $0.7 < \rho\sigma^3 < 0.9$. Their equation of state was based on a polynomial expression of the free energy. The coefficients of this expression were determined by a least-square fit of the MC results for the pressure and for the internal energy.

The perturbation theories [2.57,59] permit the equation of state of the LJ system in the liquid phase to be calculated within an error of a few percent. A complete discussion of this point is given in [2.6].

The Correlation Functions

In [2.104] VERLET tabulated the two-body correlation functions for 25 thermodynamic states. An analysis of the experimental data for the equilibrium structure factor of argon concluded that there is a large uncertainty in the validity of the x-ray experiments [2.105]. A one-to-one correspondence between the structure factor for the LJ system and the structure factor of the hard-sphere system was established.

In order to test the usefulness of the division of the LJ potential into two parts $v_0(r)$ and $\lambda w(r)$ for the computation of the functions $g(r)$ WANG et al. [2.106] studied the convergence of the Taylor expansion of $g(r)$ in the parameter λ. They published $g(r,\lambda)$ for $\rho\sigma^3 = 0.80$ and $k_BT/\varepsilon = 0.719$ and λ ranging from 0 to 1, and concluded that the expansion is quickly converging.

WANG and KRUMHANSL [2.107] calculated the triplet correlation function for $\rho\sigma^3 = 0.80$ and $k_BT/\varepsilon = 0.719$. Their conclusions about the superposition approximation are similar to those of ALDER [2.34]. The most detailed computations on g_3 were made by RAVECHE et al. [2.108] for five thermodynamic states in the domain $0.65 < \rho\sigma^3 < 0.85$ and $0.72 < k_BT/\varepsilon < 2.85$. Several expressions of the three-body correlation function in terms of the two-body correlation function are checked. The superposition approximation is in error by the same order of magnitude one finds in the hard-sphere case [2.34], and the best of the other approximations is that proposed by RAVECHE and MOUNTAIN [2.109].

Up to now, the analytical calculations of $g(r)$ have not been satisfactory because the correct thermodynamic properties are not obtained from the calculated $g(r)$. But a rather good result is found if one computes $g(r)$ for the $v_0(r)$ part of the LJ potential following the procedure of WEEKS et al. [2.57] and takes into account the $\lambda w(r)$ part following the HNC perturbation theory of LADO [2.110]. $g(r)$ for systems where the interaction is equal to $v_0(r)$, has been calculated by BARKER and HENDERSON [2.111].

b) Conclusion

A general conclusion is that the simulation results for the LJ system have certainly contributed to a real progress in the theories of the simple fluid, serving especially as a stringent test for various theoretical approximations. The simulations have also shown the inadequacy of the LJ potential to represent the equation of state of rare gases despite the good agreement for liquid argon. More realistic potentials have been considered by BARKER et al. [2.112] who use MC simulations to take into account the two-body part of these potentials and consider the three-body part by perturbation. BARKER and KLEIN [2.113] have done an extensive study of solid and liquid argon in using the potential of [2.112]; for more details we refer to [2.6].

c) The Lennard-Jones (LJ) Mixtures

Lennard-Jones mixtures have been investigated in order to compare with the mixtures of rare gases and simple fluids such as O_2, CH_4 or N_2. The main interest is to compute the change of the excess thermodynamic function upon mixing. Computations have only been done for binary mixtures. The interaction between the two components are Lennard-Jones potentials of the form

$$v_{ij}(r) = 4\varepsilon_{ij}[(\sigma_{ij}/r)^{12} - (\sigma_{ij}/r)^{6}]$$

where the indices i and j take the values 1 or 2 and where the parameters ε_{ij} and σ_{ij} satisfy the rules

$$\sigma_{12} = (\sigma_{11} + \sigma_{22})/2 \qquad \varepsilon_{12} = c(\varepsilon_{11}\varepsilon_{22})^{1/2}$$

generally c is chosen equal to 1.

J.V.L. SINGER and K. SINGER [2.114-116] have done MC simulations in the canonic ensemble, and McDONALD [2.117-119] MC simulations in the NPT ensemble. In [2.114] SINGER outlines methods by which the excess free energy, volume and enthalpy can be calculated.

In order to calculate the excess free energy SINGER proposes two procedures. We shall describe them in detail because they have been extensively used in all the works on the LJ mixtures.

In the first procedure, the free energy is calculated in three steps. The first step is to determine the variation of the free energy of a system composed of two separated amounts of each component when the parameters of the LJ interactions ε_{ii} and σ_{ii} are given identical values ε_0 and σ_0. Then the separation between the two components is removed and there is ideal mixing with a zero change in the excess free energy of the total system because ε_{12} and σ_{12} are also chosen equal to ε_0 and σ_0. In the final step, the parameters ε_0, σ_0 of the interactions between the molecules of the components are varied from ε_0, σ_0 to their values ε_{11}, σ_{11}; ε_{22}, σ_{22}; ε_{12}, σ_{12}. The change of the free energy in the first step is easily obtained from the free energy of a pure LJ system, which is assumed to be known. For the change ΔF in the final step, one method of carrying out the calculation is to consider the parameters ε_{ij}, σ_{ij} as a function of a variable s. If s has the value s_0, ε_{ij}, σ_{ij} are respectively equal to ε_0, σ_0 and if s is equal to s_1, ε_{ij}, σ_{ij} have their real values. It is easy to show that

$$\Delta F = \int_{s_0}^{s_1} ds \left[\sum_{i<j=1}^{2} \langle U_{ij}\rangle_s \, \varepsilon'_{ij}(s)/\varepsilon_{ij}(s) - \langle \Phi_{ij}\rangle_s \sigma'_{ij}(s)/\sigma_{ij}(s)\right] \qquad (2.4)$$

U_{ij} and Φ_{ij} are respectively the part of the energy and of the virial of the total system which are due to the i-j interactions. $\langle\ \rangle_s$ means that the average is to be taken in a canonic ensemble where the parameters in the LJ interactions are $\varepsilon_{ij}(s)$ and $\sigma_{ij}(s)$. $\varepsilon'_{ij}(s)$ and $\sigma'_{ij}(s)$ are the derivatives of $\varepsilon_{ij}(s)$ and $\sigma_{ij}(s)$. Finally ΔF is calculated by integration of $\langle U_{ij}\rangle_s$ and $\langle \Phi_{ij}\rangle_s$ computed by MC simulations.

In the second procedure ΔF is directly calculated by estimation of the integral

$$\Delta F = \int d\underline{r}_1 \ldots d\underline{r}_N \exp\{-\beta[U(s_1) - U(s_2)]\}$$

where U(s) is the energy of a configuration when s has a specified value.

In order to compute ΔF, J.V.L. SINGER and K. SINGER use this last procedure combined with a method described in [2.88] for the calculation of the partition function. They also use the formula (2.4) in order to check the precision of their results.

For the NPT ensemble a formula equivalent to (2.4) exists where ΔF is replaced by ΔG, the variation of the Gibbs free energy, and $\langle\ \rangle_s$ indicates that the average is to be taken in the NPT ensemble. McDONALD uses the first procedure of SINGER for the calculation of ΔG.

In the review [2.120] the results of all the simulations on LJ mixtures are summarized and compared with experimental data and also with the results of approximate theories [2.45-47]. We mention that the results [2.115,116] cover the parameter range $0.81 < \varepsilon_{11}/\varepsilon_{22} < 1.235$ and $0.88 < \sigma_{11}/\sigma_{22} < 1.12$, with $\varepsilon_{12}/k_B = 133.5$ K, $\sigma_{12} = 3.596$ Å, at the temperatures 97 K and 117 K. In [2.118,119] the results are given for choices of the parameters which permit easy comparison with experimental data. The general feature is that the excess thermodynamic properties upon mixing are very small in magnitude.

In [2.121] TORRIE and VALLEAU prove the usefulness of their importance sampling technique, or so-called umbrella sampling [2.95] by calculating ΔF for a simple mixture. As their purpose is essentially methodological, they perform simulations for the choice of the parameters $\sigma_{11} = \sigma_{22} = \sigma_{12} = \sigma$ and $\varepsilon_{11} = \varepsilon_{22} = c\varepsilon_{12} = \varepsilon$. The case c = 1.333 is investigated in detail. For this value of c the system undergoes a fluid-fluid phase transition.

Some two-body correlation functions for the LJ mixtures are plotted in [2.122].

2.3.3 The Two-Dimensional Lennard-Jones (LJ) System

The two-dimensional LJ system was studied by M. ROSENBLUTH and A. ROSENBLUTH [2.8], but more recent computations were motivated by the study of adsorbed film-gases on solid surfaces.

In [2.8] a few data are reported. FEHDER [2.123,124] published thermodynamic properties and two-body correlation functions, but the precision of the results is poor, and only qualitative conclusions seem to be possible.

TSIEN and VALLEAU [2.125] gave more precise values for the equation of state and for the two-body correlation function g(r). They found a gas-liquid phase transition and located the critical temperature at $k_BT/\varepsilon = 0.5$. The g(r)'s in [2.125] differ in many details from those published by FEHDER.

2.4 Ionic Systems

2.4.1 Generalities

a) Specific Problems

This section is devoted to classical fluids of charged particles, i.e., systems where Coulombic (electrostatic) interactions play a dominant role. Since the systems which will be considered are over-all neutral, they are made up of at least two species carrying charges of opposite sign. Let $Z_\nu e$ be the electric charge (e is the elementary proton charge) and N_ν the number of particles of species ν. If n is the total number of species, the total number of particles is $\sum_{\nu=1}^{n} N_\nu = N$, and the charge neutrality condition is

$$\sum_{\nu=1}^{n} N_\nu Z_\nu = 0 \quad .$$

The total potential energy of the system is made up of the electrostatic energy, which is, of course, pairwise additive, and of "short-range" interactions which are generally assumed to be also pairwise additive. Hence

$$V_N(1, \ldots, N) = \frac{1}{2} \sum_\nu \sum_\mu \sum_{i=1}^{N_\nu} \sum_{j=1}^{N_\mu}{}' v_{\nu\mu}(i,j) \tag{2.5}$$

$$v_{\nu\mu}(i,j) = v_{\nu\mu}^{s}(i,j) + v_{\nu\mu}^{c}(i,j) \tag{2.6}$$

$$v_{\nu\mu}^{c}(i,j) = \frac{Z_\nu Z_\mu e^2}{|\underline{r}_i - \underline{r}_j|} \tag{2.7}$$

In (2.5) the prime on the last summation means that the term $i = j$ must be left out if $\nu = \mu$. The system is characterized by $n(n+1)/2$ distinct pair potentials $v_{\nu\mu}(r)$, and consequently by the same number of partial pair distribution functions $g_{\nu\mu}(r)$ and partial structure factors $S_{\nu\mu}(k)$. A particularly important combination of the partial structure factors is the charge structure factor $S_{ZZ}(k)$

$$S_{ZZ}(k) = \sum_{\nu} \sum_{\mu} x_\nu x_\mu Z_\nu Z_\mu S_{\nu\mu}(k) \quad .$$

In Monte Carlo computations, the short-range part of the potential is treated by the usual nearest-image convention. The Coulombic part, however, poses severe problems because of the long range of the electrostatic potentials. For a two-component system, made up of equal numbers of charges $+Ze$ and $-Ze$, the importance of the Coulombic interaction can be characterized by the dimensionless parameter

$$\Gamma = \frac{(Ze)^2}{\varepsilon a k_B T} \tag{2.8}$$

where ε is the dielectric constant of the solvent ($\varepsilon = 1$ for purely ionic systems, like molten salts) and $a = (3/4\rho)^{1/3}$ is the so-called ion-sphere radius. In the weak coupling limit ($\Gamma < 1$), the charge distribution around a central particle will be essentially uniform beyond a few spacings a, and the nearest image convention is then generally sufficient to compute the electrostatic energy of a periodic system.

In the strong coupling limit ($\Gamma \gg 1$), however, account must be taken not only of the electrostatic interactions of a given particle with the nearest images of the $N - 1$ other particles in the system, but also with all the periodic replicas of these particles. The resulting infinite sums are poorly convergent; the difficulty can be overcome by using Ewald transformations similar to those used in solid-state physics to compute Coulombic lattice (or Madelung) sums [2.126-128] The Ewald procedure transforms the slowly convergent Coulomb energy sum into two rapidly convergent sums, one over direct space (Φ_1) and one over reciprocal space (Φ_2)

$$V_N^C = \frac{1}{2} e^2 \sum_i \sum_j{}' \sum_{\underline{n}} \frac{Z_i Z_j}{r_{i,j\underline{n}}} = \Phi_1 + \Phi_2 \tag{2.9}$$

$$\Phi_1 = \frac{1}{2} e^2 \sum_{i=1}^{N} \left\{ \sum_{j=1}^{N}{}' \sum_{\underline{n}} \frac{Z_i Z_j \operatorname{erfc}(\alpha r_{i,j\underline{n}})}{r_{i,j\underline{n}}} - \frac{Z_i^2 \alpha}{\sqrt{\pi}} \right\} \tag{2.10}$$

$$\Phi_2 = \frac{1}{2\pi L} e^2 \sum_{i=1}^{N} \sum_{j=1}^{N} \sum_{\underline{n}\neq\underline{0}} \frac{Z_i Z_j}{|\underline{n}|^2} \exp\left(-\frac{\pi^2|\underline{n}|^2}{\alpha^2 L^2}\right) \cos\left(\frac{2\pi}{L}\underline{n}\cdot\underline{r}_{ij}\right) \quad . \tag{2.11}$$

In these formulae $\sum_{\underline{n}}$ is a sum over all vectors of integer components, $r_{i,j\underline{n}} = |\underline{r}_i - \underline{r}_j - L\underline{n}|$, the primed summation implies $j \neq i$ if $\underline{n} = (0,0,0)$, L is the edge of the cubic cell containing N particles and the parameter α governs the convergence of both series; the sum $\Phi_1 + \Phi_2$ is independent of α, but the number of terms to be retained in the partial sums to achieve a given accuracy depends sensitively on α. If one chooses $\alpha = \sqrt{\pi}/L$, the two sums converge at the same rate [2.127]; if $\alpha > \sqrt{\pi}/L$ the direct space part converges faster and reciprocally. In any case, in order to keep truncation errors below a tolerable level, of the order of 10^2 terms must be kept all together, which renders the Ewald procedure as it stands too time consuming in MC computations. Three methods have been devised to overcome the difficulty.

The first method is a tabulation of the Ewald pair "potential" between a particle i, and a particle j plus all its periodic images on a three-dimensional grid taking full advantage of the cubic symmetry of the problem (this reduces the number of grid points by a factor of $3! \times 2^3 = 48$), and using linear interpolation between grid points [2.129,130].

A second method amounts to fitting the Ewald pair potential by a sum of Kubic harmonics [2.131] which automatically incorporate the correct symmetry properties. The coefficients of these functions of the cartesian coordinates x, y, z are chosen to be simple polynomials in the pair separation r [2.132] which are tabulated.

The third method is based on the observation by K. SINGER that the double sum of [N(N - 1)]terms over i and j in Φ_2 can in fact be trivially expressed as the square of a sum over i only (N terms), reducing the computation time by a factor of N for a given number of $\underline{n}$ vectors

$$\Phi_2 = \frac{1}{2\pi L} e^2 \sum_{\underline{n}\neq\underline{0}} \frac{\exp(-\pi^2|\underline{n}|^2/\alpha^2L^2)}{|\underline{n}|^2} \left\{\left[\sum_{i=1}^{N} Z_i \cos\left(\frac{2\pi}{L}\underline{n}\cdot\underline{r}_i\right)\right]^2 + \left[\sum_{i=1}^{N} Z_i \sin\left(\frac{2\pi}{L}\underline{n}\cdot\underline{r}_i\right)\right]^2\right\} \quad . \tag{2.12}$$

Advantage is taken of this circumstance by choosing a relatively large value of $\alpha(\alpha L \simeq 5)$, which allows a safe truncation of Φ_1 after the first term ($\underline{n} = \underline{0}$: nearest image convention). For a system of 216 ions about 10^2 terms are then retained in the $\underline{n}$ - summation of Φ_2, which is acceptable in view of the previously mentioned simplification [2.133]. If properly implemented, the three methods lead to numerical errors of less than 0.1% on the Ewald potential for any pair configuration.

For the sake of completeness we briefly mention the "particle-mesh" algorithms devised by HOCKNEY and collaborators [2.134,135], to compute the electrostatic potential from a finite-mesh solution of Poisson's equation on a periodic grid. The method has only been used in molecular dynamics simulations of 2 D plasmas and molten salts [2.136], but can be extended to 3 D, and could conceivably be applied in MC calculations.

b) Classes of Ionic Fluids

Several types of fluids of charged particles have been studied by MC computations. We shall, somewhat arbitrarily, distinguish four classes of ionic fluids, corresponding to four physically rather well defined situations:

I) Fully ionized matter characteristic of extreme temperature and pressure conditions, as encountered, e.g., in astrophysics, or laser-implosion experiments. This includes the classical electron gas, dense ion plasmas in a uniform background of degenerate electrons, mixtures of such ionic plasmas, and classical ion-electron plasmas.

II) Electrolyte solutions. A model which has been extensively studied by MC computations is the "primitive model" consisting of charged hard spheres, representing the anions and cations, in a dielectric continuum representing, rather crudely, the solvent.

III) Ionic melts or molten salts. MC computations have dealt up to now essentially with fused alkali halides and their mixtures. The primitive model of electrolytes has been extended to describe these melts. More realistic "rigid ion" potential models have been used by several authors to simulate liquid alkali halides.

IV) Liquid metals considered as a two-component system of positive ions and conduction electrons. Metallic hydrogen can be regarded as a link between classes I and II.

The following subsections will be devoted successively to ionic fluids belonging to classes I (Sect.2.4.2), II (Sect.2.4.3), III (Sects.2.4.3 and 2.4.4) and IV (Sect.2.4.5).

2.4.2 Fully Ionized Matter

For the systems considered in this subsection, i.e., dense plasmas, the interaction between particles is exclusively Coulombic. The particles are basically of two species: positively charged nuclei and electrons. Because of the very large mass ratio (m nucleus / m electron $\geqq$ 2000), the relevant temperature scales are very different for the two species. The electrons are characterized by the Fermi temperature $T_F = \hbar^2(3\pi^2\rho)^{2/3}/(2m_e k_B)$. If $T \gg T_F$, the electrons behave classically

---

$\hbar = h/2\pi$ (normalized Planck's constant)

and we are dealing with a high temperature ion-electron plasma. If on the other hand the density is sufficiently high (e.g., in degenerate stars like white dwarfs) that $T \ll T_F$, the electron gas is degenerate and can be assimilated to a rigid continuum in which the classical ions move. This latter case can be modeled in first approximation by the one component plasma (OCP).

a) The One Component Plasma

The OCP is the simplest possible model for a Coulombic fluid: N point charges Ze are immersed in a uniform neutralizing background of opposite charge; this means that the second species is not made up of discrete charges, but rather its total charge − NZe is assumed to be uniformly smeared out over the total volume V of the system. The potential energy is most compactly expressed in Fourier space

$$V_N(\underline{r}^N) = \frac{1}{2V} \sum_{\underline{k}}{}' \frac{4\pi(Ze)^2}{k^2} (\rho_{\underline{k}}\rho_{-\underline{k}} - N)$$
$$\rho_{\underline{k}} = \sum_{i=1}^{N} \exp(i\underline{k} \cdot \underline{r}_i) \tag{2.13}$$

and the prime in the $\underline{k}$ summation means that the term $\underline{k} = \underline{0}$ is left out to take proper account of the uniform background. Since the Coulomb potential is of the inverse-power type, the excess thermodynamic quantities and, more generally, all equilibrium properties of the model, depend only on the dimensionless coupling parameter Γ defined by (2.8), with $\varepsilon = 1$.

Some preliminary MC results for systems of 64 and 125 particles, averaged over $10^4 - 5.10^4$ configurations, were obtained by CARLEY [2.137], at $\Gamma = 1$ and 2.5, using a *truncated* Coulomb potential, with cut-off radii of $x = r/a = 1.87$, 3.22 and 4.03. Carley published only comparisons for the MC pair distribution functions with the numerical solutions of HNC and PY integral equations, indicating reasonable agreement between the MC and HNC results.

The pioneering MC work on the strongly coupled OCP, using the Ewald transformation described in Sect.2.4.1, is due to BRUSH et al. (BST) [2.138]. Their computations cover the range $0.05 \leq \Gamma \leq 125$, for systems of 32, 64, 108, 256 or 500 particles, with 10^5 configurations generated in most runs. Their main findings are that the excess internal energy per particle approaches within a few percent the static bcc lattice energy $\beta U_S/N = -0.895929\,\Gamma$, at large Γ, and that the pair distribution function exhibits oscillations, characteristic of "short-range order", for $\Gamma \geq 3$. The paper contains extensive tabulations of g(x), for 20 values of Γ in the range indicated above. The equation of state, which for the OCP is simply related to the excess internal energy

$$\frac{\beta P}{\rho} = 1 + \frac{1}{3}\frac{\beta U}{N}$$

goes negative for $\Gamma \geqq 3$.

The computations of BST have been improved and extended by HANSEN [2.132] who covered the range $1 \leqq \Gamma \leqq 160$ for systems of 16, 54, 128 and 250 particles and generated about 5×10^5 configurations per run. He used an accurate fit of the Ewald potential including up to 5 Kubic harmonics and investigated the N-dependence of the thermodynamic results, which turns out to be undetectable for $N \geqq 10^2$. The nonstatic part of the internal energy $\Delta\beta U/N = (\beta U/N) - (\beta U_s/N)$ is systematically larger than the results of BST (by about 15% at $\Gamma = 100$); the difference can be attributed to the better statistics and the more accurate treatment of the Ewald potential in Hansen's work. His internal energies can be fitted, within statistical errors, by a very simple function of Γ [2.139]

$$\frac{\beta U}{N} = A\Gamma + B\Gamma^{1/4} + C \tag{2.14}$$

valid for $\Gamma \geqq 1$, with $A = -0.896434$, $B = 0.86185$ and $C = -0.5551$. From the computed $g(x)$, suitably extrapolated for $x > L/2$ [2.104], HANSEN derived the structure factors $S(k)$ and direct correlation functions $c(x)$ by Fourier transformations; the $S(k)$ have been tabulated for 9 values of Γ [2.140].

Of the approximate theories for the equilibrium properties of the OCP, the HNC integral equation yields by far the best results [2.141,142]. Over the whole range investigated by the MC computations, the HNC internal energies lie systematically less than 1% below the MC results. It must, however, be stressed that this represents about 20% of $\Delta\beta U/N$ for $\Gamma \geqq 100$.

MC computations have also been carried out for the bcc crystalline phase of the OCP [2.143]. The Helmholtz free energies, obtained from integrating the equations of state of both phases, intersect at $\Gamma = 155 \pm 10$; the volume change on melting is extremely small ($\Delta V/V_{fluid} \leqq 0.05\%$).

In order to obtain an independent check of the adequacy of the Ewald summation procedure in the case of the OCP, CEPERLEY and CHESTER [2.144] have performed some MC computations on systems of 128 and 256 particles interacting via a screened potential

$$v_0(x) = \frac{\Gamma}{x}\,\mathrm{erfc}(x/\sigma)$$

and derived the pair distribution function for the OCP from that of this reference system by a perturbation scheme due to LADO [2.145]. The resulting $g(x)$ and $\Delta\beta U/N$

are rather insensitive to the cut-off parameter σ ($1 \leq \sigma \leq 1.5$), and in excellent agreement with the Ewald MC results.

b) Electron Screening

The rigid uniform background model is reasonable only in the limit of extremely high densities, e.g., in white dwarfs. At lower densities, the polarization of the degenerate electron background by the ionic charge distribution plays an important role. The resulting electron screening length is roughly equal to the Thomas-Fermi length

$$\lambda_{TF}/a = (\pi/12Z)^{1/3}\, r_s^{-1/2}$$

where $r_s = a/(a_0\, Z^{1/3})$ (a_0 is the electronic Bohr radius). Screening effects become negligible when $r_s \to 0$ (high density limit), but for $r_s \simeq 1$ (characteristic of the interior of Jupiter) they become important, and can be included through linear response theory. The Coulombic potential energy (2.13) must then be modified by the introduction of the static dielectric constant $\varepsilon(k)$ of the electron gas

$$V_N(\underline{r}^N) = \frac{1}{2V} \sum_{\underline{k}}{}' \frac{4\pi(Ze)^2}{k^2} \frac{1}{\varepsilon(k)} \left[\rho_{\underline{k}}\rho_{-\underline{k}} - N\right] \quad .$$

HUBBARD and SLATTERY [2.129] have carried out some MC computations for a H^+ plasma in such a responding electron background, using the random phase approximation for $\varepsilon(k)$ (i.e., the Lindhard dielectric function [2.146]). Their computations cover the range $10 \leq \Gamma \leq 75$, for $r_s = 0.1$ (2679 gr cm^{-3}) and 1. (2.679 gr cm^{-3}). In view of the small number of particles (27 or 32), of configurations generated in each run (12000), and of the very approximate treatment of the Ewald sums (a tabulation with a large grid spacing), the quoted results are probably affected by relatively large errors. The calculations have, however, the merit of showing that the thermodynamic properties are little changed with respect to the rigid background case, e.g., the internal energy is lowered by about 5% at $r_s = 1$ with respect to the OCP. On the other hand the effect of electron screening in reducing the degree of correlation in the H^+ plasma is clearly exhibited by the computed pair distribution functions.

c) Charged Hard Spheres in a Uniform Background

Another extension of the OCP is a model of charged hard spheres in a uniform penetrating background. Although it has no direct physical applications, this system is of great theoretical importance, since it is the simplest model which incorporates both short-range and Coulombic interactions. MC computations for this system have been performed by HANSEN and WEIS [2.147], at $\Gamma = 20$, 40 and 70 and packing fractions $\eta = 0.343$ and 0.4. The authors conclude that none of the existing perturbation theories is capable of reproducing correctly the MC pair distribution functions.

d) Ionic Mixtures

The OCP model has been generalized to treat ionic mixtures. HANSEN et al. [2.148] have performed MC computations for $H^+ - He^{++}$ mixtures in a *rigid* uniform background. The equilibrium states of such mixtures are characterized by the coupling parameter Γ [cf. (2.8)], with $\rho = (N_1 + N_2)/V$, and the concentrations x_1 and $x_2 = 1 - x_1$ of the two species; alternatively x_1 and $\Gamma' = \Gamma \bar{Z}^{-1/3}$ with $\bar{Z} = x_1 Z_1 + x_2 Z_2$, can be chosen as convenient independent variables. The MC computations of HANSEN et al. are for systems of 128 and 250 ions and cover the range $5 \leq \Gamma \leq 40$ and three concentrations, $x_1 = 0.25$, 0.5 and 0.75; averages were taken over at least 2.10^5 configurations per run. The authors find that the internal excess energy of the mixture is given very accurately (within statistical errors) by the simple linear interpolation

$$\frac{\beta U}{N}(\Gamma', x_1) = x_1 \frac{\beta U_0}{N}(\Gamma_1) + x_2 \frac{\beta U_0}{N}(\Gamma_2) \tag{2.15}$$

where $U_0(\Gamma)$ is the internal energy of the OCP [cf. (2.14)], and $\Gamma_i = \Gamma' Z_i^{5/3}$. As in the case of the OCP, the predictions of HNC theory are close to the MC energies, and satisfy also (2.15).

DEWITT and HUBBARD [2.149] have extended the work of HUBBARD and SLATTERY [2.129] to the case of metallic H - He mixtures in a *responding* background, using the Lindhard dielectric function to take into account the polarization of the electron gas. Their MC runs cover some 150 different points in the range $0 \leq r_s \leq 1$, $0 \leq \Gamma \leq 150$ and $0 \leq x_{He} \leq 0.2$; averages were taken over 40,000 configurations for systems of about 40 ions. They summarize their data in a simple equation of state, from which it appears that the deviation of the thermodynamic quantities from their rigid background values ($r_s \to 0$ limit) is essentially *linear* in r_s, and relatively small up to $r_s = 1$. Electron screening reduces the interaction energy of the ions by less than 10% whereas it raises the ionic pressure by about 2% at $r_s = 1$. These findings are in fair agreement with a simple perturbation theory which takes the rigid background model as a starting point [2.140].

e) Hydrogen Plasmas

Hydrogen plasmas in which the electrons are nondegenerate ($T > T_F$) have been considered in Monte Carlo computations by several authors in the ionization region. The pioneer work is due to BARKER [2.128] who was the first to use Ewald techniques for Coulomb energy summations. He studied a hydrogeneous plasma of 16 protons and 16 electrons at a density of 10^{18} electrons cm^{-3} and at three temperatures. At the lowest temperature ($T = 10^4$ K), he observed that the energy did not reach an asymptotic equilibrium value in 6000 steps and concluded that there was a high degree of pair formation (i.e., formation of H atoms). In a subsequent paper [2.150] BARKER pointed out the importance of quantum effects at short distances and concluded that an effective pair potential should be used when an ion and an electron approach closer than a critical distance, which depends on T, but is of the order of the Bohr radius a_0. This effective potential must reflect the quantum mechanical stability (due to the uncertainty principle) at short distances, which prevents the classical "collapse" of the proton-electron pair. Such an effective potential was constructed among others by VOROB'EV et al. [2.151] who computed the two-particle Slater sums $S_{\alpha\beta}$ for each of the three pairs (α,β = proton or electron) and identified these with the Boltzmann factors of the three effective potentials $v_{\alpha\beta}$ (r)

$$S_{\alpha\beta}(r) = c \sum_n |\psi_n(r)|^2 \exp[-(\beta E_n)] \equiv \exp[-\beta v_{\alpha\beta}(r)]$$

where c is a normalization constant. For $r \to 0$, the $v_{\alpha\beta}(r)$ remain finite, whereas for $r \to \infty$, the $v_{\alpha\beta}(r)$ go over into the purely classical Coulomb potential. VOROB'EV et al. [2.151] then used these effective potentials in grand-canonic ensemble MC computations in which T, V and the sum of the proton and electron chemical potentials, $\mu_p + \mu_e$ were kept fixed. At T = 30,000 K and $\rho_e = 6.10^{21}$ cm^{-3} they observed a degree of ionization of about 40%, which varied sensitively with density.

POKRANT et al. [2.152] have improved the simulation of the hydrogen plasma towards higher densities, by deriving density-dependent effective pair potentials which they then used in MC computations of the pair distribution functions and ionization curves at $3.10^4 \leq T \leq 5.10^4$ K and $\rho_e = 10^{20}$ and 6.76×10^{21} electrons cm^{-3}. Their MC cell contained 30 protons, 15 electrons with spin-up and 15 electrons with spin-down and between 10^5 and 3.10^5 configurations were generated in each run; using a reasonable criterion to distinguish between free electrons and those bound in the ground state of an H atom (excited states are neglected) POKRANT et al. derive the following empirical formula for the fraction f of bound electrons from their MC results

$$f = \frac{1}{2} [1 + \mathrm{erf}(b \ln T/T_0)]$$

where

$$T_0 = 10^5\ \mathrm{K}/(24.62 - 1.04\ \log_{10}\rho_e)$$

$$b = 14.35 - 0.61\ \log_{10}\rho_e$$

ρ_e being expressed in electrons cm^{-3}.

2.4.3 The Primitive Model and Its Applications

In the primitive model of electrolyte solutions, the anions and cations are represented by negatively and positively charged hard spheres, whereas the solvent is assimilated to a dielectric continuum which acts to reduce the electrostatic interaction between ions by a factor $1/\varepsilon$, where ε is the static dielectric constant of the medium. The hard-sphere diameters are chosen according to the known ionic radii, and are generally assumed to be additive. If the diameters and charges of the anions and cations are equal, the system is called the *restricted* primitive model (RPM).

a) The Restricted Primitive Model

The equilibrium properties of the RPM depend on two independent parameters which can be chosen to be Γ, as defined by (2.8) and the packing fraction $\eta = \pi\rho\sigma^3/6$, where σ is the common diameter of the ions and $\rho = (N_+ + N_-)/V = N/V$ $(N_+=N_-=N/2)$. It is often convenient to choose a density independent Coulombic coupling parameter instead of Γ, i.e.

$$q = (Ze)^2/(\varepsilon k_B T\sigma) = \Gamma/(2\eta^{1/3}) \quad . \tag{2.16}$$

A convenient reduced temperature is then $T^* = 1/q$. If c is the electrolyte concentration in moles/1, and σ is expressed in Å, the relation between η and c is simply

$$\eta = 6.3064 \times 10^{-4}\ c \times \sigma^3 \quad .$$

To illustrate the magnitudes of q and η for real systems, it may be noted that a 1 M aqueous solution of KF at room temperature ($\varepsilon = 78.5$) corresponds to $\eta = 0.012$ and $q = 2.7$ ($\Gamma \simeq 1.$), whereas $\eta \simeq 0.4$ and $q \simeq 53$ ($\Gamma \simeq 80$) for molten KF at 1173 K, provided σ is taken as the sum of the Pauling radii for K^+ and F^- [2.153].

Since the RPM is entirely symmetric with respect to the anions and cations, it is characterized by 2 (instead of 3) pair distribution functions, $g_\ell(r) = g_{++}(r) = g_{--}(r)$, and $g_u(r) = g_{+-}(r)$. The equation of state is the sum of a Coulombic and a contact term

$$\frac{\beta P}{\rho} = 1 + \frac{\beta U}{3N} + 2\eta \lim_{r \to \sigma_+} [g_\ell(r) + g_u(r)] \tag{2.17}$$

where P is the osmotic pressure in the case of electrolyte solutions; and the excess internal energy can be calculated from

$$\frac{\beta U}{N} = 12\eta q \int_1^\infty [g_\ell(x) - g_u(x)]x \, dx$$

with $x = r/\sigma$.

MC computations for the RPM have been carried out by VORONTSOV-VEL'YAMINOV et al. [2.154-157] and by VALLEAU et al. [2.158-160] in the electrolyte regime ($\eta \leq 0.1, q \leq 3$), and by LARSEN [2.161,162] in the molten salt regime ($0.15 \leq \eta \leq 0.4$; $1 \leq q \leq 100$). Representative results obtained by these various authors for the internal energy, the equation of state and the specific heat, c_V/Nk_B, are summarized in Table 2.1 as a function of η and q. VORONTSOV-VEL'YAMINOV et al. as well as VALLEAU et al. used the nearest image convention in calculating the potential energy of each configuration, for various system sizes (N = 32, 48, 64 and 128 in the former work; N = 16, 32, 64 and 200 in the latter calculations); the resulting averages were plotted by VALLEAU et al. as a function of 1/N and extrapolated linearily to 1/N = 0 in order to estimate the infinite system properties. This procedure can be expected to yield reliable results for not too strong Coulombic coupling ($\Gamma \leq 3$). LARSEN used Ewald summations in conjunction with SINGER's procedure [2.133] in his Monte Carlo computations which cover the strong coupling regime; his data are generally for 216 ion systems and he detected no systematic N-dependence of his averages. These data should be accepted with some caution, since the Ewald procedure has not been tested by an independent method in the case of the RPM under strong coupling conditions. A calculation similar to that of CEPERLEY and CHESTER [2.144] for the OCP may be worthwhile to clarify this point. From the table it appears that the pressure increases with η for fixed q, due to an increasing contribution from the contact term in (2.17), whereas it decreases with q for fixed η, due to the internal energy term in (2.17); for sufficiently high q(or Γ) the pressure goes negative, a situation similar to that of the OCP. Note that the nonideal part of the equation of state $\pi = \beta P/\rho - 1$ is affected by large errors due to the difficulty of obtaining the contact term accurately. For $\Gamma \geq 10$, $\beta U/N$ is essentially linear in Γ, as in the case of the OCP, which means that

Table 2.1 Thermodynamic properties of the restricted primitive model from Monte Carlo computations, as a function of the packing fraction η and the inverse temperature q. Γ is the plasma coupling parameter, N is the number of particles, NC is the number of configurations generated, $\beta U/N$ is the excess internal energy per particle, $\beta P/\rho$ is the equation of state, C_V/Nk_B is the specific heat at constant volume

η	q	Γ	N	NC 10^{-3}	$-\frac{\beta U}{N}$	$\frac{\beta P}{\rho}$	$\frac{C_V}{Nk_B}$	Ref.
0.000441	1.6808	0.2558	∞	40	0.1029 ± 0.0013	0.9701 ± 0.0008	0.0458	[2.158]
0.002018	1.7407	0.4400	128	16	0.211	0.946		[2.155]
0.004036	1.7407	0.5543	128	9	0.274	0.945		[2.155]
0.005023	1.6808	0.5757	∞	100	0.2739 ± 0.0014	0.9445 ± 0.0012	0.103	[2.158]
0.008072	1.7407	0.6984	128	61	0.348	0.922		[2.155]
0.02018	1.7407	0.9478	128	63	0.457	0.961		[2.155]
0.02058	1.6808	0.9212	∞	100	0.4341 ± 0.0017	0.9774 ± 0.0046	0.124	[2.158]
0.04036	1.7407	1.1942	128	51	0.582	1.049		[2.155]
0.04842	1.6808	1.2252	∞	72	0.5516 ± 0.0016	1.094 ± 0.005	0.128	[2.158]
0.06054	1.7407	1.3670	128	140	0.612	1.14		[2.155]
0.09525	1.6808	1.5353	∞	92	0.6511 ± 0.0020	1.346 ± 0.009	0.128	[2.158]
0.1498	1.8823	1.9993	216	200	0.839 ± 0.010	1.62 ± 0.15	0.16	[2.161]
0.1498	4.7055	4.9981	216	200	2.467 ± 0.017	1.43 ± 0.10	0.32	[2.161]
0.1498	9.4111	9.9963	216	200	5.465 ± 0.030	0.79 ± 0.31	0.60	[2.161]
0.1498	18.8220	19.9925	216	200	11.95 ± 0.02	0.12 ± 0.21	0.87	[2.161]
0.1498	47.0550	49.9812	216	200	33.62 ± 0.08	-0.79 ± 0.37	2.40	[2.161]
0.1498	94.1103	99.9626	216	200	70.16 ± 0.03	-3.40 ± 0.83	1.16	[2.161]
0.2507	1.5873	2.0018	216	200	0.783 ± 0.011	3.02 ± 0.21	0.09	[2.161]
0.2507	3.9684	5.0045	216	200	2.226 ± 0.032	2.70 ± 0.27	0.34	[2.161]
0.2507	7.9368	10.0091	216	200	4.876 ± 0.036	1.99 ± 0.34	1.02	[2.161]
0.2507	15.8737	20.0182	216	200	10.53 ± 0.05	1.34 ± 0.34	0.85	[2.161]
0.2507	39.6841	50.0454	216	200	28.61 ± 0.04	-0.65 ± 0.77	1.23	[2.161]
0.2507	79.3683	100.0910	216	200	59.91 ± 0.15	-3.04 ± 0.78	5.38	[2.161]
0.3503	1.4191	2.0007	216	200	0.756 ± 0.009	5.04 ± 0.08	0.05	[2.161]
0.3503	3.5476	5.0016	216	200	2.114 ± 0.012	4.69 ± 0.20	0.08	[2.161]
0.3503	7.0952	10.0033	216	200	4.601 ± 0.013	4.48 ± 0.61	0.29	[2.161]
0.3503	14.1905	20.0066	216	200	9.87 ± 0.02	3.57 ± 0.27	0.71	[2.161]
0.3503	35.4760	50.0163	216	200	26.54 ± 0.05	1.49 ± 0.73	1.56	[2.161]
0.3503	70.9522	100.0328	216	200	54.93 ± 0.05	-1.64 ± 0.72	1.95	[2.161]
0.3945	1.3634	1.9999	216	200	0.711 ± 0.026	6.71 ± 0.75	0.18	[2.161]
0.3945	3.4086	4.9998	216	200	2.067 ± 0.013	6.44 ± 0.84	0.14	[2.161]
0.3945	6.8172	9.9996	216	200	4.511 ± 0.035	6.15 ± 0.10	0.37	[2.161]
0.3945	13.6344	19.9993	216	200	9.58 ± 0.03	4.47 ± 0.26	0.44	[2.161]
0.3945	34.0859	49.9981	216	200	25.71 ± 0.02	2.61 ± 0.40	0.79	[2.161]
0.3945	68.1718	99.9963	216	200	53.26 ± 0.09	1.69 ± 0.97	1.72	[2.161]

the internal energy is essentially independent of T along an isochore in the strong coupling limit (saturation effect). In the electrolyte regime, the MC results for $\beta U/N$ and π deviate considerably from the Debye-Hückel limiting law, even at relatively low concentrations. Several approximate theories of electrolyte solutions have been tested against the MC data. The fully analytic "mean spherical" approximation (MSA) [2.163] yields good results for 1-1 aqueous ionic solutions at concentrations up to 0.5 M. The optimized cluster expansion of ANDERSEN et al. [2.164,165] yields excellent thermodynamic properties even for high concentrations (2 M), and their "exponential" approximation leads to pair distribution functions g_ℓ and g_u which are nearly undistinguishable from the results of CARD and VALLEAU [2.158] for 1-1 electrolytes. Numerical solution of the HNC integral equation [2.166] leads also to thermodynamic properties in excellent agreement with the MC data of VORONTSOV-VEL'YAMINOV et al. [2.156] and of CARD and VALLEAU [2.159]. The HNC pair distribution functions are also in good agreement with the MC results of CARD and VALLEAU; both calculations indicate a charge oscillation (or charge "inversion"), i.e., a change of sign of the radial charge distribution $g_\ell(x) - g_u(x)$, around $x = r/\sigma \simeq 2.4$ at the highest concentration considered by these authors ($c \simeq 2$ M), in agreement with a prediction of STILLINGER and LOVETT [2.167]. A complete discussion of various approximate treatments of the RPM and comparison with MC data is contained in a review by OUTHWAITE [2.168].

LARSEN fitted a semi-empirical analytic equation of state (as a function of q and η) to his MC data for the RPM [2.162]; using reasonable hard-sphere diameters he then investigated the thermodynamic properties of a number of molten alkali halides and found semi-quantitative agreement with experimental data. VORONTSOV-VEL'YAMINOV et al. [2.101] have also performed some constant pressure (NPT ensemble) MC computations on the RPM and were able to observe two coexisting phases separated by a pronounced volume discontinuity. They associate this behavior with a gas-liquid transition and locate the critical point at $T_c^* = 1/q_c \simeq 0.086$, $P_c^* = \varepsilon P_c \sigma^4/(Ze)^2 \simeq 6.7 \times 10^{-3}$ and a critical density in the range $0.18 \leqq \rho_c^* \leqq 0.28$, which corresponds to a highly concentrated electrolyte solution. The possibility of a liquid-gas transition for the RPM has been independently put forward by STELL et al. [2.169] on the basis of several distinct theories and of LARSEN's equation of state [2.169]. Although their predicted critical temperature ($T_c^* \simeq 0.085$) agrees well with that of VORONTSOV-VEL'YAMINOV et al., the critical densities differ by more than an order of magnitude ($\rho_c^* \simeq 0.01$ compared to $\rho_c^* \simeq 0.2$) and so do the critical pressures ($P_c^* \simeq 0.031 \times 10^{-3}$ instead of 6.7×10^{-3}).

VALLEAU and CARD [2.160] have applied the multistage sampling method to estimate directly the free energy and entropy of the RPM. Their results are in excellent agreement with thermodynamic integration and with various theories. It must be stressed that the method is especially suited to small systems ($N \leqq 10^2$) which are characterized by rather broad energy distributions.

b) The Dissymmetric Primitive Model

Dissymmetric versions of the primitive model of electrolytes, corresponding to different absolute charges or different diameters of the anions and cations, have been less extensively studied by MC simulations. VORONTSOV-VEL'YAMINOV and EL'YASHEVICH [2.155] studied systems with equal absolute charges but diameters in the ratios $\sigma_1/\sigma_2 = 4/3$ and 5/3. They concluded that the thermodynamic functions of such dissymmetric electrolytes were, within the limits of accuracy of their calculations, the same as those for systems composed of ions with the common diameter 1/2 $(\sigma_1 + \sigma_2)$.

An extremely dissymmetric case has been considered by GILLAN et al. [2.170] who studied mixtures of charged hard spheres and point ions ($\sigma_2 \rightarrow 0$) of opposite charge. Although most of their calculations were made in the framework of the MSA, they ran two MC computations as a check on the MSA. These were carried out on a system of 216 ions (108 anions and 108 cations), for two molten-salt like states ($\eta_1 = \pi\rho\sigma_1^3/6 = 0.3$, $q_1 = \beta(Ze)^2/\sigma_1 = 7.2$ and $\eta_1 = 0.3$, $q_1 = 45.4$). Their most interesting finding is that polymerization occurs in the second, strongly coupled state: each hard sphere is found to be surrounded by two point ions and vice-versa, leading to the formation of linear chains; such a behavior has been observed experimentally in molten BeF_2 and HF. The model may also be of relevance in the study of CuCl, since the copper ion is much smaller than the chlorine ion.

PATEY and VALLEAU [2.171] have performed an interesting MC computation to test the validity of the primitive model, which assumes that in an electrolyte solution, the only effect of the solvent is to reduce the electrostatic interaction between two hard spheres of opposite charges, immersed in a dipolar hard sphere solvent. The ions and solvent particles have the same diameter σ and the solvent particles are characterized by a reduced dipole moment $\mu^* = (\beta\mu^2/\sigma^3)^{\frac{1}{2}} = 1$, which corresponds to a dielectric constant $\varepsilon \simeq 7.8$. The MC results for the solvent-averaged potential of mean force $v_{ii}(r)$, defined by $g_{ii}(r) = \exp[-\beta v_{ii}(r)]$, indicate that the primitive model is inadequate for small ionic separations.

2.4.4 Molten Salts

a) Potentials

Molten salts have been extensively studied by MC and MD simulations since about 1970. Practically all published results are for fused alkali halides. The crudest model for such liquids is the primitive model discussed in the previous section; it is the simplest of a class of potential models which neglect the polarization of the ions by the strong internal local electric fields, and which are generally called "rigid ion models". Two rigid ion pair potentials that have been extensively used in computer simulation work are the Born-Huggins-Mayer (BHM) potential [2.172] and the Pauling potential [2.173]. Both involve four terms

$$v_{\nu\mu}(r) = R_{\nu\mu}(r) + \frac{Z_\nu Z_\mu e^2}{r} - \frac{C_{\nu\mu}}{r^6} + \frac{D_{\nu\mu}}{r^8} \qquad \nu,\mu = 1,2 \tag{2.18}$$

where R(r) is a short-range repulsive term, and the last two terms correspond to the dipole-dipole and dipole-quadrupole dispersion energies. In the BHM potential, the repulsion is exponential

$$R_{\nu\mu}(r) = B_{\nu\mu} \exp(-\alpha_{\nu\mu} r)$$

whereas in the Pauling potential, R(r) is an inverse power repulsion

$$R_{\nu\mu}(r) = \frac{A_{\nu\mu}}{r^n}$$

where the power n is of the order of 9. The various parameters occurring in both potentials are generally chosen equal to those derived by FUMI and TOSI [2.174] from the properties of solid alkali halides at 298 K. A simplified version of the Pauling potential, discarding the two dispersion terms has also been used in computer calculations

$$v_{\nu\mu}(r) = \frac{e^2}{\lambda}\left[\frac{1}{n}\left(\frac{\lambda}{r}\right)^n + Z_\nu Z_\mu \frac{\lambda}{r}\right] \quad . \tag{2.19}$$

It has the advantage of leading to a simple corresponding states law for the thermodynamic properties, if these are reduced by the related length and energy parameters λ and e^2/λ [2.175,176], where λ is essentially the sum of anion and cation radii. The model is entirely symmetrical with respect to the two ionic species; the short-range repulsive term between like ions is frequently left out since the Coulomb repulsion between like pairs is generally sufficient.

Ion Polarization is an essentially dynamic effect, and can hence only be properly studied by MD simulations. A popular model which has been extensively used in MD computations is the so-called "shell model". This model, as well as alkali halide potentials in general, is thoroughly discussed in a review paper by SANGSTER and DIXON [2.130].

b) KCl

The pioneering paper on molten salts is the MC work on KCl by WOODCOCK and SINGER [2.177]. They studied a system of 216 ions interacting via BHM pair potentials, along one isotherm in the solid phase (1045 K) and four isotherms in the liquid phase (1045, 1306, 2090, and 2874 K). The Coulomb part of the inter-ionic potentials was treated by the Ewald method, but the sums Φ_1 and Φ_2 [cf. (2.10)] were truncated after 1 and 3 terms respectively (with $\alpha = 5.714/L$), which introduces a nonnegligible systematic error on the computed energies. $2 - 5 \times 10^5$ configurations were generated per run. The main results are: I) excellent agreement between the MC internal energies (which turn out to be 90% electrostatic) and the experimental values (the difference being systematically less than 0.5%); II) almost identical $g_{++}(r)$ and $g_{--}(r)$ pair distribution functions; III) a marked charge ordering, i.e., an alternation of spherical shells of predominantly positive and negative charge which results in a very flat mean pair distribution function

$$g_m(r) = \frac{1}{2}\,[g_\ell(r) + g_u(r)]$$

where

$$g_\ell(r) = \frac{1}{2}\,[g_{++}(r) + g_{--}(r)] \quad ; \quad g_u(r) = g_{+-}(r) \quad .$$

The oscillations in the charge pair function

$$g_Z(r) = g_\ell(r) - g_u(r)$$

are, on the contrary, very pronounced; IV) a large relative volume change on melting ($\Delta V/V_{solid} \simeq 20\%$), in good agreement with experiment. WOODCOCK and SINGER determined the absolute entropies of the coexisting phases by an intuitively appealing but nonrigorous procedure; their entropy and free energy values must hence be accepted with caution [2.178].

ROMANO and McDONALD [2.179] reconsidered molten KCl, using the Pauling rigid ion potential, with a repulsive exponent $n = 8.28$. Their MC computations cover three states ($T = 1043$, 1400, and 1680 K) at atmospheric pressure, and were done for 216 ions in an NPT ensemble; averages were taken over 3×10^5 configurations. They conclude that the Pauling potential is much less satisfactory for KCl than the BHM potential, since the calculated volumes were found to be much larger than the experimental values (14% at melting, 31% at the boiling point). Subsequently ROMANO and MARGHERITIS [2.180] estimated the contribution to the internal energy of KCl due to ion polarization, using a crude model which neglects the contributions of induced dipoles to the local electric field. They find that under these conditions the polarization contribution is about 1% of the total energy. In view of the crudeness of the model and of the relatively small size of their system (64 ions), the result should be accepted with caution.

c) Other Alkali Halides

LEWIS et al. [2.181] have extended the NVT ensemble computations on KCl to eight other alkali halides: LiCl, LiI, NaCl, KBr, KF, KI, RbCl, CsF, using BHM potentials throughout. With the exception of LiI, KF and CsF, the BHM potential (with Fumi-Tosi parameters) were found to reproduce the experimental internal energies very well (better than 1%); the calculated molar volumes at zero pressure turned out to be invariably larger, by several percent (especially in the case of KF), than the experimental values. These computations confirmed the systematic tendency to a pronounced charge ordering as illustrated by the strong oscillations in $g_Z(r)$. The agreement of the mean pair distribution function $g_m(r)$ with existing x-ray and neutron diffraction data [2.182] is satisfactory, but seems to indicate that the anion-cation BHM repulsion is too soft, in agreement with the evidence from crystal phase data [2.183]. Another systematic MC study of 10 alkali halides crystallizing in the NaCl structure (three fluorides, four chlorides and three bromides) has been carried out by ADAMS and McDONALD [2.133] both at room temperature (solid phase) and at the experimental melting point (both in the solid and liquid phases). These authors work in the NPT ensemble, allowing them to obtain easily results at zero pressure. They begin by a very careful investigation of the number of terms in the Ewald sums (2.9) required to achieve a given precision. Both the Pauling and BHM potentials with Tosi-Fumi parameters are used, allowing a systematic comparison of the relative merits. The BHM potential is found to be superior to the Pauling form in virtually all respects; however, both potentials overestimate systematically the solid and liquid volumes, especially at the higher temperature. The discrepancy with experiments is largest for the liquid phase, reaching 9% with the BHM potential and 25% with the Pauling potential!

The BHM rigid ion potential has been used in a number of MD studies of alkali halides, which are mostly concerned with dynamic properties (e.g., self-diffusion). We briefly mention the work on NaCl by LANTELME et al. [2.184] and by LEWIS and SINGER [2.185], and on RbBr by COPLEY and RAHMAN [2.186]. The latter work, which studies the dynamic structure factors (density fluctuations) is of particular interest since extensive inelastic neutron scattering data under identical physical conditions are available [2.187].

As mentioned earlier, polarizable ion models, using the shell model, have been used in a number of MD simulations of NaI, NaCl, KI and Rb halides. These computations indicate substantial differences in the pair distribution functions obtained with rigid or polarizable ion models. The polarization effects lead in particular to an enhanced differentiation between g_{++} and g_{--}, in agreement with experimental data [2.182]. We refer the reader to the exhaustive review of SANGSTER and DIXON for details [2.130].

d) Corresponding States Model

ADAMS and McDONALD [2.176] have used the simple model potential (2.19) with n = 9 in a series of MC computations in the NPT ensemble to seek a corresponding states correlation of properties of molten alkali halides. Their runs cover the reduced temperature range $0.014 \leqq T^* = \lambda k_B T/e^2 \leqq 0.036$ and the reduced density range $0.182 \leqq \rho^* = \rho\lambda^3 \leqq 0.433$, at zero pressure. Although some qualitative features of experimental thermodynamic data are well reproduced, (e.g., the specific heat C_p), these authors conclude that "an interionic potential to which corresponding states theory can be rigorously applied is insufficiently flexible to account for details of the thermodynamic properties of molten alkali - metal salts". Their paper also contains some data on binary mixtures having a common anion, indicating the inadequacy of the simple potential model to account for the experimental enthalpies of mixing.

The same model potential has been used by HANSEN and McDONALD [2.188] in an extensive MD study of charge and density fluctuations in a molten alkali halide. That paper lists very accurate pair distribution functions, direct correlation functions and structure factors. Charge ordering manifests itself in a very sharp peak in the static charge structure factor.

e) Mixtures

The most extensive MC study of an alkali halide mixture is by LARSEN et al. [2.189] who studied the mixing process NaCl + KCl → (Na, K) Cl at T = 1083 and zero pressure (50-50% mixture). These authors used the BHM potential with parameters for the mixture taken essentially as averages of the Tosi-Fumi parameters for the pure salts. They find internal energies of the pure phases and the mixture in excellent agreement with experiments, while the molar volumes are about 5% too large. The volume change on mixing is found to be 1.2 ± 0.4 cm^3/mole (experiment: 0.24 ± 0.08), and the internal energy change is of the wrong sign (+ 0.8 ± 0.1 kcal/mole compared to the experimental - 0.13 ± 0.01). The entropy estimates, based on the approximate "effective free volume method" [2.190] cannot be trusted [2.178].

2.4.5 Liquid Metals

a) Generalities

Relatively few computer "experiments" on liquid metals have appeared in the literature. This is partly due to our poor knowledge of inter-ionic interactions. Liquid metals consist of two "fluids": ions and conduction electrons. Due to the large mass difference the ions behave classically whereas the electrons must be treated essentially quantum mechanically. "Pseudo-potential" theory [2.191] is used to account for the Pauli principle acting between conduction and core electrons. This ion-electron pseudopotential is then combined with theoretical models for the electron gas dielectric constant to construct an effective ion-ion pair potential in the framework of linear response theory. Such effective potentials are very sensitive to details of the theory; they are moderately reliable for alkali metals, but probably very poor for higher valence metals for which band effects become important. For that reason most published computer "experiments" deal with liquid alkali. The effective inter-ionic potential is characterized by a relatively soft repulsive core (compared to rare gas potentials) and a long-range oscillatory behavior which is essentially of the form

$$v(r) \simeq \frac{A\cos(2k_F r)}{r^3}$$

where k_F is the radius of the Fermi sphere of the electron gas (Friedel oscillations) Due to the uncertainties on the interactions, computer "experiments" have been mostly concerned with the structure (which is less sensitive to details of the potential [2.84]) rather than the thermodynamic properties of liquid metals.

The crudest model for a liquid metal is the OCP (Sect.2.4.2a), which ignores both core electrons and dielectric screening of the conduction electrons. With no adjustable parameter the MC structure factors of the OCP are in surprisingly good agreement with experimental x-ray data on liquid Na and K [2.192], except in the small k region.

b) "Realistic" Potentials

More "realistic" effective ion-ion potentials have been used both in MD and in MC simulations. The first results on the pair structure of liquid metals were obtained by PASKIN and RAHMAN [2.193] using the MD method. They obtained a radial distribution function for liquid Na in reasonable agreement with Fourier inversion of the experimental structure factor. A more systematic study of liquid Na and Al was made by SCHIFF [2.194] who used both the MC and MD methods on systems of 864 ions. He used a variety of effective ion-ion potentials to investigate the influence of the core softness and of the Friedel oscillations on the computed structure factors; the latter were obtained by Fourier transforming the computed radial distribution functions, suitably extrapolated to distances larger than half the box length [2.104]. SCHIFF found that the oscillations in S(k) corresponding to a softer repulsion are more damped, as one would intuitively expect (see also reference [2.84]), and that the effect of the Friedel oscillations on the structure factor is very weak for realistic amplitudes A.

The structure factor of liquid Na has been obtained from MC computations by FOWLER [2.195] at T = 125°C, (512 ions) and by MURPHY and KLEIN [2.196] at T = 100°C and 200°C (128 ions). Both computations use different effective pair potentials, but lead to structure factors in good agreement with the x-ray data of GREENFIELD et al. [2.197]. In both simulations, the small k behavior of S(k) is affected by large errors, because the directly determined pair distribution function was not adequately extrapolated.

MURPHY [2.198] has also determined the structure factor of liquid potassium at temperatures of 65 C and 135 C (432 ions), using an effective potential based on the same theoretical ingredients as the earlier work on Na. The agreement of his MC results with the x-ray data [2.197] is again very good, except at small k where the effects of an inadequate extrapolation of the pair distribution function are largest.

To conclude this section the important MD simulation of the dynamical properties (density fluctuations) of liquid Rb by RAHMAN [2.199] deserves special notice.

2.5 Molecular Fluids

Computer simulation studies (MC or MD) on molecular systems were undertaken for the first time ten years ago and have been pursued mainly along two lines of interest. One of these consists of selecting intermolecular potentials capable of describing real systems as accurately as possible. In view of the complexity of molecular interactions it is not surprising that so far only diatomic molecules (e.g., N_2) or water have been extensively studied. On the other hand computer simulation techniques have been employed abundantly to study the properties of highly idealized systems. Typical examples are hard spherocylinders or ellipses which have been used to investigate the orientational order-disorder phase transitions of elongated molecules and dipolar hard spheres which have proved to be more convenient than complicated potentials (such as water models) for the purpose of studying the effects of boundary conditions upon the structural and dielectric properties of systems with long range dipolar forces. Moreover, these idealized systems are often sufficiently simple to be tractable by approximate theories, and computer simulations provide an unambiguous test of the adequacy of these approximations.

To date MC (and MD) computations have been restricted to interaction potentials which are pairwise additive and depend only on the center of mass coordinates $\underline{R}_i$ and orientations $\underline{\Omega}_i$ of the molecules, (i.e., the molecules are assumed to be in their ground electronic and vibrational state). We shall distinguish (somewhat arbitrarily) between three types of potentials. These differ mainly in their treatment of the short-range forces.

Generalized Stockmayer potentials are the sum of a spherically symmetric part [e.g., a LJ or a hard-sphere (HS) potential] and an anisotropic part. The anisotropic part contains the long-range multipolar (electrostatic), anisotropic dispersive and induction forces as well as the short-range anisotropic overlap (exchange, valence) contributions, which are themselves further approximated by the first few terms of an expansion in spherical harmonics [2.200]. A potential of this form can hardly be used for a quantitative description of real molecules, because of the large number of spherical harmonics which would be needed to represent the anisotropic core of most molecules. However, the model is quite convenient for theoretical work. Computer simulations have been concerned so far only with multipolar (dipole-dipole (DD)[2.201-215], dipole-quadrupole (DQ) [2.201-203, 216], quadrupole-quadrupole (QQ) [2.202-205,216,217]) and overlap interactions [2.204,205].

A model which is expected to give a more realistic description of the anisotropic repulsive core is the atom-atom (site-site) potential. The interaction between two molecules (assumed rigid) is the sum of interactions between sites acting as force centers. In most calculations these force centers coincide with the atomic nuclei and the interactions are of the LJ or HS type. The electrostatic contribution arising from the molecular charge distribution can be taken into account either by a multipole expansion or by point charges located in such a way as to reproduce the experimental multipole moments. For linear molecules with a symmetrical charge distribution the leading term in the multipole expansion is a QQ term. Atom-atom potentials have been used to simulate diatomic molecules (N_2, O_2,...) [2.218-231] and benzene C_6H_6 [2.232].

Finally hard convex bodies such as ellipses [2.233,234] and two- or three-dimensional (3-dim) spherocylinders [2.235-239] have been studied, mainly as simple models for nematic liquids.

MC results (and MD results when they are complementary) obtained for these three types of potentials will be reviewed in the next three subsections. Other potentials such as the Kihara core model [2.240-243], the Gaussian overlap model [2.243-245] and the central force model [2.246-248], although of indeniable interest, will not be considered here as they have not yet benefitted from any MC study. Due to limits on the length of this review we refer the reader to the original papers concerned with MC [2.249-258] and MD [2.247,259-261] calculations on water-like molecules. The MD work of RAHMAN and STILLINGER prior to 1975 has been reviewed briefly by STILLINGER [2.262].

The properties of molecular fluids have been reviewed recently by EGELSTAFF et al. [2.263], GRAY [2.264], BOUBLIK [2.265] and STREETT and GUBBINS [2.266]. This last reference, concerned with linear molecules, contains also computational details.

2.5.1 Hard Convex Bodies

a) Hard Spherocylinders (3-dim.)

The simple model of hard spherocylinders is primarily of interest in the study of nematic liquid crystals formed by elongated rod-like molecules and particularly in elucidating the nature of the nematic-isotropic phase transition which occurs at high density. However, an attempt by VIEILLARD-BARON [2.236] to locate this phase transition for a system of 616 hard spherocylinders of length-to-breadth ratio $\gamma = 3$ with periodic boundary conditions failed. At the highest density for which equilibrium could be achieved ($\eta = 0.54$, where $\eta = \rho V_1$, ρ is the number density and V_1 is the volume of one spherocylinder) the system was completely melted and almost completely disoriented (isotropic phase). For higher packing fractions ($\eta > 0.54$) equilibrium could not be reached within a reasonably large number of configurations when the system was started from a perfect crystal state in which the spherocylinders had their centers arranged according to a compact hexagonal lattice with their axis all parallel.

The isotropic liquid phase is stable up to densities significantly higher than those predicted by scaled particle theory (SPT). The SPT of COTTER and MARTIRE [2.267] and the "consistent version" of COTTER [2.268] give $\eta = 0.471$ and $\eta = 0.514$, respectively, for the maximum density of the isotropic phase.

COLDWELL et al. [2.238] estimated the phase change region of 25 hard spherocylinders (with periodic boundary conditions) with a length-to-breadth ratio of 5. The centers were constrained to move in a plane but arbitrary orientations were allowed. A non-Markovian MC method, previously applied to hard disks [2.70] was used to evaluate directly the partition function in the low density region.

VIEILLARD-BARON [2.236] also shows that the value of η at which melting occurs is an increasing function of the length-to-breadth ratio. A system of 224 hard spherocylinders with $\gamma = 2$ is completely melted at $\eta = 0.57$ whereas the infinite hard-sphere system does not melt until $\eta = 0.545$ [2.15]. FEW and RIGBY [2.235], on the other hand, studying a similar system of 120 molecules ($\gamma = 2$) reached the conclusion that the freezing density is lower than that of hard spheres. These results are compatible only if the volume change on melting increases with γ.

The inconsistency in the pressure results obtained in the isotropic liquid phase by VIEILLARD-BARON [2.236] and FEW and RIGBY [2.235] motivated further MC work by BOUBLIK et al. [2.237]. These authors found pressures which are in good agreement with those of VIEILLARD-BARON (although less precise since only 96 molecules were used) and ascribe the discrepancy with the results of FEW and RIGBY to the use, by these authors, of an incorrect pressure algorithm. BOUBLIK et al. [2.237] also present tabulations of an average correlation function $g^{av}(\rho)$ which is related to the probability of finding two spherocylinders with a surface to surface distance between ρ and $\rho + d\rho$, irrespective of the mutual orientations of the two molecules and the positions of all the other molecules. This correlation function is needed to calculate the first-order perturbation correction to the free energy if one uses hard spherocylinders as a reference system [2.269,270].

Several approximate equations of state proposed for the isotropic phase of hard spherocylinders are listed below

$$(P/\rho kT)_{\text{Gibbons}} = [1 + (3\alpha - 2)\eta + (3\alpha^2 - 3\alpha + 1)\eta^2]/(1 - \eta)^3 \tag{2.20}$$

$$(P/\rho kT)_{\text{Cotter-Martire}} = [1 + (3\gamma^2 - 3\gamma + 2)\eta/(3\gamma - 1) + \eta^2(3\gamma^3 + 3\gamma^2 - 3\gamma + 1)/(3\gamma - 1)^2]/(1 - \eta)^3 \tag{2.21}$$

$$(P/\rho kT)_{\text{Boublik}} = (P/\rho kT)_{\text{Gibbons}} - \alpha^2\eta^3/(1 - \eta)^3 \tag{2.22}$$

$$(P/\rho kT)_{\text{Nezbeda}} = [1 + (3\alpha - 2)\eta + (\alpha^2 + \alpha - 1)\eta^2 - \alpha(5\alpha - 4)\eta^3]/(1 - \eta)^3 \tag{2.23}$$

Only the second equation is specific to spherocylinders; the others are applicable to any hard convex body of shape parameter $\alpha = SR/3V_1$ (R is the mean curvature $\times(1/4\pi)$, S the surface area, V_1 the volume). For spherocylinders $\alpha = (\gamma + 1)/(3\gamma - 1)$. The first three equations have been obtained by the SPT [2.271-273] and the last [2.274] from an analysis of the virial series by a method similar to that used by CARNAHAN and STARLING [2.19] to derive their hard-sphere equation of state. For this purpose NEZBEDA [2.274,275] calculates, by the MC method, the third (up to $\gamma = 8$) and fourth (up to $\gamma = 3$) virial coefficients. These coefficients had been obtained previously by RIGBY [2.276] for $\gamma = 1.4$ and 1.8. Equations (2.22) and (2.23) reduce to the CARNAHAN-STARLING equation for hard spheres in the limit $\alpha \to 1$. The four equations of state are compared in Table 2.2 with MC results. It is seen that (2.23) reproduces the MC data up to the isotropic-nematic transition within 2% which is practically within the statistical error. The SPT results are always higher than the MC values.

b) Hard Ellipsoids (3-dim.)

The third, fourth and fifth virial coefficients have been evaluated by FREASIER and BEARMAN [2.277] for different ellipsoids of revolution using a MC technique. These virial coefficients and a Padé representation of the virial series are shown to agree within a few percent with SPT up to moderate densities. A MC calculation of the pressure is more difficult [2.233], basically because the evaluation of the minimum distance between two ellipsoids is far from trivial.

Table 2.2 Compressibility factor $P/\rho kT$ of a three-dimensional system of hard spherocylinders. Comparison of MC calculations with the approximate Equations (2.20-23)

γ	η	$\left(\frac{P}{\rho kT}\right)_{MC}$	$\left(\frac{P}{\rho kT}\right)_{Gibbons}$	$\left(\frac{P}{\rho kT}\right)_{Cotter-Martire}$	$\left(\frac{P}{\rho kT}\right)_{Boubl\bar{i}k}$	$\left(\frac{P}{\rho kT}\right)_{Nezbeda}$
2	0.2454	3.23[a]	3.48	3.41	3.43	3.39
3	0.2676	4.52[a]	4.84	4.53	4.73	4.49
	0.30	5.40 ± 0.13[b]	5.95	5.51	5.78	5.41
	0.35	7.17 ± 0.11[b]	8.28	7.52	7.93	7.23
	0.40	9.60 ± 0.10[b]	11.67	10.42	11.00	9.74
	0.45	13.0 ± 0.16[b]	16.73	14.67	15.50	13.24
	0.50	18.0 ± 0.4 [b]	24.50	21.13	22.25	18.25
	0.54	23.3 ± 0.4 [b]	33.88	28.82	30.24	23.89

[a]MC results from [2.237]; [b]MC results from [2.236]

c) Hard Ellipses (2-dim.)

This two-dimensional model of a liquid crystal is clearly less realistic than the three-dimensional systems mentioned above (ellipsoids or spherocylinders) but its practical resolution is far less problematic. MC calculations by VIEILLARD-BARON [2.233] show that for sufficiently elongated ellipses this system exhibits two first-order phase transitions. At high densities one finds a solid-nematic type transition from the translationally and orientationally ordered solid phase to a nematic phase where the molecules have their centers of mass disordered while their axes stay oriented. At lower densities there is a nematic-liquid transition from the nematic phase to an orientationally isotropic liquid. For a system of 170 ellipses with axis ratio 6 (and periodic boundary conditions) melting starts at an area ratio, A_m/A_0, close to 1.15 (A_0 is the area at close packing). This value is much smaller than the area at which the hard-disk system begins to melt (A_m/A_0 = 1.266 for 870 hard disks [2.64]). The disorientation transition starts at area A_d/A_0 = 1.775 ± 0.025 and the relative area change at the transition is of the order of 1-2%. These results suggest that the nematic-liquid transition of real liquid crystals might, like the melting transition, be mainly due to geometrical factors. A quite different behavior is observed for a system of ellipses with small excentricity (axis ratio 1.01) [2.233]. As the density is decreased this system first disorients and then melts. The center of mass structure factor of a system of ellipses of axis ratio 1.44 in the fluid disoriented phase is shown in [2.234].

d) Hard Spherocylinders (2-dim.)

NEZBEDA et al. [2.239] report MC calculations for the compressibility factor and the average correlation function $g^{av}(\rho)$ (cf. Sect.2.5.1a) of a two-dimensional system of 48 hard spherocylinders in the isotropic fluid phase. The equation of state is in good agreement with SPT [2.278], though, as for the case of hard disks, at high density SPT gives results which are too high.

2.5.2 Atom-Atom Potentials

In this model, called also interaction site model, the potential energy between two molecules is the sum of site-site interactions

$$u(12) = \sum_{\alpha}\sum_{\gamma} w(|\underline{r}_1^{\alpha} - \underline{r}_2^{\gamma}|) \quad . \tag{2.24}$$

Here $\underline{r}_i^{\alpha}$ denotes the location of the αth site on the ith molecule. Most of the available MC calculations are devoted to rigid diatomic molecules.

a) Hard Diatomics (Fused Hard-Spheres, Dumbbells)

The interaction between two sites is a hard-sphere potential

$$w_{\alpha\gamma}(r) = \begin{cases} \infty & r < \sigma_{\alpha\gamma} \\ 0 & r > \sigma_{\alpha\gamma} \end{cases} \quad . \tag{2.25}$$

If the diameters are assumed to be additive

$$\sigma_{\alpha\gamma} = (\sigma_{\alpha\alpha} + \sigma_{\gamma\gamma})/2 \quad . \tag{2.26}$$

Such a model is not very realistic in the sense that it neglects the attractive forcès present in real molecules and replaces the smooth repulsive forces by an infinitely steep hard-core interaction. However, from our knowledge of atomic fluids we would expect this to be a good model to investigate. It is known that the structural properties of dense fluids are dominated by the repulsive part of the intermolecular potential. Furthermore, in real systems the repulsions are generally sufficiently harsh to be represented to a good approximation by appropriate hard-core interactions. Once the properties of the hard-core model are known, the softening of the hard core and the inclusion of the attractive forces can be achieved by use of perturbation theories [2.279,280].

MC calculations for the equation of state of homonuclear hard diatomics have been reported by FREASIER [2.218] (108 molecules, $L/\sigma = 0.6$, σ is the hard-sphere diameter, L the distance between the hard-sphere centers), FREASIER et al. [2.219] (108 molecules, $L/\sigma = 1$) and AVIRAM et al. [2.222] (256 molecules, $L/\sigma = 0.2$, 06). The results, based on two slightly different pressure algorithms, are summarized in Table 2.3.

Table 2.3 Compressibility factor $P/\rho kT$ of hard diatomics. Comparison of MC calculations with approximate theories. The theories of BOUBLIK [SPT(BB)] (2.22) with α given by (2.29), RIGBY [SPT(GR)] (2.20) with α chosen in such a way as to reproduce the exact second virial coefficient and NEZBEDA (N) [2.284] are based upon SPT. The theories of AUSLOOS (A) [2.285], PERRAM and KÖHLER (BH) [2.286] and SANDLER and STEELE (WCA) [2.287] use an effective hard-sphere diameter in conjunction with an hard-sphere equation of state. Also tabulated are the virial and compressibility equations of state given by the RISM equation of CHANDLER [2.290] and the PY equation of CHEN and STEELE [2.289] and a Padé approximant (2.27) of FREASIER et al. [2.219]

L/σ	ρd^3	MC	SPT (BB)	SPT (GR)	N	A	Padé	BH[f]	WCA	RISM[f] vir.	RISM[f] comp.	PY[f] vir.	PY[f] comp.
0.2	0.1296	1.331 ± 0.02 [a]	1.329	1.329	1.331	1.324			1.328[e]				
	0.2592	1.845 ± 0.04 [a]	1.797	1.800	1.805	1.784			1.795[e]				
	0.3888	2.596 ± 0.06 [a]	2.474	2.489	2.499	2.448			2.467[e]				
	0.5184	3.813 ± 0.12 [a]	3.474	3.523	3.542	3.427			3.453[e]				
	0.6480	5.660 ± 0.20 [a]	4.990	5.122	5.155	4.907			4.936[e]				
0.6	0.1792	1.58 ± 0.02 [a]	1.549	1.548	1.564	1.481			1.541[e]				
	0.2	1.64 [b]	1.632	1.632	1.651	1.553		1.65	1.63 [f]	1.8	1.4	1.63	1.63
	0.3584	2.614 ± 0.04 [a]	2.470	2.483	2.529	2.269			2.423[e]				
	0.4	2.84 [b]	2.766	2.786	2.842	2.518		2.79	2.64 [f]	3.2	2.2	2.70	2.74
	0.5376	4.415 ± 0.05 [a]	4.073	4.146	4.247	3.608			3.868[e]				
	0.6	5.02 [b]	4.891	5.010	5.142	4.283		4.70	4.57 [f]	5.5	3.7	4.64	4.66
	0.7168	7.56 ± 0.10 [a]	6.993	7.267	7.481	6.003			6.271[e]				
	0.8	9.24 [b]	9.149	9.628	9.928	7.750		9.43		9.8	6.4		
	0.896	13.037 ± 0.30 [a]	12.70	13.59	14.04	10.60			10.35 [e]				
1.0	0.2	1.830 ± 0.007 [c]	1.804	1.798	1.852	1.553	1.802						
	0.4	3.38 ± 0.02 [c]	3.330	3.339	3.484	2.518	3.346						
	0.6	6.37 ± 0.05 [cd]	6.312	6.440	6.751	4.283	6.418						
	0.7	8.84 ± 0.07 [c]	8.820	9.116	9.560	5.710	9.070						
	0.8	12.4 ± 0.10 [c]	12.50	13.12	13.75	7.750	13.12						
	0.9	17.5 ± 0.16 [c]	18.02	19.27	20.18	10.75	19.71						
	0.95	20.7 ± 0.16 [c]	21.83	23.57	24.66	12.78	24.70						

[a] MC results taken from [2.222]; [b] MC results taken from [2.218]; [c] MC results taken from [2.219]; [d] The compressibility factor for a 500 particle system is 6.45 ± 0.039 [2.219]; [e] results quoted from [2.222]; [f] results quoted from [2.218]

introduced into the system in this way. However, only a numerical comparison could settle how well the infinite system is approximated by the periodicity assumption. Unfortunately the only two existing MC calculations using the Ewald method are inconclusive: those of JANSOONE [2.208] in the liquid phase because of the extremely small size of the system (32 particles) and the limited information reported (Kirkwood structure factor only), those of ADAMS and McDONALD [2.212] for a polar lattice because no exact (infinite) results are available for such a system. The last authors found a difference of about 15% between the dielectric constants obtained by the RF and Ewald methods.

LADD [2.258] divided the potential energy of a given molecule in the field of all the others into two parts. The contribution of the molecules inside the nearest-image cell (centered at the selected particle) was calculated exactly (MI energy). The contribution due to the molecules in the neighboring cells surrounding the nearest-image cell was approximated by the sum over the interactions between the multipoles of the given particle and the total multipoles of all the molecules in each of the neighboring cells. In the case of a system involving only dipoles the leading term of this contribution is a dipole-octopole interaction. LADD, in his MC simulation of water [2.258] found that the contribution of this term was negligible compared to the MI term.

Another frequently used model for polar fluids is the so-called Stockmayer potential where the spherically symmetric part of the potential is a LJ interaction. Internal energies and pressures for a 108 particle system have been obtained over a wide range of densities and temperatures but small dipole moments ($\mu^2/\varepsilon\sigma^3 \leq 1$) by McDONALD [2.209]. VERLET and WEIS [2.210] gave internal energies, free energies and the angle averaged pcf for the state $kT/\varepsilon = 1.15$, $\rho\sigma^3 = 0.85$ and dipole moments up to $\mu^2/\varepsilon\sigma^3 = 4$. (864 particles; $R_c = 2.5\ \sigma$). Results for the internal energy, angle averaged pcf, mean-square force and mean-square torque have also been obtained by WANG et al. [2.204,205,307] for the state $\rho\sigma^3 = 0.8$, $kT/\varepsilon = 0.719$. Due to the small size of the system, 64 particles, these results are relatively imprecise.

Care must be taken when comparing theory with MC results. In particular, meaningful comparison for the pcf can be made only if either theoretical results can be obtained for a spherically truncated potential (as for the LHNC equation) or if the MC results are first corrected to get the infinite system behavior (e.g., by a perturbation treatment of the weak long range tail of the DD interaction [2.308]).

A theory which gives remarkably accurate results for the pcf of dipolar hard spheres is the HNC-type integral equation approach proposed by PATEY [2.303]. Basically one solves the Ornstein-Zernike equation

$$h(12) - c(12) = \rho \int h(13)c(32)d(3) \tag{2.40}$$

with boundary conditions

$$h(12) = -1 \qquad r \leq d \tag{2.41}$$

$$c(12) = h(12) - \ln g(12) - u(12)/kT \qquad r > d \quad . \tag{2.42}$$

Here d(3) stands for $1/4\pi \int d\underline{r}_3 d\underline{\Omega}_3$.

If one further limits the pcf to its projections on Δ and D, i.e.,

$$h(12) = h_S(r) + h_\Delta(r)\Delta(12) + h_D(r)D(12) \tag{2.43}$$

[and a similar relation for c(12)] and retains in ln g only terms linear in Δ and D, the boundary conditions transform to

$$h_S(r) = -1 \ , \quad h_\Delta(r) = h_D(r) = 0 \qquad r < d$$

$$\left.\begin{aligned} c_S(r) &= h_S(r) - \ln g_S(r) - u_{HS}(r)/kT \\ c_\Delta(r) &= h_\Delta(r)\,[1 - 1/g_S(r)] \qquad r > d \\ c_D(r) &= h_D(r)\,[1 - 1/g_S(r)] + \mu^2/kTr^3 \end{aligned}\right. \tag{2.44}$$

In this approximation the angle averaged pcf is independent of the dipolar strength and equal to the HNC pcf for hard spheres. At high density this solution is advantageously replaced by the exact hard-sphere pcf. This is an excellent approximation for high densities and moderate dipole moments ($\beta\mu^2/d^3 \leq 2$) [2.210,309]. At low densities where the effect of the dipole interaction is much more pronounced (or for high densities and high dipole moments) a more accurate set of equations is obtained by keeping in the expansion of ln g also terms quadratic in Δ and D [2.215]. The corresponding results for h_S (which now depends on the dipole strength), h_Δ and h_D are practically indistinguishable from the MC values (provided the equations are solved with a spherical cutoff of the dipole potential identical to that used in the MC computation).

Perturbation approaches for the angle averaged pcf proposed by PERRAM and WHITE [2.211], MADDEN and FITTS [2.310], and SMITH et al. [2.311,312] gave accurate results at high density.

Approximations for the pcf of dipolar hard spheres based upon cluster series expansions (EXP, LIN, L3) are compared by VERLET and WEIS [2.210], and STELL and WEIS [2.309]. These as well as the mean spherical approximation (MSA) of WERTHEIM [2.302] are less satisfactory than the LHNC equation.

A self-consistent Ornstein-Zernike approach by HØYE and STELL [2.313] provides a convenient and accurate representation of the pcf provided that the internal energy, dielectric constant and the contact values of the pcf are known (e.g., from LHNC) [2.303].

On the free energy level, besides the LHNC equation which gives practically exact results, a cluster series approach of VERLET and WEIS [2.210] is fairly accurate (cf. Table 2.4). Of convenient use are the Padê approximants for the series expansions of the free energy (in terms of the dipole strength) developed by RUSHBROOKE et al. [2.314] for dipolar hard spheres and by STELL et al. [2.315] for Stockmayer potentials. For dipolar hard spheres at $\rho d^3 = 0.8$ the Padê free energy differs from the LHNC result by 5% (for all dipole moments) (cf. Table 2.4). A perturbation theory by MO and GUBBINS [2.316] gives good results for small dipole moments ($\mu^2/\varepsilon\sigma^3 \leq 1$). The MSA of WERTHEIM [2.302] is poor for all dipole moments [2.206,209,210].

Various approximations for the dielectric constant are compared in Table 2.5.

Table 2.5 Dielectric constant ε of dipolar hard spheres at the density $\rho d^3 = 0.8$. From left to right: the Onsager approximation [2.324], the MSA of WERTHEIM [2.302], the LIN and L3 approximations of STELL and WEIS [2.309], the LHNC approximation of PATEY [2.303]. The MC results are taken from [2.214]

$\frac{\beta\mu^2}{d^3}$	ε_{ONS}	ε_W	ε_{LIN}	ε_{L3}	ε_{LHNC}	ε_{MC}
0.5	3.17	3.59	3.84	3.76	3.76	3.73 ± 0.1
1.0	5.62	7.80	9.27	9.66	9.66	9.0 ± 0.5
2.0	10.60	20.00	27.06	37.93	50.0	
2.75	14.36	31.86	45.62	79.17	250.	

Perturbation calculations by McDONALD [2.209] and PATEY and VALLEAU [2.317] showed that the effect of polarizability on the free energy was quite large. Lower bounds on the magnitude of the contribution from the induced interactions ranged from 35-60% according to the degree of anisotropy of the polarizability tensor [2.317]. MC calculations by PATEY and VALLEAU [2.318] indicated that these contributions are in fact considerably larger. These authors also showed that the mean polarizability and additivity assumptions are useful only for nearly isotropic molecules.

MC calculations for the surface tension of a polar fluid, modeled by a Stockmayer potential, in the Fowler-Kirkwood-Buff approximation were compared by GUBBINS et al. [2.319] with perturbation theory [2.320]. The MC results should probably not be corrected for tail effects.

b) Quadrupole-Quadrupole Interaction

PATEY and VALLEAU [2.216] investigated by the MC method the thermodynamic and structural properties of a system of hard spheres with permanent quadrupoles. The density studied is $\rho d^3 = 0.8344$ and the quadrupole strength Q^2/kTd^5 ranges from 0.02 to 1.666. As the QQ interaction is sufficiently short ranged, no problems are encountered with boundary conditions. The most striking finding is that quadrupoles have a much bigger effect on the structure than comparable dipoles. For large quadrupole moments the pcf exhibits structural details compatible with a local face-centered-cubic (fcc) structure. However, it is not impossible that these states are in a metastable region.

An expansion of the free energy in a power series of Q^2/kTd^5 is slowly convergent but, as for the dipolar case, a Padê approximant [2.216,321] is fairly accurate.

WANG et al. [2.304,205,217] reported internal energies, pcf's and mean-square forces for a LJ + QQ interaction at the density $\rho d^3 = 0.8$, temperature $kT/\varepsilon = 0.72$ and quadrupole moments up to $Q^2/kTd^5 = 1$ (64 particles).

Most of the theoretical approaches mentioned in connection with dipoles are readily applicable (with similar success) to the quadrupolar case. In particular, the LHNC theory gives results which are in complete agreement with the MC computations [2.322]. The MSA is again poor [2.322].

c) Dipole-Quadrupole Interaction

PATEY and VALLEAU [2.216] also investigated the case of a system where both dipoles and quadrupoles were present. In particular they studied the influence of the QQ interaction on the dielectric constant.

d) Anisotropic Overlap Interaction

WANG et al. [2.204, 205] studied the effect of an anisotropic overlap potential of the form

$$u_{ov}(12) = 4\delta\varepsilon(\sigma/R_{12})^{12}[\Phi^{202}(\underline{\Omega}_1) + \Phi^{022}(\underline{\Omega}_2)] \tag{2.45}$$

(δ is a dimensionless strength parameter) on the angle averaged pcf, internal energy and mean-square force of linear molecules. The strength parameter is varied from -0.2 to 0.35 which covers the range of most small molecules [2.323]. Perturbation treatments for the angle averaged pcf [2.204,312,324] and the internal energy [2.204] are satisfactory only for small values of $\delta (\delta \leq 0.1)$.

2.6 Gas-Liquid Interface

The first aim of simulations of the gas-liquid interface is the computation of the density variation, the so-called surface density profile, in the gas-liquid interfacial region. The qualitative features of the density profile are expected to be independent of the details of the interaction potential and to change only with the thermodynamic state. In order to compare with experimental results for real fluids the surface tension is also calculated. However, good agreement is only found if the intermolecular interaction of the fluid is well known.

In order to obtain the gas phase and the liquid phase in equilibrium, the system is enclosed in a rectangular box with periodic boundary conditions for two dimensions. Along the large size of the box, either the surfaces limiting the box are assumed to be rigid, or the usual periodic boundary condition is applied. In the first case, sometimes an attractive potential is associated with one of the rigid surfaces. This potential is short ranged, but its action is expected to be sufficient to stabilize the liquid phase in a well defined part of the box. In some ways this wall potential plays the role of the gravitational field for the macroscopic systems. If the periodic boundary conditions are applied in all three dimensions, at the beginning of the MC or MD simulations the particles of the system are placed in the central part of the box. The density of the particles in this part of the box corresponds to the density of the liquid state; the other part of the box is empty. During the simulation process the system relaxes under its own internal forces towards an equilibrium state formed of a two-sided film of liquid surrounded by the gas phase. Simulations with rigid boundary conditions have been done by CROXTON and FERRIER [2.325], by LIU [2.326] and by CHAPELA et al. [2.327]. Computations with the periodic boundary conditions have been done by OPITZ [2.328], LEE et al. [2.329], ABRAHAM et al. [2.330], MIYAZAKI et al. [2.331] and RAO and LEVESQUE [2.332].

The work of CROXTON and FERRIER concerns a two-dimensional system of 200 atoms interacting by a LJ potential at a temperature of 94.4 K. For the gas phase the boundary condition is a rigid surface as described above. But for the liquid phase the boundary condition is more complicated because CROXTON and FERRIER attempt to simulate in a very realistic manner the effect of an infinite bulk of liquid on the 200 atoms. The essential conclusion of this work is that the surface density profile presents large oscillations in the liquid phase before decreasing sharply towards the gas density value. From our point of view, this conclusion is not very reliable due to the boundary condition used for the liquid phase and also due to the insufficient length of the MD simulation (5000 integration steps).

For the three-dimensional systems an approximate calculation of the density profile by NAZARIAN [2.333] finds large oscillations, but the first simulation for a three-dimensional system by OPITZ [2.328] seems to invalidate this result. OPITZ investigated the equilibrium state of a liquid-gas system of 300 atoms in a rectangular box with periodic boundary conditions. For a temperature of 97.1 K and a LJ potential, OPITZ found an almost monotonic decrease from the liquid density to the gas density with a fast decrease over a domain of 8 Å. LIU [2.326] did very careful MC simulations on a LJ system of 129 atoms in a rectangular box. He used rigid boundary conditions for one of the three dimensions. He calculated the density profile for four temperatures from triple point to the critical temperature. The density profile had large oscillations in the vicinity of the rigid surface associated with the wall potential. But these oscillations were rapidly damped so that in the liquid phase the density profile had very small oscillations corresponding to 2% or 3% of the value of the liquid density. LIU concluded that these oscillations were probably real in spite of their small size. The liquid phase ended with the density decreasing rapidly towards its gas phase value. The range of this variation changed with the temperature but was typically of the order of 10 Å. In their work LEE et al. [2.329], as LIU, found an oscillatory behavior of the density profile in the liquid phase. The amplitude of the oscillations in [2.329] was 7% or 8% of the value of the liquid density.

Finally using MC and MD simulations for a system with periodic or rigid boundary conditions, ABRAHAM et al. [2.330], RAO and LEVESQUE [2.332], and CHAPELA et al. [2.327] prove definitely that the density profile is monotonic in the interfacial region and that the oscillatory behavior obtained previously is due to poor statistics in the simulations. ABRAHAM et al. simulate a system of 256 atoms at 84 K, RAO and LEVESQUE simulate systems of 1024 and 1728 atoms at 84 K and 89 K. Both calculations concern LJ systems with periodic boundary conditions. A careful choice of the initial configuration is made in order to minimize the computation time needed to obtain an equilibrium state. CHAPELA et al. made MD and MC runs for LJ systems of 255, 1020 and 4080 atoms for the temperatures $k_BT/\varepsilon = 0.7$, 0.76, 0.78, 0.83, 0.92, 1.13. They used rigid boundary conditions along the large side of the box. Their results did not show a qualitative change of the density profile with the size of the system (255,1020, or 4080 particles) at constant temperature.

The surface tension has been computed in all articles quoted for the case of the three-dimensional system, but its calculation turns out to be difficult because of large statistical fluctuations. Correction terms must be added to the simulation results. These terms are lacking in the value given in [2.332]. In the case of the rigid boundary conditions [2.327], the error on these correction terms seems to be underestimated. For LJ systems the calculated values of the surface tension do not agree with the experimental values for argon. LEE et al. obtain better agreement with a realistic potential [2.112] for argon.

In order to improve the precision of the estimation for the surface tension MIYAZAKI et al. [2.331] estimated the change in the free energy between a liquid system without a surface and a liquid system with a surface. Their computation was largely based on a method proposed by BENNETT [2.334]. The final results of MIYAZAKI stay in the error bars of the preceding evaluation [2.330] but the uncertainty is smaller.

To conclude, the simulations on gas-liquid systems have proved the monotonic character of the surface density profile. For a few thermodynamic states, precise evaluations of the density variation have been obtained. The choice of the periodic boundary conditions for all the dimensions appears to be the best choice for studying liquid surfaces, especially for the computation of surface tension.

Acknowledgements. We are grateful to K. SINGER, W.B. STREETT, D.J. TILDESLEY, T. BOUBLIK and I. NEZBEDA for sending us, prior to publication, preprints of their work. We are indebted to G.N. PATEY for a careful reading of the manuscript. One of us (J.J.W.) would like to thank G. STELL for his kind hospitality at the Department of Mechanical Engineering, SUNY at Stony Brook, where part of this work has been accomplished.

References

2.1 N. Metropolis, A.W. Rosenbluth, M.N. Rosenbluth, A.H. Teller, E. Teller: J. Chem. Phys. *21*, 1087 (1953)
2.2 W.W. Wood, J.J. Erpenbeck: Ann. Rev. Phys. Chem. *27*, (1976)
2.3 W.W. Wood: "Monte Carlo Studies of Simple Liquid Models"; in *Physics of Simple Liquids*, ed. by H.N.V. Temperley, J.S. Rowlinson, G.S. Rushbrooke (North Holland, Amsterdam 1968) Chap. 5, pp. 117-230; "Computer Studies on Fluid Systems of Hard Core Particle"; in *Fundamental Problems in Statistical Mechanics*, ed. by E.D.G. Cohen, Vol. 3 (North Holland, Amsterdam 1975) pp. 331-388
2.4 J.P. Valleau, S.G. Whittington:"A Guide to Monte Carlo for Statistical Mechanics. 1. Highways, in Modern Theoretical Chemistry", in *Statistical Mechanics*, ed. by B.J. Berne, Vol. 5 (Plenum Press, New York 1977) Chap. 4, pp. 134-168
2.5 J.P. Valleau, G.M. Torrie:"A Guide to Monte Carlo for Statistical Mechanics. 2. Byways, in Modern Theoretical Chemistry", in *Statistical Mechanics*, ed. by B.J. Berne, Vol. 5 (Plenum Press, New York 1977). Chap. 4, pp. 169-194
2.6 J.A. Barker, D. Henderson: Rev. Mod. Phys. *48*, 587 (1976)
2.7 B.J. Alder, T.E. Wainwright: J. Chem. Phys. *27*, 1208 (1957)
2.8 M.N. Rosenbluth, A.W. Rosenbluth: J. Chem. Phys. *22*, 881 (1954)
2.9 W.W. Wood, J.D. Jacobson: J. Chem. Phys. *27*, 1207 (1957)
2.10 B.J. Alder, T.E. Wainwright: J. Chem. Phys. *33*, 1439 (1960)
2.11 A. Rotenberg: New York Univ. Rpt. NYO-1480-3 (1964)
2.12 F.H. Ree, W.G. Hoover: J. Chem. Phys. *40*, 939 (1964)
2.13 F.H. Ree, W.G. Hoover: J. Chem. Phys. *46*, 4181 (1967)
2.14 W.G. Hoover, F.H. Ree: J. Chem. Phys. *47*, 4873 (1967)
2.15 W.G. Hoover, F.H. Ree: J. Chem. Phys. *49*, 3609 (1968)
2.16 G.M. Torrie, J.P. Valleau, A. Bain: J. Chem. Phys. *58*, 5479 (1973)
2.17 D.J. Adams: Mol. Phys. *28*, 1241 (1974)
2.18 L.V. Woodcock: J. Chem. Soc. Faraday Trans. II *47*, 1667 (1976)
2.19 N.F. Carnahan, K.E. Starling: J. Chem. Phys. *51*, 635 (1969)
2.20 E.J. Lefevre: Nature *235*, 20 (1972)
2.21 K.R. Hall: J. Chem. Phys. *57*, 2252 (1972)
2.22 F.C. Andrews: J. Chem. Phys. *62*, 272 (1975)
2.23 L.V. Woodcock: J. Chem. Soc. Faraday Trans. II *72*, 731 (1976)
2.24 R.J. Speedy: J. Chem. Soc. Faraday Trans. II *73*, 714 (1977)
2.25 B.J. Alder, S.P. Frankel, V.A. Lewinson: J. Chem. Phys. *22*, 881 (1954)

2.26 F.H. Ree, R.N. Keeler, S.L. McCarthy: J. Chem. Phys. *44*, 3407 (1966)
2.27 B.J. Alder, C.E. Hecht: J. Chem. Phys. *50*, 2032 (1969)
2.28 F.H. Ree, Y.T. Lee, T. Ree: J. Chem. Phys. *55*, 234 (1971)
2.29 J.A. Barker, D. Henderson: Mol. Phys. *21*, 187 (1971)
2.30 G.N. Patey, G.M. Torrie: Mol. Phys. *34*, 1623 (1977)
2.31 E.W. Grundke, D. Henderson: Mol. Phys. *24*, 269 (1972)
2.32 L. Verlet, J.J. Weis: Phys. Rev. A *5*, 939 (1972)
2.33 J.K. Percus, G.J. Yevick: Phys. Rev. *110*, 1 (1957)
2.34 B.J. Alder: Phys. Rev. Lett. *12*, 317 (1964)
2.35 B.J. Alder: J. Chem. Phys. *40*, 2724 (1964)
2.36 J.L. Lebowitz: Phys. Rev. A *133*, 895 (1964)
2.37 E.B. Smith, K.R. Lea: J. Chem. Soc. Faraday Trans. *59*, 1535 (1963)
2.38 A. Rotenberg: J. Chem. Phys. *43*, 4337 (1965)
2.39 G.A. Mansoori, N.F. Carnahan, K.E. Starling, T.W. Leland, Jr.: J. Chem. Phys. *54*, 1523 (1971)
2.40 T. Boublik: J. Chem. Phys. *53*, 471 (1970)
2.41 L.L. Lee, D. Levesque: Mol. Phys. *26*, 1351 (1973)
2.42 E.W. Grundke, D. Henderson: Mol. Phys. *24*, 269 (1972)
2.43 T.W. Melnyk, B.L. Sawford: Mol. Phys. *29*, 891 (1975)
2.44 D.J. Adams, I.R. McDonald: J. Chem. Phys. *63*, 1900 (1975)
2.45 D. Henderson, J.A. Barker: J. Chem. Phys. *49*, 3377 (1968)
2.46 W.R. Smith: Mol. Phys. *21*, 105 (1971)
2.47 P.J. Leonard, D. Henderson, J.A. Barker: J. Chem. Soc. Faraday Trans. *66*, 2439 (1970)
2.48 A. Rotenberg: J. Chem. Phys. *43*, 1198 (1965)
2.49 F. Lado, W.W. Wood: J. Chem. Phys. *49*, 4244 (1968)
2.50 B.J. Alder, D.A. Young, M.A. Mark: J. Chem. Phys. *56*, 3013 (1972)
2.51 R.W. Zwanzig: J. Chem. Phys. *22*, 1420 (1954)
2.52 M. Fisher: J. Math. Phys. *5*, 944 (1964)
2.53 L. Kadanoff: Physics *2*, 263 (1966)
2.54 D. Henderson, W.G. Madden, D.D. Fitts: J. Chem. Phys. *64*, 5026 (1976)
2.55 L. Verlet: Nuovo Cimento *18*, 77 (1960)
2.56 Y. Rosenfeld, R. Thieberger: J. Chem. Phys. *63*, 1875 (1975)
2.57 J.D. Weeks, D. Chandler, H.C. Andersen: J. Chem. Phys. *54*, 5237 (1971)
2.58 D.N. Card, J. Walkley: Can. J. Phys. *52*, 80 (1974)
2.59 J. A. Barker, D. Henderson: J. Chem. Phys. *47*, 2856 (1967)
2.60 J.A. Barker, D. Henderson: J. Chem. Phys. *47*, 4714 (1967)
2.61 W.R. Smith, D. Henderson, J.A. Barker: J. Chem. Phys. *55*, 4027 (1971)
2.62 D. Levesque, J.J. Weis: Phys. Lett. *60* A, 473 (1977)
2.63 B.J. Alder, W.E. Alley, M. Rigby: Physica *73*, 143 (1974)
2.64 B.J. Alder, T.E. Wainwright: Phys. Rev. *127*, 359 (1962)
2.65 W.G. Hoover, B.J. Alder: J. Chem. Phys. *46*, 686 (1967)
2.66 W.W. Wood: Los Alamos Scientific Laboratory Rpt. LA-2827 (Los Alamos, New Mexico 1963)
2.67 W.W. Wood: J. Chem. Phys. *48*, 415 (1968)
2.68 W.W. Wood: J. Chem. Phys. *52*, 729 (1970)
2.69 D.G. Chae, F.H. Ree, T. Ree: J. Chem. Phys. *50*, 1581 (1969)
2.70 R.L. Coldwell: Phys. Rev. A *7*, 270 (1973)
2.71 D. Henderson: Mol. Phys. *30*, 971 (1975)
2.72 A. Bellemans, J. Orban, E. De Vos: Chem. Phys. Lett. *1*, 639 (1968)
2.73 E. Dickinson: Mol. Phys. *33*, 1463 (1977)
2.74 D.A. Young, B.J. Alder: Phys. Rev. Lett. *38*, 1213 (1977)
2.75 C. Carlier, H.L. Frisch: Phys. Rev. A *6*, 1153 (1972)
2.76 F.H. Ree, T. Ree: J. Chem. Phys. *56*, 5434 (1972)
2.77 M.L. Mehta: J. Chem. Phys. *60*, 2207 (1974)
2.78 H.L. Frisch, C. Carlier: Phys. Rev. Lett. *28*, 1019 (1972)
2.79 W.G. Hoover, M. Ross, K.W. Johnson, D. Henderson, J.A. Barker, B.C. Brown: J. Chem. Phys. *52*, 4931 (1970)
2.80 J.P. Hansen: Phys. Rev. A *2*, 221 (1970)
2.81 J.P. Hansen, J.J. Weis: Mol. Phys. *23*, 853 (1972)
2.82 W.G. Hoover, S.G. Gray, K.W. Johnson: J. Chem. Phys. *55*, 1128 (1971)

2.83 M. Ross: Phys. Rev. *184*, 233 (1969)
2.84 J.P. Hansen, D. Schiff: Mol. Phys. *25*, 1281 (1973)
2.85 W.W. Wood, F.R. Parker: J. Chem. Phys. *27*, 720 (1957)
2.86 A. Rahman: Phys. Rev. *136*, A 405 (1964)
2.87 L. Verlet, D. Levesque: Physica *36*, 254 (1967)
2.88 I.R. McDonald, K. Singer: Faraday Disc. *43*, 40 (1967)
2.89 I.R. McDonald, K. Singer: J. Chem. Phys. *47*, 4766 (1967)
2.90 I.R. McDonald, K. Singer: J. Chem. Phys. *50*, 2308 (1969)
2.91 L. Verlet: Phys. Rev. *159*, 98 (1967)
2.92 J.P. Hansen, L. Verlet: Phys. Rev. *184*, 151 (1969)
2.93 D. Levesque, L. Verlet: Phys. Rev. A *2*, 2514 (1970)
2.94 G.M. Torrie, J.P. Valleau: Chem. Phys. Lett. *28*, 578 (1974)
2.95 G.M. Torrie, J.P. Valleau: J. Comp. Phys. *23*, 187 (1977)
2.96 W.B. Streett, H.J. Raveche, R.D. Mountain: J. Chem. Phys. *61*, 1960 (1974)
2.97 J.P. Hansen, E.L. Pollock: J. Chem. Phys. *62*, 4581 (1975)
2.98 H.J. Raveche, R.D. Mountain: J. Chem. Phys. *61*, 1970 (1974)
2.99 L.A. Rowley, D. Nicholson, N.G. Parsonage: J. Comp. Phys. *17*, 401 (1975) and private communication
2.100 G.E. Norman, V.S. Filinov: High Temp. *7*, 216 (1969)
2.101 P.N. Vorontsov-Vel'Yaminov, H.M. El'Y-Ashevich, L.A. Morgenshtern, V.P. Chasovskikh: High Temp. *8*, 261 (1970)
2.102 D.J. Adams: Mol. Phys. *32*, 647 (1976)
2.103 I.R. McDonald, K. Singer: Mol. Phys. *23*, 29 (1972)
2.104 L. Verlet: Phys. Rev. *165*, 201 (1968)
2.105 P.G. Mikolaj, C.J. Pings: J. Chem. Phys. *46*, 1401 (1967)
2.106 S.S. Wang, P.A. Egelstaff, K.E. Gubbins: Mol. Phys. *25*, 461 (1973)
2.107 S.S. Wang, J.A. Krumhansl: J. Chem. Phys. *56*, 4287 (1972)
2.108 H.J. Raveche, R.D. Mountain, W.B. Streett: J. Chem. Phys. *57*, 4999 (1972)
2.109 H.J. Raveche, R.D. Mountain: J. Chem. Phys. *57*, 3987 (1972)
2.110 F. Lado: Phys. Rev. A *8*, 2548 (1973)
2.111 J.A. Barker, D. Henderson: Phys. Rev. A *4*, 806 (1971)
2.112 J.A. Barker, R.A. Fisher, R.O. Watts: Mol. Phys. *21*, 657 (1971)
2.113 J.A. Barker, M.L. Klein: Phys. Rev. B *7*, 4707 (1973)
2.114 K. Singer: Chem. Phys. Lett. *3*, 164 (1969)
2.115 J.V.L. Singer, K. Singer: Mol. Phys. *19*, 279 (1970)
2.116 J.V.L. Singer, K. Singer: Mol. Phys. *24*, 357 (1972)
2.117 I.R. McDonald: Chem. Phys. Lett. *3*, 241 (1969)
2.118 I.R. McDonald: Mol. Phys. *23*, 41 (1972)
2.119 I.R. McDonald: Mol. Phys. *24*, 391 (1972)
2.120 I.R. McDonald: "Equilibrium Theory of Liquid Mixtures", in *Statistical Mechanics*, ed. by K. Singer, Vol. 1 (The Chemical Society, London 1973) Chap. 3, pp. 134-193
2.121 G.M. Torrie, J.P. Valleau: J. Chem. Phys. *66*, 1402 (1977)
2.122 K.C. Mo, K.E. Gubbins, G. Iacucci, I.R. McDonald: Mol. Phys. *27*, 1173 (1973)
2.123 P.L. Fehder: J. Chem. Phys. *50*, 2617 (1969)
2.124 P.L. Fehder: J. Chem. Phys. *52*, 791 (1970)
2.125 F. Tsien, J.P. Valleau: Mol. Phys. *27*, 177 (1974)
2.126 P.P. Ewald: Ann. Physik *64*, 253 (1921)
2.127 B.R.A. Nijboer, F.W. De Wette: Physica *23*, 309 (1957)
2.128 A.A. Barker: Australian J. Phys. *18*, 119 (1965)
2.129 W.B. Hubbard, W.L. Slattery: Australian J. *168*, 131 (1971)
2.130 M.J.L. Sangster, M. Dixon: Advan. Phys. *25*, 247 (1976)
2.131 F.C. Von Der Lage, H. Bethe: Phys. Rev. *71*, 612 (1947)
2.132 J.P. Hansen: Phys. Rev. A *8*, 3096 (1973)
2.133 D.J. Adams, I.R. McDonald: J. Phys. C *7*, 2761 (1974)
2.134 R.W. Hockney: Meth. Comp. Phys. *9*, 135 (1970)
2.135 R.W. Hockney, S.P. Goel, J.W. Eastwood: J. Comp. Phys. *14*, 148 (1974)
2.136 R.W. Hockney, S.P. Goel: Chem. Phys. Lett. *35*, 500 (1975)
2.137 D.D. Carley: J. Chem. Phys. *43*, 3489 (1965)
2.138 S.G. Brush, H.L. Sahlin, E. Teller: J. Chem. Phys. *45*, 2102 (1966)
2.139 H.E. De Witt: Phys. Rev. A *14*, 1290 (1976)

2.140 S. Galam, J.P. Hansen: Phys. Rev. A *14*, 816 (1976)
2.141 J.F. Springer, M.A. Pokrant, F.A. Stevens: J. Chem. Phys. *58*, 4863 (1973)
2.142 K.C. Ng: J. Chem. Phys. *61*, 2680 (1974)
2.143 E.L. Pollock, J.P. Hansen: Phys. Rev. A *8*, 3110 (1973)
2.144 D.M. Ceperley, G.V. Chester: Phys. Rev. A *15*, 755 (1977)
2.145 F. Lado: Phys. Rev. *135*, A 1013 (1964)
2.146 J. Lindhard: Kgl. Danke Videnskab. Selskab Mat.-Fys. Medd. *28*, No 8 (1954)
2.147 J.P. Hansen, J.J. Weis: Mol. Phys. *33*, 1379 (1977)
2.148 J.P. Hansen, G. Torrie, P. Vieillefosse: Phys. Rev. A *16*, 2153 (1977)
2.149 H.E. Dewitt, W.B. Hubbard: Astron. J. *205*, 295 (1976)
2.150 A.A. Barker: Phys. Rev. *171*, 186 (1968)
2.151 V.S. Vorob'ev, G.E. Norman, V.S. Filinov: Sov. Phys. JETP *30*, 459 (1970)
2.152 M.A. Pokrant, A.A. Broyles, T. Dunn: Phys. Rev. A *10*, 379 (1974)
2.153 L. Pauling: Z. Kristall. *67*, 377 (1928)
2.154 P.N. Vorontsov - Vel'Yaminov, A.M. El'Yashevich, A.K. Kron: Elektrokhimiya *2*, 708 (1966)
2.155 P.N. Vorontsov - Vel'Yaminov, A.M. El'Yashevich: Elektrokhimiya *4*, 1430 (1968)
2.156 P.N. Vorontsov - Vel'Yaminov, A.M. El'Yashevich, J.C. Rasaiah, H.L. Friedman: J. Chem. Phys. *52*, 1013 (1970)
2.157 B.P. Chasovskikh, P.N. Vorontsov - Vel'Yaminov: Teplefizika Vysokikh Temperatur *14*, 199 (1976)
2.158 D.N. Card, J.P. Valleau: J. Chem. Phys. *52*, 6232 (1970)
2.159 J.C. Rasaiah, D.N. Card, J.P. Valleau: J. Chem. Phys. *56*, 248 (1972)
2.160 J.P. Valleau, D.N. Card: J. Chem. Phys. *57*, 5457 (1972)
2.161 B. Larsen: Chem. Phys. Lett. *27*, 47 (1974)
2.162 B. Larsen: J. Chem. Phys. *65*, 3431 (1976)
2.163 E. Waisman, J.L. Lebowitz: J. Chem. Phys. *56*, 3086, 3093 (1972)
2.164 H.C. Andersen, D. Chandler: J. Chem. Phys. *57*, 1918 (1972)
2.165 H.C. Andersen, D. Chandler, J.D. Weeks: J. Chem. Phys. *57*, 2626 (1972)
2.166 J.C. Rasaiah, H.L. Friedman: J. Chem. Phys. *48*, 2742 (1968); *50*, 3965 (1969)
2.167 F.H. Stillinger, R. Lovett: J. Chem. Phys. *48*, 3858 (1968); *49*, 1991 (1968)
2.168 C.W. Outhwhaite: In *Specialist Periodical Reports in Statistical Mechanics*, ed. by K. Singer, Vol. 2 (The Chemical Society, London 1975)
2.169 G. Stell, K.C. Wu, B. Larsen: Phys. Rev. Lett. *37*, 1369 (1976)
2.170 M. Gillan, B. Larsen, M.P. Tosi, N.H. March: J. Phys. C *9*, 889 (1976)
2.171 G.N. Patey, J.P. Valleau: J. Chem. Phys. *63*, 2334 (1975)
2.172 M.L. Huggins, J.E. Mayer: J. Chem. Phys. *1*, 643 (1933)
2.173 L. Pauling: J. Am. Chem. Soc. *50*, 1036 (1928)
2.174 F.G. Fumi, M.P. Tosi: J. Phys. Chem. Sol. *25*, 31 (1964); *25*, 45 (1964)
2.175 M. Blander: Advan. Chem. Phys. *11*, 82 (1967)
2.176 D.J. Adams, I.R. McDonald: Physica *79* B, 159 (1975)
2.177 L.V. Woodcock, K. Singer: Trans. Faraday Soc. *67*, 12 (1971)
2.178 J.P. Valleau, S. Wittington: J. Chem. Soc. Faraday II. *69*, 1004 (1973)
2.179 S. Romano, I.R. McDonald: Physica *67*, 625 (1973)
2.180 S. Romano, C. Margheritis: Physica *77*, 557 (1974)
2.181 J.W.E. Lewis, K. Singer, L.V. Woodcock: J. Chem. Soc. Faraday II *71*, 301 (1975)
2.182 See, e.g., F.G. Edwards, J.E. Enderby, R.A. Howe, D.I. Page: J. Phys. C *8*, 3483 (1975)
2.183 F.H. Ree, A.C. Holt: Phys. Rev. B *8*, 826 (1973)
2.184 F. Lantelme, P. Turc, B. Quentrec, J.W.E. Lewis: Mol. Phys. *28*, 1537 (1974)
2.185 J.W.E. Lewis, K. Singer: J. Chem. Soc. Faraday II, *71*, 41 (1975)
2.186 J.R.D. Copley, A. Rahman: Phys. Rev. A *13*, 2276 (1976)
2.187 D.L. Price, J.R.D. Copley: Phys. Rev. A *11*, 2124 (1975)
2.188 J.P. Hansen, I.R. McDonald: Phys. Rev. A *11*, 2111 (1976)
2.189 B. Larsen, T. Førland, K. Singer: Mol. Phys. *26*, 1521 (1973)
2.190 E.M. Gosling, K. Singer: Pure Appl. Chem. *22*, 303 (1970)
2.191 N.W. Ashcroft, D. Stroud: In *Solid State Physics* (Academic Press, New York) Vol. *33*, p. 1 (1978)

2.192 H. Minoo, C. Deutsch, J.P. Hansen: J. Phys. Lett. (Paris) *38*, 191 (1977)
2.193 A. Paskin, A. Rahman: Phys. Rev. Lett. *16*, 300 (1966)
2.194 D. Schiff: Phys. Rev. *186*, 151 (1969)
2.195 R.H. Fowler: J. Chem. Phys. *59*, 3435 (1973)
2.196 R.D. Murphy, M.L. Klein: Phys. Rev. A *8*, 2640 (1973)
2.197 A.J. Greenfield, J. Wellendorf, N. Wiser: Phys. Rev. A *4*, 1607 (1971)
2.198 R.D. Murphy: Phys. Rev. A *15*, 1188 (1977)
2.199 A. Rahman: Phys. Rev. A *9*, 1667 (1974)
2.200 K.E. Gubbins, C.G. Gray, P.A. Egelstaff, M.S. Ananth: Mol. Phys. *25*, 1353 (1973)
2.201 R.J. Beshinske, M.H. Lietzke: J. Chem. Phys. *51*, 2278 (1969)
2.202 G.D. Harp, B.J. Berne: J. Chem. Phys. *49*, 1249 (1968)
2.203 B.J. Berne, G.D. Harp: Advan. Chem. Phys. *17*, 63 (1970)
2.204 S.S. Wang, C.G. Gray, P.A. Egelstaff, K.E. Gubbins: Chem. Phys. Lett. *21*, 123 (1973)
2.205 C.G. Gray, S.S. Wang, K.E. Gubbins: Chem. Phys. Lett. *26*, 610 (1974)
2.206 G.N. Patey, J.P. Valleau: Chem. Phys. Lett. *21*, 297 (1973)
2.207 G.N. Patey, J.P. Valleau: J. Chem. Phys. *61*, 534 (1974)
2.208 V.M. Jansoone: Chem. Phys. *3*, 78 (1974)
2.209 I.R. McDonald: J. Phys. C *7*, 1225 (1974)
2.210 L. Verlet, J.J. Weis: Mol. Phys. *28*, 665 (1974)
2.211 J.W. Perram, L.R. White: Mol. Phys. *28*, 527 (1974)
2.212 D.J. Adams, I.R. McDonald: Mol. Phys. *32*, 931 (1976)
2.213 K.E. Gubbins, J.M. Haile, I.R. McDonald: J. Chem. Phys. *66*, 364 (1977)
2.214 D. Levesque, G.N. Patey, J.J. Weis: Mol. Phys. *34*, 1077 (1977)
2.215 G.N. Patey, D. Levesque, J.J. Weis: Mol. Phys. *38* ,219 (1979)
2.216 G.N. Patey, J.P. Valleau: J. Chem. Phys. *64*, 170 (1976)
2.217 S.S. Wang, P.A. Egelstaff, C.G. Gray, K.E. Gubbins: Chem. Phys. Lett. *24*, 453 (1974)
2.218 B.C. Freasier: Chem. Phys. Lett. *35*, 280 (1975)
2.219 B.C. Freasier, D. Jolly, R.J. Bearman: Mol. Phys. *31*, 255 (1976)
2.220 W.B. Streett, D.J. Tildesley: Proc. R. Soc. (London) A *348*, 485 (1976)
2.221 D. Chandler, C.S. Hsu, W.B. Streett: J. Chem. Phys. *66*, 5231 (1977)
2.222 I. Aviram, D.J. Tildesley, W.B. Streett: Mol. Phys. *34*, 881 (1977)
2.223 J. Barojas, D. Levesque, B. Quentrec: Phys. Rev. A *7*, 1092 (1973)
2.224 P.S.Y. Cheung, J.G. Powles: Mol. Phys. *30*, 921 (1975)
2.225 B. Quentrec, C. Brot: Phys. Rev. A *12*, 272 (1975)
2.226 J.J. Weis, D. Levesque: Phys. Rev. A *13*, 450 (1976)
2.227 S. Romano: Z. Naturforsch. *31a*, 1108 (1976)
2.228 P.S.Y. Cheung, J.G. Powles: Mol. Phys. *32*, 1383 (1976)
2.229 W.B. Streett, D.J. Tildesley: Proc. R. Soc. (London) A *355*, 239 (1977)
2.230 K. Singer, A. Taylor, J.V.L. Singer: Mol. Phys. *33*, 1757 (1977)
2.231 G.F. Few, M. Rigby: Mol. Phys. *33*, 585 (1977)
2.232 D.J. Evans, R.O. Watts: Mol. Phys. *32*, 93 (1976)
2.233 J. Vieillard-Baron: J. Chem. Phys. *56*, 4729 (1972)
2.234 D. Levesque, D. Schiff, J. Vieillard-Baron: J. Chem. Phys. *51*, 3625 (1969)
2.235 G.A. Few, M. Rigby: Chem. Phys. Lett. *20*, 433 (1973)
2.236 J. Vieillard-Baron: Mol. Phys. *28*, 809 (1974)
2.237 T. Boublik, I. Nezbeda, O. Trnka: Czech. J. Phys. B *26*, 1081 (1976)
2.238 R.L. Coldwell, T.P. Henry, C.W. Woo: Phys. Rev. A *10*, 897 (1974)
2.239 I. Nezbeda, T. Boublik, O. Trnka: Czech. J. Phys. B *25*, 119 (1975)
2.240 T. Kihara: Rev. Mod. Phys. *25*, 831 (1953)
2.241 T. Kihara: Advan. Chem. Phys. *5*, 147 (1963)
2.242 A. Koide, T. Kihara: Chem. Phys. *5*, 34 (1974)
2.243 T.B. MC Rury, W.A. Steele, B.J. Berne: J. Chem. Phys. *64*, 1288 (1976)
2.244 B.J. Berne, P. Pechukas: J. Chem. Phys. *56*, 4213 (1972)
2.245 J. Kushick, B.J. Berne: J. Chem. Phys. *64*, 1362 (1971)
2.246 H.L. Lemberg, F.H. Stillinger: J. Chem. Phys. *62*, 1677 (1975)
2.247 A. Rahman, F.H. Stillinger, F.L. Lemberg: J. Chem. Phys. *63*, 5223 (1975)
2.248 I.R. McDonald, M.L. Klein: J. Chem. Phys. *64*, 4790 (1976)
2.249 J.A. Barker, R.O. Watts: Chem. Phys. Lett. *3*, 144 (1969)

2.250 J.A. Barker, R.O. Watts: Mol. Phys. *26*, 789 (1973)
2.251 R.O. Watts: Mol. Phys. *28*, 1069 (1974)
2.252 H. Popkie, H. Kistenmacher, E. Clementi: J. Chem. Phys. *59*, 1325 (1973)
2.253 G.N. Sarkisov, V.G. Dashevsky, G.G. Malenkov: Mol. Phys. *27*, 1249 (1974)
2.254 V.G. Dashevsky, G.N. Sarkisov: Mol. Phys. *27*, 1271 (1974)
2.255 G.C. Lie, E. Clementi: J. Chem. Phys. *62*, 2195 (1975)
2.256 H. Kistenmacher, H. Popkie, E. Clementi, R.O. Watts: J. Chem. Phys. *60*, 4455 (1974)
2.257 G.C. Lie, E. Clementi, M. Yoshimine: J. Chem. Phys. *64*, 2314 (1976)
2.258 A.J.C. Ladd: Mol. Phys. *33*, 1039 (1977)
2.259 A. Rahman, F.H. Stillinger: J. Chem. Phys. *55*, 3336 (1971)
2.260 F.H. Stillinger, A. Rahman: J. Chem. Phys. *57*, 1281 (1972)
2.261 F.H. Stillinger, A. Rahman: J. Chem. Phys. *60*, 1545 (1974)
2.262 F.H. Stillinger: Advan. Chem. Phys. *31*, 1 (1975)
2.263 P.A. Egelstaff, C.G. Gray, K.E. Gubbins: "Equilibrium Properties of Molecular Fluids"; in *Molecular Structure and Properties*, ed. by A.D. Buckingham, Vol. 2 (Butterworths, London 1975)
2.264 C.G. Gray: "Equilibrium Statistical Mechanics of Molecular Liquids"; in *Statistical Mechanics*, ed. by K. Singer, Vol. 2 (The Chemical Society, London 1975) Chap. 5, pp. 300-323
2.265 T. Boublik: Fluid Phase Equilibria *1*, 37 (1977)
2.266 W.B. Streett, K.E. Gubbins: Ann. Rev. Phys. Chem. *28*, 373 (1977)
2.267 M.A. Cotter, D.E. Martire: J. Chem. Phys. *52*, 1909 (1970)
2.268 M.A. Cotter: Phys. Rev. A *10*, 625 (1974)
2.269 T. Boublik: Collection Czech. Chem. Commun. *39*, 2333 (1974)
2.270 T. Boublik: Mol. Phys. *32*, 1737 (1976)
2.271 R.M. Gibbons: Mol. Phys. *17*, 81 (1969)
2.272 T. Boublik: Mol. Phys. *27*, 1415 (1974)
2.273 T. Boublik: J. Chem. Phys. *63*, 4084 (1975)
2.274 I. Nezbeda: Chem. Phys. Lett. *41*, 55 (1976)
2.275 I. Nezbeda: Czech. J. Phys. B *26*, 355 (1976)
2.276 M. Rigby: J. Chem. Phys. *53*, 1021 (1970)
2.277 B.C. Freasier, R.J. Bearman: Mol. Phys. *32*, 551 (1976)
2.278 T. Boublik: Mol. Phys. *29*, 421 (1975)
2.279 D. Chandler, H.C. Andersen: J. Chem. Phys. *57*, 1930 (1972)
2.280 B.M. Ladanyi, D. Chandler: J. Chem. Phys. *62*, 4308 (1975)
2.281 M. Rigby: Mol. Phys. *32*, 575 (1976)
2.282 T. Boublik, I. Nezbeda: Chem. Phys. Lett. *46*, 315 (1977)
2.283 A.J. Isahara: J. Chem. Phys. *19*, 397 (1951)
2.284 I. Nezbeda: Mol. Phys. *33*, 1287 (1977)
2.285 M. Ausloos: J. Chem. Phys. *64*, 3490 (1976)
2.286 J. Perram, F. Köhler: Unpublished, quoted from Ref. 2.218
2.287 W.A. Steele, S.I. Sandler: J. Chem. Phys. *61*, 1315 (1974)
2.288 H.C. Andersen, J.D. Weeks, D. Chandler: Phys. Rev. A *4*, 1597 (1971)
2.289 Y.D. Chen, W.A. Steele: J. Chem. Phys. *54*, 703 (1971)
2.290 D. Chandler: J. Chem. Phys. *59*, 2749 (1973)
2.291 D. Chandler: Mol. Phys. *31*, 1213 (1976)
2.292 L.J. Löwden, D. Chandler: J. Chem. Phys. *59*, 6587 (1973)
2.293 C.S. Hsu, D. Chandler, L.J. Löwden: Chem. Phys. *14*, 213 (1976)
2.294 B.C. Freasier, D. Jolly, R.J. Bearman: Mol. Phys. *32*, 1463 (1976)
2.295 T.G. Gibbons, M.L. Klein: Chem. Phys. Lett. *29*, 463 (1974)
2.296 B. Quentrec: Phys. Rev. A *12*, 282 (1975)
2.297 J.J. Weis, M.L. Klein: J. Chem. Phys. *63*, 2869 (1975)
2.298 A. Zunger, E. Huler: J. Chem. Phys. *62*, 3010 (1975)
2.299 D.J. Evans: Mol. Phys. *33*, 979 (1977)
2.300 J.C. Raich, N.S. Gillis: J. Chem. Phys. *66*, 846 (1977)
2.301 L. Blum: J. Chem. Phys. *57*, 1862 (1972)
2.302 M.S. Wertheim: J. Chem. Phys. *55*, 4291 (1971)
2.303 G.N. Patey: Mol. Phys. *34*, 427 (1977)
2.304 H.L. Friedman: Mol. Phys. *21*, 1533 (1975)
2.305 E.R. Smith, J.W. Perram: Mol. Phys. *30*, 1975 (1975)

2.306 E.R. Smith, J.W. Perram: Phys. Lett. A *53*, 121 (1975)
2.307 C.H. Twu, C.G. Gray, K.E. Gubbins: Mol. Phys. *31*, 1923 (1976)
2.308 G.N. Patey, D. Levesque, J.J. Weis: Mol. Phys. *45* , 733 (1982)
2.309 G. Stell, J.J. Weis: Phys. Rev. A *16*, 757 (1977)
2.310 W.G. Madden, D.D. Fitts: Mol. Phys. *31*, 1923 (1976)
2.311 W.R. Smith, W.G. Madden, D.D. Fitts: Chem. Phys. Lett. *36*, 195 (1975)
2.312 W.R. Smith: Chem. Phys. Lett. *40*, 313 (1976)
2.313 J.S. Høye, G. Stell: J. Chem. Phys. *67*, 524 (1977)
2.314 G.S. Rushbrooke, G. Stell, J.S. Høye: Mol. Phys. *26*, 1199 (1973)
2.315 G. Stell, J.C. Rasaiah, H. Narang: Mol. Phys. *27*, 1393 (1974)
2.316 K.C. Mo, K.E. Gubbins: J. Chem. Phys. *63*, 1490 (1975)
2.317 G.N. Patey, J.P. Valleau: Chem. Phys. Lett. *42*, 407 (1976)
2.318 G.N. Patey, J.P. Valleau: Chem. Phys. Lett. *58* , 157 (1978)
2.319 K.E. Gubbins, J.M. Haile, I.R. McDonald: J. Chem. Phys. *66*, 364 (1977)
2.320 J.M. Haile, C.G. Gray, K.E. Gubbins: J. Chem. Phys. *64*, 2569 (1976)
2.321 J.C. Rasaiah, B. Larsen, G. Stell: J. Chem. Phys. *63*, 722 (1975)
2.322 G.N. Patey: Mol. Phys. *35*, 1413 (1978)
2.323 T.H. Spurling, E.A. Mason: J. Chem. Phys. *46*, 322 (1967)
2.324 L. Onsager: Chem. Rev. *13*, 73 (1933)
2.325 C.A. Croxton, R.P. Ferrier: J. Phys. C *4*, 2447 (1971)
2.326 K.S. Liu: J. Chem. Phys. *60*, 4226 (1974)
2.327 G.A. Chapela, Graham Saville, S.M. Thompson, J.S. Rowlinson: J. Chem. Soc. Faraday Trans. II *73*, 1133 (1977)
2.328 A.C.L. Opitz: Phys. Lett. A *47*, 439 (1974)
2.329 J.K. Lee, J.A. Barker, G.M. Pound: J. Chem. Phys. *60*, 1976 (1974)
2.330 F.F. Abraham, D.E. Schreiber, J.A. Barker: J. Chem. Phys. *62*, 1958 (1975)
2.331 J. Miyazaki, J.A. Barker, G.M. Pound: J. Chem. Phys. *64*, 3364 (1976)
2.332 M. Rao, D. Levesque: J. Chem. Phys. *65*, 3233 (1976)
2.333 G. Nazarian: J. Chem. Phys. *56*, 1408 (1972)
2.334 C.H. Bennett: J. Comp. Phys. *22*, 245 (1976)

3. Phase Diagrams of Mixtures and Magnetic Systems

D. P. Landau

With 13 Figures

During the past several decades extensive experimental studies of phase transitions in magnetic materials, binary alloys, ferroelectrics, liquid crystals, and other condensed matter systems have yielded detailed information concerning phase diagrams and critical behavior which can be compared and contrasted with those in classical liquid-gas systems. Many of these physical systems are well approximated by lattice models which are relatively simple yet are not exactly soluble by present theoretical methods. In many such cases Monte Carlo computer simulations provide the most useful and accurate estimates of the location of phase boundaries as well as critical and multicritical exponents. In this chapter we shall concentrate on the use of the Monte Carlo method to study the static, bulk properties of these models (dynamic behavior will be considered in Chap.6 and surface effects in Chap.9). In the following discussion the application of the Monte Carlo method to the study of "ordinary" phase transition will be presented in Sect.3.1 and to multicritical phenomena in Sect.3.2. In Sect.3.3 we shall consider a range of "miscellaneous" systems, and conclusions and the outlook for the future will be in Sect.3.4.

3.1 Ordinary Phase Transitions in Magnets and Binary Alloys

The Monte Carlo study of phase transitions in simple systems is directed primarily towards answering two different questions: I) what is (are) the ordered state(s) of the system and over what range of "fields" is it (are they) stable; II) what is the nature of the phase transition between different phases. Extensive data exist (e.g., from specific heat, magnetization, susceptibility, neutron scattering, etc.) on a wide variety of physical systems with which the results of these calculations may be compared.

3.1.1 Ising Model

Perhaps the simplest and most versatile of all magnetic models is the Ising model [3.1] in which n = 1 dimensional spins are arrayed on a d-dimensional lattice and interact with a hamiltonian

$$\mathscr{H} = \sum_{(ij)} J s_{iz} s_{jz} + \mu H \sum_i s_{iz} + \mu H^+ \sum_i s_{iz} e^{i\underline{k}\cdot\underline{r}} \tag{3.1}$$

where J is the spin-spin interaction constant, H is the uniform magnetic field and H^+ a staggered field (generally not realizable in the laboratory) which couples to the order parameter in an antiferromagnet. In spite of the relative simplicity of the model, only a small number of special cases in spatial dimensionality d = 1 or d = 2 have been solved exactly. Besides its usefulness in magnetism, the Ising model is appropriate for studies of simple binary alloys and lattice gas models for superionic conductors. The hamiltonian of a binary alloy can be expressed in terms of local site occupation variables c_i^A, c_i^B where $c_i^A = 1$, $c_i^B = 0$ if the ith site is occupied by a type A atom and $c_i^A = 0$, $c_i^B = 1$ if it is occupied by a type B atom. The hamiltonian may then be written

$$\mathscr{H} = \sum_{(i,j)} \left[\phi_{AA}(\underline{r}_i,\underline{r}_j) c_i^A c_j^A + \phi_{BB}(\underline{r}_i,\underline{r}_j) c_i^B c_j^B + 2\phi_{AB}(\underline{r}_i,\underline{r}_j) c_i^A c_j^B\right] - \sum_i [c_i^A \mu_A(\underline{r}_i) + c_i^B \mu_B(\underline{r}_j)] + \mathscr{H}_0 \tag{3.2}$$

where the ϕ's are pair potentials, μ_A and μ_B are chemical potentials corresponding to each type of atom and $\mathscr{H}_0$ represents other interactions which will be held constant (and will thus be ignored). This hamiltonian may be easily transformed into an Ising model by making the simple transformation to spin variables

$$c_i^A = (1 + s_i)/2 \tag{3.3a}$$

$$c_i^B = (1 - s_i)/2 \quad . \tag{3.3b}$$

The resultant hamiltonian is identical to (3.1) with J and H being made up of combinations of the ϕ's and μ's. The magnetization of the Ising model ($M = \langle s_z \rangle$) corresponds to the concentration of the equivalent binary alloy and the magnetic field to the chemical potential.

The transformation shown in (3.3) is appropriate to a binary alloy which transforms to an s = 1/2 Ising model. A similar transformation to an s = 1 Ising model would apply to a ternary alloy [3.2]). Since the alloy models and Ising models are equivalent it is clear that results for one simultaneously apply to the other. The major difference lies merely in the quantities of interest, e.g., in an Ising model the field is usually held constant whereas in a binary alloy the concentration (magnetization in magnetic language) is fixed. The Ising model was first studied in the context of binary alloys by FOSDICK and co-workers [3.3-5] who used a spin-flip method to study A_3B face centered cubic binary alloys as well as square lattice, simple cubic lattice and body-centered cubic AB alloys. These calculations were repeated in more detail for the bcc lattice by GUTTMAN [3.6] using the same method. Calculations were carried out on the bcc lattice using a vacancy "spin exchange" method by FLINN and McMANUS [3.7] allowing both nearest-neighbor (nn-) and next-nearest-neighbor (nnn-) jumps. Lattices as large as 10 × 10 × 10 with cyclic boundary conditions were studied and the resultant order-parameter behavior was compared with experimental results on β-brass [3.8] and β-AgZn [3.9]. The agreement on a reduced scale is quite good (see Fig.3.1) and although there are small systematic deviations for β-brass, the simplicity of the model makes the comparison quite gratifying. This same method was applied by BEELER et al. [3.10-12] to cubic lattices, but the emphasis was on the vacancy migration.

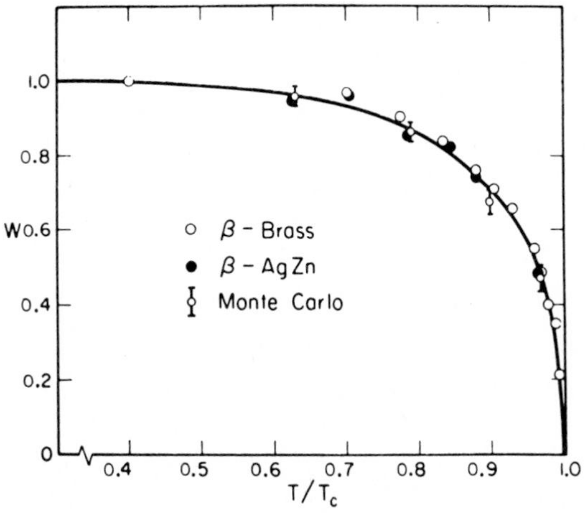

Fig.3.1 Temperature dependence of the long-range order parameter W as determined from Monte Carlo simulations and experimental measurements [3.7]

Substantial numbers of other studies of simple Ising models exist. SALSBURG et al. [3.13] applied the Monte Carlo method to the triangular Ising (lattice gas model), primarily to test the method. Similar studies were carried out on the Ising square lattice [3.14-16]. Substantially more accurate studies were later carried out on simple Ising models with nn-interactions [3.17-22] but these simulations were primarily useful for yielding information about the effects of finite lattice size and of exposed surfaces. (Results from series expansion studies [3.23] and renormalization group methods [3.24] yield more accurate estimates for critical exponents in simple systems than do Monte Carlo data.)

One of the simplest complications which can be introduced to a nn-Ising model is the addition of next-nearest-neighbor (nnn-) coupling. Monte Carlo studies have been made of nnn-square lattices [3.25-27], triangular lattices [3.28] and simple cubic lattices [3.16,29,30]. The variation of T_c with next-nearest-neighbor coupling was determined and the results were compared with those obtained using other methods. As long as the nnn-interactions stabilize the ordered structure, series expansions are quite useful for determining T_c. When the interactions are competing and a new, more complicated structure results the series are often ill behaved and analysis is unreliable. A comparison of the predicted dependence of T_c on nnn-interaction strength in the Ising square lattice as determined in several different ways [3.31-34] is shown in Fig.3.2. Here the existing series expansions were so unreliable that no estimate was made for $\alpha = K_{nnn}/K_{nn} > 0$. In the region near $\alpha = 0.5$ the results of a renormalization group study (four cell cluster approximation [3.35] are systematically low. Extended Monte Carlo data have also been obtained by BINDER and LANDAU [3.35a] who included a magnetic field as well as competing interactions. These results will be discussed in the context of phase diagrams of adsorbed surface layers (lattice gas models) in Chap.9. In the case of the simple cubic lattice the series expansion estimate extracted from 8 terms was significantly below that found from series 10 terms long [3.36]. (The results of the 10 term series appear to be approaching the Monte Carlo values). Besides shifting ordering temperatures, the addition of nnn-interaction alters values of critical parameters such as the critical entropy and critical internal energy. In the past the experimental values of these parameters were often used to suggest the effective spin or spatial dimensionality of the system; the Monte Carlo results indicate that competing interactions between different neighbors may suffice to alter these parameters significantly.

Competing interactions of another kind involve cases where there are periodic variations in the spin-spin coupling. A Kagomé net with alternating ferromagnetic and antiferromagnetic coupling (see Fig.3.3) has been studied using a spin-flip Monte Carlo method [3.37,38]. Because of the distribution of bonds there will be two inequivalent sublattices and two possible groundstates (A^+, A^-). Because the probabilities of excitations are not the same on the two sublattices, the sublattice magnetizations will be unequal in the presence of the magnetic field, even in the paramagnetic state. Fig.3.3 shows the difference in the magnetization curves corresponding to the two groundstates. This result is particularly important for understanding the experimental results on dysprosium aluminum garnet (DAG) [3.39] for which neutron scattering data taken in a field show the existence of a staggered magnetization in the paramagnetic state.

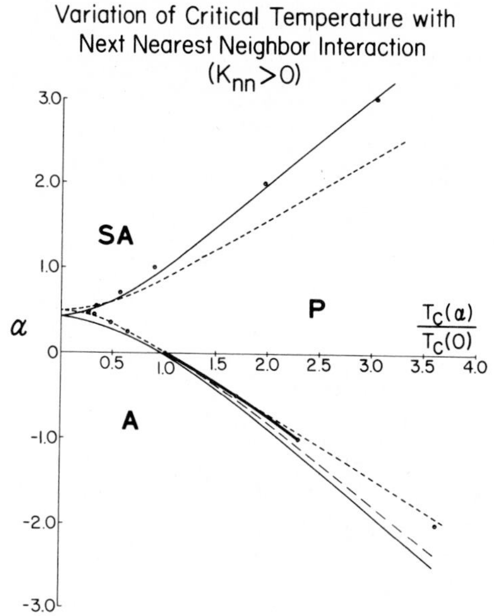

Fig.3.2 Variation of the critical temperature of the Ising square lattice with nnn-coupling ($\alpha = K_{nnn}/K_{nn}$). The circles are Monte Carlo values, the heavy solid line gives the series expansion result and the other curves show predictions obtained using perturbation theory and a many fermion representation: – – GIBBERD, —— KAWAKAMI and OSAWA, --- FAN and WU. The ordered phases are antiferromagnetic (A) and superantiferromagnetic (SA); the paramagnetic phase is P. [3.25]

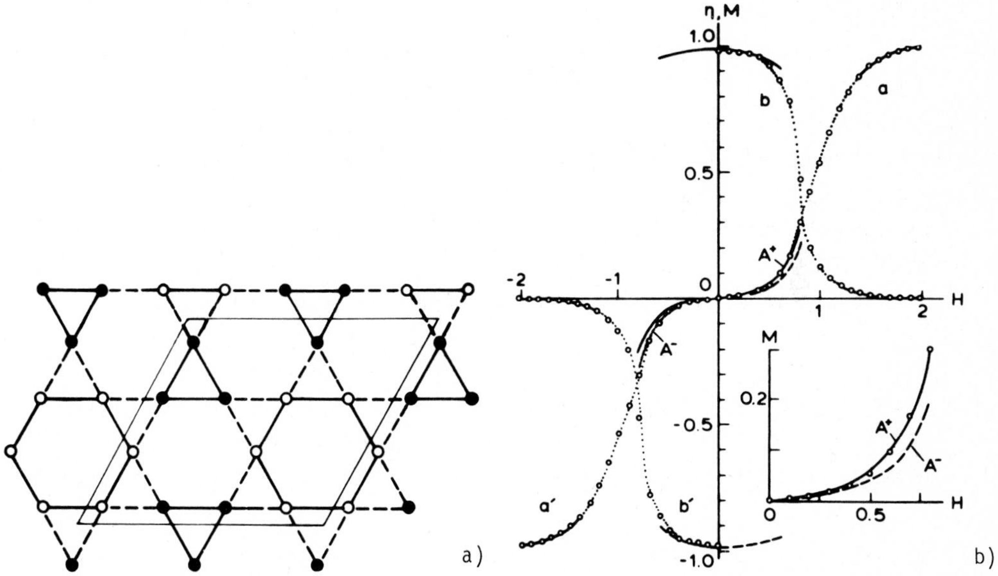

Fig.3.3 a) Two-dimensional Ising kagome net with periodic interactions: antiferromagnetic --- and ferromagnetic ——. b) Variation of magnetization M (curves a and a') and staggered magnetization η (curves b and b') with magnetic field for the net shown in (a). The curves ... were obtained by extrapolation of Monte Carlo calculations, the sizes of the σ's roughly indicating the estimated uncertainties in the extrapolations. The solid —— and broken --- curves were obtained from low temperature series expansions. Near H = 0 the solid curves correspond to the A^+ state (stable for $H > 0$), the broken curves correspond to the A^- state (stable for $H < 0$). η and M are expressed in units of their values at T = 0 K, and H in units of the critical field at T = 0 K. [3.37]

If the interactions in one direction on the lattice are much weaker than in the others, the system may behave like weakly coupled lower dimensional systems and display 3-dim critical behavior only asymptotically near to T_c. The Monte Carlo method has been used to study this "dimensionality crossover" in a 3-dim system of weakly coupled Ising chains [3.40] which interact according to the hamiltonian

$$\mathscr{H} = J_z \left[\sum_{z\text{-nn}} \sigma_i \sigma_j + \Delta \sum_{\substack{x,y-\\ nn}} \sigma_i \sigma_k \right] \quad . \qquad (3.4)$$

As the coupling between the chains weakens, the properties of the system become 1-dim-like over an ever increasing temperature range with long-range 3-dim order setting in only at quite low temperature. The Monte Carlo data in Fig.3.4 show this effect quite clearly and provide a clear understanding of the specific heat of the pseudo-one-dimensional systems such as $CoCl_2{\cdot}2NC_5H_5$ [3.41], and $CsCoCl_3{\cdot}2H_2O$ and $RbFeCl_3{\cdot}2H_2O$ [3.42].

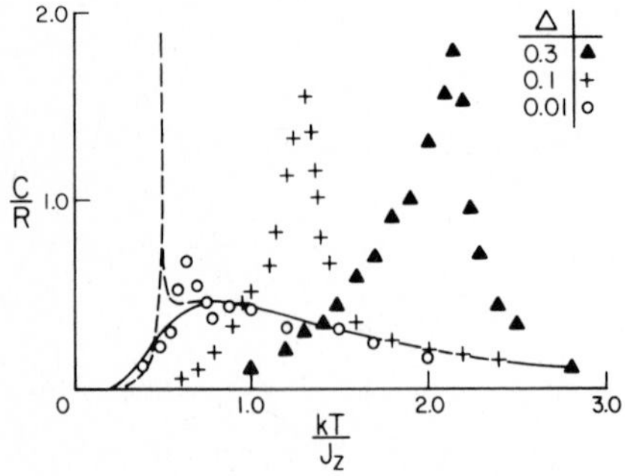

Fig.3.4 Temperature dependence of the specific heat of a 3-dim system of Ising chains interacting weakly according to (3.4) for several values of the relative interchain coupling Δ. The solid curve gives the exact 1-dim variation; the dashed line shows the exact solution for a 2-dim system with Δ= 0.01. [3.40]

Phase diagrams showing the region of stability of an antiferromagnetic state in the presence of a magnetic field are quite readily determined using Monte Carlo methods. If the interactions between spins are restricted to nearest-neighbors

$$\mathscr{H} = K \sum_{\substack{(ij)\\ \text{nn-pairs}}} \sigma_i \sigma_j + \mu H \sum_i \sigma_i \quad , \qquad (3.5)$$

we believe that in loose packed lattices the critical field curve contains no multicritical points (except at T = 0) and that, according to the universality principle the critical exponents are identical to those in zero field. The Monte Carlo technique has been used to study critical field curves of models described by (3.5) in both 2-dim [3.44,45] and 3-dim [3.46-48]. The data are quite complete and show rather convincingly that the low temperature critical field curve does indeed show the "bulge" predicted by mean-field [3.43] and other approximate theories. Existing series expansions [3.49] prove to be increasingly inaccurate as the temperature is decreased and do not correctly describe the low teperature, high field bulge. A comparison of the phase boundaries obtained in different ways is shown in Fig.3.5. The Ising model in a magnetic field has also aroused interest due to its applicability (in the lattice gas representation) to order-disorder transitions in some superionic conductors such as β-alumina [3.50,51]. A nonlinear renormalization group approach was applied to the study of Ising antiferromagnets in a field [3.52-54]. The approximate results do not reproduce the bulge in the bcc critical field curve and the temperature variation of the critical magnetization varies with the cell cluster size and shape.

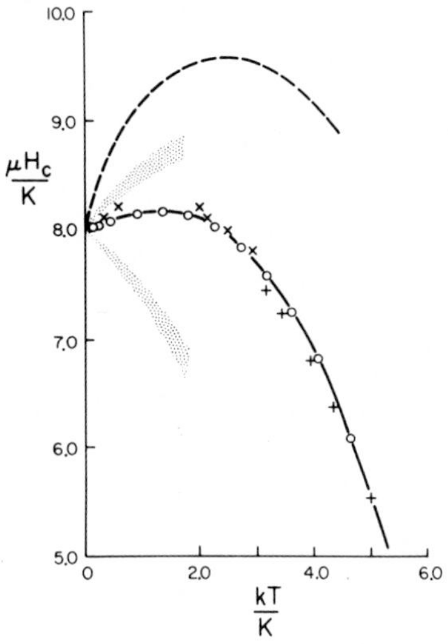

Fig.3.5 Low-temperature portion of the critical field curve. The dashed line is the mean-field prediction [3.43], the series expansion results are given by + and the infinite lattice Monte Carlo estimates by o. The thin solid line is mearly intended as a guide to the eye. Earlier N=25 Monte Carlo data [3.46] are shown by ×. The shaded area shows the crossover region between critical and noncritical behavior. [3.47]

Ising models for several specific real systems have also been studied. HARRIS [3.55] considered an S = 5/2 system of spins on an fcc lattice with periodic boundary conditions and $J_{nnn} = J_{nn}$ as a simple model for MnO. These Monte Carlo results fit the experimental data far better than the mean-field theory although the model is clearly oversimplified.

ALTMAN et al. [3.56] have simulated $FeCO_3$ using rhombohedral lattices and varying the next-nearest-neighbor interaction until both T_c as well as H_c agree with experiment. The observed temperature dependence of the order parameter cannot be reproduced unless impurities (Mn^{2+} and Mg^{2+} which are known to exist in the samples used in the neutron studies) are included. With the inclusion of these impurities into the model the Monte Carlo method reproduces the temperature dependence of the order parameter quite well. The field induced transitions in $CoCl_2 \cdot 2H_2O$ have been studied using an Ising model with nn- and nnn-interactions [3.57]. Multiple transitions were observed. Perhaps the most detailed Monte Carlo calculations which have been carried out on an Ising model corresponding to a real magnetic system are those by GEHRING et al. [3.58] and FAULHABER and HÜFNER [3.59] on DAG. In DAG a very large number of neighbors interact and theoretical treatment by any approach more sophisticated than mean-field theory is essentially impossible. Extensive data exist for DAG: neutron diffraction [3.39,60] has been used to determine the temperature dependence of the order parameter and very precise data exist for the specific heat [3.61] and optical line splittings [3.58,59]. The Monte Carlo simulations included between 50 and 106 interacting near neighbors and describe the short-range and long-range order reasonably well as shown in Fig.3.6. DAG is a very complicated system and the success of the simulations is really quite remarkable.

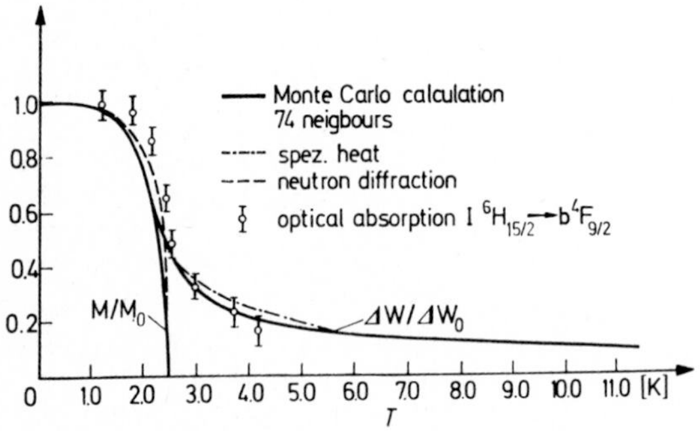

Fig.3.6 Comparision of various experimental results with a Monte Carlo calculation in DAG. The full curves come from the calculation; where the curve marked M/M_0 is the long-range order parameter and the one marked $\Delta W/\Delta W_0$ the short-range order parameter. The optical data (Φ), the specific heat results (—·—) and the sublattice magnetization (– –) from neutron diffraction experiments are also displayed. [3.59]

3.1.2 Magnetic Systems with Isotropic Interactions

Much less is known about systems with isotropic interactions between n-dimensional spins where $n \geqq 2$ and

$$\mathcal{H} = J \sum_{(ij)} \underline{s}_i \cdot \underline{s}_j + H \sum_i s_{iz} \tag{3.6}$$

where $\underline{s}_i, \underline{s}_j$ are classical vectors of length ℓ, than about the Ising systems mentioned above. Short-range spin-correlations as well as long-range order were studied in 3-dim. Heisenberg models (n = 3) [3.62-75]. Because of the extra degrees of freedom available in these models, fluctuations are much larger and computing time much longer than for comparable Ising models and the data showed substantial scatter. BINDER and MÜLLER-KRUMBHAAR [3.66-68] have developed a method which used "self-consistent-field" boundary conditions and which has been quite successful in studying Heisenberg models. MÜLLER-KRUMBHAAR and BINDER [3.69] have determined the equation of state for a simple cubic lattice and have compared the critical behavior with several real magnetic systems. The Monte Carlo data are plotted in terms of the scaling variables to determine the scaling function and are compared to the series expansion result in Fig.3.7.

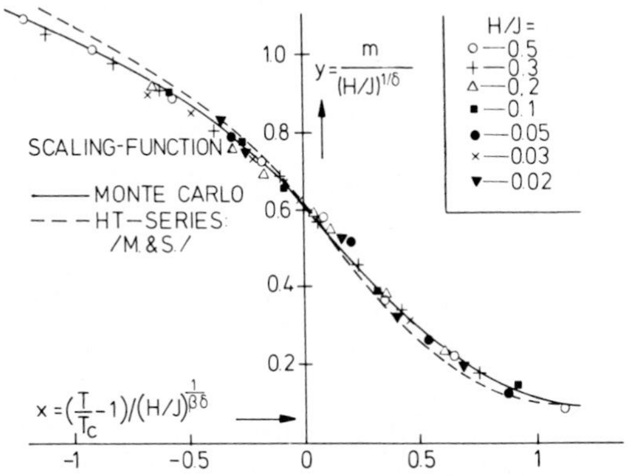

Fig.3.7 Monte Carlo results plotted in terms of the scaling variables to determine the scaling function (full curve). The dashed curve is the result of MILOSEVIC and STANLEY [3.70] obtained for the fcc lattice by series extrapolation methods. Their function has been fitted to the susceptibility amplitude C=0.966 and critical isotherm amplitude D ~ 0.61. [3.69]

In Fig.3.8 we show the critical behavior of the magnetization compared with the experimentally determined behavior of EuO. [Since Eu^{2+} has S = 7/2 we would expect that the data should be rather well described by a classical (S = ∞) model]. In addition, the spin-spin correlation functions $\langle s_0^z s_R^z \rangle$ were determined and the Fourier transform

$$\chi(\underline{k}) = \sum_{\underline{R}} e^{i\underline{k}\cdot\underline{R}} \left\langle s_0^z s_R^z \right\rangle \tag{3.7a}$$

was fitted to a critical form

$$\chi(\underline{k}) = \frac{1}{r_1^2[\varkappa_1^2 + K^2(\underline{k})]} \tag{3.7b}$$

where $\varkappa_1$ is the inverse correlation length and $K(\underline{k})$ an "effective" wave vector which becomes equal to $\underline{k}$ as $\underline{k} \to 0$. The temperature-dependence of the correlation length is shown in Fig.3.9.

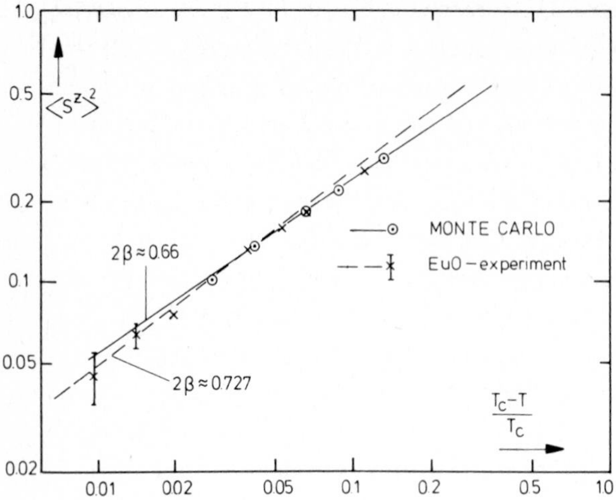

Fig.3.8 Log-log plot of the square of the spontaneous magnetization versus $1-T/T_c$ as obtained by computer simulation (×) and neutron scattering (×). [3.69]

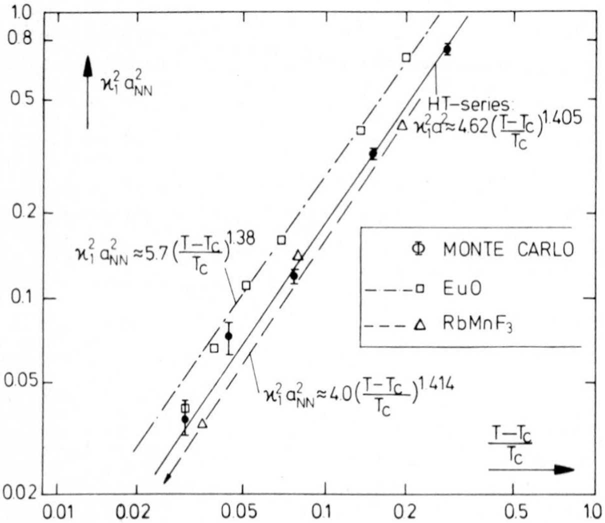

Fig.3.9 Log-log plot of the squared inverse correlation length $\varkappa_1^2 a^2$ as obtained by the computer simulation (Φ), experiments on EuO (□) and $RbMnF_3$ (△) and high temperature series expansion (full curve). [3.69]

The Monte Carlo data are in particularly good agreement with the experimental results for $RbMnF_3$, a simple cubic S = 5/2 system.

It has been suggested [3.76] that the 2-dim Heisenberg model shows a very special sort of phase transition. MERMIN and WAGNER [3.77] have shown that no long-range order can occur, but it is possible that "long-range short-range" order may be present giving rise to a divergent susceptibility (the "Stanley-Kaplan transition"). WATSON et al. [3.78] studied a Heisenberg model on square planar lattices as large as 45 × 45 with periodic boundary conditions as well as free edges. The data show substantial finite size effects (the long-range order parameter goes to zero at low temperatures only *very* slowly with increasing lattice size) and the Monte Carlo method is not able to resolve the question of whether or not the transition does occur.

The possibility of a Stanley-Kaplan transition also exists in the classical XY model. KOSTERLITZ and THOULESS [3.79] have suggested that below T_c interacting vortex pairs may be the important excitations. KAWABATA and BINDER [3.80] studied this model on square lattices as large as 30 × 30 with periodic boundary conditions. The data showed an increasing susceptibility as the temperature was lowered but the fluctuations also became so great that variations of over an order-of-magnitude occured below $kT/J \lesssim 0.5$. The results show evidence for vortex formation although upon more detailed examination it becomes clear that much of the vortex behavior is due to metastabilities. Different spin configurations are obtained depending on whether the initial state is either random or composed of all spins parallel to the z-direction. (When the spins are initially all in the x-direction clearly defined vortices do not appear [3.81]).

MOURITSEN et al. [3.81a] have studied a more complex model which is expected to be pertinent to systems such as UO_2 which have 1st order transitions in zero field. They carried out simulations on a simple cubic lattice with classical spins $\bar{s}$ interacting with hamiltonian

$$\mathcal{H} = J_1 \sum_{nn} \bar{s}_j \cdot \bar{s}_k + K \sum_{nn} (\bar{r}_{jk} \cdot \bar{s}_j)(\bar{r}_{jk} \cdot \bar{s}_k)/r_0^2 + J_2 \sum_{nnn} \bar{s}_j \cdot \bar{s}_k + P \sum (s_{jx}^4 + s_{jy}^4 + s_{jz}^4) \quad . \tag{3.8}$$

The data show clear evidence of a 1st order phase transition; both the internal energy and order parameter shown in Fig.3.10 show distinct hysteresis. This result contradicts mean-field theory which predicts a 2nd order phase transition.

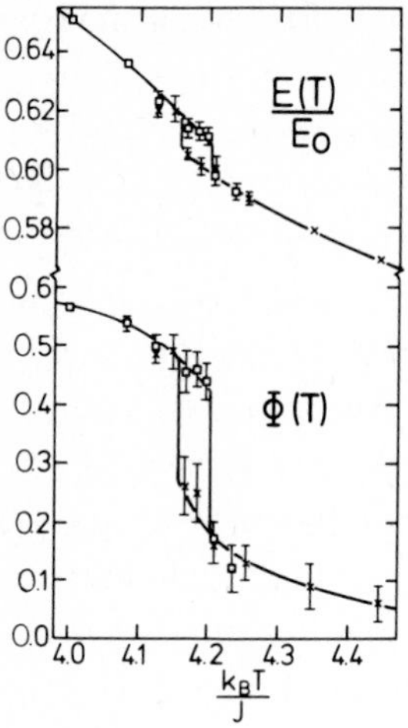

Fig.3.10 Temperature dependence of the normalized internal energy and the order parameter in the critical region for the spin system defined in (3.8). E_0 is the energy of the completely ordered state. × indicates data calculated for decreasing temperatures. □ indicates data calculated at increasing temperatures. The full line is drawn as a guide to the eye and has no further meaning. [3.81a]

3.2 Multicritical Points and Crossover Behavior

The general idea of "multicritical phenomena" was first introduce by GRIFFITHS [3.82] who pointed out that the change in the order of the transition along the phase boundaries of He^3-He^4 mixtures and many highly anisotropic antiferromagnets could be easily understood when a sufficiently complete variable space was used. In magnetic language this means that the phase boundary must be viewed not only in magnetic field-temperature space, but in a 3-dimensional (3-dim.) space which includes a staggered field H^+ as well. (The staggered field is the thermodynamic conjugate to the order parameter). In addition to the usual coexistence surface in the H-T plane two "wing"-like coexistence surfaces extend out of the H-T plane into the fuller H^+-H-T space. Each coexistence surface is bounded by a line of critical endpoints (or critical field curve) and all three phase boundaries intersect at a tricritical point T_t. The three coexistence surfaces meet in the H-T plane along the 1st order portion of the usual critical field curve. The critical behavior as T_t is approached as described by a new set of "tricritical exponents" which are in general quite different from the ordinary critical exponents found along the 2nd order phase boundaries. Asymptotically close to the 2nd order phase boundary the ordinary critical behavior dominates and is separated from the tricritical region by a "crossover" region. Other types of "multicritical behavior may be found in relatively simple systems.

3.2.1 Tricritical Phenomena

The simplest magnetic model which shows tricritical behavior is the Ising antiferromagnet with both ferromagnetic and antiferromagnetic interactions

$$\mathcal{H} = J_1 \sum_{(ij)} \sigma_i\sigma_j - J_2 \sum_{(ik)} \sigma_i\sigma_k + H \sum_i \sigma_i \quad . \tag{3.9}$$

The critical, tricritical, and crossover behavior have been studied in a nnn-antiferromagnet where J_1 and J_2 refer to nn- and nnn-coupling respectively [3.83,84]. This model has also been studied by series expansions [3.85] and the renormalization group (RG) approach [3.86]. In particular the RG studies predict that this model and the metamagnet (i.e., J_1 refers to spins in neighboring planes whereas J_2 is the coupling between nn-spins in the *same* plane) should have the same classical tricritical exponents as the gaussian model [3.87,88]. Complete Monte Carlo data have now shown this to be the case [3.84] although the interpretation of earlier data on smaller lattices [3.83] was hampered by finite size effects. The estimated locations of the phase boundary and the tricritical point are shown in Fig.3.11b. The series expansion estimates lie systematically low; this is due to the difficulties involved with analyzing a series of a quite limited number of terms along a path where crossover occurs. The shaded areas in Fig.3.11b show the approximate crossover regions as determined from the temperature dependence of the order parameter and high temperature susceptibility. The critical field curve itself shows no kink or other unusual behavior at T_t in excellent agreement with the experimental results on $FeCl_2$ [3.89] and DAG [3.90]. The Monte Carlo data have also given the first clear indication of logarithmic corrections. In Fig.3.11a both the crossover and logarithmic corrections are demonstrated for paths with $h = H/T = \text{const}$.

The second model whose tricritical behavior has been studied using the Monte Carlo method is the Blume-Capel model which is an $S = 1$ model with hamiltonian

$$\mathcal{H} = J \sum_{(ij)} s_{iz}s_{jz} + \Delta \sum_i (s_{iz}^2 - 2/3) + H \sum_i s_{iz} \quad . \tag{3.10}$$

In this model the ground state is ferromagnetic and the "crystal field" Δ plays the role of the uniform field in an antiferromagnet. Similarly, in the Blume-Capel model H is the equivalent of the ordering (or staggered) field in the antiferromagnet. Monte Carlo data are available in 2-dim [3.91,92] and in 3-dim [3.93]. The same conclusions were reached in the studies of 3-dim systems as were for the antiferromagnet: the phase boundary itself is smooth and continuous near T_t, the tricritical behavior was classical (with logarithmic corrections for the order parameter), and crossover behavior was readily observed. Generally good agreement was

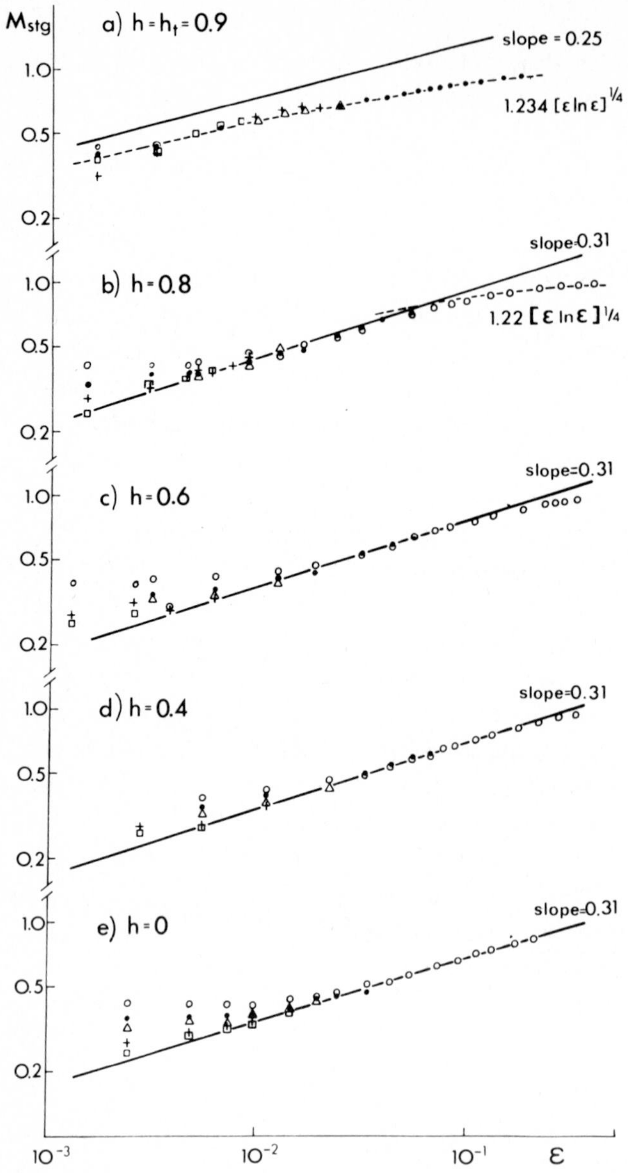

Fig.3.11a and b Multicritical behavior in an nnn-Ising antiferromagnet. (a) Critical behavior of the order parameter along paths of constant $h = H/T$. The path $h_t = 0.9$ goes through the tricritical point. Data are for $N \times N \times N$ lattices with: $N = 6$,o; $N = 8$,•; $N = 10$,△; $N = 14$,+; and $N = 20$,□. [3.84]

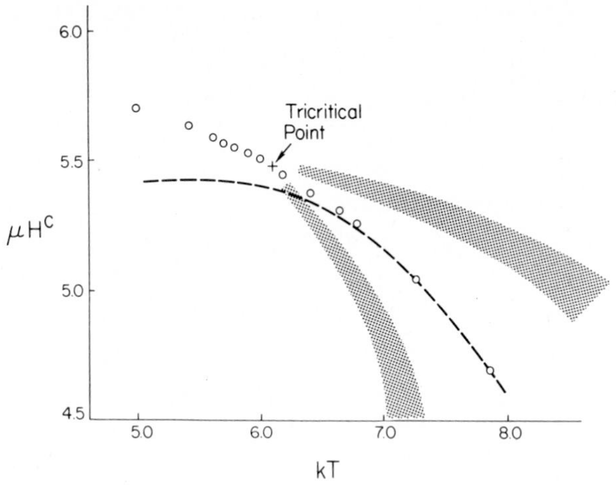

Fig.3.11b Phase boundary near the tricritical point: Dashed lines show the series expansion result and open circles give the Monte Carlo results. The shaded regions give a semiquantitative indication of the crossover regions for $T > T_t$ as determined from the order parameter and high temperature ordering susceptibility. [3.84]

found with series expansion results [3.94]. In addition, the "wings" were traced out and the critical behavior near the wings found to be the same as the ordinary ($\Delta = 0$) 3-dim Ising model.

3.2.2 Bicritical and Other Multicritical Behavior

Perhaps the simplest kind of bicritical behavior occurs in the anisotropic Heisenberg model whose hamiltonian is given by

$$\mathcal{H} = [J \sum_{(ij)} s_{iz}s_{jz} + (1 - \Delta)(s_{ix}s_{jx} + s_{iy}s_{jy})] \quad . \tag{3.11}$$

For $\Delta = 0$ the hamiltonian is identical to that of the simple Heisenberg model. For $\Delta > 0$ the asymptotic critical behavior should be Ising-like and for $\Delta < 0$ it should be XY-like. In the small Δ region crossover between Ising-Heisenberg and XY-Heisenberg behavior should occur. PATTERSON and JONES [3.95] and later BINDER and LANDAU (using much larger lattices) [3.96] simulated a classical, 2-dim anisotropic Heisenberg model with the hamiltonian given in (3.7) for $0 < \Delta \leqq 1$. The data obtained by BINDER and LANDAU indicate that spin-wave theory is valid *only* at *very* low temperatures in this model. In addition, the Monte Carlo data for small Δ provide a good estimate for the variation of T_c with Δ. The asymptotic critical behavior is consistent with Ising exponents and crossover to Heisenberg behavior for large

$(T - T_c)$ is evident. It was not possible to determine whether or not a phase transition occurs for $\Delta = 0$ (i.e., the Stanley-Kaplan transition); but if the transition is at $T = 0$, the data strongly suggest that $T_c(\Delta)$ approaches zero only logarithmically with Δ. Although the existence of real 2-dim systems which are described by (3.11) would seem unlikely, POMERANTZ [3.97] has recently used the Langmuir-Blodgett technique to prepare monolayer films of manganese stearate. Preliminary measurements of the magnetic properties of these films suggest the onset of long-range order. Since a small amount of magneto-crystalline anisotropy is not unexpected, the system probably has nonzero Δ. Moreover, the Monte Carlo result would suggest, that T_c rises very rapidly with Δ and even a minute anisotropy might be able to produce a quite "reasonable" ordering temperature. The Monte Carlo results have also guided the reanalysis of neutron scattering data on K_2NiF_4 [3.97a], with a nearly isotropic layered magnetic structure. The original rather surprising exponent estimates ($\gamma \approx 1$, $\nu \approx 1/2$) were later shown to be incorrect.

Many moderately anisotropic antiferromagnets show bicritical points with uniform fields H applied along the easy axis. From the antiferromagnetic groundstate these undergo a 1st order transition with increasing magnetic field to a canted spin-flop state at sufficiently low temperature. However, the transitions from the antiferromagnetic or spin-flop state to the paramagnetic state are 2nd order. A simple model for uniaxial antiferromagnets is the anisotropic Heisenberg model

$$\mathcal{H} = J \sum [s_{iz}s_{jz} + (1 - \Delta)(s_{ix}s_{jx} + s_{iy}s_{jy})] + H_{\parallel} \sum s_{iz} + H_{\perp} \sum s_{ix} + H_{\parallel}^{+} \sum s_{iz} e^{i\bar{k}\cdot\bar{r}} \quad . \tag{3.12}$$

Detailed Monte Carlo studies have been carried out on a classical model of this type including variations in both uniform and staggered fields in (3.12) [3.98,99]. The $H_{\parallel}^{c} - T$ and $H_{\perp}^{c} - T$ phase boundaries deduced from these studies are shown in Fig.3.12a. These results show qualitative behavior similar to the experimental data on physical systems like MnF_2, [3.100,101] and $GdAlO_3$ [3.102-104] (see Fig.3.12b), and $NiCl_2 \cdot 6H_2O$ [3.105,106]. Shapes of the phase boundaries in the region of the bicritical temperature were analyzed in terms of the skew set of scaling axes [3.107]

$$g = \Delta(H^2) - pt \tag{3.13a}$$

$$\tilde{t} = t + q\Delta(H^2) \tag{3.13b}$$

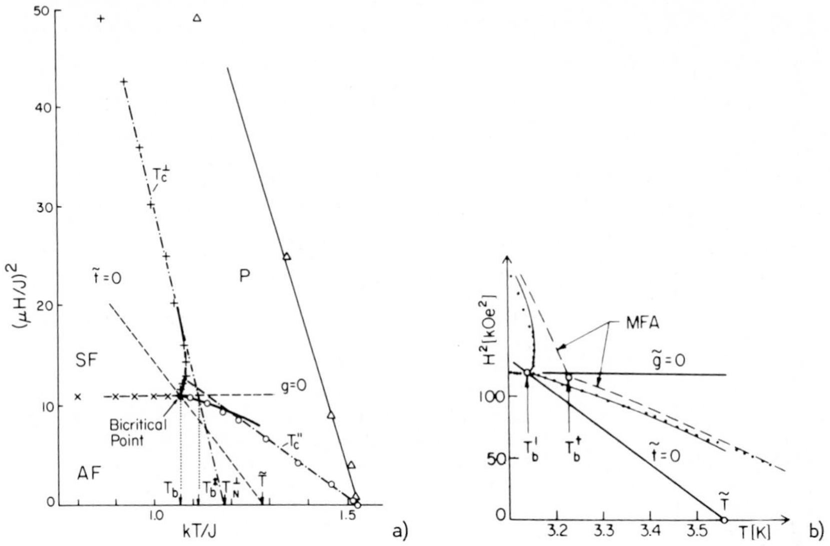

Fig.3.12a and b Bicritical phase diagrams:
(a) Monte Carlo data for a classical anisotropic Heisenberg model. Results in the H - T plane are given by ×,o, and +. The mean-field phase boundaries (-·-·-) and bicritical temperature are shown for comparison. The heavy solid curves represent the asymptotic phase boundaries resulting from a fit to (3.14). The dashed lines are the theoretical scaling axes. The triangles show the critical field data in the $H_\perp$ - T plane [(3.99].
(b) Experimental results near the bicritical point in $GdAlO_3$. The dashed lines show the extrapolations of the high- and low-field MFA phase boundaries. [3.102]

where $\Delta(H^2) = H^2 - H_b^2$ and $t = |1 - T/T_b|$. The constant p is readily determined experimentally from the slope of the spin-flop line as it approaches the bicritical point. The constant q was fixed at the value estimated from renormalization group calculations and the phase boundaries were fitted to the general form

$$\frac{g}{\tilde{t}^{\phi}} = +w_\perp \ , \ -w_\parallel \tag{3.14}$$

where $w_\perp$, $w_\parallel$ are constants describing the SF → P and AF → P phase lines respectively and ϕ is the crossover exponent. The rapid deviation of the fitted AF → P boundary from the data strongly suggests that very fine resolution is needed near the bicritical umbilicus if the asymptotic multicritical behavior is to be seen. It has been implied from extrapolated renormalization group calculations in small $H_\parallel^+$ that the SF → P phase surface in $H_\parallel$ - $H_\parallel^+$ - T space looks rather like a "bubble". The Monte Carlo data clearly indicate that the low temperature spin-flop phase spreads out into a pair of "horns" instead of a bubble, see Fig.3.13. At higher temperatures, however, the critical surface does have the shape predicted from the renormalization

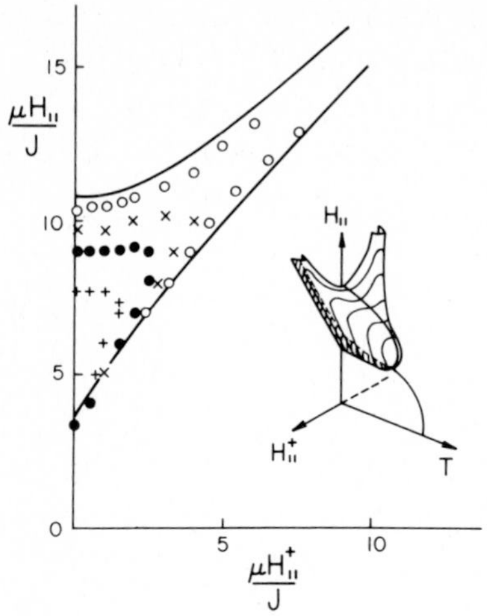

Fig.3.13 Phase boundaries of a classical anisotropic Heisenberg antiferromagnet including both staggered($H_{\parallel}^{+}$) and uniform ($H_{\parallel}$) fields along the easy axis. The solid curve shows the exact T=0 variation. The data points trace out contours at constant temperature: T=0.2 (○); T=0.4 (×); T=0.6 (•); T=0.8 (+). The insert shows a three-dimensional view of the phase diagram. The shaded area shows the approximate portion of the surface which is 1st order. [3.99]

group calculations. Fig.3.13 may be related to the tricritical "wing" diagram in the following way. A small amount of ferromagnetic nnn-coupling should not alter this phase diagram qualitatively. As the anisotropy increases the AF → SF boundary moves up until it coelesces with the upper SF → P critical field. When this happens the SF phase has disappeared and the "horns" have "flattened" out into "wings".

DOMANY et al. [3.108] have studied multicritical behavior in a generalized square lattice model of which the Baxter model and Ising model are special cases. The spins are arrayed on two interpenetrating square lattice nets and interact with hamiltonian

$$\mathcal{H} = \sum_{\substack{s.\ell. \\ \#1}} K_1^{\pm} s_i s_j + \sum_{\substack{s.\ell. \\ \#2}} K_2^{\pm} s_k s_\ell + \Lambda \sum_{(ijk\ell)} s_i s_j s_k s_\ell + G^{\pm} \sum_{(ij)} s_i s_j \tag{3.15}$$

where e.g., K^+ and K^- refer to near neighbor coupling in two opposite directions. (in the usual Ising model the coupling is symmetric and $K^+ = K^-$). The special case

$$K_1^{\pm} = K(1 - \Delta),\ K_2^{\pm} = K(1 + \Delta),\ \Lambda \neq 0,\ G^{\pm} = 0 \tag{3.16}$$

shows a tetracritical point in the (T,Δ) plane at the Baxter transition $\Delta = 0$. The finite size dependence of the specific heat maximum was used to estimate the crossover exponent. The largest lattices studied were 30 × 30 and for very small values of Δ it was not possible to separate the two peaks which appear; the variation of the specific heat amplitudes which were used to determine the crossover behavior were thus imprecise for small Δ.

A new kind of multicritical point (the Lifshitz point) has recently been identified by HORNREICH et al. [3.109] where a ferromagnetic uniformly ordered phase and a modulated ordered phase become simultaneously critical. One quite simple magnetic model which exhibits a Lifshitz point is the Ising model with ferromagnetic nn-interactions and antiferromagnetic nnn-coupling along one-direction

$$\mathcal{H} = -J_1 \sum_{nn} \sigma_i\sigma_j + J_2 \sum_{\substack{z\\ nnn}} \sigma_i\sigma_k \quad . \tag{3.17}$$

This model has been investigated by SELKE [3.110] on simple cubic lattices. For $J_2/J_1 \sim 0.27$ (the Lifshitz point) the groundstate is one in which the average value of the magnetization in a plane varies sinusoidally along the z-axis. The Monte Carlo estimates for T_c agree well with series expansion values for small J_2/J_1 all the way to the Lifshitz point. The Monte Carlo estimates for Lifshitz exponents $\beta = 0.21$, $\gamma = 1.36$ are noticeably different from Ising model values. This general trend in exponents is supported by experimental results on the structural phase transition in $BaMnF_4$ which may be near a Lifshitz point and has an effective exponent $\beta \sim 0.22$. The Monte Carlo studies in the helical phase are very difficult since the existing long-range periodicity which is incommensurate with the lattice is not compatible with periodic boundary conditions. Since the "wavelength" of the helical phase may be strongly temperature dependent any attempt to adjust the lattice dimensions to "fit" the helical phase will be quite difficult.

3.3 Phase Transitions in Miscellaneous Systems

Simple phase transitions in model systems which are different from those mentioned earlier, yet are still pertinent to real physical systems, have also been studied by the Monte Carlo method. LEBWOHL and LASHER [3.111] have simulated a lattice version of the Maier-Saupe model of a nematic liquid crystal. Rodlike molecules were placed on lattice sites of a simple-cubic lattice with periodic boundary conditions. Interactions were restricted to nearest neighbors and had the form

$$\mathcal{H} = -\varepsilon \left(\frac{3}{2} \cos\theta_{ij} - \frac{1}{2}\right) \tag{3.18}$$

where θ_{ij} is the angle between the long axes of the molecules. Lattices as large as $20 \times 20 \times 20$ were studied and rather convincing evidence of a first-order phase transition at $kT_c/\varepsilon \sim 1.1$ was found. The estimated discontinuity in the order parameter at T_c was 0.33 ± 0.04. In comparison, the MAIER-SAUPE mean-field theory [3.112] yields a value of T_c about 20% higher, a much larger discontinuity in the order parameter, and a latent heat only half as large as that obtained from the Monte Carlo data.

AVIRAM et al. [3.113] have used a Monte Carlo method to study the phase transition in a model for a displacive ferroelectric. Ions were placed on a simple cubic lattice and allowed to interact with nearest neighbors via a hamiltonian

$$\mathcal{H} = \frac{\Omega^2}{2} \sum_i x_i^2 - \frac{V}{2} \sum_{(ij)} x_i x_j + \frac{\gamma}{4} \sum_i x_i^4 \tag{3.19}$$

where x_i is a localized normal mode coordinate describing ion displacement from equilibrium. (x_i was restricted so that $|x_i| < 1$.) Data were obtained for two sets of parameters: (I) $6v/\Omega^2 = 1.5, \gamma/\Omega^2 = 3$, for which restricted mean-field theory predicts a second-order transition, and (II) $6v/\Omega^2 = 2.4$, $\gamma/\Omega^2 = 3$ for which the restricted mean-field theory predicts a first-order phase change. The Monte Carlo data indicate that the transition is 2nd order in both cases (or at least extremely "weak" first order).

Monte Carlo simulations have also been carried out on dielectric lattice models involving dipole-dipole interactions. Relatively high temperature calculations [3.114,115] on the simple cubic lattice showed that the permittivity predicted by the Onsager theory tends to be too low while the results of existing series expansions rise too rapidly. The effects of adding a magnetic field to the simple cubic lattice with dipolar interactions have also been studied [3.116]. When a small block of 343 near neighbors surrounding a dipole is considered exactly and the remaining interactions taken into account by a mean field, the data suggest a tricritical point appears at $T_t/T_N \sim 0.76$. The interpretation of all the data is complicated not only by finite size effects but also by the appearance of domains. Similar calculations were carried out on a body-centered rhombic lattice appropriate to $NaNO_2$ [3.117]. In addition to direct dipolar coupling between 558 neighbors, a "local (mean field)" contribution from the more distant spins, and a nearest-neighbor "ferromagnetic" interaction were included. Evidence for two transitions was found. A second-order transition from the paraelectric to antiferroelectric phase was seen followed by a first-order transition to the ferroelectric phase at lower temperature. Although the qualitative behavior in $NaNO_2$ is correctly described, the model itself is clearly oversimplified and must be modified in order to provide quantitative agreement with experiment.

3.4 Conclusions

Monte Carlo studies have been extremely useful for studying lattice models of a wide range of physical systems such as binary alloys, magnets, liquid crystals, ferro/antiferroelectrics, etc. With recent progress in critical and multicritical phenomena pointing out a number of new situations which may occur in systems with hamiltonians which are only slightly more complicated than in the most simplified models, there is room for extensive future use of Monte Carlo simulations. In many cases approximate methods are imprecise, series expansions intractable, and renormalization group methods not yet able to predict the locations of phase boundaries. Monte Carlo studies in these areas may then offer the only real hope for determining the applicability of various models to real physical systems. The Monte Carlo approach will also allow the accurate examination of a much wider range of hamiltonians than have yet been considered and interesting results could serve to provoke more thorough experimental examination of physical materials which are believed to be described by the models.

References

3.1 See L. Onsager: Phys. Rev. *65*, 117 (1944)
B.M. McCoy, T.T. Wu: *The Dimensional Ising Model* (Harvard University Press, Cambridge, Mass. 1973)
3.2 R. Tahir-Kehli: Phys. Rev. *169*, 517 (1968)
3.3 Lloyd D. Fosdick: Phys. Rev. *116*, 656 (1959)
3.4 J.R. Ehrman, L.D. Fosdick, D.C. Handscomb: J. Math. Phys. *1*, 547 (1960)
3.5 L.D. Fosdick: Methods Comp. Phys. *1*, 245 (1963)
3.6 L. Guttman: J. Chem. Phys. *34*, 1024 (1961)
3.7 P.A. Flinn, G.M. McManus: Phys. Rev. *124*, 54 (1961)
3.8 D. Chipman, B.E. Warren: J. Appl. Phys. *21*, 696 (1950)
3.9 L. Muldawer: J. Appl. Phys. *22*, 663 (1951)
3.10 J.R. Beeler, Jr., J.A. Delaney: Phys. Rev. *130*, 962 (1963)
3.11 J.R. Beeler, Jr.: Phys. Rev. *134*, A 1396 (1964)
3.12 J.R. Beeler, Jr.: Phys. Rev. *138*, A 1259 (1965)
3.13 Z.W. Salsburg, J.D. Jacobson, W. Fickett, W.W. Wood: J. Chem. Phys. *30*, 65 (1959)
3.14 Chen-Ping Yang: Proc. Symp. Appl. Math. *15*, 351 (1963)
3.15 R. Friedberg, J.E. Cameron: J. Chem. Phys. *52*, 6049 (1970)
3.16 D.P. Landau: AIP Conf. Proc. *18*, 819 (1974)
3.17 K. Binder: Phys. Stat. Sol. B *46*, 567 (1971)
3.18 K. Binder: Physica *62*, 508 (1972)
3.19 D.P. Landau: Phys. Lett. A *47*, 41 (1974)
3.20 D.P. Landau: AIP Conf. Proc. *24*, 304 (1975)
3.21 D.P. Landau: Phys. Rev. B *13*, 2997 (1976)
3.22 D.P. Landau: Phys. Rev. B *14*, 255 (1976)
3.23 C. Domb, M.S. Green (eds.): In *Phase Transitions and Critical Phenomena*, Vol. 3 (Academic Press, New York, 1974)
3.24 See for example J.C. LeGuillou, J. Zinn-Justin: Phys. Rev. Lett. *39*, 95 (1977)

3.25 D.P. Landau: J. Appl. Phys. *42*, 1284 (1971)
3.26 S. Takase: J. Phys. Soc. Jpn. *40*, 1240 (1976)
3.27 S. Takase: J. Phys. Soc. Jpn. *42*, 1819 (1977)
3.28 P.H.E. Meijer, G.W. Cunningham: Phys. Rev. B *15*, 3436 (1977)
3.29 P.N. Vorontsov-Vel'Yaminov, N.B. Gromova, I.A. Favorskii: Sov. Phys. Solid State *14*, 490 (1972)
3.30 D.P. Landau: Bull. Am. Phys. Soc. *17*, 276 (1972)
3.31 N.W. Dalton, D.W. Wood: J. Math. Phys. *10*, 1271 (1969)
3.32 A. Kawakami, T. Osawa: J. Phys. Soc. Jpn. Suppl. *26*, 105 (1969)
3.33 R.W. Gibberd: J. Math. Phys. *10*, 1026 (1969)
3.34 C. Fan, F.Y. Wu: Phys. Rev. *179*, 560 (1969)
3.35 M. Nauenberg, B. Nienhuis: Phys. Rev. Lett. *33*, 944 (1974)
3.36 D. Lambeth: Thesis, Massachusets Inst. of Techn., Cambridge, Mass. (1973) (unpublished)
3.37 N. Giordano, W.P. Wolf: AIP Conf. Proc. *24*, 333 (1975)
3.38 N. Giordano: Thesis (Yale University, 1977) (unpublished)
3.39 M. Blume, L.M. Corliss, J.M. Hastings, E. Schiller: Phys. Rev. Lett. *32*, 544 (1974)
3.40 D.P. Landau, T. Graim: Annals of the Israel Physical Society *2*, 582 (1978); T. Graim, D.P. Landau: Phys. Rev. B*24*, 5156 (1981)
3.41 L. De Jongh, A.R. Miedema: Advan. Phys. *23*, 1 (1974)
3.42 K. Kopinga, Q.A.G. Van Vlimmeren, A.L.M. Bongaarts, W.J.M. De Jonge: Physica B/C *86-88*, 671 (1977)
3.43 J.M. Ziman: Proc. Phys. Soc. A *64*, 1108 (1951)
3.44 D.P. Landau: Bull. Am. Phys. Soc. *15*, 1380 (1970)
3.45 B.D. Metcalf: Phys. Lett. A *45*, 1 (1973)
3.46 T.E. Shirley: Phys. Lett. A *42*, 183 (1972)
3.47 D.P. Landau: Phys. Rev. B *16*, 4164 (1977)
3.48 T.E. Shirley: Phys. Rev. B *16*, 4078 (1977)
3.49 A. Bienenstock: J. Appl. Phys. *37*, 1458 (1966)
A. Bienenstock, J. Lewis: Phys. Rev. *160*, 393 (1967)
3.50 W.J. Pardee, G.D. Mahan: J. Chem. Phys. *61*, 2173 (1974)
3.51 F.L. Lederman, M.B. Salamon: Bull. Am. Phys. Soc. *20*, 331 (1975)
3.52 K.R. Subbaswamy, G.D. Mahan: Phys. Rev. Lett. *37*, 642 (1976)
3.53 F. Claro, G.D. Mahn: J. Phys. C *10*, L73 (1977)
3.54 G.D. Mahan, F.H. Claro: Phys. Rev. B *16*, 1168 (1977)
3.55 E.A. Harris: Phys. Rev. Lett. *13*, 158 (1964)
3.56 R.F. Altman, S. Spooner, D.P. Landau: AIP Conf. Proc. *10*, 1163 (1973)
3.57 I. Ono, T. Oguchi: Phys. Lett. A *38*, 39 (1972)
3.58 K.A. Gehring, M.J.M. Leask, J.H.M. Thornley: J. Phys. C *2*, 484 (1969)
3.59 R. Faulhaber, S. Hüfner: Z. Phys. *228*, 235 (1969)
3.60 J.C. Norvell, W.P. Wolf, L.M. Corliss, J.M. Hastings, R. Nathans: Phys. Rev. *186*, 557 (1969)
3.61 B.E. Keen, D.P. Landau, W.P. Wolf: J. Appl. Phys. *38*, 967 (1967)
D.P. Landau, B.E. Keen, B. Schneider, W.P. Wolf: Phys. Rev. B *3*, 2310 (1971). This article lists extensive references to other experimental work
3.62 K. Binder, H. Rauch: Z. Phys. *219*, 201 (1969)
3.63 K. Binder: Phys. Lett. A *30*, 273 (1969)
3.64 K. Binder, H. Rauch: Z. Angew. Phys. *28*, 325 (1969)
3.65 R.E. Watson, M. Blume, G.H. Vineyard: Phys. Rev. *181*, 811 (1969)
3.66 K. Binder: Z. Angew. Phys. *30*, 51 (1970)
3.67 H. Müller-Krumbhaar, K. Binder: Intern. J. Magnetism *3*, 113 (1972)
3.68 H. Müller-Krumbhaar, K. Binder: Z. Phys. *254*, 269 (1972)
3.69 H. Müller-Krumbhaar, K. Binder: Intern. J. Magnetism *5*, 115 (1973)
3.70 S.M. Milosevič, H.E. Stanley: Phys. Rev. B *5*, 2526 (1972); B *6*, 986, 1002 (1972)
3.71 K. Binder, H. Müller-Krumbhaar: Phys. Rev. B *7*, 3297 (1973)
3.72 H. Müller-Krumbhaar, K.A. Müller: Solid State Commun. *15*, 1135 (1974)
3.73 H. Müller-Krumbhaar: Z. Phys. *267*, 261 (1974)
3.74 V. Wildpaner: Z. Phys. *270*, 215 (1974)
3.75 Th.T.A. Paauw, A. Compagner, D. Bedeaux: Physica A *79*, 1 (1975)
3.76 H.E. Stanley, T.A. Kaplan: Phys. Rev. Lett. *17*, 913 (1966)

3.77 D. Mermin, H. Wagner: Phys. Rev. Lett. *17*, 1133 (1966)
3.78 R.E. Watson, M. Blume, G.H. Vineyard: Phys. Rev. B *2*, 684 (1970)
3.79 J.M. Kosterlitz, D.J. Thouless: J. Phys. C *5*, 1124 (1972); C *6*, 1181 (1973)
3.80 C. Kawabata, K. Binder: Solid State Commun. *22*, 705 (1977)
3.81 D.P. Landau, K. Binder: Phys. Rev. B*24*, 1391 (1981)
3.81a O.G. Mouritsen, S.J. Knak Jensen, Per Bak: Phys. Rev. Lett. *39*, 629 (1977)
3.82 R.B. Griffiths: Phys. Rev. Lett. *24*, 715 (1970)
3.83 D.P. Landau: Phys. Rev. Lett. *28*, 449 (1972)
3.84 D.P. Landau: Phys. Rev. B *14*, 4054 (1976)
3.85 F. Harbus, H.E. Stanley: Phys. Rev. B *8*, 1156 (1973)
3.86 D.R. Nelson, M.E. Fisher: Phys. Rev. B *11*, 1030 (1975); *12*, 263 (1975)
3.87 E.K. Riedel, F.J. Wagner: Phys. Rev. Lett. *29*, 349 (1972)
3.88 F.J. Wegner, E.K. Riedel: Phys. Rev. B *7*, 248 (1973)
M.J. Stephen, E. Abrahams, J.P. Straley: Phys. Rev. B *12*, 256 (1975)
3.89 R.J. Birgeneau, G. Shirane, M. Blume, W. Koehler: Phys. Rev. Lett. *33*, 2078 (1974)
J.A. Griffin, S.E. Schnatterly: Phys. Rev. Lett. *33*, 1576 (1974)
3.90 N. Giordano, W.P. Wolf: AIP Conf. Proc. *29*, 459 (1976)
3.91 B.L. Arora, D.P. Landau: AIP Conf. Proc. *10*, 870 (1973)
3.92 B.L. Arora, D.P. Landau: AIP Conf. Proc. *5*, 352 (1971)
3.93 A.K. Jain, D.P. Landau: Bull. Am. Phys. Soc. *19*, 306 (1974); see also A.K. Jain: Thesis, U. of Georgia (1976), and A.K. Jain, D.P. Landau: Phys. Rev. B*22*, 445 (1980)
3.94 D.M. Saul, M. Wortis, D. Stauffer: Phys. Rev. B *9*, 4964 (1974)
3.95 J.D. Patterson, G.L. Jones: Phys. Rev. B *3*, 131 (1971)
3.96 K. Binder, D.P. Landau: Phys. Rev. B *13*, 1140 (1976)
3.97 M. Pomerantz: Bull. Am. Phys. Soc. *22*, 388 (1977)
3.97a R.J. Birgeneau, J. Als-Nielsen, G. Shirane: Phys. Rev. B*16*, 280 (1977)
The original data and analysis are given in: R.J. Birgeneau, J. Skalyo Jr., G. Shirane: Phys. Rev. B*3*, 1736 (1971)
3.98 D.P. Landau, K. Binder: AIP Conf. Proc. *27*, *29*, 461 (1976)
3.99 D.P. Landau, K. Binder: Phys. Rev. B*17*, 2328 (1978)
3.100 A.R. King, H. Rohrer: AIP Conf. Proc. *29*, 420 (1976)
3.101 Y. Shapira, C.C. Beccera: Phys. Lett. A *57*, 483 (1976)
3.102 H. Rohrer: Phys. Rev. Lett. *34*, 1638 (1975)
3.103 H. Rohrer: AIP Conf. Proc. *24*, 268 (1975)
3.104 H. Rohrer, Ch. Gerber: Phys. Rev. Lett. *38*, 909 (1977)
3.105 N.F. Oliveira, Jr., A. Paduan Filho, S.R. Salinas: Phys. Lett. A *55*, 293 (1975)
3.106 N.F. Oliveira, Jr., A. Paduan Filho, S.R. Salinas: AIP Conf. Proc. *29*, 463 (1976)
3.107 M.E. Fisher: AIP Conf. Proc. *24*, 273 (1975)
J.H. Kosterlitz, D.R. Nelson, M.E. Fisher: Phys. Rev. B *13*, 412 (1976)
3.108 Eytan Domany, K.K. Mon, G.V. Chester, Michael E. Fisher: Phys. Rev. B *12*, 5025 (1975)
3.109 R.M. Hornreich, M. Luban, S. Shtrikman: Phys. Rev. Lett. *35*, 1678 (1975); Phys. Lett. A *55*, 269 (1975); Physica A *86*, 465 (1977)
3.110 W. Selke: Z. Phys. B *29*, 133 (1978)
3.111 P.A. Lebwohl, G. Lasher: Phys. Rev. A *6*, 426 (1977)
3.112 W. Maier, A. Saupe: Z. Naturforsch. A *14*, 882 (1959); A *15*, 287 (1960)
3.113 I. Aviram, S. Goshen, D. Mukamel, S. Shtrikman: Phys. Rev. B *12*, 438 (1975)
3.114 P.N. Vorontsov-Vel'Yaminov, A.M. El'Yashevich, I.A. Favorskii: Sov. Phys.-Solid State *12*, 2984 (1971)
3.115 P.N. Vorontsov-Vel'Yaminov, I.A. Favorskii: Sov. Phys.-Solid State *15*, 1937 (1974)
3.116 P.N. Vorontsov-Vel'Yaminov, E.M. Ushakova, I.A. Favorskii: Sov. Phys.-Solid State *18*, 194 (1976)
3.117 P.N. Vorontsov-Vel'Yaminov, E.M. Ushakova, I.A. Favorskii: Izv. Akad. Nauk SSSR, Ser. Fiz. *39*, 748 (1975)

Addendum

More recent work concerns systems with dipolar interactions. KRETSCHMER and BINDER [3.118] have studied the combined effects of ferromagnetic nearest-neighbor coupling and long-range dipolar forces for Ising spins on simple cubic lattices. Various antiferromagnetic and superantiferromagnetic ordered states are observed. MOURITSEN and KNAK-JENSEN [3.119] have carried out a detailed study of classical spins with a truncated secular dipolar coupling (which simulates nuclear spin systems whose ordering can be indirectly inferred from NMR e.g. in $^{42}Ca^{19}F_2$). The same authors obtained the critical properties of four- and five-dimensional nearest-neighbor Ising models [3.120] thus vividly illustrating the ability of computer experiments to study dimensionalities other than the physical ones. This capability allows the Monte Carlo method to be used to check renormalization group predictions directly. The ordering of fcc binary alloys with nearest-neighbor repulsion was studied by PHANI et al. [3.121]. The ground-state is very degenerate and a first-order transition occurs. This transition becomes second order at a tri-critical point if a next-nearest neighbor attractive interaction is added. SELKE has extended his earlier studies of Lifshitz points to a three dimensional spherical model [3.122] and to a three dimensional two-component spin model [3.123]. A Monte Carlo investigation of the classical XY-chain has been carried out by LANDAU and THOMCHICK [3.124] and the results used as the starting point of a spin-dynamics study along the lines of [3.65] and [3.78]. VAN ROYEN and RADELAAR [3.125] have resimulated the order-disorder transition in A_3B fcc alloys using a vacancy diffusion mechanism with activation energies. Finally we mention that SHUGARD et al. [3.126] have checked the theoretical predictions for the planar model and the F-model simulating short-range correlation functions. PRELOVSEK and SEGA [3.127] studied the behavior of a classical Ising system in a transverse field.

3.118 R. Kretschmer, K. Binder: Z. Physik B*34*, 375 (1979)
3.119 O.G. Mouritsen, S.J. Knak Jensen: Phys. Rev. B*18*, 465 (1978); Phys. Rev. B*22*, 1127 (1980); Phys. Rev. B*23*, 1397 (1981)
3.120 O.G. Mouritsen, S.J. Knak Jensen: Phys. Rev. B*19*, 3663 (1979); J. Phys. A*12*, L339 (1979)
3.121 M.K. Phani, J.L. Lebowitz, M.H. Kalos, C.C. Tsai: Phys. Rev. Lett. *42*, 577 (1979)
3.122 W. Selke: Phys. Lett. *61*A, 443 (1977); J. Magn. Mag. Mater. *9*, 7 (1978)
3.123 W. Selke: J. Phys. C*13*, L261 (1980)
3.124 D.P. Landau, J. Thomchick: J. Appl. Phys. *50*, 1822 (1979)
3.125 E.W. van Royen, S. Radelaar: to be published
3.126 W.J. Shugard, J.D. Weeks, G.H. Gilmer: Phys. Rev. Lett. *41*, 1399 (1978); W.L. McMillan: To be published
3.127 P. Prelovsek, I. Sega: J. Phys. C*11*, 2103 (1978)

4. Quantum Many-Body Problems

D. M. Ceperley and M. H. Kalos

With 4 Figures

Abstract

We review methods used and results obtained in Monte Carlo calculations on quantum fluids and crystals. Available techniques are discussed for the computation of the energy and other expectation values by variational methods in which the absolute square of a trial function ψ_T is sampled by the Metropolis method. Recently developed methods for fermion systems are included. We give a more detailed exposition of the Green's Function Monte Carlo method which permits exact numerical estimates of boson ground-state properties. Our survey of results comprises applications to ^{3}He and ^{4}He, hard-sphere fluids and crystals, spin-aligned hydrogen, the one-component plasma for bosons and fermions, and simple models of neutron and nuclear matter. The reliability of the product form of ψ_T in several applications is assessed. A selected set of related topics is also taken up: low temperature excitations, results obtained by the Wigner $\hbar$ expansion, and evaluations of virial coefficients and pair correlations at finite temperatures.

In this chapter, we give a survey of the applications of Monte Carlo methods to the study of quantum phenomena in statistical physics. The emphasis is on the modelling of many-particle systems which exhibit quantum effects at the macroscopic level - systems such as liquids and crystals of helium and the electron gas. Such work is complementary to a vast literature in which more tranditional theoretical methods have been applied to these systems. Space precludes citing more than the most striking and significant treatments of that kind.

$\hbar = h/2\pi$ (normalized Planck's constant)

4.1 Introductory Remarks

Monte Carlo simulation of quantum many-body systems dates from McMILLAN's observation [4.1] that for a trial wave function having the form of a product of two-body correlation factors, the integrals required to evaluate expectations — including the total energy — can be carried out by the method of METROPOLIS et al. (cf. Chap.1 and Sect.4.2.1 below). The overwhelming bulk of research has been carried out on this basis. We describe in Sect.4.2 the range of direct applications of McMILLAN's method to a variety of physical systems as well as some extensions of the basic idea. We shall see that these simulations have been fruitful, providing useful results and insight into both physical processes and the structure of successful theories. In some cases, associated with hard-core potentials, however, product trial wave functions omit certain important properties of the ground state and the variational estimates are not accurate enough for quantitative comparison with experiment. This is true, for example, for the ground state of ^{4}He where the variational estimates of energy are 15-20% too high, precluding critical comparison of different force laws.

A more powerful — but more elaborate — method has been developed which permits in principle the Monte Carlo solution of the Schrödinger equation. It has come to be called the Green's Function Monte Carlo (GFMC) and in Sect.4.4 we will describe its general principles, its practical problems, and the results which have been obtained with its use.

Some special topics will also be discussed much more briefly: quantum corrections to classical simulations of light atoms, systems with low temperature excitations, and calculations of quantum virial coefficients.

Simulations of quantum spin lattices [4.2,3] will not be discussed as they have not yet produced many very useful results, although they are potentially interesting.

4.2 Variational Methods

The variational method has proved to be a very useful way of computing ground-state properties of many-body systems. Conceptually it is quite simple. The variational principle tells us that for any function $\psi_T(\underline{R})$, the variational energy E_T, defined as

$$E_T = \int d\underline{R}\ \psi_T(\underline{R})H(\underline{R})\psi_T(\underline{R}) / \int d\underline{R}\,|\psi_T(\underline{R})|^2 \tag{4.1}$$

will be a minimum when ψ_T is the ground-state solution of the Schrödinger equation (here $\underset{\sim}{R}$ refers to the coordinates of the N particles), and $H(\underset{\sim}{R})$ is the hamiltonian. The trial function of course must be either symmetric or antisymmetric with respect to the coordinates depending on the statistics of the particles. The variational method then consists of constructing a family of functions $\psi_T(\underset{\sim}{R},\underset{\sim}{a})$ and optimizing the parameters $\underset{\sim}{a}$ so that the energy (4.1) is minimized for $\underset{\sim}{a} = \underset{\sim}{a}^*$. The variational energy is a rigorous upper bound to the ground-state energy and if the family of functions was chosen well, $\psi_T(\underset{\sim}{R},\underset{\sim}{a}^*)$ will be a good approximation to the ground-state wave function.

The problem of constructing a good trial function for a Bose liquid at zero temperature seems to have been considered first by BIJL [4.4]. Consider a liquid of hard spheres; we shall see later that this is a good model for liquid helium. The wave function clearly must vanish if any pair of particles overlaps. The simplest way to satisfy these conditions is to make the trial function a product of two-particle correlation functions, $f(|\underset{\sim}{r}_i - \underset{\sim}{r}_j|)$. That is the BIJL wave function

$$\psi_T = \prod_{i<j} f(r_{ij}) \equiv \exp\left[-\frac{1}{2}\sum_{i<j} u(r_{ij})\right] \quad . \tag{4.2}$$

The "pseudopotential" u(r) for an isotropic liquid is a function of the radial separation only. The square of this trial function is completely equivalent to the Boltzmann distribution of a classical system with u(r) replaced by the interatomic potentials over kT (hence the name "pseudopoential"). BIJL showed how ψ_T arises naturally from perturbation theory at low densities, and with a proper choice of u(r), ψ_T will have many of the correct macroscopic properties.

This form was reinvented by DINGLE [4.5], JASTROW [4.6] and MOTT [4.7] and generalized for Fermi liquids

$$\psi_T = D(\underset{\sim}{R}) \exp\left[-\frac{1}{2}\sum_{i<j} u(r_{ij})\right] \tag{4.3}$$

where $D(\underset{\sim}{R})$ is the ideal Fermi gas wave function, i.e., a determinant of plane waves. The form of the product trial function in a quantum crystal was considered by SAUNDERS [4.8]. His form of the trial function is

$$\psi_T(\underset{\sim}{R}) = \exp\left[-\frac{1}{2}\sum_{i<j} u(r_{ij})\right] \sum_P (\pm 1)^P \phi(\underset{\sim}{r}_i - s_{P_i})\eta_{\sigma_i \sigma_{P_i}} \tag{4.4}$$

where $\phi(\underset{\sim}{r})$ is a single-particle orbital about a lattice site, usually taken to be a Gaussian function, P is a permutation of the pairing of particles to lattice sites, $\underline{s}$; η is a spinor, and the ± signs are for Bose or Fermi statistics, respectively.

The pseudopotential u(r) in these equations, in contrast to the classical situation, is varied to minimize the energy in (4.1). In practice, the pseudopotential is chosen to have some functional form with several free parameters which are then varied. For a hard-sphere system u must be infinitely repulsive for r less than the hard-sphere diameter. For the optimum pseudopotential it is easy to show that the variational energy when two particles are constrained to be a fixed distance, r, apart, $E_V(r)$, is independent of r and equal to the unconstrained variational energy in (4.1). If the interparticle potential is singular at the origin then one can determine the small r behavior of the optimum pseudopotential by requiring that $E_V(r)$ be finite as two particles approach each other. This leads to the condition

$$\lim_{r\to 0}\left\{v(r)+\frac{\hbar^2}{2m}\left[\nabla^2 u(r)-\frac{1}{2}\left(\frac{du}{dr}\right)^2+\delta_{\sigma_i\sigma_j}\frac{2}{r}\frac{du}{dr}\right]\right\}$$
$$=\text{constant}+O\left(\frac{\hbar^2}{2m}\,r\,\frac{du}{dr}\right) \tag{4.5}$$

where v(r) is the interparticle potential and σ_i and σ_j are the spin coordinates. The "spin coordinates" are in effect always different for bosons since there is no Slater determinant in their trial function. For fermions the optimum u(r) will depend on the relative spin coordinates.

The large behavior of u(r) can be derived from essentially macroscopic arguments or with the random phase approximations by demanding that the energy not change if there are long wavelength density oscillations present. For short-range interparticle potentials (i.e., not Coulomb), the macroscopic argument gives that

$$u(r) \cong \frac{mc}{\rho\pi^2\hbar r^2} \quad . \tag{4.6}$$

REATTO and CHESTER [4.9] have shown that this pseudopotential will give the correct linear behavior for small k in S(k) which is believed to be necessary [4.10] for superfluiditiy. KROTSCHECK [4.11] has shown that for the optimum u, c is an upper bound to the speed of sound in the exact ground state. For a repulsive Coulomb system, as is well known, the structure function is proportional to k^2 at small k because of the plasma modes, independent of particle statistics or temperature. BOHM [4.12] has shown that this implies the following large-r behavior of the pseudopotential

$$u(r) \cong Z \frac{e}{\hbar} \sqrt{\frac{m}{\pi\rho}} \frac{1}{r} \quad . \tag{4.7}$$

We have given some reasons for believing that the use of the product trial function can be a good approximation for quantum systems. The task now is to evaluate the multidimensional integrals to get expectation values, in particular to calculate the variational energy for (4.1). There have been many approaches to this problem, practically all of them borrowed from the theory of classical fluids: namely cluster expansions, integral equations, the Metropolis Monte Carlo method, and even molecular dynamics. The Monte Carlo algorithm [of METROPOLIS, A. ROSENBLUTH, M. ROSENBLUTH, A. TELLER, and E. TELLER [4.13] or $M(RT)^2$] is particularly well suited to this problem of calculating multidimensional integrals and has the distinct advantage over the approximate expansions that there are no approximations made that cannot be tested within the method. Molecular dynamics has been successfully applied to Bose liquids but almost all results have been obtained using Monte Carlo.

We will now describe the $M(RT)^2$ Monte Carlo algorithm for sampling the product trial function and in some detail show how it is used to find ground state properties. Then we will describe some of the many calculations that have been made with this algorithm and what has been learned from the variational studies.

4.2.1 Monte Carlo Methods with the Product Trial Function

The Monte Carlo algorithm $M(RT)^2$ [4.13] which was invented to calculate properties of classical statistical systems, is an extremely powerful way to compute multidimensional integrals. For quantum systems we want an algorithm which will produce configurations with a probability proportional to the square of the wave function. Any measurable quantity can be written as an average over such configurations. Suppose A is an operator and we wish to compute its expectation value defined as

$$\langle A \rangle \equiv \frac{\int d\underline{R}\, \psi_T^*(\underline{R})\, A\, \psi_T(\underline{R})}{\int d\underline{R} |\psi_T(\underline{R})|^2} \quad . \tag{4.8}$$

Let $\{\underline{R}_i\}$ be a set of points drawn from the probability distribution

$$p(\underline{R}) = \frac{|\psi_T(\underline{R})|^2}{\int |\psi_T(\underline{R})|^2 d\underline{R}} \tag{4.9}$$

where the integral in the denominator serves here merely to normalize $p(\underline{R})$. Then for any function $f(\underline{R})$, the central limit theorem of probability gives that

$$\lim_{M\to\infty} \frac{1}{M}\sum_{i=1}^{M} f(\underline{R}_i) = \frac{\int f(\underline{R})|\psi_T(\underline{R})|^2 d\underline{R}}{\int |\psi_T(\underline{R})|^2 d\underline{R}} \qquad (4.10)$$

and in particular

$$\lim_{M\to\infty} \frac{1}{M}\sum \psi_T^{-1}(\underline{R}_i)A(\underline{R})\psi_T(\underline{R}_i) = \langle A\rangle \quad . \qquad (4.11)$$

The $M(RT)^2$ algorithm is a biased random walk in configuration space; as usually carried out, each particle is moved one after another to a new position uniformly distributed inside a cube of side s. That move is either accepted or rejected depending on the magnitude of the trial function at the new position compared with the old position. Suppose $\underline{R}$ is the old position and $\underline{R}'$ the new. Then if $|\psi_T(\underline{R}')|^2 \geq |\psi_T(\underline{R})|^2$ the new point $\underline{R}'$ is accepted. Otherwise the new point is accepted with probability q where

$$q = \frac{|\psi_T(\underline{R}')|^2}{|\psi_T(\underline{R})|^2} \quad . \qquad (4.12)$$

It has been shown elsewhere in this book (cf. Sect.1.2) that under certain very general conditions, the points of the random walk have $|\psi_T(\underline{R})|^2$ as their density, asymptotically as the number of steps increases.

In general the algorithm is very simple to program and test, and follows very closely a Monte Carlo simulation of a classical system. However, there are a number of things specific to quantum systems which we wish to draw to the attention of the reader who is interested in doing such a calculation.

a) Finite System Size

Usually one wishes to calculate the properties of an infinite homogeneous system but practical simulations are limited to about several thousand particles with current computers. Periodic boundary conditions are used to eliminate surface effects. Quantum systems, in general, have smaller size dependence than classical systems (see below, Sect.4.2.2). The two-particle correlation function, g(r), for liquid helium looks very much like that of a classical gas near its critical point in that the correlations are of fairly short range. The energy per particle of 32 helium atoms (at equilibrium density) is only about 0.15 K lower than for 862 atoms out of a total potential energy of -18 K. The most serious long-range problem is that posed by the Coulomb potential, but that, as we shall see, seems to be satisfactorily solved by the use of the Ewald image potential, at least for single component systems. There, also 32 particles give reasonably good estimates for bulk properties and a few calculations with much larger systems, say 256 particles, give a good idea of the size dependence.

b) The Random Walk

For quantum systems, there seems to be no problem with the random walk converging to the desired probability distribution. The initial step of the random walk can be taken with particles on lattice sites, or uniformly distributed throughout the simulation box, or the last step of another random walk. Then typically 50 to 500 moves per particle are sufficient to ensure that the random walk converges to the distribution p(R). The step size of the random walk s is adjusted so that the acceptance ratio for moves is between 0.1 and 0.7. The smaller acceptance ratios are probably more desirable to insure rapid convergence and also save some computer time for Fermi trial functions. After the system has "equilibrated", the random walk continues, with various averages being kept, until either computer time is exhausted or the statistics are judged acceptable, preferably the latter.

c) Computation of the Pseudopotential

At each step of the random walk the relative change in the trial function must be computed to determine q in (4.12). If particle i is being moved and $\underline{r}_i'$ is its new trial position, then for the boson liquid trial function, (4.2), this is simply

$$q_B = \exp\left\{\sum_j \;[u(\underline{r}_i - \underline{r}_j) - u(\underline{r}_i' - \underline{r}_j)]\right\} \;, \qquad j \neq i \quad . \tag{4.13}$$

Then to move one particle takes computer time proportional to the number of particles. If u(r) is a short-ranged function, enumeration of nearest neighbors of each particle can be used to reduce this time to a constant, for large enough systems [4.14].

Care must be taken to avoid the use of pseudopotentials which are discontinuous, since the variational principle requires continuity of trial function and its derivative. A simple way of truncating the pseudopotential at the edge of the box is as follows. Suppose L is the length of the smallest side of the simulation cube. Then a continuous truncated potential is

$$u_T(r) = \begin{cases} u(r) - u(r_T) \quad , & r < r_T \\ \\ 0 \quad , & r \geq r_T \end{cases} \tag{4.14}$$

where r_T is less than or equal to L/2. This potential will have a discontinuous derivative at r_T which gives rise to an easily calculated 'tail' correction to the kinetic energy. The same remarks apply to the single particle localization function $\phi(r)$ in the solid trial function [see (4.4)]. It is important to keep $\phi(r)$ smooth if the particles are only weakly localized about their lattice sites and the system is small.

On the other hand for charged systems it is often necessary to have a long-ranged 'plasmon' trial function (4.7) and truncation is not desirable. In this case one can use the Ewald image potential where the sum in (4.13) is over all of the other particles and all of their images in the periodically extended space. (See the review by VALLEAU and WHITTINGTON [4.15] for a discussion of the theory, implementation, and difficulties of Ewald sum techniques).

d) Fermion Trial Function

The presence of the Slater determinant in the Fermion trial function (4.3) complicates the random walk somewhat since one must compute the ratio of two determinants, in addition to the pseudopotential, at each step of the random walk. The additional computation can be done in the following way. At the beginning of the random walk the inverse matrices to the Slater matrices are found (one for each spin degree of freedom). That is let

$$D_{ik} = \exp(i\underline{k}\cdot\underline{r}_i) \tag{4.15}$$

be the Slater matrix and the inverse matrix $\bar{D}$ is such that

$$\sum_{k=1}^{N} D_{ik}\bar{D}_{jk} = \delta_{ij} \quad . \tag{4.16}$$

The crucial observation is that when only one particle is moved, just one row of the Slater matrix changes, the matrix of cofactors for that row is unchanged, and one can compute the ratio of determinants by a simple scalar product. That is,

$$q = q_B \times \left|\sum_{k=1}^{N} \exp(i\underline{k}\cdot\underline{r}_i')\bar{D}_{ik}\right|^2 \quad , \tag{4.17}$$

where q_B is still given by (4.13). If the move is accepted, the inverse matrix needs to be updated and this takes roughly N^2 operations on the computer [4.16]. Because of the necessity of storing and updating the inverse matrices, it is more difficult to do a large fermion system than a large bose system. A random walk with 250 particles is a substantial calculation, but such a system is sufficiently large for particles interacting with a short-range potential [4.16]. That is, the dependence of the energy on the size of the system for 250 particles is usually comparable to the statistical error and much smaller than error of using the trial function (4.3). The electron gas in the range of metallic densities has important size dependence, even after using the Ewald image potential. But most of this error can be eliminated [4.17].

e) Computing the Trial Energy

The most important average to be done with Monte Carlo is the trial energy since it must be minimized with respect to u(r). Green's theorem can be used to cast the energy into several different forms and to check on the convergence of the random walk.

The following relation holds for continuous trial functions with periodic boundary conditions

$$- \int d\underline{R}\ \psi_T^* \ \nabla^2 \psi_T = \int d\underline{R} |\psi_T|^2 \ \ |\nabla \ln \psi_T|^2 \quad . \tag{4.18}$$

Then for the boson liquid trial function the variational energy can be rewritten as

$$E_B = \frac{\int d\underline{R} |\psi_T(\underline{R})|^2 \sum_{i<j} \left[v(r_{ij}) + \frac{\hbar^2}{4m} \nabla^2 u(r_{ij}) \right]}{\int d\underline{R} |\psi_T(\underline{R})|^2} \tag{4.19}$$

and the trial energy can be calculated once the radial distribution function, g(r), is known. For an N body system in volume V,

$$g(\underline{r}) = \frac{V(1 - 1/N) \int |\psi_T(\underline{R})|^2 \ \delta(\underline{r}_i - \underline{r}_j - \underline{r}) d\underline{R}}{\int |\psi_T(\underline{R})|^2 d\underline{R}} \quad . \tag{4.20}$$

For fermions, different distribution functions can be defined for particles of the same or opposite spins.

It does not seem to be generally realized that this transformed estimator for the energy has a serious drawback. The original form for the energy, (4.1), has a zero variance in the limit as ψ_T approaches a solution of the Schrödinger equation. That is, for any point $\underline{R}_i$, $\psi_T^{-1}(\underline{R}_i)\, H\psi_T(\underline{R}_i)$ will always have the same value and there is no sampling error. In practice, we have found that the standard error obtained using form (4.1) of the energy is at least four times smaller than that from the transformed form in (4.19) for liquid helium and the electron gas.

The second relationship, due to FEYNMAN [4.10], is applicable whenever the trial function can be split into two parts. Suppose

$$\psi_T(\underline{R}) = A(\underline{R})B(\underline{R}) \tag{4.21}$$

then

$$-2 \int d\underline{R}\; AB\underline{\nabla}A\cdot\underline{\nabla}B = \int d\underline{R}(|\underline{\nabla}A|^2 + A\nabla^2 A)B^2 \tag{4.22}$$

$$= \int d\underline{R}(|\underline{\nabla}B|^2 + B\nabla^2 B)A^2$$

Together with (4.18) we can transform the kinetic energy of the fermion trial function into many different forms but again we will in general raise the variance of the trial energy as compared to the use of (4.1). Expressions (4.18,22) are useful, however, as a check of the convergence of the random walk.

Finally we should mention that scaling can be used to relate the value of the energy at one density to its value at another density. Let 'a' stand for the set of all parameters in the trial function having units of length. Then clearly the two-particle correlation function scales in the following way

$$g(r/\ell,\rho',a/\ell) = g(r,\rho,a) \quad \text{where } \ell = (\rho'/\rho)^{1/3} \quad . \tag{4.23}$$

Now (4.19) shows that the energy for bosons can be written as an integral over $g(r)$. Of course, the optimum trial parameters at one density will not usually scale into optimum trial parameters at another density but the scaling helps locate the minimum energy.

f) The Pressure

For any equilibrium system the virial theorem gives the pressure once the kinetic energy T and the radial distribution function are known.

$$P = -\frac{1}{N}\frac{\partial E}{\partial \rho} = \frac{\rho}{3}\left[2T - (\rho/2)\int d^3r\; g(r)\; r\,\frac{dv}{dr}\right] \quad . \tag{4.24}$$

This relation has occasionally been used to check the accuracy of variational calculations. But a proof based on a scaling argument quoted by COCHRAN [4.18] shows that the virial theorem also holds for variational energies. More precisely, the two pressures in (4.24) will be equal if all lengths in the trial wave function are considered as parameters and $E(\rho)$, $T(\rho)$ and $g(r,\rho)$ are evaluated at the variational minimum of these parameters.

g) The Single Particle Density Matrix

The single-particle density matrix $n(r)$ and its Fourier transform, the momentum density $n(k)$, can be easily computed from the product trial function as sampled by the random walk. The function $n(r)$ is defined as

$$n(r) = \frac{\int d\underline{R}\ \psi_T^*(\underline{r}_1 + \underline{r}, \underline{r}_2, \ldots, r_N)\psi_T(\underline{r}_1, \underline{r}_2, \ldots, \underline{r}_N)}{\int d\underline{R}|\psi_T|^2} \quad . \tag{4.25}$$

For large r in a Bose liquid, $n(r)$ will go asymptotically to the condensate fraction [4.19], the fraction of particles in the zero momentum state. Furthermore, the probability density for finding a particle with a wave vector k is simply

$$n(k) = \rho \int d^3r\ e^{i\underline{k}\cdot\underline{r}} n(r) \quad . \tag{4.26}$$

An ingenious and very efficient way of carrying out the averages in (4.25,26) is due to McMILLAN [4.1]. A point $\underline{r}'$ is selected uniformly inside the simulation box where the random walk is taking place. That point is successively considered to be each of the particles in turn, displaced an amount $\underline{r}' - \underline{r}_i$. Then

$$n(|\underline{r}' - \underline{r}_i|) = \left\langle \frac{\psi_T(\underline{r}_1, \ldots, \underline{r}', \ldots, \underline{r}_N)}{\psi_T(\underline{r}_1, \ldots, \underline{r}_i, \ldots, \underline{r}_N)} \right\rangle \quad . \tag{4.27}$$

If the calculation is correctly arranged, only one distance and exponential need be computed for each value contributing to the average of $n(r)$.

h) Reweighting Configurations

If one has available configurations generated from one trial function it is relatively simple to calculate properties, such as the energy, of a slightly different trial function. Suppose $\{\underline{R}_i\}$ is a set of configurations drawn from the probability distribution $|\psi_T(\underline{R},\underline{a})|^2$. Then clearly one can calculate averages from a different trial function $\psi_T(\underline{R},\underline{a}')$ as follows

$$\left\langle A(\underline{a}')\right\rangle = \frac{\int \psi_T^*(\underline{R},\underline{a}')A\,\psi_T(\underline{R},\underline{a}')d\underline{R}}{\int |\psi_T(\underline{R},\underline{a}')|^2 d\underline{R}} = \frac{\lim\limits_{M\to\infty}\left\{\sum\limits_{i=1}^{M} W_i\psi_T^{-1}(\underline{R}_i,\underline{a}')A\psi_T(\underline{R}_i,\underline{a}')\right\}}{\sum\limits_{i=1}^{M} W_i} \tag{4.28}$$

where the weights W_i are

$$W_i = \frac{|\psi_T(\underline{R}_i,\underline{a}')|^2}{\psi_T(\underline{R}_i,\underline{a})|^2} \quad . \tag{4.29}$$

For this procedure to be reliable it is important that I) there be a large number of configurations, at least several hundred to eliminate bias in the estimates, and II) that all the weights be of the same order of magnitude, i.e., the largest weight be no more than that roughly 5 times larger than the mean weight. If the second condition is not satisfied it means that there is probably not enough overlap between the trial functions for the procedure to be successful. We have found this procedure allows one to compute the changes in energy due to a change in trial function parameters much more accurately and quickly than doing two separate random walks and subtracting the energies, since in the former case the estimates are positively correlated and in the latter they are not.

4.2.2 Application to Systems of Helium

The first reported use of the Metropolis random walk for a quantum system was by McMILLAN [4.1] for the ground state of liquid helium four. Independent and almost identical calculations were done at about the same time by LEVESQUE et al. [4.20]. Since that time simulations have been done for many simple quantum many-body systems. We will review most of these papers, and indicate some of what has been learned from the studies. The review is organized into three sections: hard-core boson systems, soft-core (including Coulomb) boson systems, and fermion systems.

a) Hard-Core Boson Systems

For our purposes, a hard-core system is one for which the radial distribution function is essentially zero for small r. It is generally believed that all substances in nature do not really have a hard core, but at normal densities atomic systems behave as if they did.

The physical difference between hard-core potentials and soft-core potentials is the behavior at increasing density. The energy and pressure of hard-core systems will increase very rapidly as soon as the cores begin to overlap and as a consequence they will solidify. But the potential energy of a soft-core system only increases with some power (less than one) of the density, the kinetic energy becomes dominant, and as a consequence it will melt at high density. Among atomic substances, only the lightest show significant macroscopic quantum effects at low temperatures. They are shown in Table 4.1 (from STWALLEY and NOSANOW [4.21]). The three isotopes of hydrogen, H↑, D↑ and T↑ are assumed 'spin-aligned'; a very strong magnetic field ($\sim 10^5$ G) forces all of the electrons to have the same spin, thus preventing molecules from forming. The interatomic potentials have all been fit by a Lennard-Jones 6-12 potential,

$$v(r) = 4\varepsilon[(\sigma/r)^{12} - (\sigma/r)^{6}] \qquad (4.30)$$

using virial data (ε and σ are shown in Table 4.1). The measure η was introduced by DE BOER [4.22] in his quantum theory of corresponding states and is given by

$$\eta = \hbar^2/m\varepsilon\sigma^2 \quad . \qquad (4.31)$$

Essentially η will be the ratio of the zero-point kinetic energy and the classical potential energy and thus determines the importance of quantum effects at zero temperature. It is obvious that the thermodynamic properties for any Lennard-Jones system are only determined by η, the reduced density $\rho\sigma^3$, and the reduced temperature kT/ε.

The ground-state energies of all of the atomic systems in Table 4.1 have been calculated using Monte Carlo techniques. ^{4}He is the most popular and we will discuss it first.

b) Liquid ^{4}He

McMILLAN [4.1] did the original simulation of liquid ^{4}He and many others have extended and reproduced his results.

Table 4.1 The atomic substances which show sizable quantum effects at zero temperature. The hydrogen isotopes are spin aligned, placed in a magnetic field strong enough so that all electrons are in one spin state. ε and σ are the parameters of an effective Lennard-Jones 6-12 potential, η is de Boer's quantumness parameter

Substance	Mass	ε	σ	$\eta=\hbar^2/m\varepsilon\sigma^2$	Ground state
H↑	1	6.46	3.69	0.55	Bose gas
D↑	2	6.46	3.69	0.27	Fermi gas or liquid?
T↑	3	6.46	3.69	0.18	Bose liquid
^{3}He	3	10.22	2.556	0.2409	Fermi liquid
^{4}He	4	10.22	2.556	0.1815	Bose liquid
^{6}He	6	10.22	2.556	0.1207	Crystal
H_2	2	37.00	2.92	0.0763	Crystal
D_2	4	37.00	2.92	0.0382	Crystal
Ne	20	36.2	2.744	0.0085	Crystal

I) *Direct Application of Monte Carlo*. McMILLAN used the form

$$u(r) = (b\sigma/r)^m \quad . \tag{4.32}$$

Notice that if m = 5, u(r) will satisfy, to leading order, (4.5) for the small r behavior with $b = (16/25)^{1/10} = 1.134$ (independent of density). Also note that unless m = 2 this wave function cannot give the correct long-range behavior and S(k) will not be linear at small k. The pseudopotential admits a very simple form of scaling; only the moments $\Phi_n(b) = \langle(\sigma/r_{ij})^n\rangle$ need be obtained (n = m + 2,6,12) in order to scale to any density, while for a general pseudopotential the full g (r) is needed. McMILLAN found that m = 5 gives the lowest energy, with the value of b = 1.17. This was confirmed by SCHIFF and VERLET [4.23], and by MURPHY and WATTS [4.24] in more accurate calculations.

McMILLAN found the zero pressure density to be 0.89 ± 0.01 of the experimental value and the energy at that density was (-5.9 ± 0.1) K. The experimental value is -7.14 K. The fraction of particles in the zero momentum state (n_0) was found to be 0.11 ± 0.01.

The size dependence in these numbers was found to be very small. McMILLAN could find no difference between a 32 and a 108 particle system. SCHIFF and VERLET with a much larger system (864 particles) again found the same energy within statistical errors. Combining these results we believe the size effects in the energy are probably less than 0.15 K for a 32 body system.

There are several ways these calculations have been enlarged upon: by looking at approximate integral equations, looking for better pseudopotentials and trying to invert the potential.

II) *Integral Equations*. The first comparison with Monte Carlo was done by LEVESQUE et al. [4.20]. Their results showed that the Percus-Yevick equation gives energies -2 K below the Monte Carlo and has the wrong density dependence (gets worse at higher densities). The PY2 equation gives an error of about -1 K. The hypernetted chain integral equation has energies about 1.0 K too high at ρ_0 and 3.4 K too high at $\rho = 0.028$ [4.24,25]. The equilibrium density is an extremely sensitive measure of how well the equations of state are calculated. The HNC gives an equilibrium density which is only 0.8 of the value obtained from Monte Carlo. The g(r) obtained from HNC is quite reasonable but the relatively large cancellation of potential and kinetic energy means that for helium one cannot resolve energy differences of less than 1 K with integral equations. This is a serious limitation.

III) *The Optimum Pseudopotential*. It is clear from the discussion after (4.6) that the simple form used by McMILLAN for u(r) cannot be optimal. However, attempts to improve it have generally failed. It is known that for large r, u(r) is proportional to $1/r^2$ as in (4.6). SCHIFF and VERLET [4.23] tried to add a long-range term using perturbation theory, the energy was indeed lowered, but apparently the perturbation theory is not accurate enough as the energy decreases without bound. ZWANZIGER [4.26] actually added a term like

$$u(r) = \frac{mc}{\rho\pi^2\hbar} \frac{1}{r^2 + 1/k_c^2} \tag{4.33}$$

to his trial function in the Monte Carlo calculation using the Ewald image potential. The results were somewhat inconclusive due to poor statistics but roughly, the optimum value of b was unchanged from the value at $k_c = 0$, namely 1.17. c was not varied but assumed to be the experimental speed of sound. The optimum value for k_c was found to be 0.45/Å, and the energy was lowered roughly -0.2 K. The equilibrium density and superfluid fraction remained constant within the errors. Other authors have presented other pseudopotentials with different short-range behavior. The one of de MICHELIS and REATTO [4.27] is notable in having 8 variational parameters. However, in no case is the energy lowered significantly (more than 0.2 K) [4.20,27, 28,29]. Recently CHANG and CAMPBELL [4.25] have published the best pseudopotential for the HNC equation, obtained with the use of the 'paired phonon' method. Their optimized u appears to be a short-range part as in (4.32) and the long-range part from (4.33) with the same coefficients. Here again, however, the energy in going from the simple $1/r^5$ to the most general form of u only drops by 0.14 K at ρ_0 and somewhat more at higher densities. We conclude that the 1 K difference between the Monte Carlo results and the experiment is not due to the particular form of the pseudopotential. We shall see that most of the difference can be accounted for by going beyond the product trial function to the exact ground state.

c) Solid ^4He and ^3He

We include the two isotopes of helium together since, as we shall see, the exchange effects are quite small in the crystal phase so that particle statistics have very little effect on the ground-state properties.

LEVESQUE et al. [4.20] and HANSEN and LEVESQUE [4.30] have done calculations with the solid product trial function in (4.4), with a gaussian localization orbital,

$$\phi(r) = \exp[-(A/2)r^2] \tag{4.34}$$

and only the unit permutation which appears in the sum of (4.4) is allowed. This trial function is then neither symmetric nor antisymmetric. They determined that the localized trial function has lower energy than the liquid one at high density and that the transition density is in good agreement with the experimental values for ^3He and ^4He.

For solid ^4He, a minimum density of $0.45\sigma^3$ was found, compared to an experimental value of $0.468/\sigma^3$. For ^3He the variational result was $0.42/\sigma^3$, close to the experimental value of $0.41/\sigma^3$. The other important findings were that the density distribution around a lattice site is closely approximated by a gaussian and the rms deviation divided by the nearest-neighbor distance, (Lindemann's ratio: γ) is about 0.27 at melting.

HANSEN [4.31] was able to find the structural transition HCP-BCC in solid ^3He using Monte Carlo and found $V_{HCP/BCC} \approx 21.8\ cm^3/mole$. This agrees well with experiment, $V_{HCP/BCC} = 19.8\ cm^3/mole$ but the calculated pressure is half the experimental value. Thus the liquid and solid product trial functions have been able, at least qualitatively, to account for the phase transitions in the two isotopes of helium.

HANSEN and POLLACK [4.32] have investigated a wider class of trial functions for the solid having the form

$$\psi_T = \exp\left\{-\frac{1}{2}\sum_{i<j} u(r_{ij}) + (\underline{r}_i - \underline{s}_i)\underline{\underline{G}}_{ij}(\underline{r}_j - \underline{s}_j)\right\} \tag{4.35}$$

where $\underline{s}_i$ is the set of lattice points (BCC) and $\underline{\underline{G}}$ is a tensor acting only between nearest and next-nearest neighbors. One additional parameter was introduced. However, the additional freedom in the trial function did not lower the energy significantly. They also established that the rms deviations from the lattice states were in rough agreement with the experimental values estimated from the Debye temperature, for both isotopes of helium, the variational estimates being 5 to 20% too small. This kind of comparison which is based on a harmonic model of a solid is not conclusive.

d) Interatomic Helium Potential

The interaction between the helium atoms in all of the above was assumed to be Lennard-Jones with ε and σ given in Table 4.1. Once a variational calculation has been done it is relatively straightforward to find the equation of state for a slightly different potential and thus try to determine from the variational results at zero temperature the interatomic potential.

A potential can be accepted only if the variational energies are greater than the experimental energies at all densities. In this way the Haberlandt potential [4.33] can be rejected from the liquid helium variational calculations [4.23]. Also the energy of the Lennard-Jones 9-6 potential with the Kihara parameters lies below the experimental solid helium energies [4.34]. Another test is to require that the pressure, estimated using a given potential, agree roughly with experimental pressures throughout a large density range, particularly in the high-density solid. This assumes that the errors made in using the product trial functions are roughly independent of density. We will discuss this below. Using this test, HANSEN [4.32,34] showed that a Lennard-Jones 6-12 potential with slightly changed coefficients (ε = 10.2 K, σ = 2.62 Å), the Morse potential of BRUCH and McGEE [4.35], and the Beck potential [4.36], are significant improvements. There is a significant improvement in the structure functions with these potentials as well [4.37].

The potential energy of a group of helium atoms cannot be expected to be exactly equal to a sum of two-body potentials; three and more body interactions are known to exist. It is possible to estimate the perturbational effect of a specific form. The most videly used three-particle interaction is the long-range triple dipole or AXILROD-TELLER [4.38] interaction

$$v_3(r_1,r_2,r_3) = u_0(1 + 3\cos\theta_1 \cos\theta_2 \cos\theta_3)/r_{12}^3 r_{13}^3 r_{23}^3 \quad . \tag{4.36}$$

The angles θ are the internal angles of the triangle (r_1,r_2,r_3) and u_0 = 0.324 K. MURPHY and BARKER [4.39] have calculated the energy of this term in liquid ^{4}He with the McMillan pseudopotential (4.32) and found it to be $0.14(\rho/\rho_0)^3$ K (ρ_0 is the liquid equilibrium density). Their estimate for solid ^{4}He is +0.10 $(\rho/\rho_0)^3$ K. This is a rather small energy compared with the 1 K energy difference between the various two-body potentials.

e) The Hard-Sphere Potential

The hard-sphere model has had considerable success in explaining the structure and liquid-solid transitions of classical systems. It is natural to inquire what the zero temperature properties of a system of hard spheres are. A variational calculation has been done by HANSEN et al. [4.40] using the correlation function

$$\exp\left[-\frac{1}{2}u(r)\right] = \tanh\{[(r/a)^m - 1)]/b^m\} \tag{4.37}$$

where a is the hard-core diameter. They obtained a solidification density of $0.23/a^3$ and a melting density of $0.25/a^3$. With the choice of a = 2.556 Å both the crystallization density and the structure function agree roughly with the experimental value for helium 4.

KALOS et al. [4.28] have shown that a perturbation formula from classical liquid theory [4.41] can be used to calculate the energy of a Lennard-Jones liquid from the Monte Carlo results of the hard-sphere liquid. The Lennard-Jones potential is split into an attractive part w(r) and a purely repulsive part $v_{LJ}(r) - w(r)$. Let

$$w(r) = \begin{cases} -\varepsilon & r < r_m \\ v_{LJ}(r) \quad , & r \geqq r_m \end{cases} \tag{4.38}$$

where $-\varepsilon$ is the minimum value of $v_{LJ}(r)$ and r_m is the separation at which it is attained. Then they showed that the variational energy of the Lennard-Jones liquid is given by

$$E_{LJ} = E_{HS}(a_{LJ}) + (\rho/2) \int d\underline{r}\; g_{HS}(r/a_{LJ})w(r) \tag{4.39}$$

where E_{HS} and g_{HS} are the variational hard-sphere energy and correlation function and the hard-sphere radius a_{LJ} is equal to the scattering length of the repulsive part of the Lennard-Jones potential $[v_{LJ}(r) - w(r)]$, i.e., 0.8368σ.

f) Nonuniform Helium Systems

We have seen that bulk helium can be well understood by variational calculations. LIU et al. [4.42,43] have treated two nonuniform systems with the variational method. With nonuniform problems one must generalize the product wave function to include a one-body term since otherwise a purely repulsive pseudopotential will cause the particles to fill any volume.

In the "channel" problem hard walls are placed at z = 0 and z = L, with periodic boundary conditions in the x and y directions. LIU et al. [4.42] chose to model liquid helium with hard spheres, since they were primarily interested in the structure of the system. Their trial wave function was the liquid product function times a single-particle term to make the wave function vanish at the walls.

$$\psi_T(\underline{R}) = \exp\left[-\frac{1}{2}\sum_{i<j} u(r_{ij}) \prod_{i=1}^{N} h(z_i)\right] \tag{4.40}$$

where u(r) is given by (4.37) and

$$h(z) = \tanh^{q}[z(L - z)/Ld] \quad . \tag{4.41}$$

We now have two additional variational parameters, q and d, to be optimized. The surface energy was found to be $0.24\hbar^2/ma^4$ which is in rough agreement with experiment. The most striking features of the results are the pronounced density variations across the channel. The quantum liquid forms a layered structure much like a classical liquid would in a channel. The number of particles in the zero momentum state remains at 11%.

The second nonuniform system that LIU et al. studied [4.43] was a double-sided film of helium. Since in order for the film to be stable, the interatomic potential must be attractive, the potential was assumed to be Lennard-Jones with ε and σ given in Table 4.1. At zero temperature all of the particles are bound to the film, so the wave function must vanish if any atom leaves the film. Their trial function had the form of (4.40) but now u(r) is the McMillan pseudopotential, (4.32), and h(z) is

$$h(z) = \begin{cases} 1 & |z| < z_0 \\ 2/[1 + \exp \kappa(|z| - z_0)^2] \ , & |z| \geqq z_0 \end{cases} \tag{4.42}$$

z_0 and κ are optimized to find the lowest energy. The surface energy was found to be 0.21 K/Å^2, only 25% lower than the experimental value, 0.27 K/Å^2. The density oscillations are much smaller in the film than in the fluid but still apparent.

g) Two-Dimensional Helium

If helium atoms are tightly bound to a surface then one has essentially a two-dimensional system. The simplest assumption for the effect of the substrate is that the helium atoms are confined to a plane and the potential felt by the helium atoms due to the substrate is constant. The helium-helium interaction may be taken to be Lennard-Jones, (4.30). HYMAN [4.44] using the Monte Carlo variational technique and CAMPBELL and SCHICK [4.45] using molecular dynamics (both used the McMillan r^{-5} pseudopotential) have calculated the equation of state in the liquid phase at zero temperature of this 2-D system. The two calculations are in agreement giving an equilibrium (zero presure) density of $\rho = 0.23/\sigma^2$ with a binding energy E = 0.6 K. LIU et al. [4.46] have calculated the energy in the solid phase (by MC) and determined that the system melts at $\rho = 0.46/\sigma^2$ and freezes at $0.40/\sigma^2$ in rough agreement with experiments on thin helium films. They also found that 38% of the atoms were in the zero momentum state at the liquid equilibrium density. This is a much higher fraction of condensate at the equivalent density than in the three-dimensional liquid.

h) Three-Body Pseudopotential

An interesting calculation was done for this two-dimensional Lennard-Jones system by WOO and COLDWELL [4.47] who included a three-body pseudopotential in the trial function. That is they calculated the variational energy of the trial function

$$\psi_T(\underline{R}) = \prod_{i<j} \exp\left[-\frac{1}{2} u(r_{ij})\right] \prod_{k<\ell<m} \exp\left[-\frac{1}{2} w(r_{k\ell}, r_{\ell m}, r_{mk})\right] \tag{4.43}$$

where u(r) is the usual McMillan pseudopotential and w has the form

$$w(r, s, r - s) = \{c/[r^2 + s^2 + (r - s)^2]\}^{n/2} \quad . \tag{4.44}$$

This trial function will favor configurations of particles forming equilateral triangles. Thus the potential energy will be lowered, since particles will be in the potential wells of their nearest neighbors. Configurations were sampled with the usual two-particle trial functions and then the change in energy was calculated for a variety of values of c by 'biased selection'. They find an energy drop of 0.12 ± 0.06 K but the statistical errors seem rather large. Since the exact ground-state energy is not known in two dimensions it is difficult to know whether the form of their trial function is significantly better than McMillan's trial function.

CHANG and CAMPBELL [4.25] using the convolution approximation for the three-body correlation function have optimized the $w(r_{12}, r_{13}, r_{23})$ for liquid helium in three dimensions. The energy drops by -0.44 K at the equilibrium density and more at higher density when this three-body trial function is included (see also [4.48]). As we shall see, the ground-state energy is roughly 0.6 K lower than this. Either the convolution approximation is inaccurate or higher terms than the three-body function are necessary to describe liquid helium accurately. However, it is clear that the energy can be significantly lowered by relaxing the two-body approximation.

4.2.3 Other Bose Systems

a) Spin-Aligned Hydrogen

The ground-state energy for spin-aligned hydrogen and tritium has been calculated by DUGAN and ETTERS [4.49] using Monte Carlo with the liquid trial function. The potential was taken to be the Morse form

$$v(r) = \varepsilon\{\exp[2c(1 - r/r_m)] - 2\exp[c(1 - r/r_m)]\} \tag{4.45}$$

with $\varepsilon = 6.19$ K, $c = 6.05$ and $r_m = 4.15$ Å.
The pseudopotential which they found to have the lowest energy has the form

$$u(r) = b_1 \exp(-b_2 r) \quad . \tag{4.46}$$

The optimal coefficients were within 10% of those which satisfy the small r condition of (4.5). DUGAN and ETTERS found that at low pressures, spin-aligned hydrogen is nearly an ideal Bose gas and tritium a liquid, most likely superfluid. DANILOWICZ et al. [4.50] have calculated the phase transition between gas and solid aligned hydrogen to be at 80 atm with a volume of 55 cm^3/mole. At present there are no experimental values.

The atomic hydrogen potential has been fit to a Lennard-Jones 6-12 potential by STWALLY and NOSANOW [4.21]; cf. [4.51]. They have extended the moment functions $\Phi_n(b)$ for McMillan's trial functions (4.32) with m = 5 and fitted them with polynomials. Then the optimum variational energy as a function of density can easily be obtained from those moment functions. They get good agreement with Dugan for spin-aligned hydrogen.

A series of papers by NOSANOW et al. [4.52] has treated an entire range of Lennard-Jones potentials (i.e., the η,ρ^* plane) by assuming the pseudopotential was always $(b/r)^5$. Points not available from SCHIFF and VERLET [4.23] were evaluated by additional Monte Carlo calculations.

They have determined that for $\eta < 0.456$, the zero pressure phase of Lennard-Jones bosons is a liquid, and for $\eta > 0.456$ it is a gas. There is no 'coexistence' region in this hypothetical phase diagram; that is, there is no Lennard-Jones potential which has as its zero pressure ground-state liquid and gas in equilibrium. For fermions they believe the equilibrium situation is possible.

Finally the two quantum crystals, molecular hydrogen and neon have been investigated by Monte Carlo. HANSEN [4.53] found that solid neon is well described by the Hartree theory and he obtained very good agreement with the experimental ground-state energy using a Lennard-Jones potential. BRUCE [4.54] also approximated the potential between two hydrogen molecules by a Lennard-Jones (with the ε and σ in Table 4.). He argues that this approximation is not too unrealistic for solid hydrogen. He finds rough agreement with the experimental equation of state and determines that the harmonic approximation is good at high densities. For a comparison of these Monte Carlo results with those of lattice dynamics see the review by KOEHLER [4.55].

b) Soft-Core Bose Systems

We now turn to Monte Carlo results on soft-core systems. By soft core we mean here that the interparticle potential goes as $+1/r$ for small r. In contrast to hard-core systems, the liquid or gas phase is favored at high density. There are three important systems that have been studied: the Bose one-component plasma, boson neutron matter, and the related Yukawa potential.

c) Bose Neutron Matter Calculations

Since the late 1960s it has been speculated that the interior of neutron stars and in particular pulsars is composed of neutron matter which has crystallized. See the review of BAYM and PETHICK [4.56] for background on this problem. However, different methods gave widely varying answers for the equation of state and transition density and it was proposed that a common simplified potential be chosen so that the various many-body methods could be compared. Thus the 'homework' problem was to produce the equation of state and transition density of bosons interacting with the Yukawa potential.

$$v(r) = \varepsilon \exp(-r/\sigma)/(r/\sigma)$$
$$\text{with } \varepsilon = 45389 \text{ MeV and } \sigma = 0.204 \text{ fm} \quad . \tag{4.47}$$

COCHRAN and CHESTER [4.18,57] assumed the pseudopotential had the form of a Yukawa function

$$u(r) = A \exp(-Br)/r \quad . \tag{4.48}$$

Using the potential parameters $\varepsilon = 23472$ MeV and $\sigma = 0.244$ fm, they calculated with Monte Carlo the liquid equation of state. Their solid trial function was the usual gaussian times product function (4.4,34) but they found solid energies always above the liquid. They also considered the effect of an attractive tail to the interneutron potential but again the liquid was the favored phase. PANDHARIPANDE [4.58] showed that the HNC equation reproduced the Monte Carlo energies for the homework Yukawa potential within 5%.

The Monte Carlo results were extended to the full (η,ρ^*) plane of the Yukawa potential by CEPERLEY et al. [4.59] where η is given by (4.31) and $\rho^* = \rho\sigma^3$ is the reduced density but ε and σ are now parameters of the Yukawa potential in (4.47). They found that for $\eta \geqq 0.009$, no solid will form at any density. For $\eta \leqq 0.009$ the solid is the preferred phase for intermediate densities ($\rho^* \approx 0.07$), and gas the preferred phase at high and low densities. But the 'homework' potential has $\eta = 0.022$ so that a solid will never be the preferred phase.

These authors also investigated the Yukawa solid in some detail. Two different types of trial functions were used in addition to the standard gaussian form. The first was the periodic solid trial function

$$\psi_T(\underline{R}) = \exp\left[-\frac{1}{2}\sum_{i<j} u(r_{ij}) - \frac{1}{2}\sum_i \phi(\underline{r}_i)\right] \tag{4.49}$$

where $\phi(\underline{r})$ is a function with the periodicity of the lattice. The minimum variational energy for the periodic trial functions are higher than those of the gaussian, apparently because the periodic function allows a much greater chance of double occupancy of the lattice site.

To get a feeling for the effect of exchange in the Yukawa crystal a Monte Carlo simulation of the symmetrized gaussian trial function was carried out. That is

$$\psi_T(\underline{R}) = \exp\left[-\frac{1}{2}\sum_{i<j} u(r_{ij})\right]\sum_p \exp\left[-\frac{A}{2}(\underline{r}_i - \underline{s}_{p_i})^2\right] \tag{4.50}$$

where $\underline{s}_{p_i}$ are the lattice sites with a permutation P in the pairing of particles to lattice sites. To sample this function with Monte Carlo a Metropolis random walk was done both in coordinate and permutation space. Interested readers can find details in [4.59]. It was found that at the optimum value of A, very few permutations are allowed and the energy is unchanged from the unsymmetrized gaussian trial function. However, at small values of A the energy of the permanent function is slightly larger than that of the gaussian function and very many permutations can exist. We conclude that to calculate most properties of the Yukawa crystal it is not necessary to symmetrize the trial function.

It was also found that Lindemann's quantum melting law holds in the Yukawa crystal. This states that the solid is the ground-state as long as the rms deviation from a lattice site is less than 0.26 of the nearest neighbor distance. For classical system this ratio is about 1/7.

Finally, it was found that attempts by others [4.60,61] to model neutron matter from the hard-sphere or Lennard-Jones Monte Carlo results, as in (4.39), led to a serious overestimation of the kinetic energy and therefore the total energy of neutron matter at high densities.

d) Bose One-Component Plasma

The one-component plasma is the other important Boson soft-core system. The particles interact with a Coulomb potential and a uniform background neutralizes the total charge. This is a model for helium ions in the interior of cold stars. The ground-state properties depend only on one parameter r_s

$$E_F = \frac{\int d\underline{R}\ \psi_F^* H \psi_F}{\int dR|\psi_F|^2} = E_B + \frac{\hbar^2}{2m} \frac{\int d\underline{R}\ \psi_B^2 |\nabla D|^2}{\int d\underline{R}|\psi_F|^2} \quad . \tag{4.56}$$

E_B is the exact boson ground-state energy. The second term can be expanded using methods from classical cluster series. The first term of the expansion is simply the energy of an ideal Fermi gas, $3\hbar^2 k_F^2/10m$. The second term is an integral over the exact Bose structure function, the third term an integral over the exact Bose three-particle correlation function and so on.

If E_B is replaced by the minimum Bose variational energy and the exact correlation functions are replaced in the expansion by the corresponding variational ones, then one can evaluate the first three terms of the cluster expansion with the Bose Monte Carlo method. The resulting energy will most likely be an upper bound to the ground-state energy since the replacement of the exact Bose energy by the variational is the major error for a strongly coupled Fermi liquid. This approach has been followed by NOSANOW and PARISH [4.69] for spin-aligned deuterium; and by MILLER et al. [4.52] for the full range of Fermi liquids interacting with Lennard-Jones potentials.

However, this way of treating Fermi liquids is not really variational since ψ_B is fixed to be the wave function of the Bose liquid and in general one might expect to lower the energy by allowing ψ_B to be a different trial function. If one follows this approach, as suggested by SCHIFF and VERLET [4.23], then (4.56) no longer holds and one must add more terms to the cluster expansion. SCHIFF and VERLET used this new cluster expansion for helium three [4.23] obtaining apparently good convergence. HANSEN and SCHIFF [4.70] extended these results to mixtures of helium three and helium four. MONNIER [4.63] applied the same expansion to the electron gas, and HANSEN and MAZIGHI [4.65] have recently redone his calculations.

We have mentioned only those papers using the Wu-Feenberg expansions with Monte Carlo Bose energies and correlation functions. There has been a great number of others which used integral equations to find these quantities [4.71]. Within the Wu-Feenberg theory it is impossible to judge the errors resulting from the approximations to the correlation functions and from the truncation of the perturbation expansion.

Recently we have carried out the Metropolis random walk described above for the fully antisymmetric trial function [4.16]. The liquids studied to date are ^{3}He, neutron matter, Yukawa fermions and the electron gas in two and three dimensions [4.17]. The pseudopotentials used in the trial function were identical in form to those used in the corresponding boson liquid, that is $(b/r)^5$ for helium three, the Yukawa function (4.48) for neutron matter and the Gaskell [4.72] pseudopotential for the electron gas [which is closely related to the form in (4.54)]. The variation of the energy per particle with the size of the system is small for these liquids. The largest system simulated to date contained 162 particles, but that should not be regarded as a limit for the method.

a) Helium Three

Our simulation of liquid ^{3}He gives fair agreement with experiment (-1.3 K/particle as opposed to the experimental -2.5 K) and good agreement with the SCHIFF and VERLET permutation expansion [4.23] energies at the equilibrium density. The convergence of the expansion for the potential energy and kinetic energy separately is not nearly as good (a first-order error of 0.7 K). Also at a higher density, the Wu-Feenberg expansion underestimates the variational energy by 1.1 K. Shown in Fig. 4.1 are the spin dependent structure functions for liquid helium three at equilibrium density.

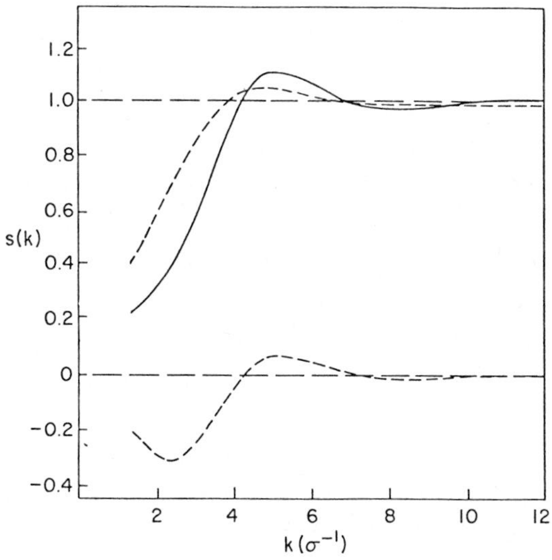

Fig.4.1 Structure functions S(k) (solid line), $S_L(k)$ (like spins upper dashed line), and $S(k)-S_L(k)$, (lower dashed line) for liquid He-3 at a density of $0.237/\sigma^3$

b) Neutron Matter

Simulations on neutron matter show that the Wu-Feenberg expansion underestimates the variational energy at all densities. Variational calculations have been carried out with the homework potential and two different Reid potentials. The fermion 'homework' system does not crystallize at any density. A recently developed integral equation called the Fermi HNC gives good agreement with these Monte Carlo results [4.73,74] at low density (< 0.3/fm^3). At higher density the three different forms for the kinetic energy give widely different results indicating that the approximation has broken down. Monte Carlo calculations have provided a crucial test of these approximations [4.74].

Figure 4.2 gives the momentum distribution. One can see the residual effect of the discontinuity at the Fermi surface caused by the determinant of the trial function. The interaction excites about 23% of the particles above the Fermi surface.

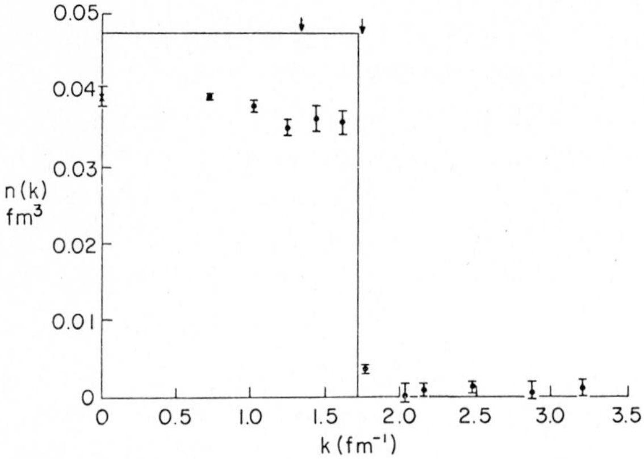

Fig.4.2 Momentum distribution, n(k) for neutron matter at a density of $0.17/fm^3$. The rectangle is the ideal Fermi gas distribution. The two arrows represent the rms values of k for the ideal gas and for this model of neutron matter

c) Yukawa Fermions

We have very roughly located the gas-solid phase boundary for a system of particles interacting with the Yukawa potential (4.47) and determined that if $\eta > 0.014$ the system is always in the gas phase. This critical value of η is very near that of the homework potential ($\eta_{HW} = 0.022$); hence it is not hard to see why other, more approximate methods, might predict that the homework system solidifies.

d) The Electron Gas

The electron gas is perhaps the most studied many-body problem. Many years ago WIGNER [4.75] predicted that the system would crystallize at low density, but a consensus of opinion on the transition density has not yet been reached [4.76]. The two-dimensional electron gas is also interesting since it appears that electrons on the surface of liquid helium are nearly a perfect realization of this system [4.77] and the Wigner transition may experimentally be seen there.

A series of Monte Carlo variational calculations for this system in two and three dimensions has been carried out [4.17].

The pseudopotentials used were of two types: Monnier's form (4.55) and GASKELL's random phase' pseudopotential [4.72]

$$u(r) = \frac{1}{\rho(2\pi)^d} \int d^dk \left[-1/S_0(k) + \sqrt{1/S_0(k)^2 + 4mv(k)/\hbar^2k^2}\,\right]\exp(i\underline{k}\cdot\underline{r}) \qquad (4.57)$$

where $S_0(k)$ is the ideal Fermi gas structure function and $v(k)$ is the Fourier transform of the interparticle potential. Gaskell's pseudopotential has energies as low as those using (4.53) but with the added advantage that there are no variational parameters and hence less computation. The best pseudopotential for the crystal had the form of (4.54).

Among the results for the electron gas [4.17] are the following:

I) The dependence of the energy on the size of the system is more important for the electron plasma since very small differences in energy are important. It was found [4.17] that this size dependence could be removed by interpolating between exact size dependence in the high and low density limits.

II) After the size dependence was corrected, the Monte Carlo correlation energies agree quite well with other calculations [4.78] in the metallic density range, $2 \leq r_s \leq 7$.

III) As BLOCH [4.79] predicted many years ago, using the Hartree Fock energies, the totally polarized liquid (all spins aligned) is the preferred phase at intermediate densities. In 3 dimensions the polarized-unpolarized transition should occur at $r_s = 26 \pm 5$ and in 2D at $r_s = 13 \pm 2$. This is at a substantially lower density than that predicted by other theories [4.80,81].

IV) The crystal energies are in good agreement with those calculated with the anharmonic crystal method [4.57] but substantially lower than those from the self-consistent phonon method [4.64].

V) The polarized liquid-crystal transition occurs at about $r_s = 67 \pm 5$ in three dimensions and about $r_s = 33 \pm 2$ in 2D. If the electrons are constrained to be unpolarized then the liquid-solid transition would occur at $r_s = 47 \pm 5$ in 3D and $r_s = 18 \pm 2$ in 2D. These variational estimates probably represent lower bounds to the true transition density since it is likely that the solid trial function is better than the liquid. For example, the 'exact' Bose Monte Carlo results for the Yukawa system [4.59] show the excess energy (above the perfect crystal energy) is overestimated 2% more in the liquid phase than in the crystal phase. If we shift the excess energy in the 3D electron plasma by the same percentage the polarized liquid-crystal transition goes from $r_s = 67$ to $r_s = 90$.

4.2.5 Monte Carlo Techniques for Low Temperature Excitations

Monte Carlo can be fruitfully used to calculate properties of systems which are close to the ground-state although this subject is largely unexplored.

As an example PADMORE and CHESTER [4.82] have calculated the energy of an excitation of a phonon or roton of wave vector $\underline{K}$ assuming the excited state wave function had the Feynman-Cohen form [4.83]

$$\psi_{\underline{K}}(\underline{R}) = \psi_0(\underline{R}) \sum_i \exp\left[i\underline{K}\cdot\left(\underline{r}_i + A \sum_j \underline{r}_{ij}/r_{ij}^3\right)\right] \quad .$$

Here $\psi_0(\underline{R})$ is the ground state wave function and $A(\underline{K})$ is a variational parameter. The energy of this trial function can be found from (4.56). PADMORE and CHESTER assumed $\psi_0(\underline{R})$ is approximated by the McMillan trial function and evaluated the integrals for the calculation of the energy by Monte Carlo. They were able to find the roton gap accurately in both 2 and 3 dimensions at several densities and reproduced within 20% the experimental excitation spectrum.

Two other examples of this sort of calculation can be cited. Recently SASLOW [4.84] computed an upper bound to the superfluid fraction in solid helium by a variational method suggested by LEGGETT [4.85]. The superfluid fraction in solid Yukawa bosons has been calculated in the same way, but using Monte Carlo results [4.59]. Unfortunately the bound computed this way is rather large (~ 15%) as compared with the experimental upper bounds (10^{-4}). Finally MEISSNER and HANSEN [4.86] have computed the sound velocity in solid neon, as a function of direction and density by using Monte Carlo.

4.3 Nearly Classical Systems

For systems which are almost classical, the WIGNER [4.87] $\hbar$ expansion can be used to correct classical results for quantum effects. For example, if A^c and $g^c(r)$ are the classical free energy and radial distribution function computed from Monte Carlo or molecular dynamics then the quantum free energy is

$$A/N = A^c/N + \rho\beta\hbar^2/24\ m \int d^3r\ g^c(r)\nabla^2 v(r) + O(\hbar^4) \quad . \tag{4.58}$$

The term in $\hbar^4$ involves averages over the 3 and 4 body classical distribution functions and can be evaluated with Monte Carlo [4.88]. This series seems to be quickly convergent for most fluids near their critical points. However, for very low temperatures and below the lambda transition in liquid helium it probably breaks down completeley. It is this purely quantum situation which we have been concerned with in this chapter. For a more complete discussion of the Wigner $\hbar$ expansion see [4.89].

4.4 The Green's Function Monte Carlo Method (GFMC)

a) Schrödinger's Equation in Integral Form

Consider the Schrödinger equation for a many-body system.

$$\left[-\sum_i \frac{\hbar^2}{2m}\nabla_i^2\,\psi(\underline{r}_i,\underline{r}_2,\ldots,\underline{r}_N) + V(\underline{r}_1,\ldots,\underline{r}_N)\right]\psi(\underline{r}_1,\underline{r}_2,\ldots,\underline{r}_N)$$

$$= E\psi(\underline{r}_1,\underline{r}_2,\ldots,\underline{r}_N) \qquad (4.59)$$

For convenience we set $\hbar^2/2m = 1$ and let $\underline{R}$ denote the point $(\underline{r}_1,\underline{r}_2,\ldots,\underline{r}_N)$ in 3N dimensional configuration space. Then (4.59) can be written succinctly as

$$[-\nabla^2\psi(\underline{R}) + V(\underline{R})]\psi(\underline{R}) = E\psi(\underline{R}) \quad . \qquad (4.60)$$

Suppose for the moment that

$$V(R) \geqq -V_0;$$

we shall note later the consequences of removing this restriction. Then

$$[-\nabla^2 + V(\underline{R}) + V_0]\psi(\underline{R}) = (E + V_0)\psi(\underline{R}) \quad . \qquad (4.61)$$

We seek an integral formulation of the Schrödinger equation. Accordingly, we consider Green's function for the operator on the left side of (4.61), namely

$$[-\nabla^2 + V(\underline{R}) + V_0]G(\underline{R},\underline{R}_0) = \delta(\underline{R} - \underline{R}_0) \quad . \qquad (4.62)$$

Appropriate boundary conditions for the problem must be contained in G. For the treatment of an isolated system of interacting particles, $G(\underline{R},\underline{R}_0)$ vanishes as the separation of any pair of particles $|\underline{r}_i - \underline{r}_j|$ increases without limit. On the other hand in calculations modelling a large system by the use of a finite number of particles in a box of side L, ψ and G must be multiply periodic in the sense that

$$\psi(\underline{R} + \underline{P}_j) = \psi(\underline{R})$$

$$G(\underline{R} + \underline{P}_j;\underline{R}_0) = G(\underline{R};\underline{R}_0)$$

and $\underline{P}_j$ is a vector all of whose 3N components are zero except the jth which is L. Finally in treating a system with hard-sphere forces,

$$\psi(R) = 0 \quad \text{if} \quad |\underline{r}_i - \underline{r}_j| \leq a, \quad \text{for any } i \neq j;$$

a is the hard-sphere diameter. In this case we require that $G(\underline{R},\underline{R}_0)$ which also vanishes when hard spheres overlap.

Substitution of the Green's function of (4.62) in (4.61) yields an integral equation

$$\psi(\underline{R}) = (E + V_0) \int G(\underline{R},\underline{R}')\psi(\underline{R}')d\underline{R}' \quad . \tag{4.63}$$

Let a succession of functions be defined for some initial $\psi^{(0)}(\underline{R})$ by

$$\psi^{(n+1)}(\underline{R}) = (E + V_0) \int G(\underline{R},\underline{R}')\psi^{(n)}(\underline{R}')d\underline{R}' \quad . \tag{4.64}$$

When the spectrum of the hamiltonian is discrete near the ground state $\psi_0(\underline{R})$ of the Schrödinger equation (4.59), then $\psi_0(\underline{R})$ is the limiting value of $\psi^{(n)}(\underline{R})$ for large n. It is possible to devise a Monte Carlo method — in the general sense of a random sampling algorithm — which produces populations drawn in turn from the successive $\psi^{(n)}$. Suppose that a set of configurations $\{R^{(0)}\}$ is drawn at random from the given function $\psi^{(0)}(\underline{R})$. Then for each configuration $\underline{R}_k^{(0)}$, let new configurations $\underline{R}_\ell^{(1)}$ be selected at random from the density function $(E+V_0)G\left(\underline{R}_\ell^{(1)}, \underline{R}_k^{(0)}\right)$ conditional on $\underline{R}_k^{(0)}$. Note that the number of configurations is not conserved. The expected number of configurations appearing in a unit neighborhood of R, averaged over all possible $\underline{R}_k^{(0)}$ is

$$(E + V_0) \int G(\underline{R},\underline{R}')\psi^{(0)}(\underline{R}')d\underline{R}' = \psi^{(1)}(\underline{R}) \quad , \tag{4.65}$$

which is identical with (4.64) for n = 1. Thus the sampling of $G\left(\underline{R}_\ell^{(1)},\underline{R}_k^{(0)}\right)$ produces a population of configurations $\{\underline{R}^{(1)}\}$ drawn from $\psi^{(1)}$. Clearly, repetition leads to a population whose density is $\psi^{(2)}$ and further iteration to samples drawn from $\psi^{(n)}$ for any n. We call the population drawn from $\psi^{(n)}$ the nth "generation".

Unfortunately, E_0, the exact eigenvalue of the ground state, is not known in advance. Equation (4.64) requires the correct eigenvalue to yield $\psi^{(n)}$ asymptotically constant. On the other hand it is equally clear from (4.64) that if one uses, for computational purposes, a trial eigenvalue E_t larger than E_0, the $\psi^{(n)}$ will grow in normalization reflected in a growth in the population of configurations. If E_t is too small the population declines. Thus the distribution of configurations has the correct marginal distribution and, in addition, the eigenvalue E may be estimated from the change in population size

$$E + V_0 = \frac{(E_t + V_0) \int \psi^{(n)}(\underline{R})d\underline{R}}{\int \psi^{(n+1)}(\underline{R})d\underline{R}} \quad . \tag{4.66}$$

A Monte Carlo estimator for E_0 is then $\bar{E}_1$ given by

$$\bar{E}_1 + V_0 \cong (E_t + V_0)N_n/N_{n+1} \tag{4.67}$$

where N_n is the number of configurations in the population drawn from $\psi^{(n)}$ according to (4.64) and the symbol $\cong$ indicates that the estimates is biased. A source of bias is a consequence of the asymptotic character of the result. Another expresses the fact that the expected value of the quotient of two integrals is not the quotient of the expected values. An estimator, $\bar{E}_2$ for which the second kind of bias is smaller is

$$(\bar{E}_2 + V_0) \cong (E_T + V_0) \sum_{k=n_1}^{n_2} N_k \Bigg/ \left(\sum_{k=n_1}^{n_2} N_{k+1}\right) \quad . \tag{4.68}$$

In any case the magnitude of the bias must be evaluated or bounded in obtaining practical results. Including the size of more generations decreases the bias as a consequence of the increased statistical correlation and reduced fluctuation of numerator and denominator.

The second bias can be eliminated completely and the first frequently reduced in magnitude by the following method. Multiply (4.60) by some known $\psi_T(\underline{R})$ which satisfies the boundary conditions. It has been found effective to use a trial function which minimized the energy in a variational study of the same problem. Upon integrating over the full space of $\underline{R}$ and using the Hermiticity of the hamiltonian, one obtains

$$E = \frac{\int \psi_0(\underline{R})[-\nabla^2 + V(\underline{R})]\psi_T(\underline{R})d\underline{R}}{\int \psi_0(\underline{R})\psi_T(\underline{R})d\underline{R}} \quad . \tag{4.69}$$

If one samples a population $\underline{R}_i,\ldots,\underline{R}_N$ from the density function $\psi_0(\underline{R})\psi_T(\underline{R})$ then

$$\bar{E}_3 \cong \frac{1}{n}\sum_{k=1}^{n} \psi_T(\underline{R}_k)^{-1}(-\nabla^2 + V)\psi_T(\underline{R}_k) \tag{4.70}$$

is an unbiased estimator of E. It has the property that as ψ_T approaches ψ_0, $\bar{E}_3$ becomes exact, independent of the values of $\underline{R}_k$. Thus one expects that for "reasonable" ψ_T, the estimator $\bar{E}_3$ will be less biased than $\bar{E}_2$ before the convergence of $\psi^{(n)}$ to $\psi^{(\infty)}$ and it may well have less variance. We will discuss practical results below. Estimator $\bar{E}_2$ is called a "growth" estimate, $\bar{E}_3$ a "variational" estimate.

The iteration process (4.64) does not converge to $\psi_T\psi_0$ but, formally at least, it is easy to change it so that it will do so. Simply multiply (4.64), through by $\psi_T(R)$ and introduce $\psi_T(\underline{R}')/\psi_T(\underline{R}')$ into the integral. The sequence

$$\psi_T(\underline{R})\psi^{(n+1)}(\underline{R}) = (E_t + V_0) \int [\psi_T(\underline{R})G(\underline{R},\underline{R}')/\psi_T(\underline{R}')] \times \psi_T(\underline{R}')\psi^{(n)}(\underline{R}')d\underline{R}' \tag{4.71}$$

can be sampled randomly as before and converges to $\psi_T\psi_0$ where ψ_0 is the lowest state not orthogonal to ψ_T. The estimator (4.68) may also be used with N_k referring to the size of the population generated with the bias ψ_T. Clearly this is a practical thing to do when the sampling is carried out according to (4.71). We shall see below that this is theoretically advantageous.

We now consider the question of treating infinite attractive potentials such as occur in a two-component plasma. Suppose one writes

$$V(\underline{R}) = V_+(\underline{R}) - V_-(\underline{R}) \tag{4.72}$$

and with V_- unbounded from above, and V_+ has lower bound $-V_0$. Then (4.61) becomes

$$[-\nabla^2 + V_+(\underline{R}) + V_0]\psi(\underline{R}) = V_-(\underline{R})\psi(\underline{R}) + (E + V_0)\psi(\underline{R}) \quad . \tag{4.73}$$

Using Green's function for the operator on the left, we have

$$\psi(\underline{R}) = \int G(\underline{R},\underline{R}')V_-(\underline{R}')\psi(\underline{R}')d\underline{R}' + (E + V_0) \int G(\underline{R},\underline{R}')\psi(\underline{R}')d\underline{R}' \tag{4.74}$$

or

$$V_-(\underline{R})\psi(\underline{R}) = \int V_-(\underline{R})G(\underline{R},\underline{R}')V_-(\underline{R}')\psi(\underline{R}')d\underline{R}' + (E + V_0) \int V_-(\underline{R})G(\underline{R},\underline{R}')V_-^{-1}(\underline{R}')V_-(\underline{R}')\psi(\underline{R}')d\underline{R}' \tag{4.75}$$

an integral equation for the new dependent variable $V_-(\underline{R})\psi(\underline{R})$ which must be integrable. In some applications it is best to arrange the decomposition so that V_0 is zero and transpose $E\psi$ to the left to be incorporated into Green's function.

b) Sampling Green's Function by Random Walks

The development of the preceding section rests upon a crucial technical assertion, that it is possible to sample $(E_t + V_0)G(\underline{R},\underline{R}')$ or $(E_t + V_0)\psi_T(R)G(\underline{R},\underline{R}')/\psi_T(\underline{R}')$ for $\underline{R}$ conditional on $\underline{R}'$. Now for some simple problems - e.g., a particle in a box - one can construct the analytic form of G and hence an algorithm for sampling it. For interesting problems of statistical physics this is not possible, but we now sketch how it is possible to sample $G(\underline{R},\underline{R}')$ without knowing it explicitly. This algorithm is, of course, the heart of the method.

Let D be the full domain in configuration space in which the particles move. Consider some $D_0(\underline{R}_0) \subset D$ and suppose that

$$U_0 \geqq V(\underline{R}) + V_0 \quad \text{for} \quad \underline{R} \in D_0 \quad . \tag{4.76}$$

We introduce the Green's function on the domain D_0, a "partial" Green's function

$$(-\nabla^2 + U_0)G_U(\underline{R}_1,\underline{R}_0) = \delta(\underline{R}_1 - \underline{R}_0) \quad \text{for} \quad \underline{R}_1,\underline{R}_0 \notin D_0 \tag{4.77}$$

with the boundary condition

$$G_U(\underline{R}_1,\underline{R}_0) = 0 \quad \text{for} \quad R_1,R_0 \notin D_0 \quad . \tag{4.78}$$

In principle U_0 could be a function of $\underline{R}$, but in practice only constant values have been used. The Green's functions introduced here are symmetric because of the boundary conditions. Equation (4.62) can be rewritten for a source at $\underline{R}_1$ as

$$[-\nabla_1^2 + V(\underline{R}_1) + V_0]G(\underline{R},\underline{R}_1) = \delta(\underline{R} - \underline{R}_1) \quad . \tag{4.79}$$

On multiplying (4.77) by $G(\underline{R},\underline{R}_1)$, (4.79) by $G_U(\underline{R}_1,\underline{R}_0)$, subtracting and integrating, one finds

$$\begin{aligned} G(\underline{R},\underline{R}_0) = G_U(\underline{R},\underline{R}_0) &+ \int_{\partial D_0(\underline{R}_0)} [-\nabla_{n_1} G_U(\underline{R}_1,\underline{R}_0)]G(\underline{R},\underline{R}_1)d\underline{R}_1 \\ &+ \int_{D_0(\underline{R}_0)} \{[U_0 - V(\underline{R}_1) - V_0]/U_0\}U_0 G_U(\underline{R}_1,\underline{R}_0)G(\underline{R},\underline{R}_1)d\underline{R}_1 \quad . \end{aligned} \tag{4.80}$$

This last equation shows how the full Green's function is related to a "partial" Green's function satisfying (4.76-78) on a subdomain. This relation may be understood in the following way. First observe that if (4.77) is integrated over $D_0(\underline{R}_0)$, we have

$$\int_{\partial D_0(\underline{R}_0)} [-\nabla_{n'} G_U(\underline{R}',\underline{R}_0)] d\underline{R}' + U_0 \int_{D_0(R_0)} G_U(\underline{R}',\underline{R}_0) d\underline{R}' = 1 \quad . \tag{4.81}$$

Since $G_U(\underline{R}',\underline{R}_0)$ and therefore $-\nabla_{n'} G_U(\underline{R}',\underline{R}_0)$ are non-negative, they may be interpreted as probability density functions for a move in a random walk which may go either to some $R' \in D_0(\underline{R}_0)$ or to $\underline{R}' \in \partial D_0(\underline{R}_0)$, respectively. Equation (4.80) may be related to expectations of random processes of this kind. In particular it expresses the fact that $G(\underline{R},\underline{R}_0)$ is the sum of $G_U(\underline{R},\underline{R}_0)$ for the subdomain plus the expected value of $G(\underline{R},\underline{R}_1)$ taken over an ensemble of random events. These events comprise one of the following two possibilities: I) a move at random with density $-\nabla_{n_1} G_U(\underline{R}_1,\underline{R}_0)$ to $\underline{R}_1$ on the boundary $\partial D_0(\underline{R}_0)$; or II) a move at random with density $U_0 G(\underline{R}_1,\underline{R}_0)$ to $\underline{R}_1$ on the interior of $D_0(\underline{R}_0)$ with the averaging of $G(\underline{R},\underline{R}_1)$ carried out with probability $[U_0 - V(\underline{R}_1) - V_0]/U_0$. It is true that in either case $G(\underline{R},\underline{R}_1)$ is not known but we may now construct a domain $D_0(\underline{R}_1)$ containing $\underline{R}_1$. Then $G(\underline{R},\underline{R}_1)$ may be expressed as in (4.80) as the sum of $G_U(\underline{R},\underline{R}_1)$ plus the average of $G(\underline{R},\underline{R}_2)$ taken over points $\underline{R}_2$ reached by random steps to the boundary or interior of $D_0(\underline{R}_1)$. The process may be iterated leading to a sequence of $\underline{R}_n$ with each chosen from its predecessor $\underline{R}_{n-1}$ by a move to $\underline{R}_n$ on the boundary of $D_0(\underline{R}_{n-1})$ drawn from $-\nabla_n G_U(\underline{R}_n,\underline{R}_{n-1})$ or to $\underline{R}_n$ on the interior of $D_0(\underline{R}_{n-1})$ drawn from $G(\underline{R}_n,\underline{R}_{n-1})$ and propagated with probability $[U_0-V(\underline{R}_n)-V_0]/U_0$. $G(\underline{R},\underline{R}_0)$ is the expected value of the sum of all the $G_U(\underline{R},\underline{R}_n)$ for the $\underline{R}_n$ that are generated in this random walk. Each of these terms potentially makes a contribution to the next "generation" as in (4.64). That this procedure yields $G(\underline{R},\underline{R}_0)$ can be proved formally, but we justify it here by a "physical" argument. Equation (4.62) has as its solution the expected density for observing at $\underline{R}$ an object which was started at $\underline{R}_0$ and which diffuses subject to an absorption process whose rate is $V + V_0$. Now at any stage of the diffusion process a domain $D_0(R) \subset D$ can be constructed. Green's function defined in (4.77,78) describes a diffusion in $D_0(\underline{R})$ subject to an absorption rate that is too large in the domain - cf. (4.76) - and to perfect absorption at the boundary. Diffusion continues until the first passage across that boundary. To describe the diffusion in the full domain such a first passage merely defines a surface source for subsequent diffusion. In addition the excess absorption owing to U_0 is compensated by the reintroduction of a fraction $\{[U_0-V(R)-V_0]/U_0\}$ of those objects absorbed in the interior.

The details of the construction of the $D_0(\underline{R}')$ and the estimation of the upper bound $U_0(\underline{R}')$ are rather technical and depend upon the problem at hand. It has been found useful to construct $D_0(\underline{R})$ as a cartesian product of subspaces, one for each particle. Taking each subspace to be a sphere is especially convenient. Then separation of variables permits the explicit calculation of G_U as a product and this, in turn, leads to an explicit algorithm for sampling G_U.

The treatment of [4.28] does not include the absorption U_0, but the generalization is trivial. Let

$$(-\nabla^2 + \frac{\partial}{\partial t})G_0(\underline{R},\underline{R}',t) = \delta(\underline{R} - \underline{R}')$$

$$(-\nabla^2 + U_0 + \frac{\partial}{\partial t})G_U(\underline{R},\underline{R}',t) = \delta(\underline{R} - \underline{R}') \tag{4.82}$$

with G_0 and G_U vanishing outside $D(\underline{R}')$ as before. Then

$$G_U(\underline{R},\underline{R}') = \int_0^\infty G_U(\underline{R},\underline{R}',t)dt = \int_0^\infty \exp(-U_0 t)G_0(\underline{R},\underline{R}',t)dt \quad . \tag{4.83}$$

By sampling t from $U_0 \exp(-U_0 t)$ as well as the times described (3.11) in [4.28] the sampling of G_U is accomplished.

c) Importance Sampling

The sampling algorithm embodied in (4.80) yields a population drawn from $G(\underline{R},\underline{R}_0)$ for any $\underline{R}_0$. It must be altered to sample $\psi_T(\underline{R})G(\underline{R},\underline{R}_0)/\psi_T(\underline{R}_0)$ as indicated in (4.71). Again, formally this presents no particular problem. Equation (4.80) can be transformed by multiplying by $\psi_T(\underline{R})$ to produce

$$\begin{aligned}\psi_T(\underline{R})G(\underline{R},\underline{R}_0)/\psi_T(\underline{R}_0) &= \psi_T(\underline{R})G_U(\underline{R},\underline{R}_0)/\psi_T(\underline{R}_0)\\ &+ \int_{\partial D_0} [-\psi_T(\underline{R}_1)G_U(\underline{R}_1,\underline{R}_0)/\psi_T(\underline{R}_0)]\psi_T(\underline{R})G(\underline{R},\underline{R}_1)/\psi_T(\underline{R}_1)d\underline{R}_1\\ &+ \int_{D_0} \{1-[V(\underline{R}_1) + V_0]/U_0\}U_0\psi_T(\underline{R}_1)G_U(\underline{R}_1,\underline{R}_0)/\psi_T(\underline{R}_0)\\ &\times \psi_T(\underline{R})G(\underline{R},\underline{R}_1)/\psi_T(\underline{R}_1)d\underline{R}_1\end{aligned} \tag{4.84}$$

to produce an integral equation for $\psi_T(\underline{R})G(\underline{R},\underline{R}_0)/\psi_T(\underline{R}_0)$. The new equation can also be sampled by a random walk in which the kernels which describe the passage from $\underline{R}'$ to $\underline{R}$ are modified by factors $\psi_T(\underline{R}')/\psi_T(\underline{R})$ - cf. the discussion following (4.80). We now show that the use of ψ_T is of very great potential in increasing the efficiency of the method.

To see how this may be possible, we define $E_T(\underline{R})$ by

$$(-\nabla^2 + V + V_0)\psi_T(\underline{R}) = [E_T(\underline{R}) + V_0]\psi_T(\underline{R}) \quad . \tag{4.85}$$

Combining this with (4.77) which defines Green's function G_U, we obtain

$$\int_{\partial D_0(\underline{R}_0)} \{\psi_T(\underline{R}')[-\nabla_{n'}G_U(\underline{R}',\underline{R}_0)]/\psi_T(\underline{R}_0)\}d\underline{R}'$$

$$+ \int_{D_0(\underline{R}_0)} \{[U_0-V(\underline{R}')-V_0]\psi_T(\underline{R}')G_U(\underline{R}',\underline{R}_0)/\psi_T(\underline{R}_0)\}d\underline{R}' \qquad (4.86)$$

$$= 1 - \int_{D_0(\underline{R}_0)} \{[E_T(\underline{R}') + V_0]\psi_T(\underline{R}')G_U(\underline{R}',\underline{R}_0)/\psi_T(\underline{R}_0)\}d\underline{R}'$$

Comparing with (4.80) we see that the first integral on the left is the expected number of steps to the boundary of $D_0(\underline{R}_0)$ which will be made when such steps are generated with density $\psi_T(\underline{R}')[-\nabla_{n'}G_U(\underline{R}',\underline{R}_0)]/\psi_T(\underline{R}_0)$ in a random walk which generates $\psi_T(\underline{R})G(\underline{R},\underline{R}_0)/\psi_T(\underline{R}_0)$. The second integral on the left gives the expected number of corresponding steps to the interior of $D_0(\underline{R}_0)$.

When $E_T(\underline{R}) + V_0 \geqq 0$ (a weak condition since $V(\underline{R}) \geqq V_0$), the right side cannot exceed one. Thus the expected number of steps to boundary or interior of $D_0(\underline{R}_0)$ made in the Green's function random walk is generally less than one and the random walk terminates at some stage.

Now we recall that the nth step in the random walk makes a contribution of $\psi_T(\underline{R})G_U(\underline{R},\underline{R}_n)/\psi_T(\underline{R}_n)$ to the full weighted Green's function $\psi_T(\underline{R})G(\underline{R},\underline{R}_0)/\psi_T(\underline{R}_0)$ and that each such partial contribution potentially contributes to the next generation of configurations. If $E_T(\underline{R})$ is replaced by E_t, then the integral on the right side of (4.86) is exactly the expected number of configurations in the next generation which results from the contribution of $G_U(\underline{R},\underline{R}_0)$ for domain $D_0(\underline{R}_0)$ to the full $G(\underline{R},\underline{R}_0)$. Now if ψ_T were ψ_0, the lowest eigenfunction, then $E_T = E_0 > V_0$, and we have the result that three possible events have probabilities that add up to one; the events are a move to the boundary, a move to the interior so as to continue the Green's function random walk, and the event that produces a configuration in the next generation. All three may then be sampled as mutually exclusive and exhaustive events. Thus one may arrange the algorithm so that the random walk terminates when and only when a next generation configuration is produced. Under these circumstances, viz., $\psi_T = \psi_0$ and $E_t = E_0$, the random walk produces exactly one new configuration and is guaranteed to terminate.

In addition, if (4.85) is combined with (4.62) defining the full Green's function, and the periodic and any other boundary conditions for ψ_T are used, the result is

$$\int \{[E_T(\underline{R}) + V_0]\psi_T(\underline{R})G(\underline{R},\underline{R}_0)/\psi_T(\underline{R}_0)\}d\underline{R} = 1 \qquad (4.87)$$

Again, if $\psi_T(\underline{R}) = \psi_0(\underline{R})$ and $E_T(\underline{R}\ ) = E_0$,

$$\left|\int [\psi_0(\underline{R})G(\underline{R},\underline{R}_0)/\psi_0(\underline{R}_0)]d\underline{R}\right|^{-1} = E_0 + V_0 \qquad (4.88)$$

The integral on the left in (4.88) is simply the expected size of the total population, say N_2, which results in the next generation after one configuration at $\underline{R}_0(N_1 = 1)$ as in (4.68). With $E_t + V_0 = 1$ in (4.68) we see that the energy estimate from growth of generations is identically E_0, independent of $\underline{R}_0$ or any distribution used in sampling $\underline{R}_0$. The estimate of E then has zero variance, as does (4.69). Of course this ideal result requires knowing ψ_0, but we expect that "reasonable" ψ_T, such as those which prove useful in variational calculations will reduce the variance significantly. Much experience has borne this out although it is not necessarily true that the ψ_T in some class of trial functions which minimizes the energy also minimizes the variance. An important example will be discussed below.

d) Quantum Mechanical Expectations

In the preceding sections we have shown how the energy of the ground state can be estimated efficiently. But other expectations are also of considerable interest. These have the form

$$<F> = \frac{\int \psi^*(\underline{R}) F \psi(\underline{R}) d\underline{R}}{\int |\psi(\underline{R})|^2 d\underline{R}} \tag{4.89}$$

$$= \frac{\int \psi_T \psi f(\underline{R}) \psi/\psi_T d\underline{R}}{\int (\psi_T\psi)(\psi/\psi_T) d\underline{R}} \quad . \tag{4.90}$$

The second of these assumes a real eigenfunction, that F is merely multiplication by $f(\underline{R})$ over some domain and casts the result in the form suitable for evaluation by Monte Carlo given a population of configurations drawn from $\psi_T(\underline{R})\psi(\underline{R})$. The extra factor or "weight" $\psi(\underline{R})/\psi_T(\underline{R})$ must be included. Now the completeness of the eigenfunctions of the hamiltonian implies

$$\sum_k \psi_T(\underline{R})\psi_k(\underline{R})\psi_T^{-1}(\underline{R}')\psi_k(\underline{R}') = \delta(\underline{R} - \underline{R}') \quad . \tag{4.91}$$

If one uses $\delta(\underline{R} - \underline{R}') = \psi_T(\underline{R})\psi^{(0)}(\underline{R})$ in the iteration (4.71) one sees easily [4.90] that the coefficient of the asymptotic value of $\psi_T\psi^{(n)}$ contains, aside from constants, the factors

$$\psi_T(\underline{R})\psi_0(\underline{R})\psi_T^{-1}(\underline{R}')\psi_0(\underline{R}')$$

so that the asymptotic generation size, conditional upon $\underset{\sim}{R}'$, is $\psi_0(\underset{\sim}{R}')/\psi_T(\underset{\sim}{R}')$. Thus for each configuration, say $\underset{\sim}{R}_k$, drawn from $\psi_T(\underset{\sim}{R}_k)\psi(\underset{\sim}{R}_k)$, further sampling can, in principle, yield statistically independent estimates of the weight to be given to $\underset{\sim}{R}_k$.

Unfortunately, although importance sampling significantly accelerates this process, the computations have had to be carried to substantially greater length for reasonable estimates of quantities other than the energy by this method. That is, the variance of the weight is large for large numbers of generations.

If one assumes that $\psi_T(\underset{\sim}{R})$ is close to $\psi_0(\underset{\sim}{R})$, then a convenient estimate from a perturbation theory is possible. Let us define a "mixed expectation" as

$$<F>_M = \frac{\int \psi_0\psi_T\cdot\psi_T^{-1}F\psi_T d\underset{\sim}{R}}{\int \psi_T\psi_0 d\underset{\sim}{R}} = \frac{\int \psi_0 F\psi_T d\underset{\sim}{R}}{\int \psi_T\psi_0 d\underset{\sim}{R}} \tag{4.92}$$

and write

$$\psi_0(\underset{\sim}{R}) = \psi_T(\underset{\sim}{R}) + \varepsilon\phi(\underset{\sim}{R}) \tag{4.93}$$

$$\begin{aligned} <F>_M &= \frac{\int \psi_T F\psi_T d\underset{\sim}{R}}{\int \psi_T^2 d\underset{\sim}{R}} + \frac{\varepsilon}{\int \psi_T^2 d\underset{\sim}{R}}\left|\int \phi F\psi_T d\underset{\sim}{R} - \frac{\int \phi\psi_T d\underset{\sim}{R} \int \psi_T F\psi d\underset{\sim}{R}}{\int \psi_T^2 d\underset{\sim}{R}}\right| + O(\varepsilon^2) \\ &\equiv <F>_T + \varepsilon F_1 + O(\varepsilon^2) \end{aligned} \tag{4.94}$$

where $<F>_T$ is the purely variational estimate. A similar calculation yields for the true ground-state expectation

$$<F>_0 = <F>_T + 2\varepsilon F_1 + O(\varepsilon^2)$$

from which we derive an "extrapolated estimate"

$$<F>_X = 2<F>_M - <F>_T = <F>_0 + O(\varepsilon^2) \quad . \tag{4.95}$$

This method applies as well to the calculation of off-diagonal matrix elements (cf. Sect.4.4.1 below) and therefore to the treatment of condensate fractions and momentum distributions.

This estimate has been tested by comparing with values estimated from the asymptotic method, based on (4.91) and, more convincingly in certain cases, by carrying out calculations with different ψ_T and finding consistent extrapolated estimates for expectations. Where the "extrapolated estimate" is unambiguous, it is preferable to the asymptotic results since it requires substantially less computation.

e) Implementation

The sampling of G_U [except for the extra step of sampling $U_0 \exp(-U_0 t)$] in cartesian product spaces is discussed in [4.28]. In particular it is shown that if each D_0 is taken as a product of one sphere for each coordinate then the algorithm reduces to the joint sampling of Green's functions for three dimensional spheres. The calculations of [4.28] were concerned with hard spheres only; the sphere radii were set simply by the condition that any pair of particles moving in or to the surfaces of their spheres could not approach closer than the hard-sphere diameter.

For calculations with a continuous potential, at least in part repulsive, the problem is somewhat more complicated. Although, in principle, any choice of domains $D_0(\underline{R})$ that can cover D may be used, the choice of size of domain and corresponding upper bound U_0 can strongly affect the efficiency. In treating potentials of the Yukawa type [4.59,91] the dominating factor is the r^{-1} singularity of the potential. There it was found reasonable to set $U_0 = \zeta V(\underline{R}_0)$ where ζ is a constant greater than one and to tabulate radii (as a function of nearest neighbor separation) such that moves in such spheres would not permit V(R) to exceed U_0. Some experiments showed the computational efficiency not sensitive to ζ in the neighborhood of $\zeta = 2$.

In later computations of the Lennard-Jones potential, the method was somewhat more involved; U_0 was set to $V(\underline{R}_0)+ V_1$, V_1 a constant determined from computer experiments. The average computing time required to sample completely a single iterate of the homogeneous (4.64) depends upon constants such as ζ and V_1 although average answers do not. These parameters may be varied in short auxiliary calculations to determine values which approximately minimize the computer time. In addition the value of V_0 was set to be about twice the minimum potential observed in practise, rather than a rigorous lower bound, with no complications arising.

Some discussion is worth giving here on the effect of the importance function upon the sampling of moves, that is, sampling $\psi_T(\underline{R})G_U(\underline{R},\underline{R}_0)/\psi_T(\underline{R}_0)$ rather than $G_U(\underline{R},\underline{R}_0)$. If one expands $\psi_T(\underline{R})$ about $\underline{R}_0$ through the first term, one has

$$\psi_T(\underline{R})/\psi_T(\underline{R}_0) = 1 + (\underline{R} - \underline{R}_0) \cdot \nabla\psi_T(\underline{R}_0)/\psi_T(\underline{R}_0) \quad . \tag{4.96}$$

The unmodified Green's function G_U is isotropic for each sphere. Thus the effect of (4.96) upon the marginal distribution of all variates except direction is nil. But the gradient of the log of ψ_T indicates a relative preference for some directions over others, e.g., a preference for a close pair to move apart.

In later evolutions of the method the practical effect of (4.86) on the sampling was recognized: for sampling steps which generate G from G_U, left side of (4.86), the factor $[E_T(\underline{R}) + V_0]$ on the right acts as an effective "absorption" rate in decreasing the probability that the random walk continues when it is at $\underline{R}_0$.

Finally we note that the periodicity of the Green's function and of the wave function are ensured by the usual computational device [4.92] of moving by ± L any particle which leaves the domain at ∓ L/2. In this context this corresponds to using an infinite series of images (in the sense of potential theory) of the point $\underline{R}_0$.

We find that convergence to the ground state takes very roughly 50 generations starting from a population of configurations sampled from $|\psi_T(R)|^2$. The convergence does depend upon the accuracy of ψ_T to some extent and care must be taken to assure that convergence is complete. Indeed the principle complication and consumer of computer time is the necessity for computing accurate eigenvalues and expectations well beyond the point of apparent convergence to be sure of convergence to the required precision. With programs improved to the current state 64 body problems with Lennard-Jones forces require about 30 hours of time on a CDC 6600 to converge and give an average energy of better than 1%. For Yukawa forces the time is about a factor of 10 less.

4.4.1 Results

Early applications of the GFMC method included very simple few-body nuclear problems [4.93] and the helium atom [4.94]. A calculation [4.95] of 32 particles with Lennard-Jones forces of the de Boer-Michels type was somewhat more interesting. It used no importance sampling and was consequently very inefficient; it was the failure of that particular program to give anything reasonable for a system of 256 particles that gave the impetus to the development of importance sampling. But even with 32 particles, the result—later amply confirmed—that the energy of an ensemble of bosons with Lennard-Jones forces is substantially deeper than found variationally was first obtained.

In accord with the insightful suggestion of VERLET, the GFMC method was next applied to a hard-sphere quantum system, and the methodology considerably enhanced in the process.

The results [4.28] for the energy were somewhat (3-5%) deeper than found variationally, [4.40] and the radial distribution somewhat more structured. The improvement of agreement with experimentally measured structure functions was striking. Crystal calculations were made using ψ_T for the form given by (4.4). With the help of the perturbation theory also developed in [4.28], the energy for a Lennard-Jones system was estimated. A minimum of about -6.8 ± 0.2 K was obtained at a density of ρ = 1.0 ± 0.1 of the experimental density. The observed energy is -7.14 K.

The study of an inhomogeneous system of hard-sphere bosons was undertaken by LIU et al. [4.90]. Their system was periodic in two directions, but boundary conditions appropriate to hard parallel walls were applied in the third. This problem had also been studied using variational methods [4.42], where in the interior of the system a weakly layered density profile was found. These layers were strikingly enhanced in the GFMC results, as shown in Fig.4.3. In addition the interior structure was found to depend sensitively upon the width of the channel. New layers appear whenever the channel becomes wide enough to accommodate another peak of the ordinary radial distribution function. Interestingly the authors investigated a classical channel system at an appropriate density and found similar effects. PERCUS [4.95] was able to use a perturbation theory to relate quantitatively the channel results to the structure seen in a homogeneous system.

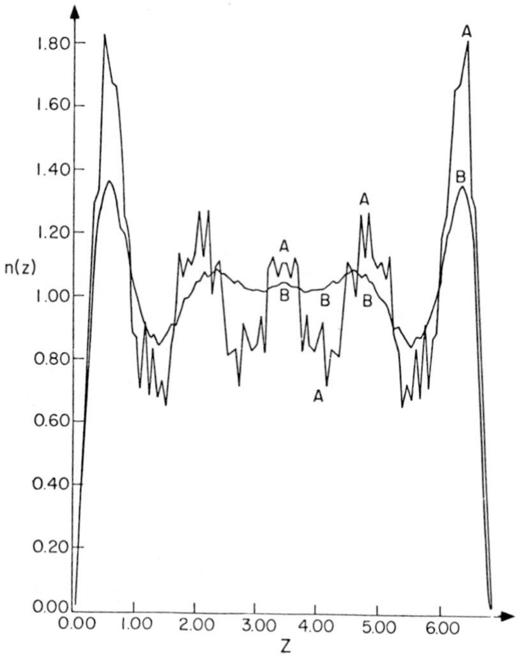

Fig.4.3 Density profiles for quantum hard spheres in a channel at average density $\rho a^3 = 0.2$. The width of the channel is 6.8a. The line B shows the results of a variational calculation, A the result of GFMC

LIU and KALOS [4.96] investigated the density profile of a film of bosons with two free surfaces using a hard-sphere plus square-well potential. This too showed fairly convincing evidence of layered structure near the surface. Unfortunately the model potential was not well suited for ^{4}He and the statistical errors in the density profile left something to be desired. This was a consequence mostly of the poor importance function ψ_T used to describe the homogeneous system.

Recently, WHITLOCK et al. [4.97] have returned to a direct calculation of ^{4}He with the de Boer-Michels form of the Lennard-Jones potential. This work is still in progress so only preliminary data are available. One important result is clear already: the potential used does indeed give a substantially deeper minimum than the variational results, confirming the previous results of KALOS [4.92] and of KALOS et al. [4.28], and in particular the perturbation theory of the latter paper. The newer results have much improved accuracy; a preliminary result for the equilibrium energy of liquid ^{4}He is -6.85 ± 0.05 K. This is uncorrected for three-body forces and correlations introduced by zero point motion of long wave-length phonons. These effects are being evaluated and appear to be about equal and of opposite sign.

The discrepancy in the energy as compared with the predictions of the product wave function (4.2) is even larger than one would expect from the error in the kinetic energy (as seen in the hard-sphere results) and the error amplifying effect of the cancellation by the negative potential energy. It seems likely that the neglect of three-body and higher correlations is the root of the problem [4.25].

The newer results on ^{4}He show again a more structured S(k) as compared with variational results – see Fig.4.4 – and reasonable agreement with recent experimental data for the momentum density in liquid ^{4}He obtained by MARTEL et al. [4.98] and by WOODS and SEARS [4.99]. The latter authors estimated the condensate fraction, n_0, at 1.1 K to be 6.9 ± 0.8% and extrapolated to 10.8 ± 1.3% at T = 0. The result obtained with GFMC [4.97] was 11.2 ± 0.2%. These are to be contrasted with the experimental results of 0.024 ± 0.01 and 0.018 ± 0.01 reported by MOOK et al. [4.100] and by MOOK [4.101], respectively.

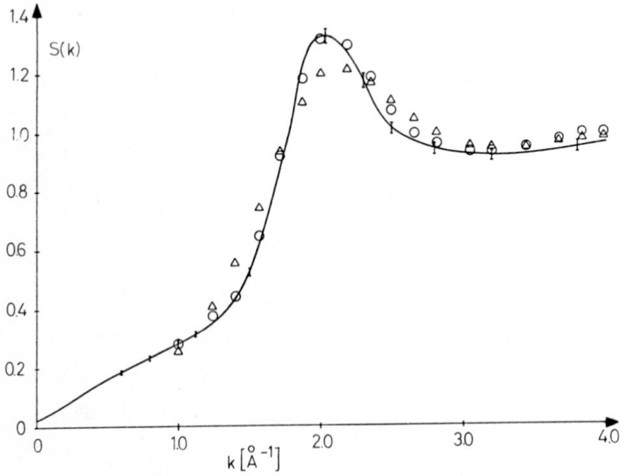

Fig.4.4 Comparison of structure functions for He-4 at equilibrium density. The solid line shows the smoothed experimental data of ACHTER and MEYERS [4.111] with bars indicating one standard deviation. S(k) computed variationally (Lennard-Jones 6-12; McMillan trial function) is shown by triangles. The circles show the results computed using GFMC

In the recent calculations of WHITLOCK et al. certain technical questions have been carefully investigated. In one case it was shown that the estimate of g(r) obtained from asymptotic weights [cf. discussion following (4.91)] is consistent with those obtained by extrapolation, (4.95). Also it was found that consistent energy values could be obtained when parameters of ψ_T were changed. Even more notable was a series of calculations at a density of $0.3648/(2.556\ \text{Å})^3$, the experimental equilibrium. At this density, product trial functions with pseudopotentials of very different character are available. GFMC calculations have been made with ψ_T having the McMillan form [4.1] and with the "α" and "β" functions suggested by de MICHELIS and REATTO [4.27]. It was found that use of the β function (constructed so as to agree well with experimental S(k) at the expense of a high energy value) reduced the variance of the energy estimates of the GFMC by a factor of 7 as compared with results obtained with McMillan's ψ_T. Thus the variational optimal is not necessarily the minimum variance ψ_T, although in no case was a ψ_T derived from variational results a poor choice. All three give consistent energy values. Most significant are that "extrapolated" estimates of g(r), S(k), $\langle V\rangle$ and n_0 (the condensate fraction) agree very well although the variational and "mixed" estimates, (4.94,92), respectively, are rather different. The validity of the extrapolation procedure is amply confirmed here.

CEPERLEY et al. have reported two investigations [4.59,91] on boson systems with Yukawa potentials. The variational calculations of these papers were described above in Sect.4.4. GFMC calculations were performed as well as a test of the accuracy of the energies and other quantities calculated. In general the variational energies are accurate; in the earlier paper [4.91] they are less than 1% above the GFMC results. Other expectations do not agree as well. The radial distribution is of the order of 10% more structured as calculated more exactly. In the second paper somewhat larger discrepancies were noted. One generally important point emerges from these results. Generally speaking, the energies deduced variationally for a crystal are more reliable than those found for the liquid, presumably owing to the simplification in the character of the wave function which results from crystal order. Thus the estimation of the location of melting and freezing transitions from variational energies is likely to be systematically biased.

In these papers also, additional verification of the independence of GFMC results to changes in ψ_T is given. In particular [4.59] compares calculations of energy with ψ_T of the gaussian type (4.34) or periodic (4.49). These give rather different variational energies but the GFMC results with the different ψ_T agree.

4.5 Virial Coefficients and Pair Correlations

FOSDICK and co-workers have published several papers [4.102-104] in which path integrals formulated to give the density matrix of a quantum system have been approximated by discrete paths and the resulting many dimensional integrals carried out by Monte Carlo. A review of the theory and an introduction to this class of approximations has been given also by BRUSH [4.105].

JORDAN and FOSDICK [4.102] calculated the second virial coefficient and pair correlation for a Lennard-Jones interaction. They regarded 2 K as a lower limit for the practical application of the method. Virial coefficients agreed within 2-3% of numerical values obtained from phase shifts. They calculated exchange contributions as well as direct, but at 2 K the exchange contribution to the pair correlation function showed very poor statistics.

In a second paper [4.103] the same authors included the effect of a third particle upon the pair correlations and estimated the third virial coefficient. This paper is technically interesting in that a version of the $M(RT)^2$ method [4.13] was used to sample paths representative of those required for the potential and the temperature involved. They reported results from 5 to 273 K but no permutation effects were included. The truncation errors (owing to replacement of continuous by straight paths) were made small compared to the Monte Carlo errors on say the pair distribution which ranged from a few to of the order of 20% in 20,000 samples. Interestingly, they found fair agreement with HENSHAW's data [4.106] for the pair distribution in liquid helium at 5 K.

Very recently WHITLOCK and KALOS [4.107] have taken up this class of problems, using the GFMC method to generate the quantum mechanical density matrix. They have formulated the general problem of treating a many-body system at finite temperature (including permutations to treat Bose or Fermi statistics) and treated in detail the pair correlation and second virial of a hard-sphere system. The general methods of Sect.4.4 above are followed except that an additional variable, the "pseudo time" = $1/k_BT$, is included in the full and partial Green's functions. An importance sampling transformation is introduced: there is an optimal importance function, namely the correct density matrix, which guides the random walk to the right place at the right "time". They found that approximate trial forms for the density matrix make the computation feasible but that improvements in that function improve the Monte Carlo efficiency in a drastic way. They were able to calculate direct and exchange pair correlations and virial coefficients from about 0.2 to about 10 K. The direct contributions were computed to about 7600 K. It is noteworthy that at high temperatures the difference between the quantum and classical pair correlations and virials is statistically significant. The worst error in the direct virial coefficients was obtained at low temperature — the standard error was 3% for 10^4 samples. The exchange virial grows worse at high temperatures and the best data at

8 K with the same number of samples give an error of 3%. All results were within statistical errors of those obtained by LARSEN [4.108] and by BOYD et al. [4.109] using partial wave expansions.

4.6 Conclusions

What, then, has been learned from these studies of quantum systems? First, it has been found that the product trial function is a good first approximation to the ground-state wave function of quantum liquids, and, with the gaussian localization, a good description of quantum solids. The variational method is able to predict qualitatively all of the zero temperature phase changes of helium and the quantitative energies are fairly close to the experimental values. The actual form of the pseudopotential is not very critical for computing ground-state energies but is important for other ground-state properties. The 'optimum' pseudopotential resulting from the 'paired-phonon' calculation of CHANG and CAMPBELL [4.25] is quite simple, consisting of the solution of the two-body Schrödinger equation for small r and the phonon pseudopotential of CHESTER and REATTO [4.9] for large r. The hypernetted chain integral equations are relatively accurate for quantum systems but are usually not reliable enough to predict phase transitions. In addition they are seriously inconsistent in estimating the kinetic energy.

The variational calculations have verified the intuitive picture of a crystal; the atoms move about the lattice sites with a gaussian density distribution. The ground-state energy is influenced very little by exchange effects.

Hence for most purposes one can regard the particles in crystals as distinguishable. The crystal becomes unstable if the rms displacement around a lattice site is greater than 0.27 of the nearest neighbor distance, this ratio depends only weakly on the interparticle potential and particle statistics.

There are many remaining unexplained problems for which quantum Monte Carlo is a powerful tool. First, it has been proposed [4.110] that all groups compute the equation of state of a more complicated neutron matter potential involving spin and tensor interactions. Monte Carlo can again provide a check on the various integral equation and cluster expansion techniques. Second, there are a number of interesting problems concerning quantum fluids and solids in restricted geometries which can be treated variationally as was done by LIU et al. [4.43,46]. Mixtures are also largely unexplored. Related to this are finite body problems such as nuclei, atoms, molecules and droplets which can in principle be approximated by product trial functions and for which Monte Carlo is particularly well suited since there are no periodic boundaries and hence no finite system size problems. The possibilities for the fermion Monte Carlo in molecular systems are interesting since it is radically different from present methods of calculation.

The variational method can be used to explore excited states. The product wave function in (4.3) is valid for excited states, where D is now an excited state of the ideal gas or harmonic crystal. The reweighting technique can be used to calculate accurately small differences between excited states.

Monte Carlo can also be used to explore a wider class of trial functions as was done by WOO and COLDWELL [4.38] for two-dimensional helium. FEYNMAN [4.83] has suggested a trial function incorporating back flow in Fermi liquids which is possible but perhaps tedious to sample with Monte Carlo.

The utility of GFMC is only now becoming apparent. It is true that it is not yet as convenient and economical a tool as variational studies. In careful studies of systems of ^{4}He such as the assessment of the effect of different potentials on the equation of state and other observable properties, it is likely to prove indispensable. In other research it should continue to be used to provide at the least critical "benchmarks" for the validity of less accurate methods.

As far as this review goes, the major lacuna is the absence of an exact numerical method, possibly of the GFMC type, for treating fermion systems. The fact that fermion wave functions have nodal surfaces whose character cannot usually be specified in advance is a major stumbling block but not necessarily an insuperable one.

We believe that the extension of GFMC to the treatment of equilibrium many-body problems at temperatures other than zero will be feasible and extremely useful. For the distant future the most remote speculations are for the development of a method capable of treating quantum many-body systems evolving in time.

Acknowledgements. We are grateful to K. BINDER, P.A. WHITLOCK, and G.V. CHESTER for helpful suggestions and critical review of the manuscript. The preparation of this review was begun while the authors were visiting the Laboratoire de Physique Thêorique et Haute Energies at Orsay and they would like to thank B. JANCOVICI and D. LEVESQUE for their hospitality and the Centre National de la Recherche Scientifique for support. The work was also supported in part by the U.S. Department of Energy under contract. No. EY-76-C-02-3077.

References

4.1 W.L. McMillan: Phys. Rev. A *138*, 442 (1965)
4.2 D.C. Handscomb: Proc. Camb. Phil. Soc. *58*, 594 (1962); *60*, 115 (1964)
4.3 M. Suzuki, S. Miyashita, A. Kuroda, C. Kawabata: Phys. Lett. A *60*, 478 (1977)
M. Suzuki: Prog. Theor. Phys. *56*, 1454 (1976); Commun. Math. Phys. *51*, 182 (1976)
4.4 A. Bijl: Physica *7*, 869-886 (1940)
4.5 R.B. Dingle: Phil. Mag. *40*, 573 (1949)
4.6 R. Jastrow: Phys. Rev. *98*, 1479-1483 (1955)
4.7 N.F. Mott: Phil. Mag. *40*, 61 (1949)
4.8 E. Saunders: Phys. Rev. *126*, 1724-1736 (1962)
4.9 L. Reatto, G.V. Chester: Phys. Lett. *22*, 276 (1966)
4.10 R.P. Feynman: Phys. Rev. *94*, 262 (1954)

4.11 E. Krotscheck: Phys. Rev. A *15*, 397 (1977)
4.12 D. Bohm, D. Pines: Phys. Rev. *92*, 609 (1953)
4.13 N. Metropolis, A.W. Rosenbluth, M.N. Rosenbluth, A.M. Teller, E. Teller: J. Chem. Phys. *21*, 1087 (1953)
4.14 L. Verlet: Phys. Rev. *159*, 98 (1967)
4.15 J.D. Valleau, S.G. Whittington: "A Guide to Monte Carlo for Statistical Mechanics I", in *Modern Theoretical Chemistry,* Vol. V, Statistical Mechanics, ed. by B. Berne (Plenum Press, New York 1977)
4.16 D. Ceperley, G.V. Chester, M.H. Kalos: Phys. Rev. B *16*, 3081 (1977)
4.17 D. Ceperley: Monte Carlo study of the electron gas in two and three dimensions. Phys. Rev. B (in press)
4.18 S.G. Cochran: Ph. D. Thesis, Cornell University, Ithaca, New York (1973)
4.19 O. Penrose, L. Onsager: Phys. Rev. *104*, 576 (1956)
4.20 D. Levesque, Tu Khiet, D. Schiff, L. Verlet: Orsay report, 1965 (unpublished)
4.21 W.C. Stwalley, L.H. Nosanow: Phys. Rev. Lett. *36*, 910 (1976)
4.22 J. de Boer: Physica (Utrecht) *14*, 139 (1948)
4.23 D. Schiff, L. Verlet: Phys. Rev. *160*, 208 (1967)
4.24 R.D. Murphy, R.O. Watts: J. Low Temp. Phys. *2*, 507 (1970)
4.25 C.C. Chang, C.E. Campbell: Phys. Rev. B *15*, 4238 (1977)
4.26 J. Zwanziger: Ph. D. Thesis, Cornell University, Ithaca, New York (1972)
4.27 D. de Michelis, L. Reatto: Phys. Lett. A *50*, 275 (1974)
4.28 M.H. Kalos, D. Levesque, L. Verlet: Phys. Rev. A *9*, 2178 (1974)
4.29 R.D. Murphy: Phys. Rev. A *5*, 331 (1972)
4.30 J.P. Hansen, D. Levesque: Phys. Rev. *165*, 293 (1968)
4.31 J.P. Hansen: Phys. Lett. A *30*, 214 (1969)
4.32 J.P. Hansen, E.L. Pollack: Phys. Rev. A *5*, 2651 (1972); see also J.P. Hansen: J. Phys. C (Paris) *31*, Suppl. C *3*, 67 (1970)
4.33 R. Haberlandt: Phys. Lett. *14*, 197 (1965)
4.34 J.P. Hansen: Phys. Lett. A *34*, 25 (1971)
4.35 L.W. Bruch, I.J. McGee: J. Chem. Phys. *52*, 5884 (1970)
4.36 D.E. Beck: Mol. Phys. *14*, 311 (1969)
4.37 R.D. Murphy, I.J. McGee: Phys. Lett. A *45*, 323 (1973)
4.38 B.M. Axilrod, E. Teller: J. Chem. Phys. *11*, 293 (1943)
4.39 R.D. Murphy, J.A. Barker: Phys. Rev. A *3*, 1037 (1971)
4.40 J.P. Hansen, D. Levesque, D. Schiff: Phys. Rev. A *3*, 776 (1971)
4.41 J.D. Weeks, D. Chandler, H.C. Anderson: J. Chem. Phys. *54*, 5237 (1971)
4.42 K.S. Liu, M.H. Kalos, G.V. Chester: J. Low Temp. Phys. *13*, 227 (1973)
4.43 K.S. Liu, M.H. Kalos, G.V. Chester: Phys. Rev. B *12*, 1715 (1975); see also erratum, Phys. Rev. B
4.44 D.S. Hyman: Ph. D. Thesis, Cornell University, Ithaca, New York (1970) unpublished
4.45 C.E. Campbell, M. Schick: Phys. Rev. A *3*, 691 (1971)
4.46 K.S. Liu, M.H. Kalos, G.V. Chester: Phys. Rev. B *13*, 1971 (1976)
4.47 C.W. Woo, R.L. Coldwell: Phys. Rev. Lett. *29*, 1062 (1972)
4.48 C.E. Campbell: Phys. Lett. A *44*, 471 (1973)
4.49 J.V. Dugan, Jr., R.D. Etters: J. Chem. Phys. *59*, 6171 (1973)
R.D. Etters, J.V. Dugan, R.W. Palmer: J. Chem. Phys. *62*, 313 (1975)
4.50 R.L. Danilowicz, J.V. Dugan, Jr., R.D. Etters: J. Chem. Phys. *65*, 498 (1976)
4.51 M.D. Miller, L.H. Nosanow: Phys. Rev. B *15*, 4376 (1977)
4.52 L.H. Nosanow, L.J. Parish, E.J. Pinski: Phys. Rev. B *11*, 191 (1975)
M.D. Miller, L.H. Nosanow, L.J. Parish: Phys. Rev. Lett. *35*, 581 (1975)
M.D. Miller, L.H. Nosanow, L.J. Parish: Phys. Rev. B *15*, 214 (1977)
L.H. Nosanow: J. Low Temp. Phys. *26*, 613 (1977)
4.53 J.P. Hansen: Phys. Rev. *172*, 919 (1968)
4.54 T.A. Bruce: Phys. Rev. B *5*, 4170 (1972)
4.55 T.R. Koehler: Meth. Comp. Phys. *15*, 277 (1976)
4.56 G. Baym, C. Pethick: Ann. Rev. Nucl. Sci. *25*, 27 (1975)
4.57 S.G. Cochran, G.V. Chester: Preprint, Cornell University, Ithaca, New York (1973)
4.58 V.G. Pandharipande: Nucl. Phys. A *248*, 524 (1975)
4.59 D. Ceperley, G.V. Chester, M.H. Kalos: Phys. Rev. B *17*, 1070 (1978)

4.60 D. Schiff: Nat. Phys. Sci. *245*, 130 (1973)
4.61 R.G. Palmer, P.W. Anderson: Phys. Rev. D *9*, 3281 (1974)
4.62 J.P. Hansen, B. Jancovici, D. Schiff: Phys. Rev. Lett. *29*, 991 (1972)
4.63 R. Monnier: Phys. Rev. A *6*, 393 (1972)
4.64 H.R. Glyde, G.H. Keech, R. Mazighi, J.P. Hansen: Phys. Lett. A *58*, 226 (1976)
4.65 J.P. Hansen, R. Mazighi: Phys. Rev. A, to be published
4.66 L.L. Foldy: Phys. Rev. *124*, 649 (1961); *125*, 2208 (1962)
4.67 W.J. Carr: Phys. Rev. *122*, 1437 (1961)
4.68 F.Y. Wu, E. Feenberg: Phys. Rev. *128*, 943 (1962)
4.69 L.H. Nosanow, L.J. Parish: Ann. Acad. Sci. N.Y. *224*, 226 (1973)
4.70 J.P. Hansen, D. Schiff: Phys. Rev. Lett. *23*, 1488 (1969)
4.71 D.K. Lee, F.H. Ree: Phys. Rev. A *6*, 1218 (1972)
V.R. Pandharipande, H.A. Bethe: Phys. Rev. C *7*, 1312 (1973)
4.72 T. Gaskell: Proc. Phys. Soc. London *77*, 1182 (1961)
4.73 J.G. Zabolitzky: Phys. Rev. A *16*, 1258 (1977)
4.74 B. Day: Nuclear matter calculations. Rev. Mod. Phys. *50*, 495 (1978)
4.75 E.P. Wigner: Phys. Rev. *46*, 1002 (1934); Trans. Faraday Soc. *34*, 678 (1938)
4.76 C.M. Care, N.H. March: Advan. Phys. *24*, 101 (1975)
4.78 F.A. Stevens, Jr., M.A. Pokrant: Phys. Rev. A *8*, 990 (1973)
K.S. Singwi, A. Sjolander, M.P. Tosi, R.H. Land: Phys. Rev. B *1*, 1044 (1970)
D.L. Freeman: Phys. Rev. B *15*, 5512 (1977)
4.77 M. Cole: Rev. Mod. Phys. *46*, 451 (1974)
4.79 F. Bloch: Z. Physik (Leipzig) *57*, 549 (1929)
4.80 S. Mizawa: Phys. Rev. *140*, 1645 (1965)
4.81 A.K. Rajagopal, J.C. Kimball: Phys. Rev. B *15*, 2819 (1977)
4.82 T. Padmore, G.V. Chester: Phys. Rev. A *9*, 1725 (1974)
T.C. Padmore: Phys. Rev. Lett. *32*, 826 (1974)
4.83 R.P. Feynman, M. Cohen: Phys. Rev. *102*, 1189 (1956)
4.84 W.M. Saslow: Phys. Rev. Lett. *36*, 1151 (1976)
4.85 A.J. Leggett: Phys. Rev. Lett. *25*, 1543 (1970)
4.86 G. Meissner, J.P. Hansen: Phys. Lett. A *304*, 61 (1969)
4.87 E. Wigner: Phys. Rev. *40*, 729 (1932)
4.88 J.P. Hansen, J.J. Weis: Phys. Rev. *188*, 314 (1969)
4.89 J.P. Hansen, I.R. McDonald: Theory of Simple Liquids (Academic Press, New York 1976) p. 200
4.90 K.S. Liu, M.H. Kalos, G.V. Chester: Phys. Rev. A *10*, 303 (1974)
4.91 D.M. Ceperley, G.V. Chester, M.H. Kalos: Phys. Rev. D *13*, 3208 (1976)
4.92 M.H. Kalos: Phys. Rev. A *2*, 250 (1970)
4.93 M.H. Kalos: Phys. Rev. *128*, 1791 (1962); Nucl. Phys. A *126*, 609 (1969)
4.94 M.H. Kalos: J. Comp. Phys. *1*, 127 (1966)
4.95 J.K. Percus: J. Stat. Phys. *15*, 423 (1976)
4.96 K.S. Liu, M.H, Kalos: Unpublished
4.97 P.A. Whitlock, D.M. Ceperley, G.V. Chester, M.H. Kalos: Submitted to Phys. Rev.
4.98 P. Martel, E.C. Svensson, A.D.B. Woods, V.F. Sears, R.A. Cowley: J. Low Temp. Phys. *23*, 285 (1977)
4.99 A.D.B. Woods, V.F. Sears: Phys. Rev. Lett. *39*, 415 (1977)
4.100 H.A. Mook, R. Scherm, M.K. Wilkinson: Phys. Rev. A *6*, 2268 (1972)
4.101 H.A. Mook: Phys. Rev. Lett. *32*, 1167 (1974)
4.102 H.F. Jordan, L.D. Fosdick: Phys. Rev. *143*, 58 (1966)
4.103 H.F. Jordan, L.D. Fosdick: Phys. Rev. *171*, 128 (1968)
4.104 L.D. Fosdick, R.C. Jacobson: J. Comp. Phys. *7*, 157 (1971)
4.105 S.G. Brush: Rev. Mod. Phys. *33*, 79 (1961)
4.106 P.G. Henshaw: Phys. Rev. *119*, 14 (1960)
4.107 P.A. Whitlock, M.H. Kalos: Submitted to J. Comp. Phys.
4.108 S.Y. Larsen: J. Chem. Phys. *48*, 1701 (1968)
4.109 M.E. Boyd, S.Y. Larsen, J.E. Kilpatrick: J. Chem. Phys. *45*, 499 (1966)
4.110 H.A. Bethe: Workshop on Condensed and Nuclear Matter, May 3-6, 1977, Urbana, Ill., USA
4.111 E.K. Achter, L. Meyers: Phys. Rev. *188*, 291 (1969)

5. Simulation of Small Systems

H. Müller-Krumbhaar

With 9 Figures

Two aspects of small systems are considered. The first is the simulation of systems of small extent and given shape, where surface properties can no longer be neglected in comparison with bulk properties. The second aspect is the appearance of small subsystems as parts of the model system being considered. These latter "clusters" play an essential role for the dynamics of phase transitions. Various aspects such as finite-size scaling behavior, relation between surface and bulk properties and interactions between clusters are discussed. Effective methods for the evaluation of clusters statistics are described in an appendix.

5.1 Introductory Remarks

This chapter is concerned with the properties of a variety of systems which have the common feature of "smallness". More precisely, we will discuss the effect of finiteness or variable system size with respect to static and dynamic properties. There are two different aspects to be considered. First, the smallness of the system itself may be of physical interest, such as liquid (or solid) droplets, superparamagnetic particles, two-dimensional adsorption clusters on a substrate or the clusters occuring in spinodal decomposition of binary mixtures. Second, there is the question, to what extent a simulation in a relatively small system can be used to make predictions about the behavior of a very large system. These two groups of problems can be characterized by the boundary conditions used, which in the first case are given by the physics involved, while in the second case they are usually periodic or using some self-consistent field (to account for extrapolation of the system size). A very frequent task of course is the combination of both cases, such as the evaluation of cluster statistics in large systems. There, one has to choose the basic system size so large, that a further increase does not change the results significantly.

For the simulation of physical clusters there exist two different numerical methods, which are complementary to each other: "Molecular Dynamics" (MD) and

"Monte Carlo" (MC) simulation. The first method is conveniently used if the short-time dynamics of the systems have to be investigated, such as phonon spectra in particular. Being interested in static quantities at thermodynamic equilibrium, the second method has the advantage, that configurational parts of the free energy are simulated directly.

Being interested only in thermostatic quantities at equilibrium, the Monte Carlo method therefore is more effective than the Molecular Dynamics method, because the time scale (which controles the computing time) is not related to the very short intrinsic time of molecular motion. The danger of being trapped in some metastable state of the system is thus reduced by the MC method.

This summary therefore only accounts for MC simulations. A few remarks, however, will be made, whenever a specific problem has been treated by MD only.

The organization of the article is as follows. The first part deals with static properties of systems at thermodynamic equilibrium. In Sect.5.2.1, clusters in continuous space are briefly discussed. There is not much Monte Carlo work available on this subject, partly due to limitations in computing time and partly due to problems of appropriate cluster definition.

Both problems are less severe in lattice models, to which the rest of Sect.5.2.2 on static quantities is addressed. Both Ising-type and Heisenberg-type models are discussed. The first of the three subsections deals with superparamagnetism of small particles and with finite-size behavior of computer-simulated models. The second part gives a summary of cluster statistics (= small subsystems) in large lattices, and the last part deals with the role of clusters in the percolation problem. The various thermodynamic properties of clusters are discussed as well as geometric relations and shapes.

Section 5.3 is concerned with time-dependent processes in small systems. Particular emphasis is given to the cluster dynamics in various models. In the first part the role played by cluster dynamics in first-order phase transitions is outlined and the relation to nucleation theory is sketched. The second part discusses cluster dynamics near second-order phase transitions in two and three-dimensional systems.

An appendix provides some hints for the construction of economic cluster counting programs and a few remarks concerning special applications.

5.2 Statics

5.2.1 Clusters in Continuous Space

The appearance of "clusters" of atoms is an important effect during processes of nucleation, phase separation and percolation. If the clusters are formed as liquid droplets from the gas phase, all atoms may change their position relative to each other on a continuous length scale. In contrast to the clusters in continuous space, clusters may also be imbedded in a host lattice, such as impurities in a crystal, where the atoms then are preferably located at discrete positions relative to the host lattice.

The first problem concerning atoms clustering in continuous space is the proper definition of a "cluster", i.e., the question which atoms are part of the cluster and which ones are not. The usual definition [5.1-3] of a cluster involves a physical constraint. All atoms, namely, are considered to be part of a cluster, if they all lie within a certain predefined volume V around the center of mass. This definition may be used at sufficiently low temperatures and low densities of atoms, where the clusters are very compact and sufficiently far separated from each other.

Since the definition of the constraining volume requires some a priori knowledge about the global properties of the cluster (size, shape), it might be better to use a local definition, which allows one to decide for each atom whether it is part of a cluster without knowing the total size. Such a (somewhat less ambiguous) definition [5.4,5] would be given by marking a constraining volume around every single atom. All atoms then are considered to be part of a cluster which have at least one (or more) other atom(s) within their close neighborhood (= constraining volume). The latter range would be associated with the interaction range of the attractive pair potential. (Fig.5.1).

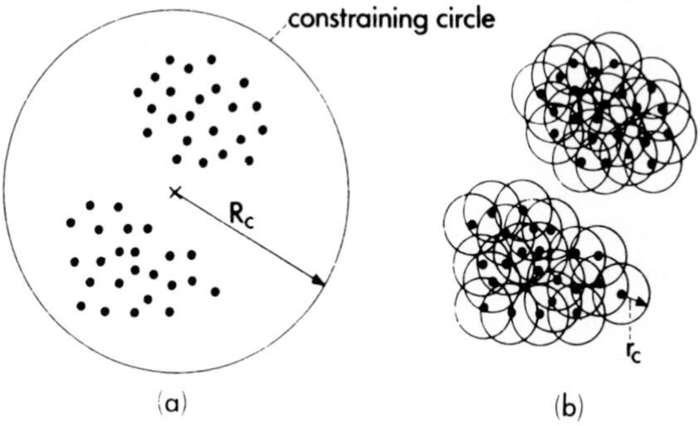

Fig.5.1 Schematic (two-dimensional) illustration of cluster definitions. Part (a) used a constraining volume of radius R_C, part (b) uses a local radius r_C of neighborhood. The same agglomeration of molecules is treated as one cluster in (a) and as two clusters in (b) [5.5]

At low temperatures and low concentrations the cluster properties derived with the two different definitions are expected to be essentially the same. The second definition, however, has the advantage that it can be used also at higher particle densities up to the percolation threshold, where the first definition obviously makes no sense.

LEE et al. [5.1] probably did the first sound simulations in this field using the constrained cluster volume. They investigated the Helmholtz free energy and the radial density distribution for clusters of atoms interacting with a 12-6 Lennard-Jones potential, corresponding to argon droplets of up to 100 atoms.

The main result is the calculation of the free energy for clusters of various sizes, at different remperatures. At sufficiently low temperatures ($T \leq 80$ K) the result depends only weakly upon the constraint. The dependence is largest for the small clusters.

The radial density distribution function is difficult to analyze. As was recently pointed out [5.6] this function depends markedly upon whether the center of mass of the cluster coincides with an atom or not. The strong fluctuations near the center of the cluster therefore are not surprising.

An extension of these simulations to the more complicated clusters of water molecules was performed subsequently [5.2,3]. The main difference to the Lennard-Jones particles here is the nonspherical symmetry of the water molecule and the long-ranged interaction potential. The radial density distribution function shows a pronounced dipole-ordering near the center of the cluster [5.2]. This was confirmed in a later study [5.3] which in addition indicated that near the surface of the cluster the dipoles are oriented preferenitally parallel to the surface. The main conclusion of these simulations is, that for very small water clusters the traditional "capillarity approximation" [5.6] is not adequate.

Most other cluster properties were investigated by means of molecular dynamic techniques, such as the existence of a critical cluster size of $N = 55$ atoms, above which crystalline structure is observed for sufficiently low temperatures [5.7]. Concerning nucleation theory it was demonstrated [Ref. 5.6, footnote on page 150] that rotational and translational modes of clusters in continuous space have to be included if one wants to calculate nucleation rates from a microscopic model, at least for clusters consisting of only a few atoms.

Microcrystalline clusters give contributions to the cluster partition functions from their phonon excitations which also were calculated by MD techniques [5.6,8]. It is, however, important first to include all *configurational* parts in the cluster free energy. Because of the long lifetime of metastable states this could be more safely done by using MC techniques. MD methods are usually limited to simulation times below some 10^{-10} seconds, while the simulation time in MC methods has no direct connection with real time in these models and it is therefore easier to include the important configurations by MC than by MD methods.

In fact, it turns out that these configurational contributions are as important for nucleation theory as they are difficult to calculate, since it is not so clear under what conditions the motion of the atoms in a cluster can be separated into harmonic parts (phonons) and anharmonic parts, which lead to configurational reordering of the microclusters. For that reason it seems adequate to resort to models, where phonons are suppressed by definition, as is described in the next sections of this chapter. In these models one then uses MC simulation to obtain rather precise statistics of clusters up to several thousand particles, which provide an excellent test for analytical theories. These models even may be rather realistic for the interpretation of physical clusters in host lattices and substrates, as long as elastic deformations of the host lattice are not important, and hence may have applications to clusters in metallic alloys etc.

5.2.2 Lattice Models

a) General Remarks

The small lattice systems to be considered in the following can be grouped into different classes according to their boundary conditions.

The first group, being discussed in Sect.5.2.2, consists of systems, which have a given finite extent and given shape. In addition one can discriminate between periodic boundary conditions and free (or fixed) boundaries. The first type is used to preserve translational symmetry for extrapolation to infinite system size, the second type is physically realized, e.g., in superparamagnetic particles.

The second group, the "clusters", is entirely different. Here, the size and the shape of the system or cluster is an essential result of the simulation rather than an input. They have direct physical importance and hence will be discussed in more detail than the above mentioned cases (Fig.5.2).

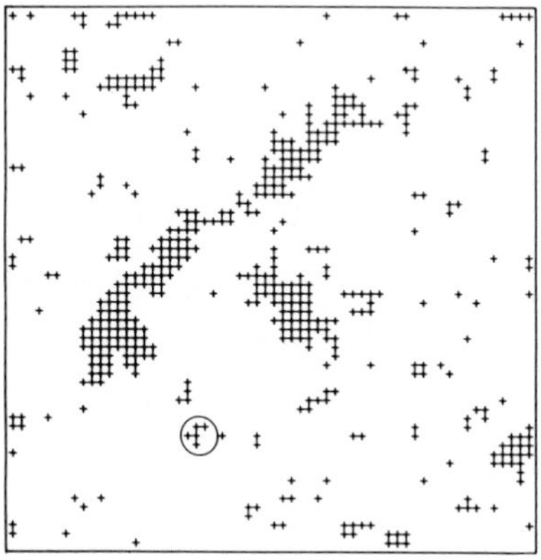

Fig.5.2 Equilibrium clusters in a two-dimensional (60 × 60) lattice-gas model. The circle surrounds an (ℓ = 5)-particle cluster [5.37]

A particular aspect of small systems has been completely neglected in this chapter due to limitations of space and expertise, namely the simulation of chain-molecules and polymerisation. As an introduction, however, the reader is referred to [5.9;1.12].

b) Finite-Size Behavior and Superparamagnetism

The investigation of finite-size effects is an important task for the application of MC methods in statistical mechanisms, since there one usually wants to make predictions about the behavior of systems with some 10^{23} particles.

The most frequently used lattice systems in studies of cooperative phenomena are the Ising model and — partly — the Heisenberg model. The first model has many applications as lattice-gas model, both models are also used to interpret ferromagnetism. They are both described in some detail elsewhere (Chap.6; also [1.4,68]).

Finite systems do not exhibit phase transitions. But sufficiently large systems show rounded peaks instead of singularities for certain derivatives of the free energy such as, e.g., the specific heat. The change in shape of these peaks with system size and other properties was investigated by BINDER [5.10,11] and LANDAU [5.12-14].

The general thermodynamics of small systems are treated in [5.15,16]. Using the general similarity of second-order phase transitions, FISHER and FERDINAND [5.17,18] gave an analysis of the scaling behavior for finite systems. The specific heat $C_\infty(T)$ in an infinite system should vary with temperature as

$$C_\infty(T) \sim (|\varepsilon|^{-\alpha}-1)/\alpha + \ldots$$
$$\varepsilon = (T - T_c)/T_c \tag{5.1}$$

near the critical temperature T_c, with a critical exponent α. In a finite system with volume N^d (d: dimensionality, N: length in lattice units), the specific heat should vary [5.17,18] as

$$C_N(T) \sim A(T)(|\varepsilon_N|^{-\alpha}-1)/\alpha + B(T) \tag{5.2}$$

where A and B are slowly varying functions,

$$\varepsilon_N^2 = [(T - T_{max})/T_c]^2 + c^2/N^{2/\nu} \tag{5.3}$$

and

$$T_{max} = T_c(1 - b/N^{1/\nu}) \quad . \tag{5.4}$$

The dimensionless constants b and c are of order unity. We see here that the maximum of the specific heat is shifted in temperature (5.4) and rounded (5.3). In (5.3) appears the critical exponent ν of the correlation length. This illustrates the idea, that the rounding of the anomaly sets in when the correlation length $\xi \sim |\varepsilon|^{-\nu}$ becomes of the order of the system size N.

More difficult to understand is the temperature shift of the maximum in C(T). For free boundaries one can argue that the missing correlation across the boundaries facilitates fluctuations and thus reduces the critical temperature. This is in fact supported by the MC results [5.10-14]. For periodic boundary conditions, however, the situation is not so clear. In a three-dimensional Ising model the "critical" temperature is also reduced for finite N [5.1,4], while in the two-dimensional case it is increased [5.13].

The quantitative behavior of the two- and three-dimensional Ising model is in good agreement with the conclusions drawn from the exactly solvable two-dimensional Ising model [5.17,18]. In particular, the finite size scaling behavior of the specific heat equation (5.2) is nicely reproduced by the Monte Carlo simulations [5.11,13,14]. The order parameter m ("magnetization" in magnetic language or "concentration" in lattice-gas terminology), which vanishes at T_c in the infinite system as

$$m \sim |\varepsilon|^{\beta} \tag{5.5}$$

[ε from (5.1)], is expected to scale with the system size N as

$$\tilde{m}_N N^{\beta/\nu} = X^0(\varepsilon N^{1/\nu}) \tag{5.6}$$

where $X^0(x)$ is a smooth (scaling) function in $x = \varepsilon N^{1/\nu}$ [with $X^0(x) \sim x$ for $x >> 1$], and $\tilde{m}_N$ denotes the expectation value of the absolute magnetization, since m is zero in a finite system at zero field. This also is in excellent agreement with MC results [5.13,14], where one finds that all points $\tilde{m}(T,N)$ lie on a single smooth curve if plotted according to (5.6). For the susceptibility χ the finite-size scaling behavior is predicted to be

$$\chi_N T\, N^{-\gamma/\nu} = Y^0(\varepsilon N^{1/\nu}) \tag{5.7}$$

[with $Y^0(x) \sim x$ for $x \gg 1$], where γ is the usual critical exponent for χ_∞, and Y^0 is a scaling function. Again, this scaling behavior is nicely reproduced, both above and below T_c, even without fitting any parameters (Fig.5.3).

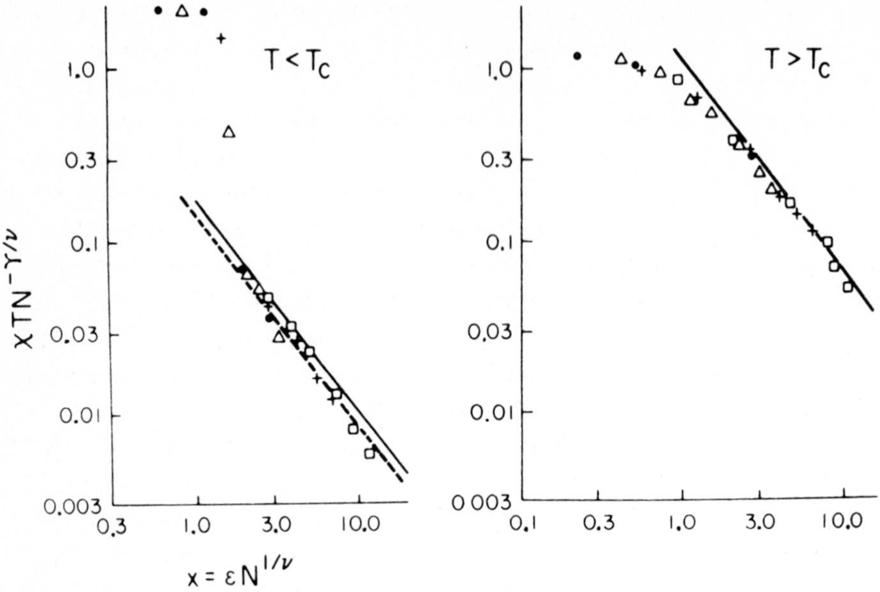

Fig.5.3 Finite-size scaling plots for the low- and high-temperature susceptibilities for a three-dimensional Ising lattice with periodic boundary condition. The dashed line is a best fit to the data while the solid lines are theoretical predictions [5.14]

A special point is the discrimination between bulk and surface properties for lattices with free boundaries. Here the influence of the boundaries is considerably stronger than in the cases with periodic boundary conditions and it can be compared with predictions about surface scaling [5.19].

The scaling function $X^0(x)$ for the order parameter accordingly should have the form

$$X^0(x) \approx Bx^{\beta} + B_s x^{\beta_s} \tag{5.8}$$

with amplitude factors B and B_s for bulk and surface contributions and critical exponents β and β_s where $\beta_s = \beta - \nu$.

Since the bulk contributions are known with good precision one may subtract those and then plot the resulting surface contribution $\tilde{m}_N^{(s)}$ appropriately normalized versus the scaling variable $x = \varepsilon N^{1/\nu}$

$$\tilde{m}_N^{(s)} N^{\beta/\nu} \sim B_s x^{\beta_s} \quad . \tag{5.9}$$

Similar relations hold for other quantities such as susceptibility and specific heat. All relations are well reproduced by the simulations [5.14] for scaling variables larger than about $x \approx 3$.

The size dependence of magnetization, susceptibility, specific heat and the probability distributions P(m) and P(E) for magnetization and energy in the Heisenberg model were analyzed by PAAUW et al. [5.20]. The scaling behavior near T_c was also well reproduced. The multicritical behavior of a class of models containing the Baxter model was investigated by DOMANY et al. [5.21]. They used finite-size scaling theory to extrapolate finite system to infinity in order to calculate critical crossover exponents.

A magnetic system which is composed of many noninteracting clusters of atomic magnets (spins) may exhibit the phenomenon of superparamagnetism [5.22]. The spins inside a cluster interact which each other and lead to an effective magnetic moment of the cluster. For noninteracting clusters these moments are oriented at random and thus the system as a whole has no spontaneous magnetization but responds strongly to an applied field. The properties of such clusters of given shape are difficult to analyze analytically and, therefore, are ideal objects for MC simulations. A superparamagnetic cluster may be modelled by a Heisenberg model for ferromagnetism in a finite lattice with free boundaries [5.10,23-25]. At first a shift in "critical" temperature is observed equivalent to the behavior of the three-dimensional Ising model. The fluctuations in the Heisenberg model of course are much larger than in the Ising model due to the continuous symmetry of the spin space. Another effect is the strong dependence of the absolute magnetization $|m| = \sqrt{\langle m^2 \rangle}$ upon shape [5.23]. This can be attributed mainly to the relative number of surface sites, since a sphere-like system shows values for $|m|$ which are much closer to the (extrapolated) value for an infinite system than a rectangular system of about the same number of sites (Fig.5.4). The magnetization $|m|$ also has a strong radial dependence [5.25]. It decays from the center towards the surface. The resulting effective magnetic radius of the cluster which is obtained by the usual experimental applications therefore can be considerably smaller than the radius of the particle, in particular for small particles and near the critical temperature of the corresponding infinite system. Many details of the cluster magnetization depending on next-nearest neighbor exchange, temperature, cluster size and external field were evaluated in [5.25].

A boundary condition particularly useful for the extrapolation of Heisenberg models to infinite system size is the self-consistent boundary condition [5.26-28]. Here the free boundary conditions are completed with an additional field that acts only in the boundary layer in such a way, that the gradient of the order parameter at the boundary layer vanishes. This is the self-consistency condition in the spirit of the Bethe-approximation [5.26], from which the field is determined iteratively. The result is a considerably reduced dependence of the systems properties

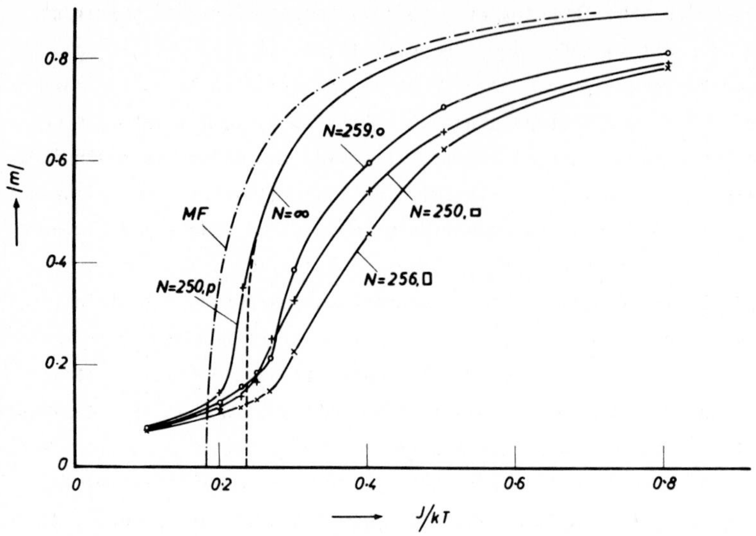

Fig.5.4 Absolute values of the magnetization of superparamagnetic particles with different shapes (as indicated by the symbols) and about the same number of spins. The spherical shape has the largest magnetization [5.23]

upon size. This is mainly due to the fact that these boundary conditions produce a true phase transition in contrast to periodic boundary conditions, where no singularity occurs.

The influence of "rounding" of the critical properties in experimental situations finally may be estimated using the three-dimensional Ising model with free boundaries [5.14]. With the calculated amplitudes for the surface contributions one obtains the result that rounding effects may be observed at values of the reduced temperature $|\varepsilon| \approx 10^{-3}$ for particles with a diameter of several thousand lattice units. A further complication in real systems, namely the mixture of particles of different sizes, may at least partly be taken into account by calculating weighted averages with a suitable distribution. Such distributions usually depend on the way in which the material was prepared. Some examples are given in Sect.5.3.3, on cluster growth during phase separation.

c) Equilibrium Cluster Statistics in Systems with Interaction

The evaluation of the "free energy" of a small cluster of particles (or more precisely: its contribution to the free energy of the system) from a microscopic model is not an easy task, as mentioned already in Sect.5.2.1 and in the book by ZETTLEMOYER [5.6]. For that reason it seems adequate not to study isolated clusters in some constraining volume in order to evaluate some cluster-free energy and then insert this into some clusters probability functional, but to study the cluster probabilities directly. The cluster probabilities are the mean number of clusters of a given size (as seen in Fig.5.2) per lattice site.

These cluster probabilities in fact are the essential ingredient for all cluster theories of dynamics of phase transitions [5.6,8,29-32]. Several groups therefore have made Monte Carlo simulations on Ising lattices, with and without conservation of the order parameter (= concentration of particles in the lattice). There are various ways to evaluate the cluster statistics. Economic cluster counting methods are described in the appendix and in [5.33].

The definition of clusters in such a lattice system is obviously less ambiguous than in a continuous space. In most cases an Ising model with nearest-neighbor interaction on a simple quadratic or cubic lattice was analyzed. A cluster then can simply be defined as the set of occupied sites in the lattice, which are mutually connected by at least one nearest-neighbor bond.

The clusters then may be analyzed with respect to several criteria. A natural quantity to investigate is the probability $p(\ell,s)$, that a cluster of ℓ particles with surface s occurs in a unit volume. This function of course does not give any detailed information about the shape and topology of the cluster, except for the trivial limiting cases that s is at its limiting values for a very compact or very thin wire-like cluster. A different measure has been proposed recently [5.34], namely the cyclomatic number $c = m - \ell + 1$, where ℓ is the number of particles and m the number of bonds in the cluster. This number is related to the surface bonds s by $c(\ell,s) = \ell[(1/2)Q - 1] -(1/2)s + 1$, where Q is the coordination number of the lattice.

Let us now first look at the size distribution of clusters. For this purpose one defines a cluster probability n_ℓ

$$n_\ell = \sum_s p(\ell,s) \qquad (5.10)$$

for the occurence of a cluster with ℓ particles in a unit volume. The total number $\tilde{M}$ of particles in the system then is given by

$$\tilde{M} = N^d \sum_\ell \ell n_\ell \quad . \qquad (5.11a)$$

Since most of the MC investigations are using magnetic terminology ("spins" instead of "particles") we will adapt this in the following sections. Clusters then are defined as sets of connected "down"-spins in a surrounding of "up"-spins. The total magnetization $M(-1 \leqq M \leqq +1)$ of the system is according to (5.11a)

$$M = 1 - 2 \sum_\ell \ell n_\ell \quad . \qquad (5.11b)$$

At sufficiently low temperatues there are only a few small clusters present in an Ising model at thermal equilibrium. According to classical droplet-model-theories [5.6,29,30,35] the probability n_ℓ should decay exponentially with the surface area

$$\log(n_\ell) \sim -\ell^{(d-1)/d} \tag{5.12}$$

in zero "magnetic field" corresponding to the liquid-gas coexistence curve for fluids.

This has direct implications for nucleation theory which states that clusters above a critical size grow and shrink in the other case. Quantitative agreement between nucleation theory and critical clusters size was observed in [5.36], for an essentially two-dimensional model, while apart from undetermined prefactors a good fit of cluster occurences to the classical formula (5.12) was obtained [Ref. 5.37, Fig.3b] in the three-dimensional Ising model.

At very low temperatures the main contributions of clusters to the free energy come from the surface energy of the clusters while fluctuations and hence contributions to the entropy are less important. The surface energy of a compact cluster is easily calculated and it is therefore not surprising that (5.12) gives a good representation of cluster probabilities at low temperatures.

The situation is different at high temperatures, i.e., sufficiently close to the critical point T_c. Here, the fluctuations control the systems behavior and are the essence for the understanding of critical phenomena at continuous phase transitions. Considerable effort therefore has been invested into MC simulation near T_c [5.4, 37-40]. Because of the sum rule (5.11) one should expect, that the critical behavior, which is expressed by some nonanalytical behavior in the thermodynamic variables at T_c [see (5.1,5)], also shows up in the cluster properties. Since one knows [5.41] that the free energy and its derivatives fulfill certain conditions of homogeneity near T_c one should expect an analogous homogeneity to hold for the cluster probabilities.

With the two variables, the reduced temperature $\varepsilon = (T_c - T)/T_c$ and the reduced external field $h = \mu H/k_B T$ (where μ is the magnetic moment of a spin) one is led to the scaling form [5.37,42]

$$n_\ell = \ell^{-\tau}\tilde{n}_\ell(h\ell^y, \varepsilon\ell^z) \tag{5.13}$$

where τ, y and z are the critical exponents and $\tilde{n}_\ell$ is the scaling function. The relations with the usual critical exponents then are easily obtained, using the sum rule (5.11) and then taking derivatives $\partial M/\partial\varepsilon$, $\partial M/\partial h$ along the coexistence line ($h = 0$) or along the critical isochore ($\varepsilon = 0$). Results then are [5.42]

$$\beta = (\tau - 2)/z \quad , \tag{5.14a}$$

$$\beta\delta = y/z \quad , \tag{5.14b}$$

$$\delta = (2 - \tau + y)/z \quad . \tag{5.14c}$$

The critical properties are determined by two independent exponents (e.g., β and δ) which determine the others (e.g., γ). Thus one has for the cluster model of (5.13) with three exponents one additional exponent (e.g., γ) to be chosen freely, while the others then are fixed. This choice in fact represents a transformation from the cluster size variable ℓ to an "effective" cluster size $\ell' = \ell^y$. A physical interpretation of this substitution is, that only the relatively "compact" parts of a cluster, which may be connected by thin bridges of particles ("down" spins), are thermodynamically relevant and that the diffuse noncompact parts only contribute to the background magnetization and not to critical fluctuations.

MC simulation results with rather high precision are available for the two-dimensional [5.37,40] and three-dimensional [5.37,38] case. In two dimensions the cluster probabilities can be very well approximated by FISHER's droplet model [5.43]. This semiphenomenological model assumes the cluster probability n_ℓ to be given by

$$n_\ell = q_0 \ell^{-\tau} \exp(-a\varepsilon\ell^{\sigma} - h\ell) \tag{5.15}$$

with some constants q_0 and a. It obviously has the scaling structure (5.13) with the special value $y = 1$. With (5.14a,b) this gives $\sigma = 1/\beta\delta$, $\tau = 2 + 1/\delta$.

It is seen in Fig.5.5a that this relation holds in two dimensions [5.40] over a temperature range of $0.02 < |\varepsilon| < 0.12$. In three dimensions the unmodified droplet model (5.15) is not very well reproduced [Ref.5.37, Fig.3]. This could be expected since it was observed [5.39], that already about 5% below the critical temperature the "down" spins form a giant cluster that goes in all three directions through the whole system. This "percolating" cluster contains a finite fraction of all down spins in the system, in an infinite system therefore an infinite number.

Already in the neighborhood of the percolation threshold the tendency to form giant clusters will be reproduced in the cluster statistics. But since many of the connections between the relatively compact parts of the cluster are just thin bridges of a few spins it can be expected that this effect may at least partly be compensated by setting the cluster exponent y not equal to unity: $y < 1$. Using $y = 0.5$ a considerably better fit to a scaling form could be achieved [5.37] (Fig. 5.5b).

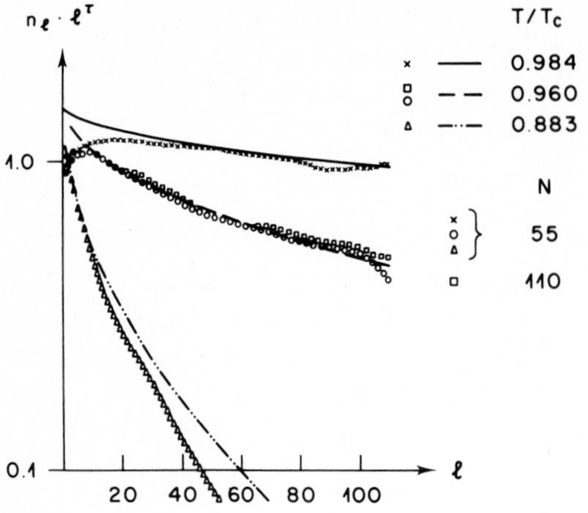

Fig.5.5a Comparison of the cluster probabilities n_ℓ in the two-dimensional Ising-model between Monte Carlo simulation (dots) and Fisher's droplet model near T_c. Very good agreement is achieved [5.40]

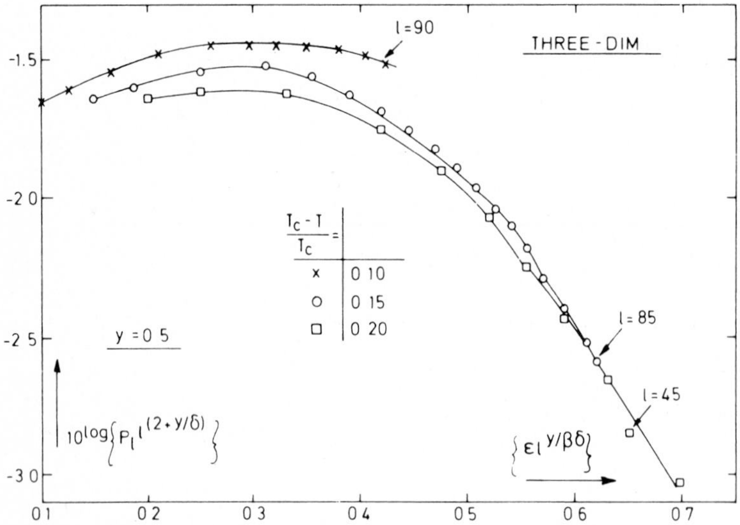

Fig.5.5b Scaling behavior of the cluster probabilities n_ℓ $(=p_\ell)$ in the three-dimensional Ising model below T_c, involving the additional scaling exponent y (5.13). Even with this generalization scaling cannot be achieved near the percolation temperature $T_p \approx 0.95\ T_c$ (see Fig.5.7) [5.37]

But ultimately it is necessary to avoid percolation, if one wants to use the cluster picture for the interpretation of the behavior of the system near the critical point. This can be achieved by a redefinition of what is meant by a cluster. Part of a cluster, e.g., would be a "down" spin, which has all neighbors as down spins. In this way only the compact parts of the clusters which apparently represent the correlations would contribute to the cluster statistics. A disadvantage of this definition of course is that the sum rule (5.11) is violated and that percolation for the ordered phase (a necessary condition for long-range order) then might not be found in the cluster picture. On the other hand, the scaling formalism (5.13,14) would still be applicable, since this is based on the existence of large compact fluctuations of the diameter of a correlation length $\xi \sim |\varepsilon|^{-\nu}$. We will leave this point for the moment and return to it in Sect.5.2.2d.

The next relevant parameter for the analysis of cluster statistics is usually considered to be the cluster surface, s, or the number of bonds between the "down" spins in a cluster and the surrounding "up" spins. MC simulations studying cluster surfaces are reported in [5.4,34,42,44] (concerning clusters with interaction between the spins). Fitting the cluster size ℓ and then varying the shape using a detailed balance principle, the probability distribution $p_e(\ell,s')$ for the *external* surface s' was calculated in the two- and three-dimensional Ising model [5.4,42]. s' here describes the length of the external contour of a cluster, while s in addition also contains the contributions from holes inside the clusters. For not too large clusters $\ell \leqq 200$ at not too high temperatures there is not much difference between s' and s and hence between $p_e(\ell,s')$ and $p(\ell,s)$.

The average cluster surface $\bar{s}'$ accordingly is expected to scale [5.42] as

$$\bar{s}' \sim \ell^{1-1/d}\, \varepsilon^{\beta\delta-1-\nu} \tag{5.16}$$

in the critical region. The simulation results are reasonably consistent with this scaling relation in two dimensions, while stronger deviations in three dimensions again probably can be attributed to the percolation problem.

The total cluster surface s should increase somewhat more rapidly with ℓ than s', because the possibility of having holes inside a cluster increases with the cluster size. One therefore would expect an exponent x to be closer to 1 in a scaling relation $s \sim \ell^x$ compared with (5.16), for sufficiently large ℓ. The above scaling theory (see 5.13-16), suggests values of x = 0.56 ... 0.69 (d = 2 dimensions) and x = 0.7 ... 0.77 (d = 3), depending upon the actual values of the exponent y [5.42].

Monte Carlo simulations on the two-dimensional system [5.44] seem to give even larger values for the averaged total surface $\bar{s}(\ell) = \sum_s sp(\ell,s)/\sum_s p(\ell,s)$. Above T_c and approaching $T = \infty$, x converges to 1, as should be the case for a noninteracting system. This value there is good for clusters with $\ell > 50$. As the temperature is lowered, the asymptotic value x = 1 is still obtained, but the range of validity is restricted to larger clusters. Below T_c an effective value of $x \approx 0.75$ seems to hold over the accessible range of cluster sizes and temperatures, but of course it is not clear, whether one is really in the asymptotic limit of large clusters, above which there is no further change.

Another attempt, finally, to parametrize the surface properties of clusters was made [5.34] using the cyclomatic number $c(\ell,s)$ mentioned above. The average number of cycles (of bonds within a cluster) per lattice site $\langle c\rangle_s$ should contain some information about the compactness (or inversely: "ramification") of the clusters in the system. It was defined as [5.34]

$$\langle c \rangle_s = \sum_{\ell,s} c(\ell,s)p(\ell,s) \quad . \tag{5.17}$$

This value undergoes a drastic change around T_c. Directly at T_c it is finite and was calculated [5.45] from the relation with the Whitney polynomial [5.46-48].

From all these investigations it is indicated, that the definition of clusters in a purely geometrical sense, i.e., a group of sites connected by nearest-neighbor bonds, causes problems in the neighborhood of the critical point in three and certainly above T_c in two dimensions. On the other hand, this geometric description seems just to be appropriate to describe percolation (see the next section), which in fact is a purely geometric phenomenon. Only an improved definition of "clusters" in connection with MC simulations might finally lead to a coherent description of incompact clusters near the critical point.

d) Cluster Statistics in the Percolation Problem

The existence of an infinite cluster in a lattice system (where "connectivity" may easily be defined) is called percolation. Here again, a cluster is a group of occupied sites connected by nearest-neighbor bonds between particles, where the occupation with particles is at random with a probability p. Since the general MC results are treated in Chap. 8.3 in this book and recent review articles exist [5.62,63], only a summary of the state of the art is given. (Again we use the analogy, that empty sites correspond to "up" spins and occupied sites to "down" spins).

From the definition of percolation [5.49,50] it is clear, that the geometry of the clusters is essential for the process. In critical phenomena it is not important for the macroscopic thermodynamic behavior whether two large clusters of "down" spins are connected by a thin bridge of "up" spins. But in the percolation problem these thin branches of the "ramified" [5.44] clusters finally connect the more compact parts of the clusters to the giant percolating cluster. In this case, therefore, one expects that the cluster models which make use of the nearest-neighbor bond definition (see Sect.5.2.2c), provide the appropriate definition for percolation. Expansion of macroscopic quantities of the system in polynomials of the concentration c of "down" spins (without interaction between spins) namely may be expressed equivalently by expansions in these geometrically defined clusters [5.50,51].

If there is no interaction between the spins, the Monte Carlo process is trivial insofar, that the only probabilistic decision to be made then concerns the average concentration of "down" spins in a lattice of "up" spins. The main computational effort then is the cluster analysis. Programming methods are described in the appendix.

The first large-scale cluster analysis on a computer has been performed by DEAN and BIRD [5.52] for clusters in noninteracting systems. They evaluated cluster probabilities n_ℓ (5.10) for various concentrations of "down" spins and various lattice systems. Detailed analyses of their data and comparison with theoretical model predictions, however, have only come up later (see [5.37,53-55] and in particular [5.56]. Recent MC simulations are concerned with the critical concentration $p_c (0 < p_c < 1)$ for percolation and corresponding critical exponents [5.57,58].

The cluster statistics can be analyzed in much the same way as described in Sect.5.2.2c, evaluating the occuring clusters with respect to size ℓ and surface s. The total number of clusters $N_c(p)$ per lattice site depending upon p was analyzed is [5.59]. The surface s was studied in some detail in [5.55,60]. (Note the varying definitions of cluster "surface" as mentioned in Chap.8.3). Percolation in a continuum model was studied in [5.61] and good agreement with experimental data on alkali-tungsten bronzes was found.

The most complete picture of percolation is obtained using a scaling theory [5.53,56] similar to conventional critical phenomena, but now for the cluster distribution $n_\ell(p)$

$$n_\ell(p) \sim \ell^{-\tau} f(\varepsilon \ell^{\sigma}) \tag{5.18}$$

where $\varepsilon = p - p_c$, τ and σ are two critical percolation exponents and f is a scaling function. (Details will be published in [5.56]). Invoking in addition finite-size scaling theory [5.17,18] one can also interpret the dependence upont system size [5.55] of the simulation results and obtain rather precise values for the critical percolation density p_c.

The highest precision of MC data so far seems to have been obtained in a recent analysis of a triangular lattice with $(4000)^2$ sites [5.56]. From scaling (5.18) one obtains at the percolation point $p = p_c$ the relation

$$\sum_{\ell'=\ell}^{\infty} n_{\ell'} \sim \ell^{1-\tau} \tag{5.19}$$

after summation. This relation was found to be fulfilled asymptotically for $\ell \approx 1000 \ldots 2000$ $(\tau \approx 2.055)$. For larger values of ℓ finiteness of the system causes remarkable deviations, as is clearly seen in Fig.5.6. The deviations for much smaller values of $\ell \leqq 100$ have to be considered as corrections to scaling.

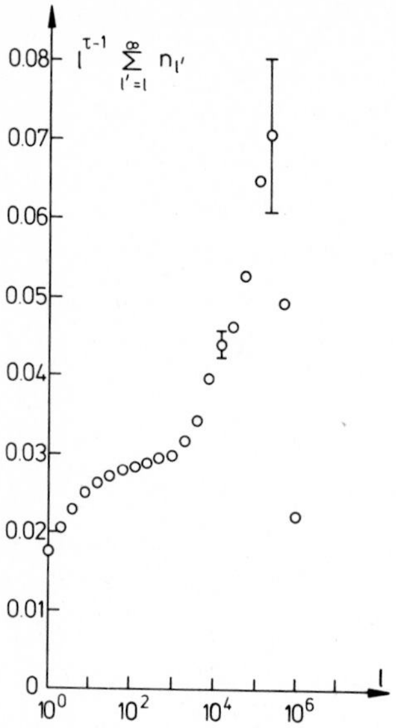

Fig.5.6 Test of the asymptotic power-law behavior of clusters in the percolation problem (5.19), on a 4000^2 two-dimensional triangular lattice. Near a cluster size of 10^3 particles the asymptotic behavior of an infinite system is well reproduced, while above $2\cdot10^3$ particles in the cluster the finiteness of the system causes deviations [5.56]

A more difficult situation is encountered when the particles (or spins) in the lattice interact with each other in such a way, that like spins attract each other while opposite spins repel each other (ferromagnetic interaction). The number of reversed spins as well as their clustering properties depend upon temperature, assuming thermodynamic equilibrium to be maintained.

Below the critical temperature T_c the ordered phase ("up" spins) then must percolate to maintain long-range order. The reversed spins then cannot percolate in a two-dimensional system at the same time. An exception is the critical temperature, where both phases are just at the percolation limit $p_c = 1/2$. The critical percolation density p_c was shown to increase above T_c in two dimensions [5.62] so that for $T > T_c$ there is no percolation at all.

The situation is quite different in three dimensions [5.39]. There both phases percolate at and above T_c and in fact even below T_c until at a certain temperature $T_p \approx 0.95\ T_c$ the ordered phase percolates, while the other phase only forms finite clusters below T_p. The observed lowering of the critical percolation density p_c (Fig.5.7) in the system with interaction compared to the one without interaction was confirmed by exact inequalities [5.63] and series expansions [5.64]. Other properties have not been analyzed in detail and therefore the interrelations between criticality and percolation cannot be quantified at the moment.

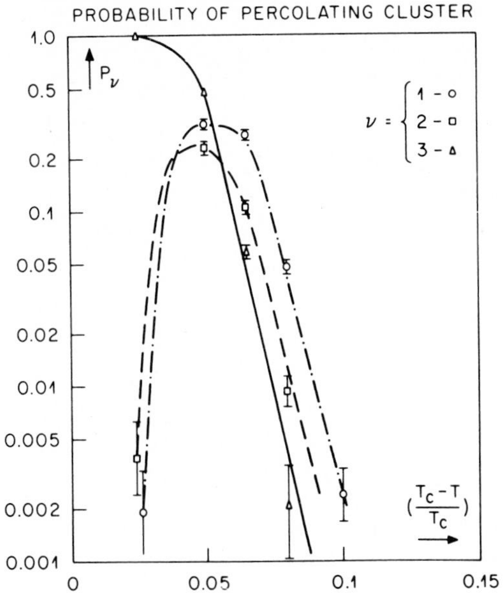

Fig.5.7 Probability of occurrence of one- two- and three-dimensional percolating clusters in the ferromagnetic Ising-model (30^3 sites) near T_c. About 5% below T_c a giant cluster percolates in all three directions through the system. This explains the problems in Fig.5.5b. [5.39]

5.3 Cluster Dynamics

5.3.1 First-Order Phase Transitions

A phase transition is called to be of first order or discontinuous, if the order parameter before and after the transition differs by a finite amount.

If the system originally was in a two-phase state where the two coexisting phases were separated by an interface, the transition occurs locally by a mere displacement of the interface.

The more interesting situation is realized when originally there is only one phase present. An interface between the old and the new phase has to be generated by thermal fluctuations at the cost of a corresponding amount of free energy. According to nucleation theory in this case a supercritical cluster of the new phase has to be formed. Once a cluster has become supercritical it will grow in the average and finally fill the whole system. The variation with time of the state of the system can be associated with a variation of the cluster distribution functions. Since the general aspects of the dynamics at phase transitions are discussed in Chapt.6 of this volume we shall concentrate here on the cluster behavior in particular.

The description of such a transition in terms of time-dependent cluster distributions at present seems to be the only formalism, which allows one to discuss the full transition from the coexistence line through the nucleation region and the spinodal region, and finally into the new phase [5.32]. Other approaches, like time-dependent Ginzburg-Landau theory and Cahn-Hilliard-Langer theory for spinodal decomposition, show a nonphysical spinodal line, while nucleation theory without the consideration of coagulation does not show the experimentally observed breakdown of the metastable states (for comparison of the various approaches see [5.32]).

Large-scale computer simulations [5.31,65-69] form the basis for a cluster reaction theory [5.31,32,68,70]. Clusters grow, coagulate and disintegrate by single spin-flips or by pair exchange of neighboring spins (the combined mechanisms have not been studied sofar). Most of the simulations have been carried out on the two-dimensional Ising model. The qualitative behavior of the system has been pictured by cine films [5.65,67]. Quantitatively the system's behavior can be described by nonlinear relaxation functions

$$\phi_A(t) = (\langle A(t)\rangle - \langle A(\infty)\rangle)/(\langle A(o)\rangle - \langle A(\infty)\rangle) \tag{5.20}$$

for macroscopic observables A. Their qualitative behavior during a first-order phase transition is shown in Fig.7.5.

The initial flat part corresponds to nucleation processes, where thermal fluctuations slowly generate supercritical nuclei, which again grow slowly. The faster decaying part corresponds to the coagulation region, where already formed small clusters agglomerate to larger clusters. The latter behavior should lead to maxima in the cluster probabilities $n_\ell(t)$ as a function of time, where the maxima occur at later times for larger cluster size ℓ. This is clearly observed in Fig.5.8 [5.68], where the cluster probabilities n_ℓ for large ℓ first increase due to coagulation of smaller clusters ($\ell' + \ell'' \rightarrow \ell$) and then decrease again because of further coagulation ($\ell + \tilde{\ell} \rightarrow \tilde{\tilde{\ell}}$).

In three dimensions the changeover from nucleation to coagulation seems to be related to the percolation problem [5.71]. But this is probably an accidental coincidence, since the relaxation functions have an equivalent shape in two dimensions and there percolation cannot occur for the phase which does not occupy the majority of sites in the lattice.

In these systems only single spin-flip processes have been considered. Inclusion of pair-exchange would modify the behavior insofar, that a supercritical cluster could grow faster by incorporating the monomers diffusing around in the neighborhood. Accordingly, there would be fewer large clusters and cooperative effects between the clusters would become important because of their overlapping diffusion fields.

Computer simulations of dynamic processes are very costly since the various time scales involved require either many simulation runs or very large systems. It could be shown, in fact, that the relaxation function $\phi(t)$ obtained for both alternatives becomes equivalent [Ref.5.68, Fig.9], provided the size of the smallest system used is large compared with the correlation length between sites and large compared with the critical nucleus.

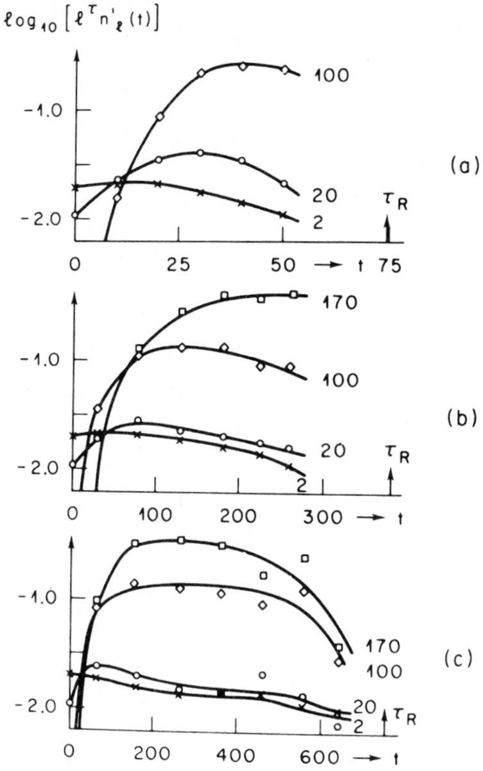

Fig.5.8 MC simulation of cluster probabilities varying with time during a first-order phase transition. The external field driving the phase transition decreases from (a) to (c). Parameter of the curves is the cluster size ℓ. First, the probability of small clusters increases and then goes down again, later the probabilities of larger clusters follow this behavior [5.68]

For these applications in first-order transitions it is more advisable to make a single run on a large system than to make several runs on smaller systems. In small systems the boundaries may cause problems as was shown in [5.72]. For a very large square Ising model the lifetime of the metastable states was insensitive to a variation of the system size since several supercritical clusters were already present when coagulation became important. For systems smaller than the critical nucleus the time τ needed for the transition is inversely proportional to the probability that a cluster covering half the system size is formed. In the intermediate range, where the system is larger than the critical nucleus but smaller than the average distance between several critical nuclei, the probability for finding a critical nucleus increases with the size of the system. The transition time there is then inversely proportional to the system size.

Closely related to the first-order phase transition is the phenomenon of spinodal decomposition. This effect is discussed in sufficient detail in Chapt.6 and, therefore, will not be treated here.

5.3.2 Second-Order Transitions

The dynamics of clusters near a second-order phase transition are completely different from the behavior at a first-order transition. No precursor effects are known to occur near first-order transitions. The cluster distribution function n_ℓ, therefore, has to be studied as a function of time. At a second-order transition it is well known [5.69] that fluctuations have very long correlation times. A similar effect of kinetic slowing down, therefore, is expected to occur for the cluster dynamics in thermal equilibrium.

The cluster distribution function at equilibrium is constant in the dynamic sense of detailed balance. Individual clusters grow and shrink by single-particle processes or by coagulation and disintegration. But in the average the inverse to every single process occurs with the same frequency so that the equilibrium distribution remains constant in the average. This holds for all Hermite models [5.73] where no propagating modes exist.

The state of the system (i.e., whether it is close to a critical point) then shows up in the different time scales of the cluster dynamics. As a starting point both for theory and for computer simulations one defines a cluster reaction matrix $W(\ell,\ell')$ which contains the probabilities for the coagulation of a cluster of size ℓ with a cluster of size ℓ'. This matrix of course is symmetric in ℓ and ℓ'. A cluster of size ℓ now in the average has a certain activity for the reaction with clusters of different sizes. This is conveniently expressed by the reaction constant R_ℓ [5.31,68,74]

$$R_\ell = n_\ell^{-1} \sum_{\ell'=0}^{\infty} W(\ell,\ell')(\ell' + 1)^2 \quad . \tag{5.21}$$

In the space of the cluster variable ℓ this R_ℓ then is a diffusion-"constant" (ℓ-dependent). It appears as rate determining factor in the cluster reaction equation, known from nucleation theory [5.30,74]

$$\frac{\partial}{\partial t} g_\ell = R_\ell \frac{\partial^2}{\partial \ell^2} g_\ell(t) + \frac{\partial}{\partial \ell}\{\ell n[R_\ell n_\ell(\infty)]\} \frac{\partial}{\partial \ell} g_\ell(t) \tag{5.22}$$

where $g_\ell = [n_\ell(t) - n_\ell(\infty)]/n_\ell(\infty)$ is an instantaneous deviation of the cluster distribution $n_\ell(t)$ from equilibrium $n_\ell(\infty)$.

The evaluation of the cluster reaction matrix of course is extremely time consuming (see Appendix). The creation and annihilation of isolated particles (reversal of isolated spins) was neglected. In order to have meaningful statistics the reaction matrix has to contain a sufficient number of uncorrelated events. The reaction of two clusters (coagulation, disintegration in thermal equilibrium follows from detailed balance) therefore was only counted, after a characteristic correlation time τ_μ for the magnetization had elapsed after the last reaction. But of course effective autocorrelation times for large clusters may be much larger than τ_μ, since the growth and shrinkage of a cluster represents a diffusion in the one-dimensional space of the cluster-variable ℓ.

Since every spin in the cluster has a nonzero chance to flip, one expects the cluster-reaction constant R_ℓ to increase for large ℓ as [5.31]

$$R_\ell \sim \ell^r \quad , \quad r \approx 1 \tag{5.23}$$

with some proportionality constant which depends on the microscopic details of the process. It is this exponent r which is related to the dynamic critical exponents $\Delta_{\mu\mu} = \Delta_{\mu E} = \Delta_{EE}$ of magnetization and energy relaxations

$$\Delta_{\mu\mu} = (2 - r)\beta\delta \quad . \tag{5.24}$$

The Monte Carlo simulations in a 30^3 simple cubic [5.31b] and a 110^2 square-lattice [5.31a] are consistent with the assumption (5.23). This seems to hold even better in the three-dimensional case as seen in Figs. 5.9a and b. The slope of the straight lines is taken from renormalization group results, using (5.24). The cluster definition was identical to the one used in Sect.5.2. This of course causes the same problem as for static quantities, that namely close to T_c the occurrence of percolating clusters (in three dimensions, almost percolating in two dimensions) obscures the critical behavior. Again it may be possible to circumvent this problem by an appropriate cluster definition, but corresponding MC simulations do no exist.

Only a few simulations have been performed for systems with conserved quantities [5.66] and [6.54,55,39]). Data on cluster diffusion constants (see Sect.6.2.3, Fig.6.4) were not sufficiently close to T_c in order to make a comparison with theoretical predictions on critical dynamics.

It may finally well be that the cluster reaction theory based on the present definition of clusters is the right approach to the kinetic models for percolation. Such a model might represent the diffusional movement of the atoms of the percolating phase in a host lattice. Since at the percolation threshold the surface of clusters increases proportional to the number of sites involved one should expect in that case a reaction exponent r = 1 (5.23) to describe the asymptotic reaction rate of large clusters.

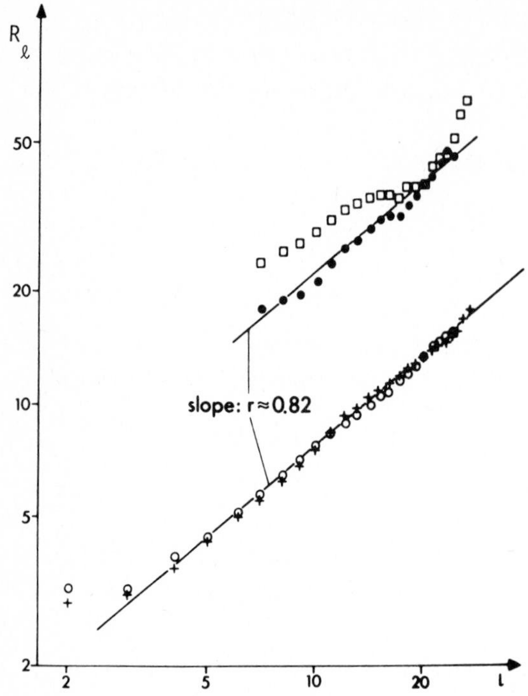

	R_ℓ	J/k_BT
o	$W(\ell,1)/n_\ell^F$	0.46
●	$\sum_{\ell'=1}^{10} (\ell'+1)^2 W(\ell,\ell')/n_\ell^F$	
+	$W(\ell,1)/n_\ell^F$	0.45
□	$\sum_{\ell'=1}^{10} (\ell'+1)^2 W(\ell,\ell')/n_\ell^F$	

Fig.5.9a Log-log plot of the reaction rate R_ℓ (5.21) versus number ℓ of spins in a cluster for a 110^2 square Ising lattice. For sufficiently large ℓ, R_ℓ follows approximately a power law equation (5.23). (The critical point here is at $J/k_BT = 0.440$) [5.31a]

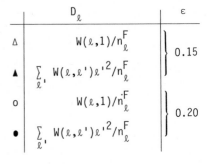

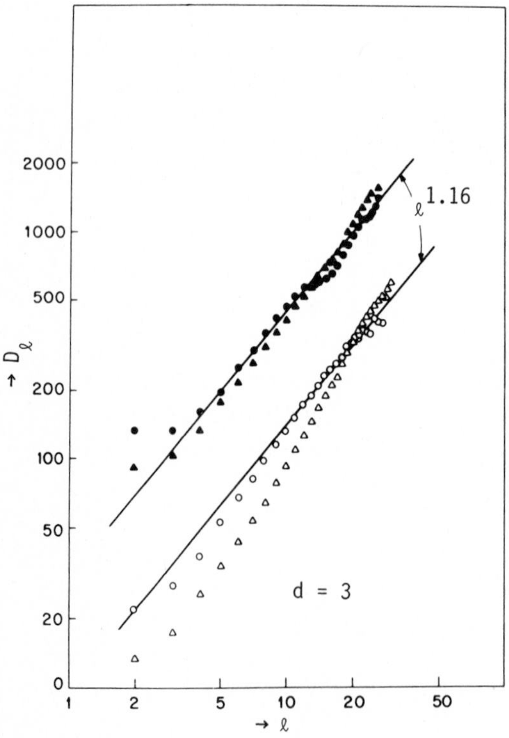

	D_ℓ	ε
Δ	$W(\ell,1)/n_\ell^F$	0.15
▲	$\sum_{\ell'} W(\ell,\ell')\ell'^2/n_\ell^F$	
o	$W(\ell,1)/n_\ell^F$	0.20
●	$\sum_{\ell'} W(\ell,\ell')\ell'^2/n_\ell^F$	

Fig.5.9b The same as in Fig.5.9a but for the slightly changed definition of D_ℓ instead of R_ℓ, now in a three-dimensional kinetic Ising-model with 30^3 sites [5.31b]

Appendix

Cluster Counting Alogorithm

As an example of a cluster counting computer program, this appendix describes our methods used, e.g., in [5.38,39].

Complete information about the clusters in a system is obtained if every particle (= site in a lattice system) has been labelled with a cluster identification number (CLNR) such that all particles within a specific cluster are assigned to the same unique number. This also allows one to calculate the total cluster surface by simply adding up the contributions of all particles to the surface. The investigation of the topology of the clusters (e.g., simple or multiple connectivity) is not straightforward and requires additional analysis.

In the following a fast cluster counting algorithm is described which produces the above mentioned assignment between CLNR and lattice site in a lattice system of arbitrary dimensionality. An algorithm in a continuous space may be constructed analogously if one first produces a matrix containing the information about "neighborhood" between particles in the system.

Assume the lattice system to be given in a d-dimensional matrix MA1($\underline{x}$), where $\underline{x}$ is a (discrete) lattice vector and MA($\underline{x}$) = {0,1} indicates an empty or filled site respectively. One then wants to construct a matrix MA2($\underline{x}$), which contains the CLNR for every lattice site.

There are several ways to realize this task. The fastest procedure would be to construct an inverted matrix, MINV(CLNR; L = 1 ... LMAX) which contains for every CLNR the position vectors of all particles in that cluster. During the counting procedure the lattice-matrix MA1 is investigated row by row, layer by layer etc., whether neighboring sites are both filled. Then they are part of a cluster, a CLNR is assigned and the positions of the sites are stored in MINV. This works as long as two neighbors are not already part of two different clusters. In that case the inverted matrix is used to find all the sites with the higher CLNR and relabel them with the lower CLNR. From then on the originally assumed two clusters are treated correctly as one cluster.

This fast procedure, however, generally requires more storage for the inverted matrix MINV than is available on most machines and is, therefore, usually not applicable (A storage-saving possibility would be to increase the size of the matrix MINV dynamically according to the instantaneous needs of the cluster analysis, instead of defining a rigid matrix structure at the beginning of the program. But this is not an easy task in a programming language like FORTRAN).

At the cost of increased computing time one can circumvent this storage limitation with an alternative procedure. If a connection between two clusters is encountered (two neighbors having different cluster number), the counting procedure is interrupted and restarted at the beginning with the following modification. At every site only the cluster identification number (CLNR) is checked. If it corresponds to the larger one of the two connected clusters mentioned above, it is replaced by the lower one, until the interruption point has been reached. From there on the original examination of neighborhood between particles is continued. For very diffuse clusters in large systems, however, this method is not very effective since the frequent relabelling runs are rather time consuming.

A very advantageous method with respect to both low computing time and moderate storage requirement was therefore constructed in the following way. When a connection between two cluster parts is encountered (two neighbors with different CLNR), no relabelling is performed, but the information of the connection between the two CLNR's is stored in an additional matrix. The fastest way to do this would be to define a two-dimensional matrix with rows and columns numbered with the CLNR. The matrix elements are set equal to zero except at intersections of rows and columns, where a connection between the CLNR's exists. But since the CLNR's easily may exceed values of 1000 in a three-dimensional $30 \times 30 \times 30$ Ising lattice this (very sparse) matrix again requires enormous storage space. Taking advantage of this sparseness one therefore better defines a two-dimensional matrix MCROSS for the cross-relations between the cluster parts CLNR, such that each row with index CLNR is successively filled with the cluster identification numbers (CLNR) of the cluster parts connected with the first CLNR. This required in no case matrices MCROSS larger than 2000×40.

In this way a single run through the lattice MA1 is sufficient to fill the matrix MA2 with preliminary cluster numbers (CLNR) and store the information about connections in the matrix MCROSS. Then the matrix MCROSS is analyzed. For this purpose one first defines a vector IRL for the relabelling of the cluster numbers, which initially contains the current number of its elements as elements [IRL(i) = i]. Then one goes through the connectivity matrix MCROSS and for every connection between a higher number CLNR = ih and a lower number CLNR = il one replaces the element of the vector IRL as IRL(ih) = il. This procedure is repeated on the matrix MCROSS, until no further change in IRL has occured (usually after one to three iterations).

The matrix of cluster numbers MA2 then is relabelled in a single run, by replacing the old cluster number CLNR = MA2($\underline{x}$) by the new one: MA2($\underline{x}$) = IRL(CLNR). All other properties, such as number of particles within a cluster, are obtained by trivially adding up the cluster parts, using the relabelling vector IRL.

One more run through the lattice is needed, of course, if one wants a printout of all particle coordinates associated with a particular cluster.

Percolation of a cluster is also easily investigated. One defines a vector VP(i) whose indices correspond to the cluster numbers (CLNR) and initializes it with zeros. Starting in the first layer, one inserts a "one" for each cluster number occuring. For all successive layers it is checked, if a particular cluster occurs, whether the respective element of VP is equal to the layer number minus one. In this case the element is set equal to the layer number. (This means that, e.g., in the third layer only elements of VP will be changed, if CLNR occurs in layer three and if VP(CLNR) is equal to "two").

After the last layer has been analyzed percolating clusters are identified by having "VP(CLNR)" equal to the total number of layers. Rotating the lattice by permuting the indices allows for a separate analysis of percolation in all spatial directions independently.

Cluster reaction rates are obtained in an analogous way. If a lattice point is changed from "empty" to "occupied" (or vice versa) one first checks, whether there is already a neighbor present. If yes, the lattice is twice analyzed with respect to its cluster distribution, once before and once after the occupation. The difference in cluster occurences immediately shows the type of reaction.

A final remark concerns the problem of how to test the program. There are two main sources of possible errors in this case. The one is the correct treatment of clusters that touch the boundary. This can be checked by inserting clusters of known properties. In three dimensions one should check the three types of boundary-crossing clusters separately: clusters going through a plane, an edge and a corner. The correct treatment of the bulk properties then is tested with a consistency argument. After a 90° rotation of the lattice, the cluster counting program cannot recognize that the system is identical. Identical results for all six possible orientations at both high and low cluster densities suggest the program to be correct.

References

5.1 J.K. Lee, J.A. Barker, F.F. Abraham: J. Chem. Phys. *58*, 3166 (1973)

5.2 F.F. Abraham: J. Chem. Phys. *61*, 1221 (1974)

5.3 J. Miyazaki, G.M. Pound, F.F. Abraham, J.A. Barker: J. Chem. Phys. *67*, 3851 (1977)
K. Nishioka: Phys. Rev. A *16*, 2143 (1977)

5.4 K. Binder, D. Stauffer: J. Stat. Phys. *6*, 49 (1972)

5.5 K. Binder: J. Chem. Phys. *63*, 2265 (1975)

5.6 J.J. Burton, C.L. Briant: "Atomistic Models of Microclusters", in *Nucleation Phenomena,* ed. by A.C. Zettlemoyer (Elsevier, Amsterdam 1977) p. 131

5.7 M.R. Hoare, P. Pal: Nature (Phys. Sci.) *236*, 35 (1972); J. Crystal Growth *17*, 77 (1972)

5.8 J.J. Burton: In *Modern Theoretical Chemistry,* Vol. 6, ed. by B. Berne (Plenum Press, New York 1977)

5.9 J.G. Curro: J. Chem. Phys. *61*, 1203 (1974)
D.C. Rapaport: J. Phys. A *9*, 1521 (1977)

5.10 K. Binder: Phys. Stat. Sol. *46*, 567 (1971)

5.11 K. Binder: Physica *62*, 508 (1972)

5.12 D.P. Landau: Phys. Lett. A *47*, 41 (1973)

5.13 D.P. Landau: Phys. Rev. B *13*, 2997 (1976)

5.14 D.P. Landau: Phys. Rev. B *14*, 255 (1976)
5.15 T.L. Hill: J. Chem. Phys. *36*, 3182 (1962)
5.16 T.L. Hill: *Thermodynamics of Small Systems* (Benjamin, New York 1963)
5.17 M.E. Fisher, A.E. Ferdinand: Phys. Rev. Lett. *19*, 169 (1967)
5.18 A.E. Ferdinand, M.E. Fisher: Phys. Rev. *185*, 832 (1969)
5.19 K. Binder, P.C. Hohenberg: Phys. Rev. B *9*, 2194 (1974)
5.20 Th.T.A. Paauw, A. Compagner, D. Bedeaux: Physica *79*, 1 (1975)
5.21 E. Domany, K.K. Mon, G.V. Chester, M.E. Fisher: Phys. Rev. B *12*, 5025 (1975)
5.22 E. Kneller: "Theorie der Magnetisierungskurve kleiner Kristalle", in *Ferromagnetismus*, Hrsg. H.P.J. Wijn, Handbuch der Physik, Bd. XVIII/2 (Springer, Berlin, Heidelberg, New York 1966), S. 438
I.S. Jacobs, C.P. Bean: In *Magnetism*, ed. by G.T. Rado, H. Suhl (Academic Press, New York 1963)
5.23 K. Binder, H. Rauch, V. Wildpaner: J. Phys. Chem. Sol. *31*, 391 (1970)
5.24 K. Binder: Z. Angew. Phys. *30*, 51 (1970)
5.25 V. Wildpaner: Z. Phys. *270*, 215 (1974)
5.26 H. Müller-Krumbhaar, K. Binder: Z. Phys. *254*, 269 (1972)
5.27 K. Binder, H. Müller-Krumbhaar: Phys. Rev. B *7*, 3297 (1973)
5.28 H.C. Bolton, C.H. Johnson: Phys. Rev. B *13*, 3025 (1976)
5.29 J. Feder, K. Russel, J. Lothe, M. Pound: Advan. Phys. *15*, 117 (1966)
5.30 A. Zettlemoyer (ed.): *Nucleation* (Dekker, New York 1969); *Nucleation II* (Marcell Dekker, New York 1977)
5.31 K. Binder, D. Stauffer, H. Müller-Krumbhaar: a) Phys. Rev. B *12*, 5261 (1975) b) Phys. Rev. B *10*, 3853 (1974)
5.32 K. Binder: Phys. Rev. B *15*, 4425 (1977)
5.33 J. Hoshen, R. Kopelman: Phys. Rev. B *14*, 3438 (1976)
5.34 C. Domb, E. Stoll: J. Phys. A *10*, 1141 (1977)
5.35 R. Becker, W. Döring: Ann. Phys. *24*, 719 (1935)
5.36 C. van Leeuwen, J.P. van der Eerden: Surface Sci. *64*, 237 (1977)
5.37 H. Müller-Krumbhaar, E.P. Stoll: J. Chem. Phys. *65*, 4294 (1976)
5.38 H. Müller-Krumbhaar: Phys. Lett. A *48*, 459 (1974)
5.39 H. Müller-Krumbhaar: Phys. Lett. A *50*, 27 (1974)
5.40 E. Stoll, K. Binder, T. Schneider: Phys. Rev. B *6*, 2777 (1972)
5.41 H.E. Stanley: *Introduction to Phase Transitions and Critical Phenomena* (Clarendon Press, Oxford 1971)
5.42 K. Binder: Ann. Phys. (N.Y.) *98*, 390 (1976)
5.43 M.E.Fisher: Physics *3*, 255 (1967)
5.44 C. Domb, T. Schneider, E. Stoll: J. Phys. A *8*, L90 (1975)
5.45 H.N.V. Temperley: J. Phys. A *9*, L113 (1976)
5.46 H.N.V. Temperley, E.H. Lieb: Proc. R. Soc. A *322*, 251 (1971)
5.47 R.J. Baxter: J. Phys. C *6*, L445 (1973)
5.48 J.W. Essam: Discrete Math. *1*, 83 (1971)
5.49 S. Kirkpatrick: Rev. Mod. Phys. *45*, 574 (1973)
V.K.S. Shante, S. Kirkpatrick: Advan. Phys. *20*, 325 (1971)
5.50 J.W. Essam: In *Phase Transitions and Critical Phenomena*, ed. by C. Domb, M.S. Green, Vol. 2 (Academic Press, New York 1972) p. 197
5.51 M.F. Sykes, D.S. Gaunt, J.W. Essam: J. Phys. A *9*, L43 (1976) and references
5.52 P. Dean, N.F. Bird: National Laboratory, Teddington, England, Rpt. No. MA61 (1966)
5.53 D. Stauffer: Phys. Rev. Lett. *35*, 394 (1975)
5.54 M.M. Bakri, D. Stauffer: Phys. Rev. B *14*, 4215 (1976)
5.55 A. Sur, J.L. Lebowitz, J. Marro, M.H. Kalos, S. Kirkpatrick: J. Statist. Phys. *15*, 345 (1976)
5.56 J. Hoshen, D. Stauffer, G.H. Bishop, R.J. Harrison, G.D. Quinn: Preprint, submitted to J. Phys. A *12*, 1285 (1979)
5.57 K.J. Duff, V. Canella: AIP Conf. Proc. *10*, 541 (1972)
5.58 M.E. Levinshtein, B.I. Shklovskii, M.S. Shur, A.L. Efros: Zh. Eksp. Theor. Fiz. *69*, 386 (1975)
5.59 G.D. Quinn, G.H. Bishop, R.J. Harrison: J. Phys. A *9*, L9 (1976)
5.60 P.L. Leath: Phys. Rev. Lett. *36*, 921 (1976); Phys. Rev. B *14*, 5046 (1976)
5.61 I. Webman, J. Jortner, M. Cohen: Phys. Rev. B *13*, 713 (1976) and references
5.62 T. Odagaki, N. Ogita, H. Matsuda: J. Phys. Soc. Japan *39*, 618 (1975)

5.63 A. Coniglio, C.R. Nappi, F. Peruggi, L. Russo: J. Phys. A *10*, 205 (1977) and references
5.64 M.F. Sykes, D.S. Gaunt: J. Phys. A *9*, 2131 (1976)
5.65 N. Ogita, A. Ueda, T. Matsubara, H. Matsuda, F. Yonezawa: J. Phys. Soc. Jpn. (Suppl.) *26*, 145 (1969)
5.66 A. Sur, J. Lebowitz, J. Marro, M. Kalos: Phys. Rev. B *15*, 3014 (1977)
5.67 T. Schneider, E. Stoll: In *Anharmonic Lattices, Structural Transitions and Melting*, ed. by T. Riste (Nordhoff, Leiden 1971) p. 275
5.68 K. Binder, H. Müller-Krumbhaar: Phys. Rev. B *9*, 2328 (1974)
5.69 E. Stoll, K. Binder, T. Schneider: Phys. Rev. B *8*, 3266 (1973)
5.70 P. Mirold, K. Binder: Acta Met. 25, 1435 (1977)
5.71 H. Müller-Krumbhaar: Unpublished
5.72 E.P. Stoll, T. Schneider: Physica B *86-88*, 1419 (1977)
5.73 K. Kawasaki: In Ref. 5.50, Vol. II, p. 443
5.74 K. Binder, D. Stauffer: Advan. Phys. *25*, 343 (1976)

Addendum

Further information on the structure of clusters is given by DOMB [5.75], discussing the cyclomatic number in the Ising model, and LEATH and REICH [5.76], who discuss internal and external surface for percolation clusters. KLEIN et al. [5.77] investigated the typical cluster radius for interacting systems. The finite size effects for the percolation problem were further studied by KIRKPATRICK [5.78], by REYNOLDS et al. [5.79], and by HOSHEN et al. [5.80].

5.75 C. Domb: Ann. Israel Phys. Soc. *2*, 225 (1978)
5.76 P.L. Leath, G.R. Reich: J. Phys. C *11*, 4017 (1978)
5.77 W. Klein, H.E. Stanley, P.J. Reynolds, A. Coniglio: Phys. Rev. Lett. *41*, 1145 (1978)
5.78 S. Kirkpatrick: in *Ill-Condensed Matter*, ed.by R. Balian, R. Maynard, and G. Toulouse (North-Holland, Amsterdam 1979)
5.79 R.J. Reynolds, H.E. Stanley, W. Klein: J. Phys. A *11*, 199 (1978)
5.80 J. Hoshen, R. Kopelman, E.M. Monberg: J. Statist. Phys. *19*, 219 (1978)

6. Monte Carlo Studies of Relaxation Phenomena: Kinetics of Phase Changes and Critical Slowing Down

K. Binder and M. H. Kalos

With 15 Figures

Monte Carlo calculations of the dynamic properties of various stochastic models are reviewed, concentrating upon applications of kinetic Ising models to biopolymers, anisotropic magnets, and binary metallic alloys. Dynamic phenomena at critical as well as tricritical points are treated, and simulations are described which elucidate the kinetic mechanisms of first-order phase transitions (like nucleation, spinodal decomposition, etc.). Consequences of these simulations both for experiment and for analytic theories are outlined.

6.1 Introductory Remarks

The suitability of the Monte Carlo method for the study of relaxation phenomena rests on the fact that Monte Carlo sampling is a *kinetic process in itself by construction* (cf. Chap.1) [6.1-3]. One may attempt to choose the details of this kinetic process in such a way that a description of the desired system, as realistic as one may wish, is obtained. Systems which can be suitably studied by this method include: binary and ternary alloys, adsorbed surface layers, highly anisotropic magnets, uniaxial order-disorder ferroelectrics, molecular crystals like NH_4Cl, cooperative phenomena in biopolymers, problems related to vacancy or interstitial diffusion, etc. An important feature is that the simulation method is *not* restricted to small deviations from thermal equilibrium, as most well-established analytical methods are; on the contrary, the "initial condition" used for the Monte Carlo sample is usually a state far from thermal equilibrium [6.1-3]. Thus nonlinear relaxation phenomena, like nucleation or spinodal decomposition, are easily accessible to a study. Hence Monte Carlo sampling is a powerful method in materials science, a fact which is far from being exhaustively exploited.

In the present review we intend to give a broad survey only, referring the reader to the various original papers for more details. The phenomena considered may be classified into two kinds: first we take up relaxation processes close to equilibrium and afterwards nonlinear phenomena at phase transitions. Although the applications are extremely diverse (ranging from polymer kinetics to metallurgy), all topics are discussed in the framework of kinetic Ising models [6.4] in one, two and three dimension. Each of the following sections will begin with a brief description of the physical system involved, and a demonstration of its equivalence to an Ising system. The technical side of simulations of this kind is discussed at length in Chap.1 and will not be repeated here. But we do give a formal exposition of the kinetic master equation which governs all of the models involved. In all sections the common theme is the model predictions of "relaxation functions" which are related to physically observable quantities like inelastic neutron cross sections, NMR linewidths, etc. Where relevant experimental data exist these are compared with predictions, but in many cases we are as much interested in the physical insight that can be derived from the simulations better than from any other means. One can also mediate between controversial theories where neither is contradicted by experiment.

Additional applications of simulations of interface kinetics and crystal growth based upon Ising models may be found in the next chapter of this book. In order to present a better understandinq of the theoretical concepts underlying these applications we recall (Chap.1) that in the course of the Monte Carlo simulation one generates a "Markov chain" of phase space points $\underline{R}_\nu$. In a magnet, $\underline{R}_\nu \equiv (\mu_1,\ldots,\mu_N)$, where μ_i is the magnetic moment associated with the ith atom, in a binary alloy (AB) $\underline{R}_\nu = (c_1, \ldots\, c_N)$. where $c_i = 1$ if the ith lattice site is occupied by an A-atom and zero otherwise, etc. It is useful to describe the state of the system in terms of the probability $P(\underline{R},t)$ that the system is in state $\underline{R}$ at the νth step, introducing a time variable t by $t = \tau_s\nu/N$, where τ_s is an arbitrary scale factor (which is unity if we measure the time in "Monte Carlo steps/atom") [6.1-3] (MCS). The evolution of the system is then described in terms of a Markov master equation for the probability $P(\underline{R},t)$ to which we associate a vector $|P(t)\rangle$ so that $P(\underline{R},t) \equiv \langle\underline{R}|P(t)\rangle$. Treating t as a continuous variable, we have

$$\frac{d|P(t)\rangle}{dt} = -\hat{L}|P(t)\rangle, \quad \frac{d}{dt}P(\underline{R},t) = -\left\langle\underline{R}|\hat{L}|P(t)\right\rangle \quad , \tag{6.1}$$

where the Liouville operator $\hat{L}$ is expressed in the R representation in terms of a transition probability $W(\underline{R}\rightarrow\underline{R}')$ per unit time

$$-\left\langle\underline{R}|\hat{L}|P(t)\right\rangle = -\sum_{\underline{R}'} W(\underline{R}\rightarrow\underline{R}')P(\underline{R},t) + \sum_{\underline{R}'} W(\underline{R}'\rightarrow\underline{R})P(\underline{R}',t) \quad . \tag{6.2}$$

The physical significance of (6.1,2) is just that of a rate equation: the first term in (6.2) describes the loss of probability due to all transitions away from the state $\underline{R}$, while the second term describes the gain of probability due to all transitions from neighboring states into that state. The nature of the transitions $\underline{R}\rightarrow\underline{R}'$ depends on the physics of the problem: in a magnet a transition may be a single spin flip ($\mu_i \rightarrow -\mu_i$), in an alloy an interchange of neighboring atoms

$(c_i \to c_j, c_j \to c_i)$, etc. As a general condition, however, we require that detailed balance is obeyed with the equilibrium distribution

$$P_0(\underline{R}) \equiv P(\underline{R},\infty) = (1/Z)\exp(-\mathcal{H}/k_BT) \tag{6.3}$$

where $\mathcal{H}$ is the hamiltonian of the system and $Z = \mathrm{Tr}\exp(-\mathcal{H}/k_BT)$ is its partition function. That is

$$W(\underline{R} \to \underline{R}')P_0(\underline{R}) = W(\underline{R}' \to \underline{R})P_0(\underline{R}'); \quad \hat{L}|P_0\rangle = 0 \quad . \tag{6.4}$$

Apart from the restriction imposed on $W(\underline{R} \to \underline{R}')$ by (6.3,4), its precise behavior is not determined; one may choose $W(\underline{R} - \underline{R}')$ so that it simulates the physical system as realistically as possible. Of course, since $\hat{L}$ defined in (6.1,2) is a hermitean operator, the dynamics described is purely relaxational, one cannot describe any propagating modes like phonons, magnons, etc. [6.5]. Furthermore the dynamics is that of classical and not quantum statistics, of course (no tunneling phenomena can be described, etc.). Nevertheless a large variety of systems is accessible to study, as will become evident below.

The averages taken in Monte Carlo sampling are time averages, defined as

$$\left\langle \hat{A}(t) \right\rangle = \left\langle \underline{R}|\hat{A}|P(t) \right\rangle = \sum_{\underline{R}} A(\underline{R})P(\underline{R},t) \quad , \tag{6.5}$$

where $A(\underline{R}) \equiv \langle\underline{R}|\hat{A}|\underline{R}\rangle$ is any observable quantity of interest for the system. It turns out that it is of central importance to study the quantities A_c for which conservation laws exist, i.e.,$(d/dt)\langle\hat{A}_c(t)\rangle \equiv 0$ for every state. Since first of all the total probability is unity and hence conserved, we have

$$\sum_{\underline{R}} P(\underline{R},t) = \sum_{\underline{R}} \left\langle \underline{R}|P(t) \right\rangle = 1 \quad , \tag{6.6}$$

and hence

$$\sum_{\underline{R}} \frac{dP(\underline{R},t)}{dt} = -\sum_{\underline{R}} \left\langle \underline{R}|\hat{L}|P(t) \right\rangle = 0 \quad ; \quad \langle\underline{R}|\hat{L} = 0 \quad . \tag{6.7}$$

Using (6.1,7), the equation of motion for $\langle\hat{A}(t)\rangle$ is cast into the canonic form [6.5]

$$\frac{d}{dt}\left\langle \hat{A}(t) \right\rangle = -\left\langle R|\hat{A}\hat{L}|P(t) \right\rangle = \left\langle R|[\hat{L},\hat{A}]|P(t) \right\rangle \quad , \tag{6.8}$$

and thus we find that for a conserved quantity $\hat{A}_c$ the commutator vanishes

$$[\hat{L},\hat{A}] = \hat{L}\hat{A} - \hat{A}\hat{L} = 0 \quad . \tag{6.9}$$

Hence we note that conservation laws for models with "stochastic" dynamics can be cast in a notation formally analogous to systems with "hamiltonian" dynamics. The important physical consequence of (6.9) is the "hydrodynamic slowing down" [6.4]; correlation functions of conserved quantities will for small wave vector $\underline{k}$ and large times behave as $\langle\hat{A}(\underline{k},t)\ \hat{A}\ (-\underline{k},0)\rangle \propto \exp(-D_A k^2 t)$, where D_A is some appropriate kinetic coefficient.

In binary alloys, for instance, the average concentration of either constituent is a conserved quantity, in adsorbed surface layers the coverage of adatoms is conserved, etc. Hence we will draw a clear distinction between models with and without conservation laws in the following applications.

6.2 Kinetics of Fluctuations in Thermal Equilibrium

6.2.1 Dynamics of Models for Chain Molecules

Here we are concerned with the helix-coil transition of a biopolymer [6.6]. In the case of homopolypeptides one has N identical peptide groups. The hydrogen of a base-NH in each peptide may form a hydrogen bond with a base-C = O in a peptide which is four units away from the base-NH. These hydrogen bonds lead to a spiral arrangement of the polymer (the α-helix [6.6]). If a peptide unit does not take part in this hydrogen binding, then the structural arrangement (with respect to the neighboring units) is not specified. Therefore the polymer may also be in a random structure (the coil [6.6]). Introducing a structure-variable $\mu_k(t) = +1$ if the kth unit is in a helical state at time t,.$\mu_k(t) = -1$ if it is an a coil state, it is clear that the helix-coil transition can be discussed in terms of an Ising model [6.6-9] in which the sites constitute a one-dimensional chain. In the hamiltonian of the system one often takes into account only interaction energies between neighboring units [6.7-9]

$$\mathcal{H} = -J\sum_{j=1}^{N-1}\mu_j\mu_{j+1} - H\sum_{j=1}^{N}\mu_j \ ; \quad \sigma = \exp[-(4J)/k_BT] \ ; \quad s = \exp(2H/k_BT) \ . \tag{6.10}$$

Thus the μ_j are formally identifiable as the "spins" of an Ising-model. Either the interaction parameters J,H or the "statistical weights" σ,s which are often used have to be fitted to experiments where one studies the variation of the helix fraction $\theta = (\langle\mu_j\rangle + 1)/2$ with temperature [6.6-9]. "Abnormal" end-effects [6.10] in (6.10) have been neglected, although one could have chosen different parameters involving the spins μ_1,μ_N in (6.10).

The kinetic properties of such molecules in aqueous solution have been studied experimentally by various techniques: ultrasonic attenuation, temperature jump, nuclear magnetic resonance, etc. [6.7-9]. In the simplest theoretical description, one assumes that the solution acts as a heat bath on the polymer, inducing random transitions of the peptides from the helix state to the coil state and vice versa. This is precisely the situation described by (6.1-4), and no conservation law comes into play. A very simple choice of transition probability consistent with (6.4) is that of the GLAUBER kinetic Ising model [6.11]

$$W(\mu_i \to -\mu_i) = \frac{1}{2\tau_s}\,[1 - \tanh(\delta\mathcal{H}/2k_BT)] \quad . \tag{6.11}$$

Here τ_s is an adjustable parameter which fixes the time scale of the relaxation process, and $\delta\mathcal{H}$ the cost of energy required by the "spin flip" [computed from (6.10)] More complicated choices of W which are still consistent with (6.4) but which involve up to *three* time-scale constants τ_{s1}, τ_{s2}, τ_{s3} are conceivable and probably necessary to obtain a more realistic description of the kinetics [6.7-9]. We disregard this complication here for simplicity. Experimental interest focuses [6.7-9] on the relaxation function $\phi_\theta(t)$ of the helix fraction

$$\phi_\theta(t) = [\theta(t) - \theta(0)]/[\theta(\infty) - \theta(0)] \quad . \tag{6.12}$$

This quantity was readily obtained from computer simulations [6.12]. There was considerable discussion [see, e.g., [6.8,9] of the extent of the agreement between the initial and final relaxation times, $\tau_i[d\phi_\theta(t)/dt|_0 = \tau_i^{-1}]$ and $\tau\{\phi_\theta(t) \propto \exp(-t/\tau)$ as $t \to \infty\}$, respectively. For the conditions studied by OGITA et al. [6.12] these times differed by less than by a factor of two.

Another interesting aspect of the Glauber model (6.10,11) is that it is exactly soluble in the special case H = 0, where the spin correlations $\langle\mu_j(t_0)\mu_{j-k}(t_0 + t)\rangle$ can also be obtained for arbitrary sites j,k [6.11]. Thus this model is very suitable to test the quantitative accuracy of the Monte Carlo simulation. It turns out that the Monte Carlo results [6.2,3,13] agree with the exact results very nicely (see, e.g., [Ref.6.2, Fig.1a]), if one takes both an average over t_0 (at $k_BT/J \leqq 1$ an average over several hundred MCS/spin is necessary) and over the sites j of a chain with N = 220 spins.

Finally we mention the extension to the case of hetero-polymers. Of practical importance is the case where the chain is composed of two constituents (A,B) in (quasi-)random succession (e.g., in nucleic acids one may have either adenin and thymin or guanin and cytosin) [6.6]. This case can be transformed to a generalization of (6.10) where a term $-\sum_{j=1}^{N} h_j\mu_j$ is added to the hamiltonian; the field $h_j = +h$ if site j is occupied by an A-molecule, $h_j = -h$ for a B-molecule [6.6].

This term then accounts for the fact that the two consituents involve different binding enery parameters. The treatment of an Ising chain in random fields is cumbersome even with respect to static properties, and thus Monte Carlo methods are valuable for obtaining θ vs. T-curves [6.14,15]. Some results on the relaxation function $\phi_\theta(t)$ for this case can be found in [6.15].

6.2.2 Critical Slowing Down in Systems Without Conservation Laws

Here we consider a highly anisotropic ferromagnet or antiferromagnet with magnetic moments localized at lattice sites. Such a system is again described by a hamiltonian of the type of (6.10), where now J represents the exchange interactions between the magnetic moments, and H represents the magnetic field (in units of the Bohr magneton). The important difference between the present case and (6.10) is, of course, that now the spins are arranged on a three-dimensional lattice rather than a one-dimensional chain. However, for some layered substances the exchange perpendicular to the layers is negligible in comparison with the in-layer exchange, and thus it makes sense to consider two-dimensional lattices. Lattice vibrations induce random spin flips through the coupling between the lattice and the magnetic moments (spins), which can be modelled again by (6.11). At H = 0 a second-order phase transition occurs at some critical temperature T_c, where also some relaxation times of the model diverge ("critical slowing down"). The relaxation of the model is described conveniently in terms of relaxation functions $\phi_{B,C}(\underline{q},t)$ for observables B,C, which are defined as normalized fourier transforms of correlation functions of the observables

$$\phi_{BC}(\underline{q},t) = S_{BC}(\underline{q},t)/S_{BC}(\underline{q},0) \; ; \qquad S_{BC}(\underline{q},t) = \sum_j \exp(i\underline{q}\cdot\underline{x}\ )[\langle\hat{B}(0,0)\hat{C}(\underline{x}_j,t)\rangle - \langle\hat{B}\rangle\langle\hat{C}\rangle]. \tag{6.13}$$

Here B,C may be the local magnetization μ_j, or the local energy $\mathscr{H}_j = -J\sum_i \mu_i\mu_j$, etc. Relaxation times are then defined as integrals, $\tau_{BC}(\underline{q}) = \int_0^\infty \phi_{BC}(\underline{q},t)dt$. One considers also the autocorrelation function

$$\phi_{BC}^A(t) = \sum_q S_{BC}(\underline{q},t) / \sum_q S_{BC}(\underline{q},0) = [\langle\hat{B}(0,0)\hat{C}(0,t)\rangle - \langle\hat{B}\rangle\langle\hat{C}\rangle]/[\langle\hat{B}\hat{C}\rangle - \langle\hat{B}\rangle\langle\hat{C}\rangle]$$

and its relaxation time $\phi_{BC}^A = \int_0^\infty \phi_{BC}^A(t)dt$. While such relaxation times are accessible (at least in principle) through linewidth measurements in resonance experiments, full information on the relaxation function of the magnetization itself can be obtained from inelastic neutron scattering. The character of the critical slowing down is expressed by

$$\tau_{BC}(0) \propto |T - T_c|^{-\Delta_{BC}} \quad ; \quad \tau_{BC}^{A} \propto |T - T_c|^{-\Delta_{BC}^{A}} \quad .$$

Monte Carlo studies of this model are available for the case of three-dimensional (3d) simple cubic $8 \times 8 \times 8$ lattices with periodic boundary conditions [6.1,16], two-dimensional square 55×55 lattices with periodic boundary conditions [6.13,17]. Due to the small size of the systems considered no estimates for the exponents Δ_{BC}, Δ_{BC}^{A} introduced above were obtained in the 3d-case. This case will not be discussed further, nor the results of [6.17] where higher moments of the relaxation functions $\Delta_{BC}(g,t)$ were studied. But information relevant for critical exponents was obtained in [6.13] (for detailed reviews see [6.3,18]). The results are

$$\Delta_{\mu\mu} \approx \Delta_{\mu\mathscr{H}} \approx \Delta_{\mathscr{H}\mathscr{H}} \approx 1.9 \pm 0.1,\ \Delta_{\mu\mu}^{A} \approx 1.6 \pm 0.1,\ \Delta_{\mu\mathscr{H}}^{A} \approx 0.95 \pm 0.1,\ \Delta_{\mathscr{H}\mathscr{H}}^{A} \approx 0 \quad ,$$

corresponding exponents below T_c and above T_c being equal. These results are in full accord with an *extended* dynamic scaling hypothesis [6.19]: (I) magnetization fluctuations and energy fluctuations decay with the same relaxation time, in contrast to conjectures of SUZUKI [6.20]; (II) the self-correlation exponents are consistent with the scaling laws [6.13] connecting them to static exponents α,β [6.21] $\Delta_{\mu\mu} - \Delta_{\mu\mu}^{A} = 2\beta$, $\Delta_{\mu\mathscr{H}} - \Delta_{\mu\mathscr{H}}^{A} = 1 - \alpha$, $\Delta_{\mathscr{H}\mathscr{H}} - \Delta_{\mu\mu}^{A} = 2(1 - \alpha)$, since in the 2d Ising model the order parameter exponent $\beta = 1/8$ and the specific heat exponent $\alpha = 0$; (III) the exponent $\Delta_{\mu\mu}$ exceeds slightly the susceptibility exponent $\gamma(= 7/4$ [6.21]), in contrast to the conventional theory of critical slowing down [6.4] and approximations based on a mode-mode coupling approach [6.22]. All the results (I) - (III) are consistent with high temperature series expansions [6.23] and with renormalization group treatments of a continuum version of the kinetic Ising model [6.24]. But for some time there was considerable discussion of the validity of the conventional theory in the literature [6.4,13,20], and the simulations described above have served to clarify this matter.

Apart from this impact on the theory of dynamic critical phenomena, such investigations may contribute to interpret corresponding experiments on anisotropic magnets and ferroelectrics. Since at T_c dynamic scaling implies that $\tau_{\mu\mu}(q) \propto q^{z}$, $z = \Delta_{\mu\mu}/\nu$ where ν is the critical exponent of the correlation length [6.19], $\Delta_{\mu\mu}$ could in principle be measured by inelastic neutron scattering. Such a measurement exists for the uniaxial 3d antiferromagnet FeF_2 [6.25], but is is hardly accurate enough to show unambiguously the difference between $\Delta_{\mu\mu}$ and γ, which is very small in three dimensions. More promising in this respect would be the study of anisotropic quasi-two-dimensional magnets (like K_2CoF_4, Rb_2CoF_4 [6.26]). It is possible to measure $\tau_{\mu\mu}^{A}$ and hence $\Delta_{\mu\mu}^{A}$ with better precision, since $\tau_{\mu\mu}^{A}$ is proportional to the NMR linewidth. Using the above scaling relation $\Delta_{\mu\mu} - \Delta_{\mu\mu}^{A} = 2\beta$, one in fact obtains [6.24] a slight difference between $\Delta_{\mu\mu}$ and γ

using data for FeF_2 [6.27]. Another consequence of the simulations [Ref.6.28, Fig.2] that the Fourier transform $\tilde{\phi}_{BC}(\omega)$ of $\phi_{BC}(0,t)$ is slightly nonlorentzian [i.e., $\phi_{BC}(0,t)$ deviates from a simple $\exp(-t/\tau_{\mu\mu})$ decay], is in good accord with other theoretical approaches ("cluster dynamics" [6.29], renormalization group calculations [6.30]), but experimental verification is still lacking.

The kinetic Ising model can also be used to represent a ferroelectric, if one associates a "pseudospin" with the electric dipole moment of each unit cell. Due to the longer range of the interaction between the pseudospins representing the electric dipole moments, layered ferroelectric crystals (like square acid or $SnCl_2 \cdot 2H_2O$ [6.31]) are expected at best to be in qualitative accord with the simulations for nearest-neighbor kinetic Ising models. Nevertheless it seems encouraging that preliminary results on $SnCl_2$ seem to indicate [6.32] $\Delta_{\mu\mu} \approx 2$ and nonlorentzian line shapes ("polydispersive relaxation"). Clearly Monte Carlo simulations of kinetic Ising models with spins interacting with the dipole-dipole interaction would be desirable.

Finally we mention that the kinetic Ising model considered here may be used also as a model of a monolayer of adatoms, where the prefered sites of the substrate layer form a square lattice [6.33,34]. If an occupied site corresponds to spin up and an empty site to spin down, the "coverage" is $\theta = (1 + \langle\mu_i\rangle)/2$ [6.33]. Random spin flips correspond to random evaporation and condensation of the adatoms, and one considers thermal equilibrium between the adsorbed layer and the adsorbate gas phase. Since surface diffusion of adatoms in neglected, this model is perhaps too unrealistic a description of the kinetics, and we are unaware of experiments which correspond to it.

6.2.3 Relaxation in Systems with Conserved Quantities

Here we consider simple lattice models of binary alloys (AB), in which atoms A or B are placed on the sites of a perfect crystal. If we assume pairwise interactions $v_{AA}(\underline{r}_i - \underline{r}_j)$, $v_{AB}(\underline{r}_i - \underline{r}_j)$ and $v_{BB}(\underline{r}_i - \underline{r}_j)$, we may write the hamiltonian in terms of a local occupation variable c_i ($c_i = 1$ if the ith site is occupied by an A-atom, $c_i = 0$ for B-atoms, vacancies being neglected)

$$\begin{aligned}\mathscr{H} = \sum_{i \neq j} \{c_i c_j v_{AA}(\underline{r}_i - \underline{r}_j) + [c_i(1 - c_j) + c_j(1 - c_i)]\, v_{AB}(\underline{r}_i - \underline{r}_j) \\ + (1 - c_i)(1 - c_j) v_{BB}(\underline{r}_i - \underline{r}_j)\} + \sum_i [c_i g_A + (1 - c_i) g_B] + \mathscr{H}_0 \quad .\end{aligned} \tag{6.14}$$

Here g_A, g_B are the chemical potentials of the two constituents, and $\mathscr{H}_0$ represents the other degrees of freedom of the lattice (e.g., lattice vibrations, etc.). Kinetics is introduced in this model by random interchange of neighboring atoms

($c_i \to c_j$, $c_j \to c_i$). Introducing a pseudospin μ_i by $\mu_i = 2\,c_i - 1$, (6.14) transforms again into an Ising hamiltonian of the type (6.10), with $J = (2\,v_{AB} - v_{AA} - v_{BB})/4$, $H = [g_B - g_A + \sum_{j(\neq i)}(v_{BB} - v_{AA})]/2$. An important difference concerns the dynamics, however: rather than single spin flips (6.11) one has spin exchanges $\mu_i \to \mu_j$, $\mu_j \to \mu_i$ [6.4], and hence the magnetization $\langle\mu_i\rangle$ which corresponds to the concentration $\langle c_i\rangle = (1 + \langle\mu_i\rangle)/2]$ is conserved.

Two cases have to be distinguished: (I) if $J > 0$, the system undergoes "ferromagnetic ordering" below T_c, which corresponds to an unmixing transition; the spontaneous magnetization $\langle\mu_i\rangle$ below T_c is just one half of the concentration difference of the A-rich and B-rich phases at the coexistence curve (cf., e.g., [Ref.6.35, Fig.1]. In this case the order parameter of the transition is conserved. (II) if $J < 0$, the system undergoes "antiferromagnetic ordering" below T_c, which corresponds to an order-disorder transition. While the concentration is conserved, the sublattice concentration (which is the order parameter of the transition) is not.

While in case (II) one expects [6.24] a behavior of the order parameter relaxation function rather similar to the kinetic Ising model without any conservation law (Sect.6.2.2), a different behavior is expected [6.24,35] in case (I); due to the order parameter conservation, the relaxation function is monodispersive: $\phi_{\mu\mu}(\underline{q},t) = \exp(-Dq^2t)$ for $t \to \infty$, $q \to 0$ and q^2t finite, where D is a diffusion constant, and $D \propto S^{-1}_{\mu\mu}(0,0) \propto |1 - T/T_c|^{\gamma}$. As a consequence, the energy fluctuations are expected to decay nonexponentially [6.36-38]: $\phi_{\mathcal{H}\mathcal{H}}(0,t) \propto t^{-d/2}$ as $t \to \infty$ for d-dimensional lattices [6.35].

A determination of D as a function of $T \to T_c$ was obtained in a study [6.39] where the exchanges were restricted so that the energy as well as the order parameter were conserved. Due to the second conservation law, this Kadanoff-Swift model [6.5] allows the determination of a second transport coefficient, the heat conduction coefficient κ. While estimates for both D and κ could be obtained with reasonable precision at infinitely high temperature, where agreement with analytic approximations was obtained [6.39], the results for finite temperatures are rather unreliable due to "computational difficulties" [Ref.6.39, Fig.11]. This failure probably was due to inappropriate choice of initial states, namely states with specified energy and $\langle\mu\rangle = 0$ but otherwise completely random spin configuration; at $T \neq 0$ the system does not come close to equilibrium at large wavelengths due to the slow relaxation required by the conservation laws, except if one would use prohibitively large computing time. It would be valuable to repeat such investigations, by dividing the computation in two steps [6.3]; in a first run without any conservation laws, one generates equilibrium configurations characteristic of a chosen temperature which are used as starting configurations in a second series of runs where conservation laws are employed. A simulation of this type was performed recently to study the diffusion of sodium in β"-alumina [6.40].

It must be remarked that the kinetics used above is a good model of no real material, where atoms are exchanged via vacancies rather than directly. The latter mechanism was simulated by moving one vacancy through a 64 × 64 square lattice of an ordering alloy at 50% concentration [6.41]. Figure 6.1 shows that the relaxation time $\tau_{\mu\mu}$ could be obtained as a function of temperature with reasonable precision. But unfortunately $\tau_{\mu\mu}$ was not accurate enough to see a difference between its exponent $\Delta_{\mu\mu}$ and γ. Clearly it would be interesting to extend this work to the 3d-case, in order to compare it to the available experimental data [6.42].

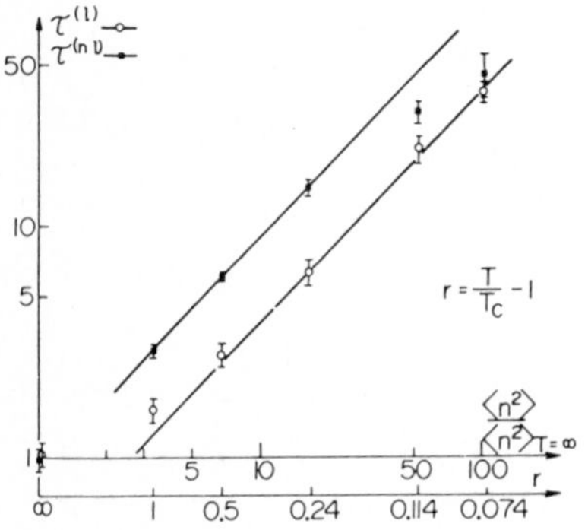

Fig.6.1 Log-log plot of the relaxation times $\tau^{(\ell)}$ (for linear response) and $\tau^{(n\ell)}$ (for nonlinear response) of a kinetic Ising model with vacancy mechanism plotted vs. the order-parameter susceptibility (denoted by $\langle n^2 \rangle / \langle n^2 \rangle_{T=\infty}$).[6.41]

6.2.4 Dynamics at a Multicritical Point

We consider a very anisotropic antiferromagnet in a magnetic field. The Néel temperature T_N below which spontaneous sublattice ordering occurs decreases if the field is increased, until one reaches a tricritical point at H_t, $T_N(H_t) = T_t$: For $H > H_t$ the transition from the antiferromagnetic state to the paramagnetic state is of first order rather than of second order [6.43]. In a three-dimensional system critical exponents describing the approach to the tricritical point for static quantities (specific heat $C \propto |T - T_t|^{-\alpha t}$, sublattice magnetization $m_s \propto |T - T_t|^{-\beta t}$, staggered susceptibility $\chi_s \propto |T - T_t|^{-\gamma t}$, etc.) have their "classical" mean-field values ($\alpha_t = 1/2$, $\beta_t = 1/4$, $\gamma_t = 1$), apart from logarithmic correction factors [6.43] In addition to this tricritical point in the so-called "metamagnet", tricritical and other multicritical points appear in many other physical systems (see [6.43] and Chap.3 for more detailed discussions of static multicritical phenomena). Of course, it is an intriguing question to ask what is known about dynamic multicritical phenomena - a question on which little experimental and theoretical information is available. This question was treated by Monte Carlo methods [6.44]. An Ising hamiltonian with antiferromagnetic nearest-neighbor interaction but ferromagnetic next-nearest neighbor interaction was chosen, and 20 × 20 × 20 lattices with periodic boundary conditions were treated. The kinetic evolution was again determined by (6.11), i.e., with no conservation law. This model simulates the "metamagnet" described above. The same relaxation times as for ordinary critical kinetic Ising models (Sect.6.2.2) were obtained, but a rather different behavior was found: (I) The exponents of the relaxation times satisfy the conventional theory [6.4], i.e., $\Delta^t_{\mu\mu} = \Delta^t_{\mu\mathcal{H}} = \Delta^t_{\mathcal{H}\mathcal{H}} = \gamma_t = 1$ (see Fig.6.2). (II) The relaxation function $\phi_{\mu\mu}(t)$ is a simple exponential (Fig.6.2), and hence one has a lorentzian line-shape. Dynamic scaling is found to hold if one uses everywhere the tricritical exponents. (III) The critical amplitudes $\tau^0_{BC}(\tau_{BC} = \tau^0_{BC}\,|1 - T/T_t|^{-\Delta^t_{BC}})$ are all the same: $\tau^0_{\mu\mu} \approx \tau^0_{\mu\mathcal{H}} \approx \tau_{\mathcal{H}\mathcal{H}}$ [6.44].

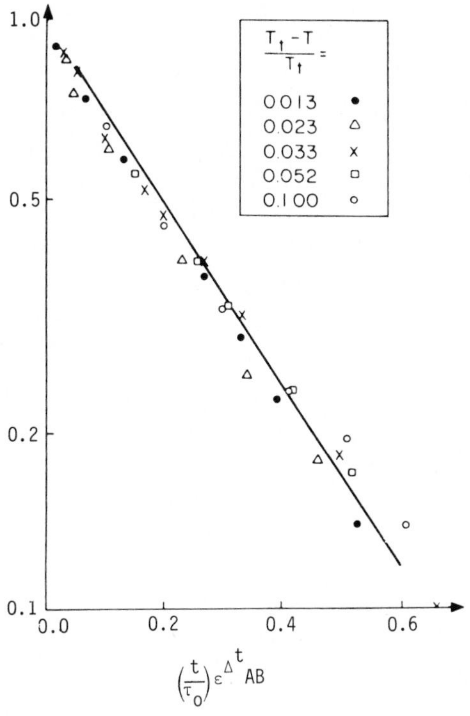

Fig.6.2 Scaled relaxation function of the order parameter at the tricritical point of a three-dimensional kinetic Ising model of a metamagnet, demonstrating simple exponential decay. [6.44]

6.2.5 Dynamics of "Clusters": Their Reaction Rate and Diffusion Constant

In uniaxial systems close to T_c the fluctuations of the local order parameter can be considered as "clusters" of various size, just as the critical fluctuations of gas density near the gas-liquid critical point may be described as microscopically small liquid droplets [6.45-6.47,29]. For example, in an antiferromagnetic material a cluster would most naturally be defined as a localized region with a well-defined order. These clusters may have rather irregular shapes [6.48] but their typical linear dimension is the correlation length ξ [6.47,29]. A more detailed discussion of cluster properties can be found in Chap.5. It has been suspected that the slow relaxation of these clusters may be responsible for the "central peak" at structural phase transitions [6.49] and related phenomena in ferroelectrics [6.50], and attempts have been made to account for the linewidth of light scattering at liquid-gas critical points in terms of clusters [6.51]. While most of this work is of doubtful validity [6.47], it is clear that clusters play an important role in nucleation phenomena at first-order transitions [6.52], for example, metastable alloy phases containing clusters are important in metallurgy. Thus there is a broad motivation to study both the static [6.46] and dynamic [6.28,29] properties of clusters.

In order to give this concept of a cluster a precise meaning, one may again use an Ising ferromagnet and consider an ℓ-cluster as a group of exactly ℓ minus-spins connected to each other by nearest-neighbor bonds. (We suppose that either we have a positive spontaneous magnetization or a field $H > 0$, such that the plus-spins forms a "percolating" [6.53] background, and clusters of plus-spins need not be considered [6.46]). While static critical phenomena may then be traced back to the concentration function n_ℓ of the clusters [6.46], dynamic phenomena in the kinetic Ising model (Sect.6.2.2) are related to a reaction matrix $W(\ell,\ell')$, which gives the average number of cluster splitting reactions ($\ell + \ell' \rightarrow \ell,\ell' - 1$) in thermal equilibrium per unit time. Due to a detailed balance condition [cf. (6.4)] the nature of the reverse reaction is implied. It turns out [6.19] that the time evolution pattern is basically described in terms of an effective cluster reaction rate R_ℓ, defined by [6.29] $R_\ell = \sum_{\ell'=1}^{\infty} (\ell' + 1)^2 W(\ell,\ell')/n_{\ell'}$, which gives the rate of change (by growing or shrinking) of ℓ-clusters. In the case where the order parameter is conserved, one has also to consider an effective cluster diffusion constant D_ℓ [6.35].

The reaction rate R_ℓ was obtained close to T_c for both 110×110 square [6.29] and $30 \times 30 \times 30$ simple cubic lattices [6.28]. A power-law behavior $R_\ell \propto \ell^r$ was found in both cases (Fig.6.3), with $r \cong 0.82$ ($d = 2$) and $r \cong 1.16$ ($d = 3$). Both values roughly satisfy the scaling relation [6.28] $\Delta_{\mu\mu} = (2 - r)(\beta + \gamma)$ (cf. Sect. 6.2.2). Figure 6.3 also shows that growing or shrinking of clusters in single steps ($\ell' = 1$) accounts only for about 1/4 of the observed total rate.

The cluster diffusivity D_ℓ was obtained so far only in the case of the $d = 2$ spin-exchange model, both from equilibrium [6.54] and nonequilibrium [6.55] studies. Figure 6.4 shows that also D_ℓ follows a power law, with $D_\ell \propto \ell^{-1.05}$, which disagrees with the prediction [6.56] $D_\ell \propto \ell^{-(1+1/d)}$. The reasons for this discrepancy are not yet clear [6.35]. Note that the behavior of D_ℓ enters crucially in the kinetics of the initial stages of phase separation (Sect.6.3.3).

The authors are not aware of any experiments to which these results could be compared directly, although studies of cluster kinetics are conceivable, e.g., in adsorbed monolayers. But these simulations yield valuable background information for the studies of relaxation far from equilibrium, which are discussed in the following section.

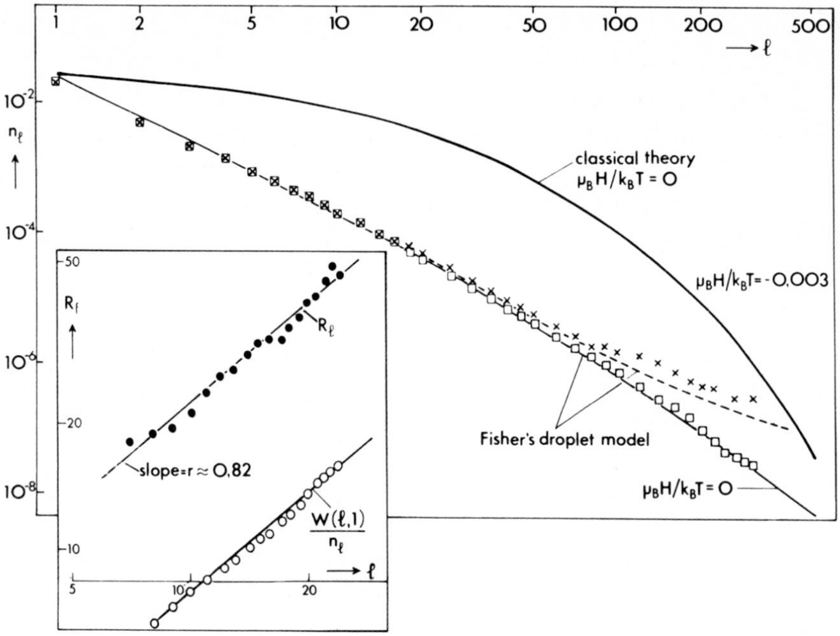

Fig.6.3 Concentration n_ℓ of clusters containg ℓ reversed spins plotted vs. ℓ for the square Ising lattice at $T/T_C \approx 0.96$ and two values of the field. Predictions of both the classical nucleation theory and the Fisher droplet model [6.45] (neither of which contain any adjustable parameter) are included. The inset shows the reaction rate (for zero magnetic field) and includes also the contribution $W(\ell,1)/n_\ell$ of single flip processes. [6.75]

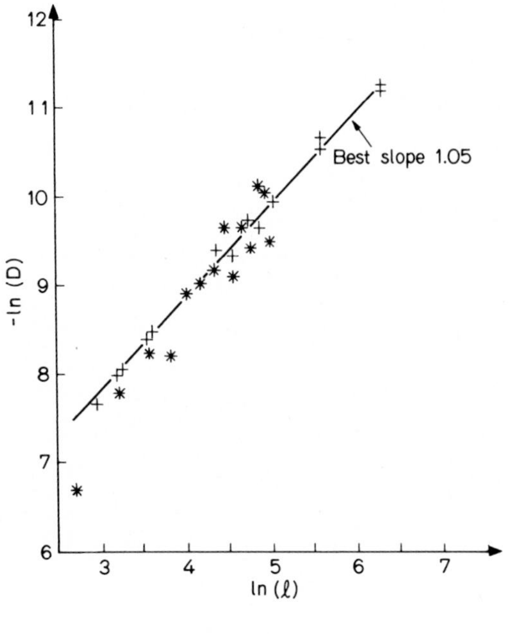

Fig.6.4 Log-log plot of the cluster diffusion constant D_ℓ versus ℓ, for a two-dimensional spin-exchange Ising model at $T/T_C = 0.6$. [6.54]

6.3 Kinetics of Nonlinear Relaxation

Suppose that a system was kept in thermal equilibrium at a temperature T_0 for times $t < 0$, and the temperature is changed to T abruptly at $t = 0$. Then nonconserved quantities $\langle A(t)\rangle$ will relax from $\langle A(0)\rangle = \langle A\rangle_{T0}$ to $\langle A(\infty)\rangle = \langle A\rangle_T$ (the subscripts denote the temperature at which an ensemble average is taken). Thus one may introduce a nonlinear relaxation function $\phi_A(t)$ by [6.57,58]

$$\phi_A(t) = [\langle A(t)\rangle - \langle A(\infty)\rangle]/[\langle A(0)\rangle - \langle A(\infty)\rangle] \quad , \tag{6.15}$$

and define a nonlinear relaxation time $\tau_A = \int_0^\infty dt\, \phi_A(t)$ [cf. (6.12) for an example].

6.3.1 Nonlinear Critical Slowing Down

The critical behavior of the nonlinear relaxation function defined above has been investigated especially in the case of kinetic Ising models without conservation laws. RACZ [6.59] has suggested, on the basis of a Ginzburg-Landau theory, that the critical behavior of τ_A is given by

$$\tau_A^{(n\ell)} \propto |T - T_c|^{-\Delta_A^{(n\ell)}} \quad , \quad \Delta_A^{(n\ell)} = \Delta_{\mu\mu} - a \quad , \tag{6.16}$$

if the static critical behavior of $\langle A\rangle$ is described by $\langle A\rangle_T - \langle A\rangle_{T_c} \propto |T - T_c|^a$. Equation (6.16) results also from a more general scaling theory [6.47], from cluster dynamics [6.47,29], from renormalization group treatments [6.30,60], or series expansion [6.61].

Several Monte Carlo estimates for $\Delta_A^{(n\ell)}$ have been obtained in two dimensions [6.13,41,47,48,62]. Reference [6.13] suggested $\Delta_\mu^{(n\ell)} = 1.85 \pm 0.10$. Unfortunately the accuracy of this estimate and the estimate for $\Delta_{\mu\mu} \approx 1.9$ (6.14) does not allow a significant discussion of the difference, which should be $\Delta_{\mu\mu} - \Delta_\mu^{(n\ell)} = \beta = 0.125$ according to (6.16) in this case. The same lack of accuracy hampers the study [6.41] using the vacancy model (see Fig.6.1). Even more inaccurate were the earlier predictions [6.48] $\Delta_\mu^{(n\ell)} \approx 1.7$ and $\Delta_{\mathcal{H}}^{(n\ell)} \approx 0.3$, which disagree with (6.16): $\Delta_{\mu\mu} - \Delta_{\mathcal{H}}^{(n\ell)} = 1 - \alpha = 1$. Also the recent study [6.62], where the first application of the "self-consistent-effective field boundary condition" [6.63] (cf. Chap.1) to dynamic phenomena was given, suggested $\Delta_{\mathcal{H}}^{(n\ell)} \approx 0.4$ in disagreement with the scaling law. We think that these estimates are unreliable since they were taken too close to T_c where good statistical accuracy could not be reached (cf. Chap.1 for accuracy limitations near critical points). In fact, another study [6.47] where data not so close to T_c were analyzed gave $\Delta_{\mathcal{H}}^{(n\ell)} \approx 0.9$ (Fig.6.5), in good agreement with the scaling law, (6.16).

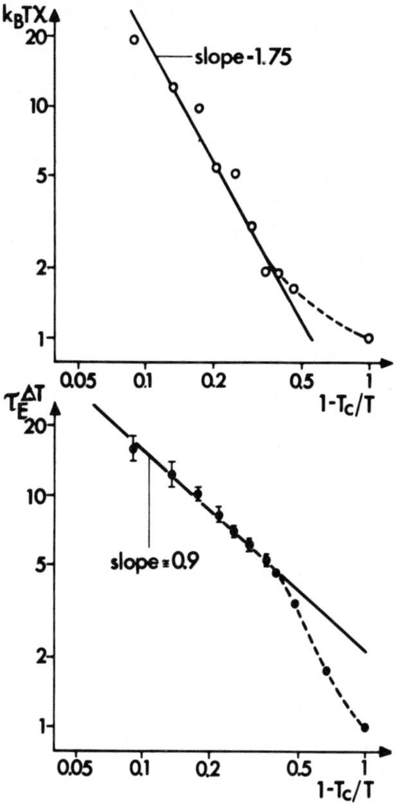

Fig.6.5 Log-log plot of nonlinear energy relaxation time (lower part) and static susceptibility (upper part) vs. temperature. This simulation of a 80 × 80 square lattice with classical Ising spins (i.e., $S \to \infty$) shows that the expected exponents are indeed observed for $0.1 \lesssim 1 - T_C/T \lesssim 0.5$. While corrections to scaling become dominant for higher temperatures, finite size rounding and inaccuracy due to finite time averaging make meaningful calculations closer to T_C very difficult. [6.47]

Experimental studies of this nonlinear slowing down are available in ordering alloys. For Ni_3Mn $\phi_\mu^{(n\ell)}(t)$ was obtained by recording the neutron superlattice Bragg peak intensity as a function of time after a temperature jump [6.64]. Similarly, the relaxation function $\phi_\rho^{(n\ell)}(t)$ of the electrical resistivity ρ has been measured for $AuCu_3$ [6.65]. The critical behavior of $\phi_\rho^{(n\ell)}(t)$ should be the same as that of $\phi_{\mathscr{H}}^{(n\ell)}(t)$ [6.66]. However, no attempt has been made to find the relaxation times defined above by numerical integration. Thus the interpretation of these experiments in terms of (6.16) is hampered by difficulties similar to those for the computer experiments.

6.3.2 Nucleation Kinetics at First-Order Phase Transitions

If an intensive thermodynamic variable of a system (temperature T, pressure p, magnetic field H, ...) is suddenly changed by a finite amount (ΔT, Δp, ΔH, ...), it may happen that one crosses a line of first-order phase transitions in the phase diagram. The system is now unstable and another phase develops in the course of the nonlinear relaxation process. Usually this phase change occurs via nucleation and growth, i.e., small "droplets" or "clusters" of the new phase appear on the background of the old instable phase due to thermal fluctuations, and then start to grow to form macroscopic domains. Apart from this "homogeneous" nucleation, nucleation at inhomogeneities (impurities, surfaces, etc.), i.e., "heterogeneous nucleation" is important. However, rather preliminary computer simulations of the latter process are available. The kinetics of adsorption at surfaces was simulated [6.67-69], where first the adsorption isotherm (i.e., coverage θ as a function of adsorbate gas pressure) was determined [6.67,69], and the formation of adatom clusters was studied [6.68], the cluster size ℓ range being in the range $\ell \leqq 12$. Since such extremely small clusters are hardly relevant to test the concepts of nucleation theory, it would be very desirable to extend such studies to larger cluster sizes. For that purpose, lattices much larger than 20×20 (as used in [6.68]) would be necessary in order to avoid unwanted finite size effects (see Chap.1) and obtain better statistics. Therefore, we will discuss only some more extensive studies of homogeneous nucleation in two-dimensional lattice gas systems [6.70-73,52,54,55] in the following. Since some of this work has been recently reviewed in detail [6.74-76], we will again give only a brief discussion, but we emphasize that studies of homogeneous nucleation in two dimensions may again be considered as nucleation on a surface if the adatoms are restricted to be within the first monolayer. Studies without conserved order parameter [6.70,71] correspond to cases where diffusion of adatoms within the layer is negligible in comparison with evaporation-condensation from the layer, while the reverse is true for studies with conserved order parameter [6.73,54,55]. Similarly these studies may also be transformed to models of crystal growth, if the start of the growth of a new layer is permitted only after total completion of the preceding lawer (cf. Chap.7).

In order to elucidate the significance of these simulations, we briefly sketch the theoretical background (Fig.6.6) [6.75]. The formation of a cluster of size ℓ of the new phase is determined by two factors: (I) the growth rate R_ℓ, which was already considered in Sect.6.2.5, (II) the free energy of formation F_ℓ (i.e., the excess free energy in comparison with the old phase). While F_ℓ increases steadily with ℓ, when the old phase is a stable equilibrium state, F_ℓ has a maximum in ℓ if the old phase is metastable or instable. The maximum occurs at $\ell = \ell^*$, the critical cluster size. In contrast to the mean-field theory of nonequilibrium relaxation for non conserved order parameter [6.58] and the related Cahn-Hilliard theory for conserved order parameter [6.77], there is no sharp distinction between metastable and unstable states [6.52]. States are metastable if $F_{\ell^*}/k_BT \gg 1$ while they are unstable if $F_{\ell^*}/k_BT \lesssim 1$. The maximum can be understood as a result of a competition between a negative bulk term and a positive surface term; for very small ℓ, surface energy dominates and hence F_ℓ is positive and increases with ℓ. For large ℓ, bulk energy dominates and F_ℓ is negative and decreasing. Neglecting nonlinear terms describing coalescence of clusters (coagulation), nucleation is then described by a continuity equation for the cluster concentrations $\bar{n}_\ell(t)$ [6.52,75]

$$\frac{\partial \bar{n}_\ell}{\partial t} + \frac{\partial J_\ell}{\partial \ell} = 0 \; ; \quad J_\ell = - R_\ell \frac{\partial \bar{n}_\ell(t)}{\partial \ell} - R_\ell \bar{n}_\ell(t) \frac{\partial}{\partial \ell} \frac{F_\ell}{k_B T} \quad . \tag{6.17}$$

The cluster current J_ℓ consists of two terms: the first is a diffusive term, and the second ($\propto \partial F_\ell/\partial \ell$) is a drift term, which changes sign for $\ell = \ell^*$. Hence the drift acts against the diffusion for subcritical clusters ($\ell < \ell^*$), while it acts in the same direction as diffusion for supercritical clusters ($\ell > \ell^*$). Cluster growth is thus described as a Brownian motion in the space of cluster sizes $\{\ell\}$ by (6.17). For further analysis it is convenient to transform from F_ℓ to the equilibrium cluster size distribution n_ℓ via $n_\ell = \exp(-F_\ell/k_B T)$ (for a metastable state, this is a fictitious quantity only, of course), Fig.6.6. It is then easy to see that $J_\ell = - R_\ell n_\ell (\partial/\partial \ell)\{\bar{n}_\ell(t)/n\}$ and hence a steady state solution $\{\partial \bar{n}_\ell(t)/\partial t = 0\}$ of (6.17) is obtained for $J_\ell = J$ independent of ℓ(J is the so-called nucleation rate). From $-(\partial/\partial \ell)\{\bar{n}_\ell(t)/n_\ell\} = J(R_\ell n_\ell)^{-1}$ one obtains, using the boundary conditions $\lim_{\ell \to 0} \{\bar{n}_\ell(t)/n_\ell\} = 1$, $\lim_{\ell \to \infty} \{\bar{n}_\ell(t)/n_\ell\} = 0$

$$1 = J \int_0^\infty d\ell (R_\ell n_\ell)^{-1} \quad , \quad \frac{\bar{n}_\ell(t)}{n_\ell} = J \int_\ell^\infty d\ell (R_\ell n_\ell)^{-1} \quad . \tag{6.18}$$

It is seen that knowledge of R_ℓ and n_ℓ is sufficient to calculate both the nucleation rate J and the steady-state cluster size distribution $\bar{n}_\ell(t)/n_\ell$. Since it is very hard to derive both R_ℓ and n_ℓ from experimental quantities, no significant experimental test of this theory has been possible so far. Since this Becker-Döring theory [6.78] is based on many simplifying approximations (use of only one "coordinate" ℓ to describe the clusters, neglect of coagulation effects, etc. [6.52]), a test of this theory would be extremely desirable, since it is applied in many branches of chemistry and physics [6.78]. For this problem, computer simulations are of help. Both R_ℓ and n_ℓ have been determined (cf. Sect.6.2.5, Fig.6.3). For a more detailed discussion of static cluster properties see Chap.5. It turns out that the actual cluster size distribution is fairly well accounted for by the droplet model of FISHER [6.79], while the classical formula [6.78,75] for $d = 2$, $n_\ell/n_0 = \exp[-2\ell (H/k_B T) - (2f_s/k_B T)\sqrt{\pi}\, \ell^{1/2}]$, where f_s is the bulk surface tension, is a bad approximation, at least for ℓ not too large [6.52]. Note that in our case this classical formula contains no adjustable parameters - f_s is known exactly [6.80] and the coexistence between spin-up phase and spin-down phase occurs precisely at $H = 0$. Hence the energy of overturning ℓ spins from $\mu = +1$ to $\mu = -1$ ($\Delta\mu = 2$) is $2H\ell$. This is a huge advantage of the 2d lattice gas model in comparison to 3d Lennard-Jones models of fluids, where neither the location of the coexistence line nor the surface tension is known accurately enough. Hence we feel that the molecular dynamics computer simulations of [6.81] devoted to a test of nucleation theory in

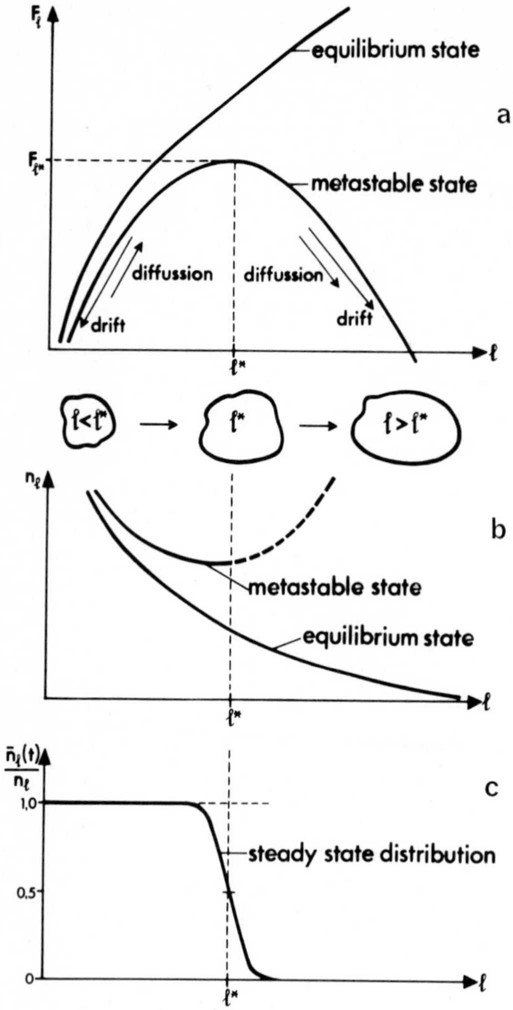

Fig.6.6a-c Schematic plot of the formation energy F_ℓ of a cluster plotted vs. cluster size ℓ and the associated mechanism of cluster growth over the critical size ℓ^* (a), cluster concentration n_ℓ for equilibrium states and its analytic continuation for metastable states (b), and the nonequilibrium cluster size distribution in a steady-state nucleation process. [6.75]

that case are necessarily inconclusive [6.52]. In our case, however, it is possible both to compute J and $\bar{n}_\ell(t)/n_\ell$ unambiguously from Fig.6.3 via (6.18), and to determine the *actual* steady-state values $\bar{n}_\ell(t)/n_\ell$ observed in the simulation, Figs. 6.7,8. In Fig.6.7 the actual (nonstationary) $\bar{n}_\ell(t)$ are plotted versus time for various values of ℓ and three values of changes ΔH of the field. After a short time (which is comparable to the relaxation of equilibrium fluctuations) the $\bar{n}_\ell(t)$-curves become flat. No true steady state is reached, of course, since the fraction

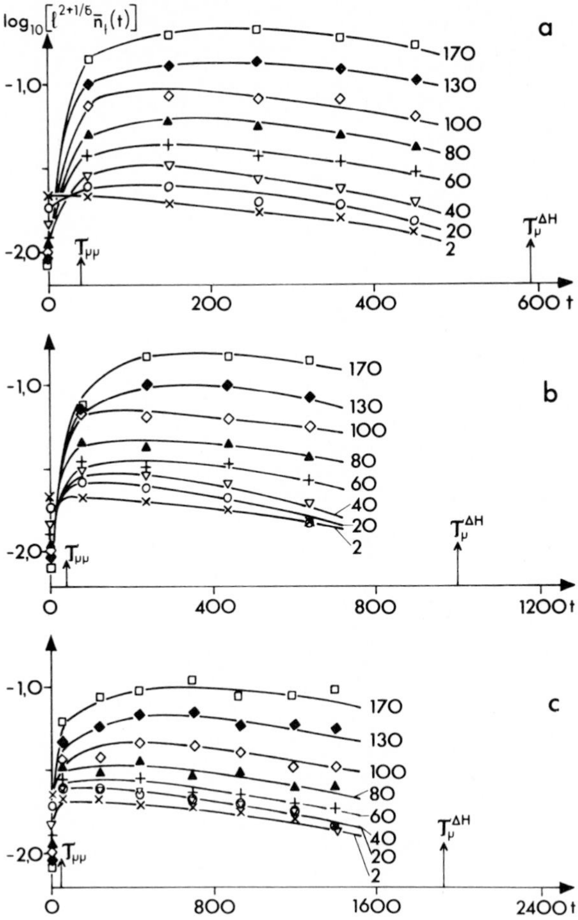

Fig.6.7a-c Nonequilibrium cluster concentrations $\bar{n}_\ell(t)$ plotted vs. times as observed in simulations of a 110 × 110 square Ising lattice for $T/T_c \approx 0.96$ and $\mu_B H/k_B T = -0.015$ (a), -0.012 (b) and -0.008 (c). Parameter of the curves (which are only drawn to guide the eye) is ℓ. Arrows indicate the values of both the equilibrium relaxation time $\tau_{\mu\mu}$ and the nonlinear relaxation time $\tau_\mu^{(n\ell)}$ (denoted by $\tau_\mu^{\Delta H}$ in the figure). [6.52]

of metastable phase is not kept constant, as artificially assumed in (6.18), but decreases on a time scale set by the nonlinear relaxation time $\tau_\mu^{\Delta H}$. If we nevertheless extrapolate the flat regions of $\bar{n}_\ell(t)$ back to t = 0 to estimate the "experimental" steady state values, we obtain Fig.6.8. The error bars indicate the uncertainty of this extrapolation. Dashed curves are the prediction of (6.18). Good agreement is obtained except for cluster sizes in the vicinity of ℓ_ξ, the typical size of clusters in the initial thermal equilibrium state. Since there are no adjustable parameters whatsoever, this agreement constitutes the first significant test of the validity of nucleation theory.

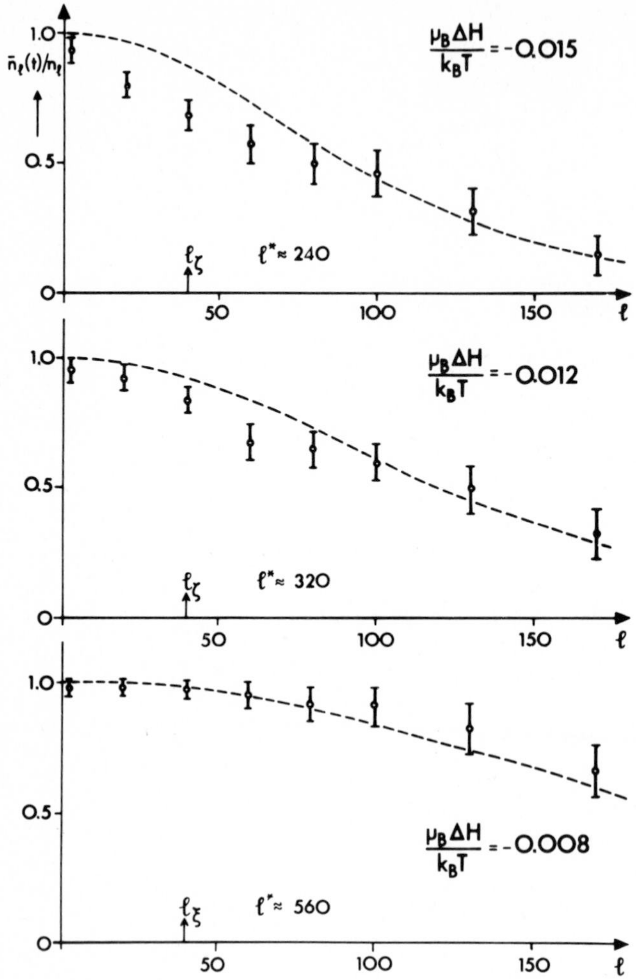

Fig.6.8 Relative cluster concentration $\bar{n}_\ell(t)/n_\ell$ of steady-state nucleation plotted versus ℓ for various fields. The points are Monte Carlo estimates based on the data of Fig.6.7, the curves are the prediction of nucleation theory, using the data of [Ref.6.18, Fig.6.3]. Arrows show the typical cluster size ℓ_ξ in thermal equilibrium at that temperature. [6.75]

Fig.6.9a-l Snapshot pictures of the development of a 55 × 55 square Ising lattice being in thermal equilibrium at $T/T_c \approx 0.73$ until t = 0 (first picture), after which a switch of the field from zero to $\mu_B H/k_B T = -0.16$ takes place. Snapshot pictures are shown at time intervals of 5 Monte Carlo steps/spin. Reversed spins are indicated by black dots, up spins are not shown. Clusters with $\ell > 5$ are set off by contours. The reactions ($\ell' \to \ell$) indicate the development of a cluster with ℓ spins from the present (ℓ) to the next (ℓ') shown configuration. Later on the two largest clusters (with $\ell = 42$ on the last picture) form stable domains of the new phase. [6.52]

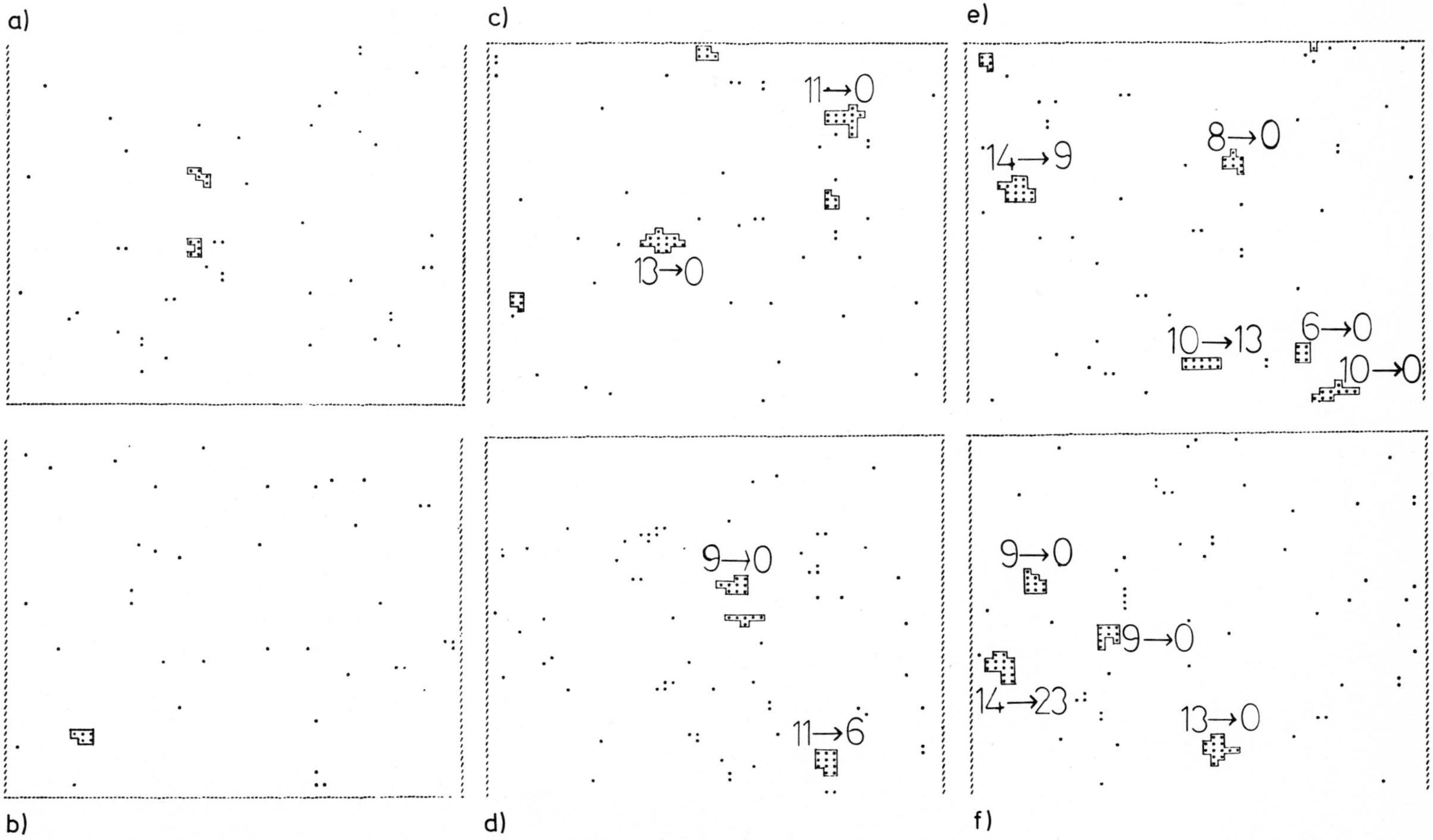
a)
b)
c)
11→0
13→0
d)
9→0
11→6
e)
14→9
8→0
10→13
6→0
10→0
f)
9→0
9→0
14→23
13→0

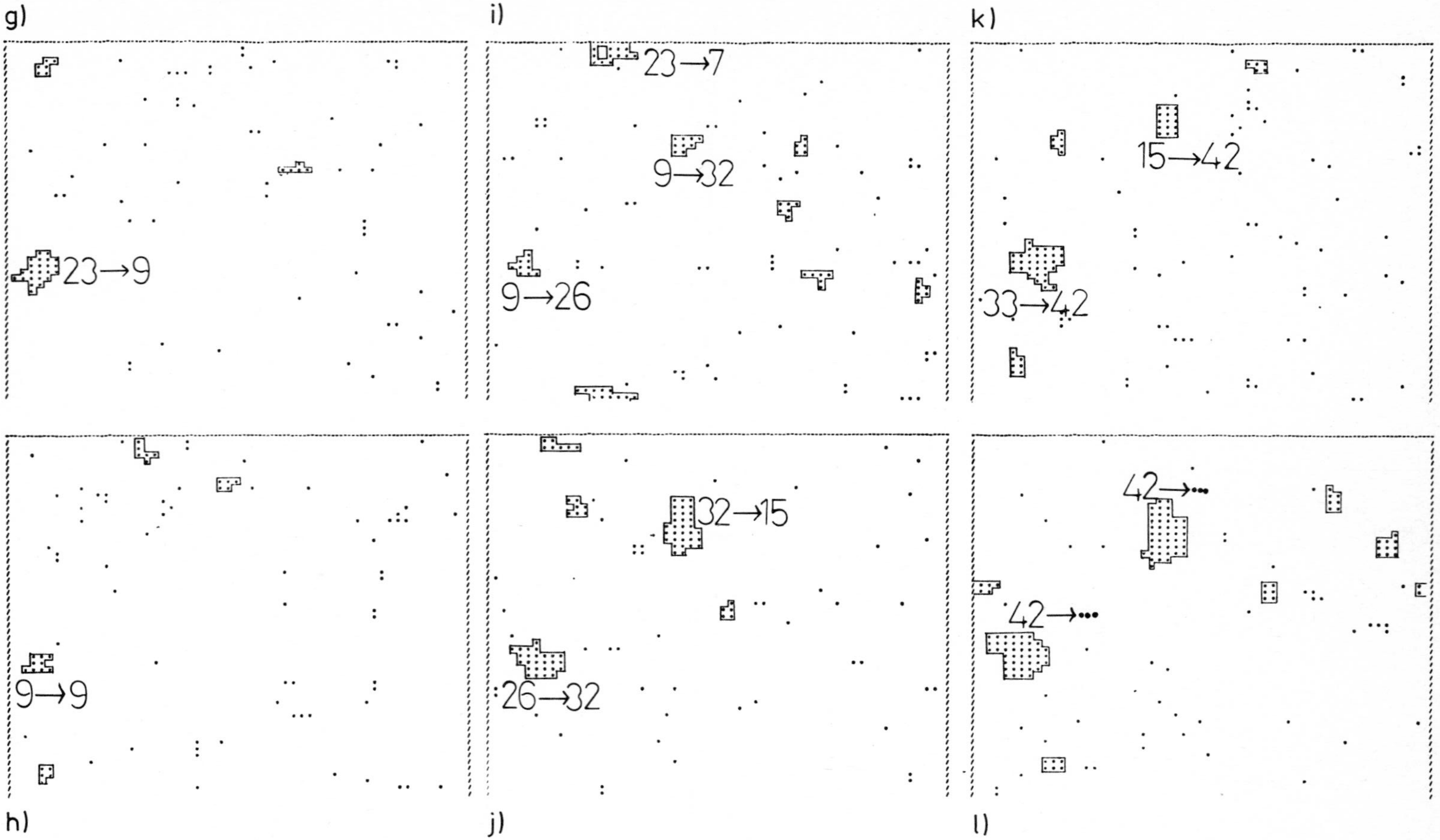

Fig.6.9 g-l (Caption see page 244)

One may use these simulations to get a detailed insight into the processes occurring on microscopic scales both in space and in time, of course. As an example we show in Fig.6.9 a series of snapshot pictures taken at successive time intervals Δt = 5 Monte Carlo steps/spin after the change of the field. Minus spins are shown as black dots while plus spins are not shown. The changes from picture to picture are indicated as reaction equations $\ell_1 \rightarrow \ell_2$. It is seen that small (subcritical) clusters are formed due to thermal fluctuations, and disintegrate again. Even clusters which come close to ℓ^* tend to shrink again (cf. Fig.6.9g,j) before a new fluctuation takes them over the potential barrier. This series of pictures [6.52] refers to a rather small "supersaturation" ΔH of the metastable phase. Other series of pictures at high ΔH were the initial phase is extremely unstable are described in more detail in [6.74,75].

In simulations with conserved order parameter at d = 2 [6.37,55,82] no evidence of nucleation phenomena could be detected so far. These findings are easily interpreted by the fact, however, that the critical cluster size ℓ^* is close to unity (which is the "limit of metastability" [6.70]) except in an extremely narrow concentration interval Δc close to the coexistence curve, cf. [Ref.6.70, Fig.21]. No data within Δc have been taken so far. In three dimensions Δc is much larger, cf. [Ref.6.70, Fig.22], and hence it was easier to actually observe nucleation phenomena in this case [6.73]. The qualitative behavior is very similar to the description presented above. Since a comparable quantitative analysis is still lacking we do not go into the details here. But we mention that in the two-dimensional case with conservation some indirect evidence for the validity of nucleation concepts was obtained by a different kind of simulation [6.54]. There a cluster of size ℓ was put in a box containing supersaturated "vapor" (Δc), and equilibrium between the cluster (at size ℓ^*) and surrounding gas (with supersaturation Δc^*) was established and studied [see also Chap.5]. The relation between ℓ^* and Δc^* obtained in this way is consistent with what one expects from nucleation theory.

All the above simulations apply to fairly high temperatures, where fluctuations are important already in thermal equilibrium, and rather high "supersaturations". Hence the models for n_ℓ, R_ℓ checked by these simulations should be applicable to treat nucleation close to a critical point. In fact a treatment along these lines [6.52] gives a much better account of data on the nucleation of CO_2 close to T_c [6.83] than the conventional theory of nucleation [6.78], see [Ref.6.52, Fig.13]. But discrepancies still remain due to incomplete knowledge of n_ℓ, and further studies probing nucleation theory over a wider range of temperatures and "supersaturations" would be valuable.

6.3.3 Kinetics of Spinodal Decomposition and Grain Growth in Alloys

We consider binary (AB) alloys, such as ZnAl, which are suddenly quenched from a high temperature to a lower one within its miscibility gap (Fig.6.10). Then the initially homogeneous phase is thermally unstable; thermal equilibrium requires coexistence of two phases, one A-rich and one B-rich. We are here concerned with the time evolution of this phase separation process which has to follow the quench. This is a problem of great practical interest in metallurgy, and also presents a challenge to the theorist. Typical experimental findings on this kinetics are shown in Fig.6.11. We start by summarizing the main theoretical ideas on this process.

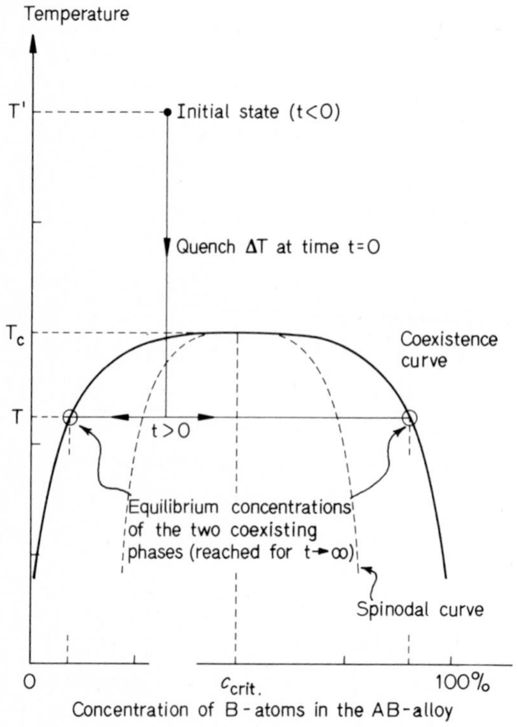

Fig.6.10 Schematic phase diagram of a solid binary alloy with a miscibility gap, into which a quench at time t = 0 is performed. [6.88]

According to the "classical" theory of CAHN-HILLIARD [6.77] two regimes must be distinguished; between the coexistence curve and the spinodal curve (Fig.6.10) the system is "metastable", i.e., it is thought to be stable with respect to *weak* fluctuations. The kinetics there should be initially as described in Sect.6.3.2. Within the spinodal curve, however, the system is thought to be unstable with respect to *weak* delocalized (i.e., long-wavelength) fluctuations. The growth of these fluctuations into zones of the two coexisting phases is called the spinodal decomposition mechanism. Experimentally one then observes [6.84] that the structure

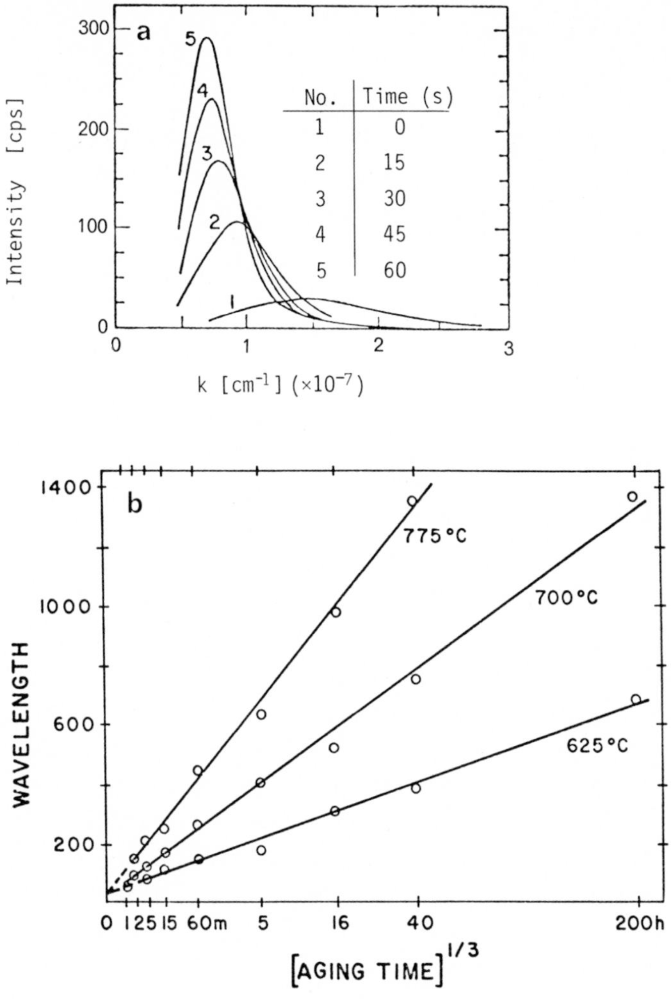

Fig.6.11 (a) X-ray scattered intensity $S(\underline{k},t)$ from quenched ZnAl-alloys plotted versus wave vector k at various times. [6.84]. (b) Plot of the "characterisitc wavelength" of the precipitated zones of a quenched Cu-Ni-Fe alloys versus $t^{1/3}$. [6.85]

factor $S(\underline{k},t)$ develops a peak at small wave vectors, which increases in height and whose location $k_m(t)$ shifts to smaller and smaller wave vectors as time goes on (Fig.6.11a). One finds that the data are consistent with a $k_m(t) \propto t^{-a}$ behavior at large times, where in most cases a = 1/3 (e.g., [6.85]) and in some cases a is even smaller [6.86]. A qualitatively similar behavior is expected, of course, if microdomains have been built up by the nucleation mechanism and then grow further ("coarsening" of grains). The explicit predictions of the Cahn-Hilliard theory can be explained in terms of the equation of motion of $S(\underline{k},t)$, which is the fourier transform of the (spatial) correlation function of concentration fluctuations

$$S(\underline{k},t) = \sum_{r} e^{i\underline{k}\underline{r}} \left\langle [c(\underline{r},t) - \bar{c}][c(\underline{r}' - \underline{r},t) - \bar{c}] \right\rangle \quad , \tag{6.19}$$

where $c(\underline{r},t)$ is the fractional concentration of the B-component at position $\underline{r}$ and time t, and $\bar{c} \equiv \langle c(\underline{r},t)\rangle$. Note that $S(\underline{k},t)$ in (6.19) is a correlation at equal times [in contrast to (6.13)], but it nevertheless depends on time since we consider a strongly nonlinear relaxation process (Fig.6.10) rather than thermal equilibrium. The equation of motion then is [6.85,87]

$$\frac{\partial S(\underline{k},t)}{\partial t} = -2\,Mk^2\,[\{Kk^2 + [A + 3B(\bar{c} - c_{crit.})^2]\}S(\underline{k},t) - k_BT] \quad , \tag{6.20}$$

where M is a mobility, and the constants $K > 0$, $B > 0$, $A(T < T_c) < 0$ can be related to expansion coefficients of a coarse-grained free energy extrapolated into the nonequilibrium region (see [6.88] for a more detailed discussion). If now the term $K\xi^{-2} = A + 3B(\bar{c} - c_{crit})^2 > 0$, then (6.20) has a steady-state solution $S(\underline{k},\infty) = k_BT/\,K(k^2 + \xi^{-2})$, which is the typical Ornstein-Zernike result [6.75] for one-phase equilibrium, ξ being the correlation length of concentration fluctuations. Deviations of $S(\underline{k},t)$ from this equilibrium value decay with time exponentially fast. The correlation length ξ diverges at the spinodal curve; nucleation is neglected in (6.20) and thus the "metastable states" between coexistence curve and spinodal curve have infinite lifetime, they appear in the theory in just the same way as true equilibrium states. Spinodal decomposition occurs for $K\xi^{-2} < 0$, however: $S(\underline{k},t)$ increases exponentially fast with time for wave vectors $k < k_c = \sqrt{|\xi^{-2}|}$. The maximum growth rate occurs for $k_m = k_c/\sqrt{2}$. Thus one would expect that $S(\underline{k},t)$ develops a peak for $k = k_m$, which grows exponentially with time. No coarsening (shift of k_m to smaller values as time goes on) is obtained from the theory. Clearly, this prediction is unphysical and due to neglected nonlinear effects [6.88-90,76, 35]. While a discussion of the asymptotic kinetics of a nonlinear version of (6.20) yielded $k_m(t) \propto t^{-1/3}$ [6.89], a complete theory describing the crossover from (6.20) [it was assumed that (6.20) is reasonable during the initial stages of the phase separation [6.89]] was lacking. The computer simulations [6.36-38,54,55,73,82,88, 91,92] to be described below have stimulated the formulation of a nonlinear theory [6.90] of spinodal decomposition, which is simular to (6.20) but ξ^{-2} is not a constant but depends on time (its time-dependence is determined self-consistently). While this theory predicts that nonlinear effects are important already during the initial stages, and hence (6.20) is never valid for systems with short-range interactions, the meaning of a spinodal line within this theory remains obscure. Furthermore it predicts (as shown in [6.35]) that $k_m(t) \propto t^{-1/4}$. This contrasts to the above prediction $k_m(t) \propto t^{-1/3}$, which also results from a somewhat different treatment, the linearized diffusion theory of LIFSHITZ and SLYOZOV [6.93]. The latter theory has recently been rederived [6.35] as a special case of a generalized nucleation theory, taking properly the condition of constant average concentration $\bar{c}$ into account. There it was shown [6.35,56] that at intermediate times values of the exponent $a = 1/(d + 3)$ instead of 1/3 are possible due to cluster coagulation events. This theory [6.35,94], provides a treatment of spinodal decomposition in terms of "cluster diffusion" and cluster growth and thus implies a completely gradual transition between nucleation and (nonlinear) spinodal decomposition. While it predicts the time evolution of the cluster size distribution, it does not yield the structure factor $S(\underline{k},t)$ which is accessible experimentally by neutron or x-ray scattering. In summary, then, we are far from a complete and consistent theory of the phase separation process.

The Monte Carlo method can yield valuable insight into this problem, adopting the description of the alloy in terms of the appropriate kinetic Ising model (which was introduced already in Sect.6.2.3). In the simulations [6.36-38,55,73,88] one usually starts with a completely random distribution of the two species (A,B), which corresponds to choosing the initial temperature $T' \to \infty$ (Fig.6.10). Then a run at the desired temperature T is made, and one may record both $S(\underline{k},t)$ and the time evolution of the cluster size distribution. While one obtains thus very complete information, the method is in practice limited to rather early stages of the process; the very slow relaxation would require to go to prohibitively large computing times in order to observe later stages. In addition, one then would have to study very large systems in order to avoid finite-size effects. It has been argued [6.95] that even the finite size of macroscopic systems may prevent one from reaching the regime where the theory of LIFSHITZ-SLYOZOV [6.93] becomes strictly valid. Thus it is clear that the computer simulations will not be able to check the validity of these theories [6.93,95], but one is able to study both the initial and intermediate stages of the process. Since the simulations have recently been reviewed in detail [6.88] we here give only a brief account of them.

Figure 6.12 shows three typical examples of $S(\underline{k},t)$ plotted vs. t, with $\underline{k}$ being the parameter of the curves, showing the early time evolution of a three-dimensional model system [6.38]. Fig.6.12a and b refer to two different temperatures at $\bar{c} = 0{,}5$, while Fig.6.12c refers to $\bar{c} = 0.2$. It is seen that $S(\underline{k},t)$ increases with t more slowly than linear, i.e., one does not see any indication of the predicted exponential increase, (6.20). The same observation applies to the two-dimensional case [6.37], implying that the linearized Cahn-Hilliard theory [6.77,87] is not at all valid during the initial stages.

A plot of $S(\underline{k},t)$ vs. k at various times (Fig.6.13a) [6.38] reveals a striking similarity with the experimental data Fig.6.11a, however. Furthermore it turns out that the temperature dependence of the data can be understood in terms of a dynamic scaling representation [6.90]

$$S(\underline{k},t) = |1 - T/T_c|^{-\gamma}\tilde{S}(q,\tau), \quad q = k|1-T/T_c|^{-\nu}, \quad \tau = t|1-T/T_c|^{\gamma} \quad , \qquad (6.21)$$

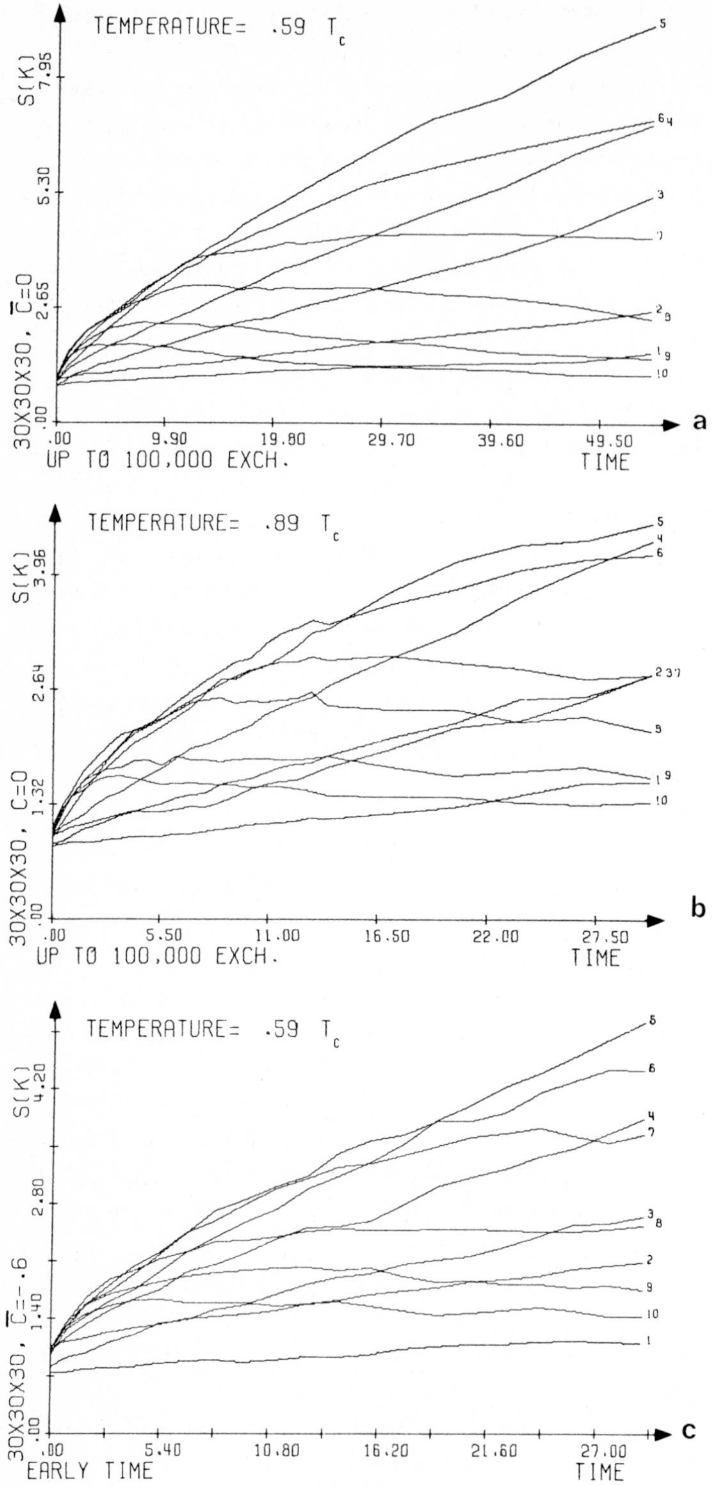
TEMPERATURE= .59 T_c
S(K)
30X30X30, C̄=0
UP TO 100,000 EXCH.
TIME
a
TEMPERATURE= .89 T_c
S(K)
30X30X30, C=0
UP TO 100,000 EXCH.
TIME
b
TEMPERATURE= .59 T_c
S(K)
30X30X30, C̄=-.6
EARLY TIME
TIME
c

see Fig.6.13b [6.38]. The data agree fairly well with the predicted scaling function [6.90]. While it has been suggested [6.70,71,52] that the dynamic scaling hypotheses [6.19] can be extended to phenomena far from thermal equilibrium as implied by (6.21), an experimental verification of (6.21) is so far available for liquid binary mixtures only [6.96].

Figure 6.14a [6.97] illustrates the observed behavior of $\ln[k_m(t)]$ as a function of $\ln(t)$. The theoretical slopes $-a = -1/(d + 3) = -1/6$ 6.35,56 and $-a = -1/3$ are included for comparison. At earlier times the data are indeed consistent with the former prediction and show a crossover to a somewhat larger exponent at later times. As expected, one cannot really tell if the asymptotic value $a = 1/3$ is actually reached. In the two-dimensional simulations at $c = 0.2$ also the cluster size distribution $n_\ell(t)$ was recorded and the time evolution of the average grain (or "cluster") size $\bar{\ell}(t)$ determined (Fig.6.14b) [6.55]. The theoretical prediction [6.35,56] is $\bar{\ell}(t) \propto t^{d/(d+3)} = t^{0.4}$, which is fairly consistent with the observed $t^{0.359}$-law. Due to difficulties with the "percolation" problem [6.53] (i.e., a finite fraction of particles is within ramified infinite clusters as $N \to \infty$) no such cluster observations are possible for $c = 0.5$ or $d = 3$, respectivley. But in the case $d = 2$, $c = 0.2$ a direct observation of the configurations of the system [6.55] beautifully reveals the coarsening of the grains (Fig.6.15).

6.4 Conclusions and Outlook

In this chapter a survey has been given on applications of the Monte Carlo method to relaxation phenomena. Since the Monte Carlo sampling is a stochastic process described by a markovian master equation (see Chap.1), no kinetic phenomena in systems described by the newtonian equations of motion can be modeled. For the latter purpose, the "molecular dynamics" method [6.98] would have to be used which is outside the scope of this book. Thus Monte Carlo studies of kinetic processes are restricted to phenomena where the system under study is coupled to some other system outside of explicit consideration (reservoir of particles, or heat bath, etc.), which induces random changes in the system variables. It turns out, however, that a rich variety of phenomena belongs to this class: biopolymers in aqueous solution, where the latter causes random transitions of peptide units from the helix to the coil state (and vice versa); highly anisotropic ferro- and antiferromagnets,

---

Fig.6.12a-c Plot of $S(k,t)$ vs. t for a three-dimensional $30 \times 30 \times 30$ model system at $\bar{c} = 0.5$, $T = 0.59\ T_c$ (a), $\bar{c} = 0.5$, $T = 0.89\ T_c$ (b) and $\bar{c} = 0,2$, $T = 0.59\ T_c$ (c). Parameter of the curves is $\rho = 30\ k/2\pi$. [6.38]

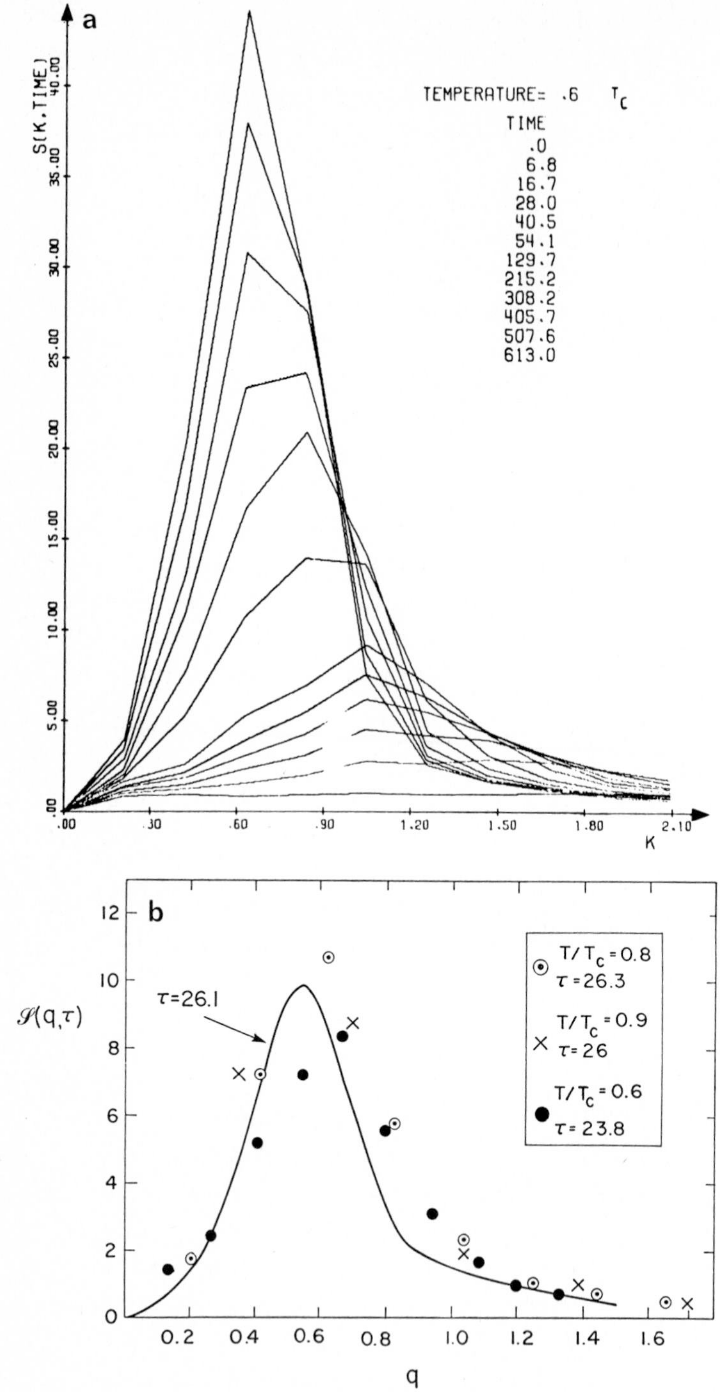

Fig.6.13 (a) Structure factor $S(\underline{k},t)$ plotted vs. K (in units of the reciprocal lattice spacing) at various times. (b) Monte Carlo data at various temperatures are compared to the predicted scaling functions $\tilde{S}(q,\tau)$ [6.90]. [6.38]

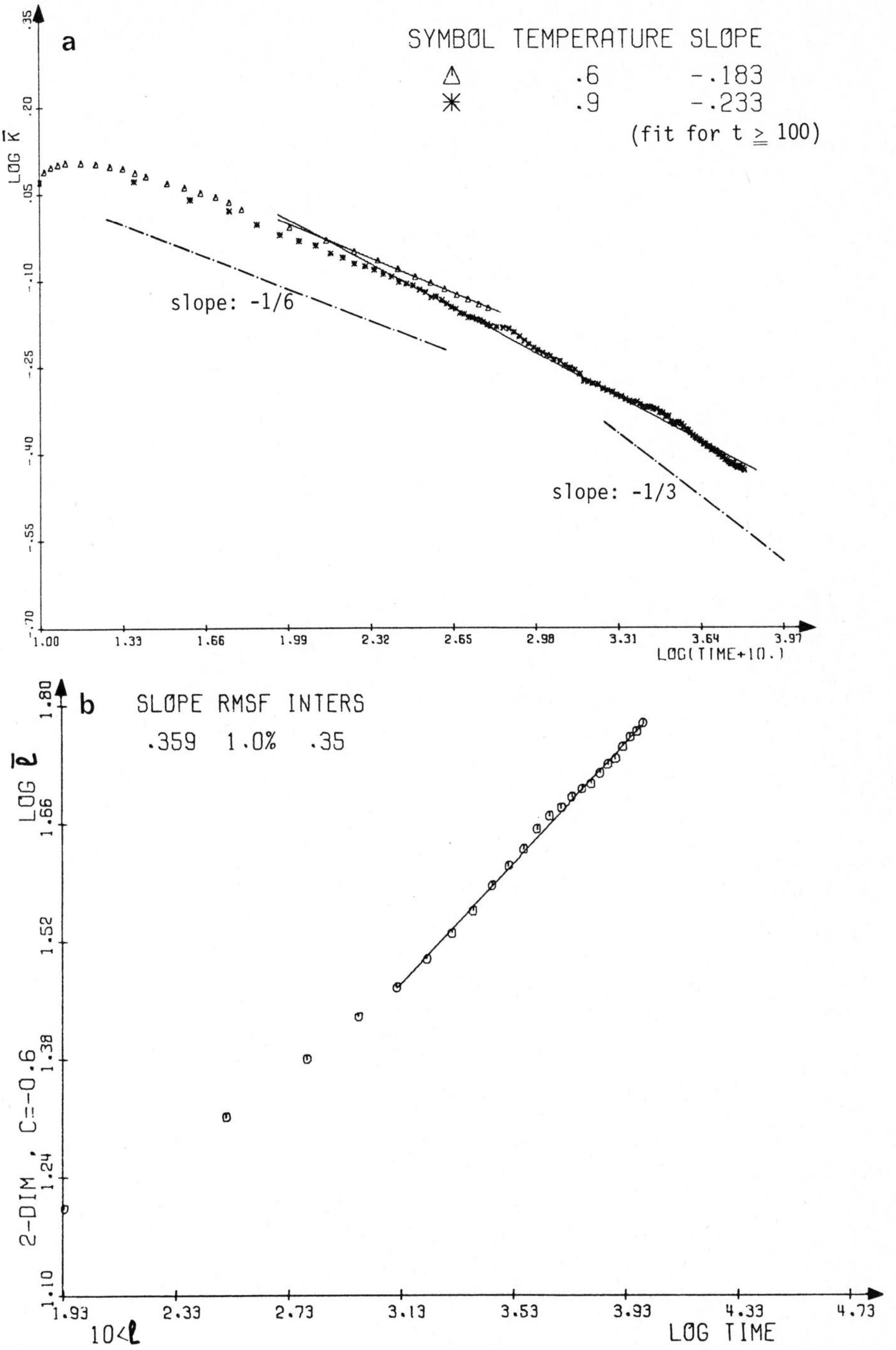

Fig.6.14 (a) Typical wave vector $k_m(t) \equiv \sum k\, \underline{S}(\underline{k},t) / \sum k\, S(\underline{k},t)$ plotted versus time for a three-dimensional model system at $\bar{c} = 0.5$ and two temperatures [6.38]. (b) Average grain size $\bar{\ell}$ plotted versus time for a two-dimensional model system at $T/T_c = 0.6$, $\bar{c} = 0.2$, [6.55]

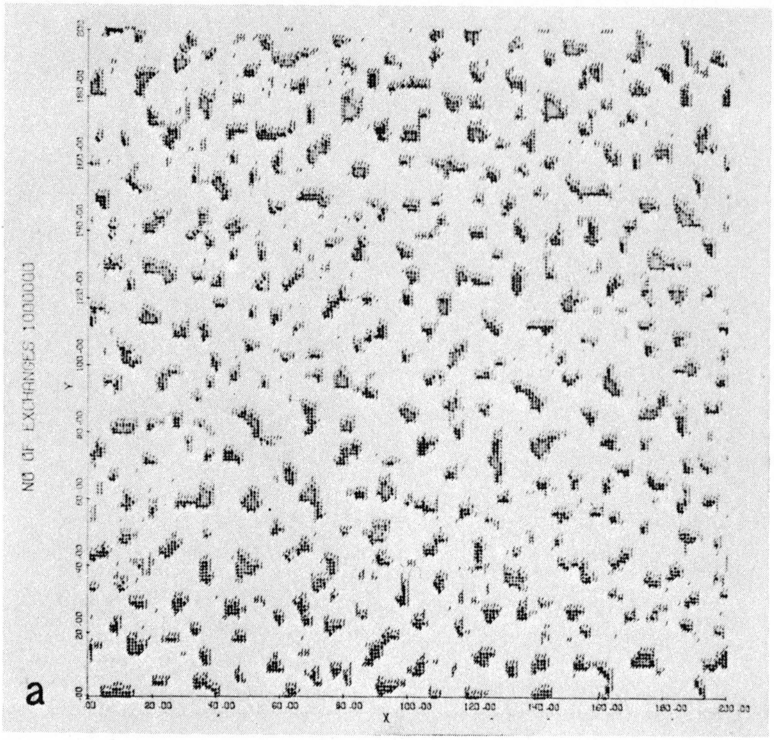
NO OF EXCHANGES 1000000
Y
X
a

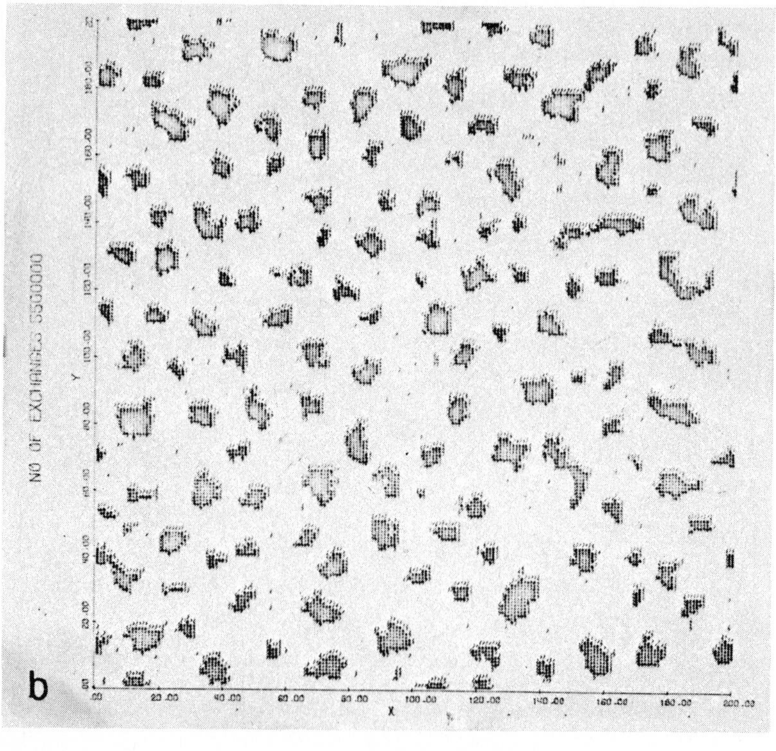
NO OF EXCHANGES 5500000
Y
X
b

where the coupling to lattice vibrations leads to random flipping of the spins; uniaxial ferro- or antiferroelectrics, where the coupling to lattice vibrations leads to random changes in orientation of the atomic dipoles; molecular crystals where similarly the orientation of nonpolar molecule groups changes randomly; diffusion problems on surfaces and in bulk metallic alloys and nonmetallic systems like superionic conductors [6.97]; random evaporation and condensation of atoms at planar surfaces, crystal growth, "cluster" kinetics (nucleation problems) etc. We have shown that most of these processes can be approximated by a description in terms of appropriate kinetic Ising models, which are immediately suitable for Monte Carlo studies. Discussing a few examples, it was demonstrated that these Monte Carlo studies have in fact produced useful results: the relaxation functions of both the "helix fraction" in biopolymers and the magnetization of anisotropic magnets (near an ordinary critical and near multicritical points) has been calculated, clarifying some problems of recent theories on these subjects; the temperature dependence of various relaxation times for nonlinear critical slowing down was obtained to check proposed scaling relations; nucleation theory was checked in detail, obtaining both the time dependence of cluster concentrations in a nucleation process and the equilibrium cluster concentrations and reaction rates, which are a necessary input for a theoretical prediction of the nucleation rate; the phase separation kinetics and grain growth in alloys was simulated and compared both to experiments and to recent theories of spinodal decomposition and coarsening. We feel that in these metallurgical problems of phase transformation kinetics there are particular good opportunities for future applications of the Monte Carlo method. Of course, there are other topics of diffusion problems unrelated to phase transformations, which are also very well suited for an application of simulation techniques. For instance, interesting results have been obtained [6.40,99] on the correlation factor f for diffusion in disordered alloys. This correlation factor is defined in terms of the diffusion constant D, atomic jump frequency Γ and jump distance λ by the generalized Einstein relation $D = \Gamma\lambda^2 f/6$. Other subjects where we expect dynamic Monte Carlo simulations to be helpful in the future are relaxation effects in molecular crystals and superionics, biological systems (transport and reactions at membranes), macromolecules, and phenomena far from thermal equilibrium, like chemical reactions, etc.

Fig.6.15a and b "Snapshot pictures" of the configuration of a 200 × 200 system with $\bar{c}$ = 0.2 at times t = 500 and t = 8925 Monte Carlo steps/atom after the quench to T/T_c = 0.6. B-atoms are indicated as black dots, A-atoms are not shown [6.55]

References

6.1 H. Müller-Krumbhaar, K. Binder: J. Statist. Phys. *8*, 1 (1973)

6.2 K. Binder: Advan. Phys. *23*, 917 (1974)

6.3 K. Binder: In *Phase Transitions and Critical Phenomena*, ed. by C. Domb, M.S. Green, Vol. 5b (Academic Press, London 1976) p. 1

6.4 For a review of earlier work on kinetic Ising-models, see K. Kawasaki: In *Phase Transitions and Critical Phenomena*, ed. by C. Domb, M.S. Green, Vol. 2 (Academic Press, London 1972) p. 443

6.5 L.P. Kadanoff, J. Swift: Phys. Rev. *165*, 310 (1968)

6.6 D. Poland, H.A. Scheraga: *Theory of Helix-Coil Transitions in Biopolymers* (Academic Press, New York 1970)

6.7 T. Tanaka, K. Soda, A. Wada: J. Chem. Phys. *58*, 5707 (1973)
T. Tanaka, A. Wada, M. Suzuki: J. Chem. Phys. *59*, 3799 (1973)

6.8 G. Schwarz, J. Engel: Angew. Chem. Intern. Ed. Engl. *11*, 568 (1972)
J. Engel, G. Schwarz: Angew. Chem. Intern. Ed. Engl. *9*, 389 (1970)

6.9 A. Baumgärtner, K. Binder: J. Statist. Phys. *18*, 423 (1978)

6.10 See e.g. K.S.J. Nordholm, S.A. Rice: J. Chem. Phys. *59*, 5605 (1973)

6.11 R.J. Glauber: J. Math. Phys. *4*, 294 (1963)

6.12 N. Ogita, A. Ueda, T. Matsubara, F. Yonezawa, H. Matsuda: In *Synergetics*, ed. by H. Haken (Teubner, Stuttgart 1973) p. 177

6.13 E. Stoll, K. Binder, T. Schneider: Phys. Rev. B *8*, 3266 (1973)

6.14 D.P. Landau, M. Blume: Phys. Rev. B *13*, 287 (1976)

6.15 I. Morgenstern, K. Binder: J. Chem. Phys. *69*, 253 (1978)

6.16 H. Müller-Krumbhaar, K. Binder: Intern. J. Magnetism *3*, 113 (1972)

6.17 T. Schneider, E. Stoll: Phys. Rev. B *10*, 959 (1974)

6.18 T. Schneider: In *Local Properties at Phase Transitions*, ed. by K.A. Müller, A. Rigamonti (North-Holland, Amsterdam 1976)

6.19 B.I. Halperin, P.C. Hohenberg: Phys. Rev. *177*, 952 (1969); Rev. Mod. Phys. *49*, 435 (1977)

6.20 M. Suzuki: Progr. Theoret. Phys. *43*, 882 (1970)

6.21 B.M. McCoy, T.T. Wu: *The Two-Dimensional Ising Model* (Harvard University Press, Cambridge, Mass. 1973)

6.22 K. Kawasaki: Progr. Theoret. Phys. *40*, 706 (1968)

6.23 H. Yahata: J. Phys. Soc. Jpn. *30*, 657 (1971)

6.24 B.I. Halperin, P.C. Hohenberg, S.-K. Ma: Phys. Rev. B *10*, 139 (1974)

6.25 M.T. Hutchings, M.P. Schulhof, H.J. Guggenheim: Phys. Rev. *135*, 154 (1972)

6.26 L.J. de Jongh, A.R. Miedema: *Experiments on Simple Magnetic Model Systems* (Taylor and Francis, London 1974)
H. Ikeda, K. Hirakawa: Solid State Commun. *14*, 529 (1974)
E.J. Samuelsen: J. Phys. Chem. Sol. *35*, 785 (1974)

6.27 A.M. Gottlieb, F. Heller: Phys. Rev. B *3*, 3615 (1971)

6.28 K. Binder, D. Stauffer, H. Müller-Krumbhaar: Phys. Rev. B *10*, 3853 (1974)

6.29 K. Binder, D. Stauffer, H. Müller-Krumbhaar: Phys. Rev. B *12*, 5261 (1975)

6.30 R. Bausch, H.K. Janssen: Z. Phys. B *25*, 275 (1976)

6.31 J. Feder: Ferroelectrics *12*, 71 (1976)

6.32 E.R. Monaschi, A. Rigamonti, L. Menafra: Phys. Rev. B *14*, 2005 (1976)

6.33 G. Doyen, G. Ertl, M. Plancher: J. Chem. Phys. *62*, 2957 (1975)

6.34 K. Binder, D.P. Landau: Surf. Sci. *61*, 577 (1976)

6.35 K. Binder: Phys. Rev. B *15*, 4425 (1977)

6.36 J. Marro: Thesis, Yeshiva University, New York (1975), unpublished

6.37 A.B. Bortz, M.H. Kalos, J.L. Lebowitz, M.H. Kalos: Phys. Rev. B *10*, 535 (1974)

6.38 J. Marro, A.B. Bortz, M.H. Kalos, J.L. Lebowitz: Phys. Rev. B *12*, 2000 (1975)

6.39 A. Sadiq: Phys. Rev. B *9*, 2299 (1974)

6.40 G.E. Murch, R.J. Thorn: Phil. Mag. *35*, 493 (1977)

6.41 Z. Racz, M.F. Collins: Phys. Rev. B *11*, 2564 (1975)

6.42 H. Yamauchi, D. de Fontaine: In *Order-Disorder Transformations in Alloys*, ed. by H. Warlimont (Springer, Berlin, Heidelberg, New York 1974) p. 148

6.43 For reviews, see E.K. Riedel: AIP Conf. Proc. *18*, 865 (1973)
M.E. Fisher: AIP Conf. Proc. *24*, 273 (1975)
6.44 H. Müller-Krumbhaar, D.P. Landau: Phys. Rev. B *14*, 2014 (1976)
6.45 M.E. Fisher: Physics *3*, 255 (1967)
L.P. Kadanoff: In *Critical Phenomena*, ed. by M.S. Green (Academic Press, New York 1971)
6.46 K. Binder: Ann. Phys. (N.Y.) *98*, 390 (1976)
6.47 R. Kretschmer, K. Binder, D. Stauffer: J. Stat. Phys. *15*, 267 (1976)
6.48 N. Ogita, A. Ueda, T. Matsubara, H. Matsuda, F. Yonezawa: J. Phys. Soc. Jpn. S *26*, 145 (1969
6.49 T. Schneider, E. Stoll: J. Phys. C *8*, 283 (1975)
6.50 G.J. Adriaenssens: Phys. Rev. B *12*, 5116 (1975)
6.51 B.J. Ackerson, C.M. Sorensen, R.C. Mockler, W.J. O'Sullivan: Phys. Rev. Lett. *34*, 1371 (1975)
6.52 K. Binder, D. Stauffer: Advan. Phys. *25*, 343 (1976)
6.53 J.W. Essam: In *Phase Transitions and Critical Phenomena*, ed. by C. Domb, M.S. Green, Vol. 2 (Academic Press, New York 1972) p. 197
6.54 K. Binder, M.H. Kalos: J. Stat. Phys. *22*, 363 (1980); see also
M. Rao, M.H. Kalos, J.L. Lebowitz, J. Marro: In *Computer Simulation for Materials Applications*, ed. by R.J. Arsenault, J.R. Beeler, J.A. Simmens (NBS, Gaithersburg, Md, 1976) p. 180
6.55 M. Rao, M.H. Kalos, J. Marro, J.L. Lebowitz: Phys. Rev. B *13*, 4328 (1976)
6.56 K. Binder, D. Stauffer: Phys. Rev. Lett. *33*, 1006 (1974)
6.57 M. Suzuki: Intern. J. Magn. *1*, 123 (1971)
6.58 K. Binder: Phys. Rev. B *8*, 3419 (1973)
6.59 Z. Racz: Phys. Rev. B *13*, 263 (1976)
M.E. Fisher, Z. Racz: Phys. Rev. B *13*, 5039 (1976)
6.60 K. Kawasaki: Progr. Theoret. Phys. *56*, 1705 (1976)
6.61 M.F. Collins, Z. Racz: Phys. Rev. B *13*, 3074 (1976)
6.62 H.C. Bolton, C.H. Johnson: Phys. Rev. B *13*, 3025 (1976)
6.63 H. Müller-Krumbhaar, K. Binder: Z. Phys. *254*, 269 (1972)
K. Binder, H. Müller-Krumbhaar: Phys. Rev. B *7*, 3297 (1973)
6.64 M.F. Collins, H.C. Teh: Phys. Rev. Lett. *30*, 781 (1973)
6.65 T. Hashimoto, T. Miyoshi, H. Ohtsuka: Phys. Rev. B *13*, 1119 (1976)
6.66 K. Binder, D. Stauffer: Z. Phys. B *24*, 407 (1976)
6.67 F.F. Abraham, G.M. White: J. Appl. Phys. *41*, 1841 (1970)
6.68 A.I. Michaels, G.M. Pound, F.F. Abraham: J. Appl. Phys. *45*, 9 (1974)
6.69 R. Gordon: J. Chem. Phys. *48*, 1408 (1968) and references contained therein
6.70 K. Binder, H. Müller-Krumbhaar: Phys. Rev. B *9*, 2328 (1974)
6.71 K. Binder, E. Stoll: Phys. Rev. Lett. *31*, 47 (1973)
6.72 E. Stoll, T. Schneider: Physica B&C *86-88*, 1419 (1977)
6.73 A. Sur, J.L. Lebowitz, J. Marro, M.H. Kalos: Phys. Rev. B *15*, 535 (1977)
6.74 K. Binder: Advan. Colloid Interface Sci. *7*, 279 (1977)
T. Schneider, E. Stoll: In *Anharmonic Lattices, Structural Transitions and Melting*, ed. by T. Riste (Noordhoff, Leiden 1974)
6.75 K. Binder: In *Fluctuations, Instabilities and Phase Transitions*, ed. by T. Riste (Plenum Press, New York 1975) p. 53 and references contained therein
6.76 K. Binder: In *Proc. Intern. Conf. on Statistical Physics, Budapest, Sept. 1975*, ed. by P. Szepfalusy (Akademiai Kiado, Budapest 1976) p. 219
6.77 J.W. Cahn, J.E. Hilliard: J. Chem. Phys. *31*, 688 (1959)
J.W. Cahn: Acta Met. *9*, 795 (1969)
6.78 A.C. Zettlemoyer: *Nucleation* (Marcel Dekker, New York 1969)
6.79 M.E. Fisher: Rpt. Progr. Phys. *30*, 615 (1967); see also 6.45
6.80 L. Onsager: Phys. Rev. *65*, 117 (1944)
6.81 F.F. Abraham: *Homogeneous Nucleation Theory* (Academic Press, New York 1974)
6.82 J.L. Lebowitz, M.H. Kalos: Scripta Met. *10*, 9 (1976)
6.83 J.S. Huang, W.I. Goldburg, M.R. Moldover: Phys. Rev. Lett. *34*, 639 (1975)
6.84 K.B. Rundman, J.E. Hilliard: Acta Met. *15*, 1025 (1967)
K.B. Rundman: Ph. D. Thesis, University of Illinois (1965) unpublished
6.85 E.P. Butler, G. Thomas: Acta Met. *18*, 947 (1970)
6.86 M. Bouchard, G. Thomas: Acta Met. *23*, 1485 (1975)
6.87 H.E. Cook: Acta Met. *18*, 297 (1970)

6.88 K. Binder, M.H. Kalos, J.L. Lebowitz, J. Marro: In *Nucleation III*, ed. by A.C. Zettlemoyer: Adv. Coll. Interf. Sci., *10*, 173 (1979)
6.89 J.S. Langer: Ann. Phys. (N.Y.) *65*, 53 (1971); Acta Met. *21*, 1649 (1973)
6.90 J.S. Langer, M. Bar-On, H.D. Miller: Phys. Rev. A *11*, 1417 (1975)
6.91 P.A. Flinn: J. Statist. Phys. *10*, 89 (1974)
6.92 K. Binder: Z. Phys. *267*, 213 (1974)
6.93 I.M. Lifshitz, V.V. Slyozov: J. Phys. Chem. Sol. *19*, 35 (1961)
6.94 P. Mirold, K. Binder: Acta Met. *25*, 1435 (1977)
6.95 M. Kahlweit: Advan. Colloid Interface Sci. *5*, 1 (1975)
6.96 N.C. Wong, C.M. Knobler: Bull. Am. Phys. Soc. *22*, 298 (1977)
6.97 G.D. Mahan: Phys. Rev. B *14*, 780 (1976)
6.98 A. Rahman: Phys. Rev. A *136*, 405 (1964)
6.99 H.J. de Bruin, G.E. Murch: Phil. Mag. *27*, 1475 (1973)
H.J. de Bruin, G.E. Murch, H. Bakker, L.P. van der Mey: Thin Solid Films *25*, 47 (1975)
H. Bakker, N.A. Stolwijk, L. van der Meij, T.J. Zuurendonk: Nucl. Metallurgy *20*, 96 (1976)
J.R. Manning: Nucl. Metallurgy *20*, 109 (1976)

7. Monte Carlo Simulation of Crystal Growth

H. Müller-Krumbhaar

With 17 Figures

The properties of crystal surfaces are frequently described in terms of lattice models. The dynamics of crystal growth are introduced as stochastic processes for adsorption, evaporation and surface migration. Monte Carlo methods directly simulate these dynamic processes. They have sofar been applied to short-ranged interactions between atoms, elastic forces are just about to be considered. Equilibrium properties of crystal surfaces such as surface energy, correlation functions, interface width and step structure have been analyzed. The dynamics of crystal growth has been studied with respect to different growth laws which were observed experimentally. They are traced back to the various microscopic mechanisms such as the roughening transition, screw dislocations, layer growth, surface diffusion etc. Particular emphasis is given to the relation between these models and general concepts of phase transitions such as critical fluctuations, nucleation and metastability.

7.1 Introductory Remarks

The simulation of crystal growth [7.1.2] offers a broad range of applications for the Monte Carlo method [7.3,4] for several reasons.

The growing crystal defines a natural lattice system onto which the noncrystalline growth units (atoms, molecules) are condensed. For growth from solution or from the vapor one may assume with sufficient generality, that the growth units impinge on the solid-vapor (-solution) interface at random with an average frequency. It is therefore natural to describe the adsorption-desorption mechanism as a stochastic process, for which a master equation may be formulated.

A large variety of phenomena makes the simulation particularly interesting. Solidification in general is a first-order phase transition, which is theoretically much less well understood than second-order transitions [7.2]. During growth the crystal surfaces may show many structural details such as clusters of atoms, surface steps, screw dislocations, all of which are originated on an atomic length scale but may lead to cooperative phenomena of macroscopic dimensions. These structures of course depend on the underlying lattice symmetry and the type and range of interactions between atoms.

Crystals of nonpure substances bring additional degrees of freedom into play such as surface ordering, segregation of impurities, adsorption on foreign substrate, etc. Inclusion of grain boundaries finally makes the system inhomogeneous in all three spatial directions. These processes then may depend very sensitively upon the growth conditions, such as growth rate, chemical composition, temperature distribution, and are of fundamental importance for the final properties of the grown crystal.

Many of these phenomena are not well understood at present. Kinetic theories suffer from the spatial inhomogeneity due to the interface, which makes it necessary to treat coupled systems of differential-difference equations (or nonlinear partial differential equations). Experiments suffer from the large number of experimental parameters which have to be controlled simultaneously.

The Monte Carlo method (Chap.1) now can handle a large number of variables in an inhomogeneous many-particle system, including possibly important fluctuations in space and time. It may serve as a test for the theories as well as a first model-study in order to find out what the experimentally relevant parameters are. It may therefore advantageously be used to fill the present gaps between theory and experiment.

The main concern of crystal growth simulation sofar has been the investigation of low-indexed surfaces of monatomic crystals, assuming very short-ranged interaction between the atoms. Since the crystal surface advances during growth it is better called [7.5] an "interface" in a two-phase solid-liquid or solid-vapor system, in contrast, e.g., to the "surface" of a magnet, which defines the fixed external boundary condition. A small deviation from thermodynamic equilibrium will cause the interface to move. Since at sufficiently low temperatures the low-indexed crystal faces are atomically flat the surface structure must undergo drastic changes as the interface advances to the next layer. (In the following, "surface" will be used synonymously with "interface").

The study of equilibrium structures (Sect.7.2) is, therefore, only a first step in studying growth mechanisms. Since one always deals with a rather small model-system on the computer, one generally uses periodic boundary conditions to simulate an infinitely extended crystal surface of prescribed fixed crystallographic orientation. In this way one of course cannot simulate the crystal morphology but assumes, that the morphological (overall) changes occur much slower than the relaxation towards local equilibrium within the particular interface. The same method, i.e., fixing the average direction by boundary conditions, is also used to study the properties of surface steps.

A special problem, which has attracted much attention again recently, is the surface roughening transition. It is expected to show up in several properties of crystal surfaces and therefore will be discussed in a special paragraph.

The main part of this chapter is contained in Sect.7.3. First, a few remarks on the dynamics of the Monte Carlo procedure relate the present application to Chap.1. Then follows the survey over kinetics on low-index faces. Particular emphasis is given to the interpretation of the layerwise growth as first-order phase transition, involving nucleation and metastability. Surface diffusion as an accelerating factor is discussed in the context of both plane surfaces and stepped surfaces. The important role of surface steps is summarized in the next subsection with particular attention to cooperative effects, such as step-bunching and screw-dislocation spirals. Rather unclear is the situation in the roughening problem, in particular the way how growth changes from two-dimensional nucleation to normal growth. The final paragraph of Sect.7.3 is on more component crystals and segregation of impurities. The first Monte Carlo simulation in crystal growth, performed by CHERNOV and LEWIS [7.6,7], in fact, had started already with this topic. And it forms the bridge to the probable future development as will be noted in the outlook in the final Sect.7.4.

A special guide to the literature is finally provided by Table 7.1. There, all the presently available publications on Monte Carlo simulation in crystal growth [7.6-64] have been analyzed with respect to 15 different topics. The rows are marked with the reference numbers. A cross (+) at the intersection of a row and a column indicates, that the respective subject has been analyzed in the reference.

7.2 Crystal Surfaces at Equilibrium

7.2.1 Singular Faces

a) Kossel Model (Simple Cubic-(100) Face), Two-Dimensional Spin-s-Ising Model, Thin Films

The simplest model for a growing crystal is the Kossel model or SOS (Solid-on-Solid) model. It consists of a simple cubic lattices which may be partly filled with atoms (or molecules) in such a way, that no empty spaces are allowed in the bulk of the crystal. This is achieved by the so-called SOS restriction [7.65,66] which allows only columns to be built normal to the interface (between the filled and the empty part of the lattice), thus excluding overhanging structures at the interface. There is a hard-core repulsion between atoms which allows at most one atom per lattice point and a short-ranged attractive interaction (usually to nearest neighbors only) which causes condensation. Due to the SOS restriction the model is not realistic near the melting point but may well be used to simulate crystal growth from the vapor or from a solution. The model does not aim at a realistic description of the solid and its surface (neither does it allow for surface reconstruction effects nor for phonons). But it contains the essential configurational degrees of freedom for building up new crystal layers.

The hamiltonian of the model is conveniently written as

$$\mathcal{H} = \varepsilon \sum_{<i,j>} |h_i - h_j|^p \quad ; \quad p = 1 \quad , \qquad h_j = \text{integer} \tag{7.1}$$

where $\varepsilon > 0$ is the interaction energy per bond, the sum goes over nearest-neighbor pairs in a plane and the exponent p in the usual case is $p = 1$. The local variables h_i may vary between $\pm\infty$ and indicate the local deviation of the interface from some reference plane $h_i = 0$. An alternative form is given by [7.67]

$$\mathcal{H} = -\frac{1}{2}\varepsilon \sum_{<i,j>_{3d}} S_i S_j + V(\{S_i\}) \qquad S_i = \pm 1 \tag{7.2}$$

where the sum now goes over a three-dimensional lattice, $S_i = +1$ stands for "filled" and $S_i = -1$ for "empty" lattice point. The potential $V(\{S_i\})$ is constructed in such a way that it is infinite for any overhanging structure at the interface and zero otherwise, thus providing the SOS restriction. In case that $V = 0$ for all configurations the model then is identical to the three-dimensional Ising model [7.68,69] (spin-½), where an interface may be introduced by appropriate boundary conditions.

The Monte Carlo simulation of the equilibrium properties of these interface models is straightforward (Fig.7.1a,b). One defines conditional probabilities $W(h_i \to h_i + \delta)$ for the local increase or decrease ($\delta = \pm 1$) in height h_i which fulfill the detailed balance condition (1.4). Local changes in height then are made with a random number generator, where the probability for the transition depends on the energy change in the usual way. Details will be discussed in Sect.7.3.1.

Fig.7.1 (a) Computer simulation of a "Kossel" (SOS) model at low temperatures. The surface shows two steps and a few isolated clusters and holes. (b) The same model at high temperatures. Clusters of adatoms and holes over the surface. The crystal appears to be locally rough

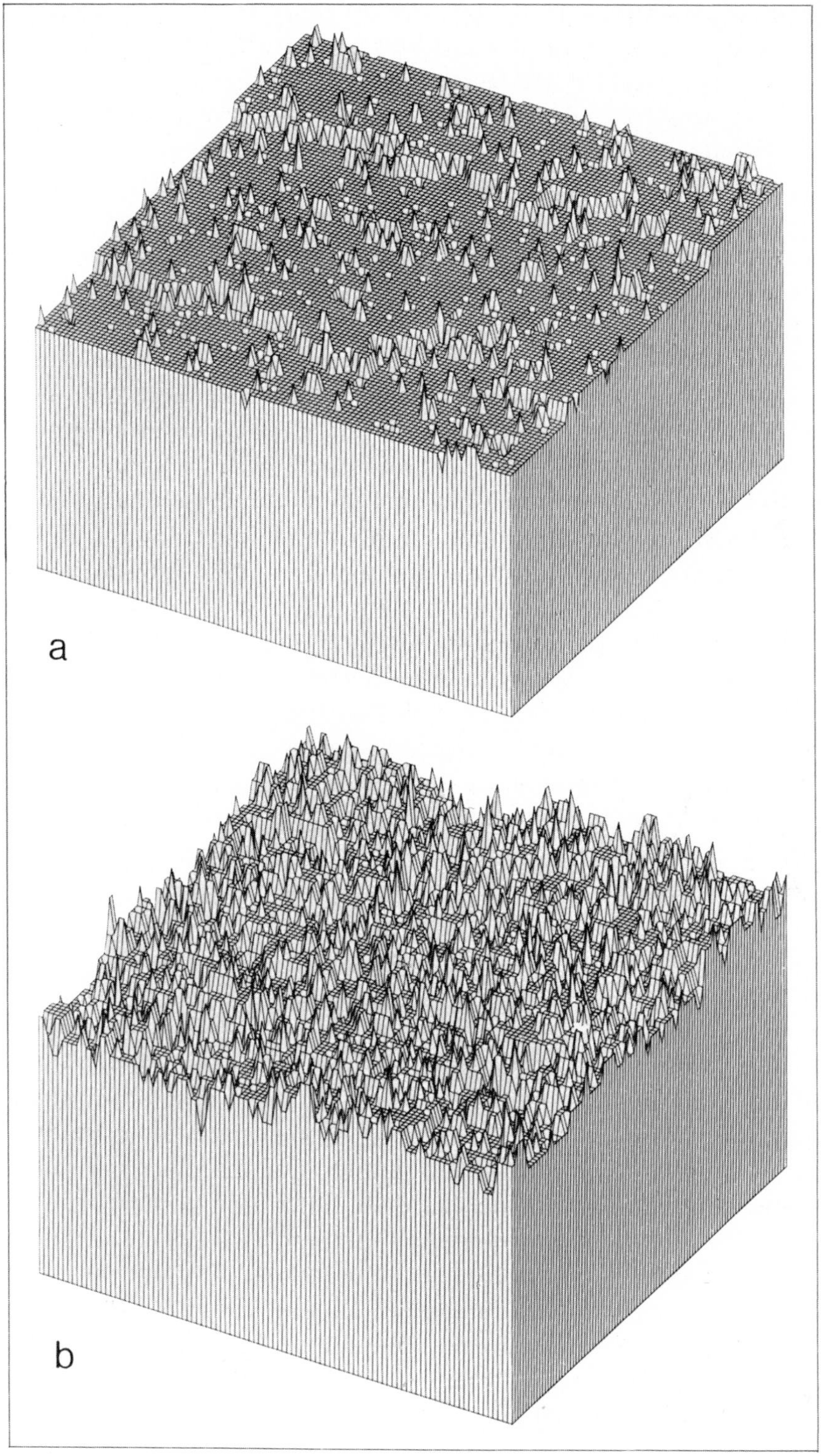
a
b

At sufficiently low temperatures the interface is a flat plane (Fig.7.1.a) with only a few isolated atoms sitting on the plane and a few holes within the first solid layer [7.13-15]. As long as the temperatures are low enough, i.e., as long as clustering of adatoms and holes is not important, the properties of this interface are well described by a layerwise mean-field approximation (MFA) [7.66,67,70]. As the temperature increases towards the critical temperature T_{2d} of the two-dimensional Ising model, fluctuations become important and the mean-field approach (including Bethe-approximation and related theories [7.34]) ceases to be reliable. For temperature T larger than the roughening temperature $T_R(\geq T_{2d})$ a column-wise mean-field approximation [7.71] may be appropriate, but a detailed comparison has not been made. All Monte Carlo calculations [7.13,14,16,17,46] show a rapid increase in interface energy around T_{2d}, analogous to the two-dimensional Ising model.

Early interface models (N-level model) [7.72] limit the local deviation of the interface from its average position ($h_i = 0$) to $h_i = \pm (N-1)/2$, N = odd; $h_i = \pm N/2$; N = even. They can be mapped onto spin $s = (N-1)/2$ Ising models in two dimensions. The maximum energy density (= broken bonds projected into a plane parallel to the interface) there is finite and proportional to the maximum interface thickness.

For certain questions such as the problem of interface roughening [7.73] these models are not appropriate. But they may be used to explain structures of thin films adsorbed on a flat substrate. The two-dimensional spin-½ Ising model, e.g., may represent a monomolecular adsorption layer, the spin-1 Ising model accordingly may represent a mono- or bimolecular layer etc.

If the restriction in height fluctuations is only partly given up, such that $0 \leq h_i < \infty$, the model represents a crystal grown on a foreign substrate, where additional interactions may be introduced in order to formally account for lattice mismatch.

In both of the above cases a number of phase transitions within the layers may occur as they become partly filled [7.65].

The SOS model with the full range of discrete values $-\infty < h_i < +\infty$ allows the interface thickness and its associated energy to go to infinity even locally if $T \to \infty$. This limit clearly is unphysical and a result of the SOS restriction, since it produces columns of atoms and holes of arbitrary height (depth), while a real crystal would tend to melt.

Giving up the SOS restriction the interface would vanish at the critical temperature T_{3d} of the resulting three-dimensional Ising model [7.68] . This second-order transition again is clearly not to be understood as "melting" and hence the solid-liquid transition of a pure substance is not contained within this class of models. For the solid-vapor or solid-solution transitions, however, these models are suitable and various quantities like the interface energy (7.1) or "local roughness" and moments $\langle[h(\underline{0}) - h(\underline{r})]^{2n}\rangle$, n = 1,2,3, ... ($\underline{r}$ = distance between two points in a plane parallel to the interface) are available from Monte Carlo simulations and accessible to experiments [7.74,75].

b) Other Low-Index Faces

In the simple cubic lattice with nearest-neighbor interaction the (100) face clearly is the one with minimum energy. Because of their limited importance for crystal growth, not much Monte Carlo work has been done on sc-(h,k,ℓ) faces with $|j|+|k|+|\ell|<5$, (but see Sect.7.2.2). For sc-faces not in (100) direction in addition there is the problem of ground-state degeneracy. In Monte Carlo work a sc-(110) face, for example [7.26] is realized by prescribing the boundary condition that after N lattice parameters from one boundary to the next the crystal has increased its height by N lattice parameters. Calculating the energy differences across the (periodic) boundaries, one therefore substracts a number of N height units from all the h_i's at the higher end of the crystal model before taking the differences (7.1). The states of minimum energy then can be either a series of N straight steps of unit height or a single straight step of N units high or any number of steps of arbitrary height, provided that they add up to N height units over N lattice parameters between the boundaries. This energetic degeneracy of these possible ground states has the consequence, that the entropy of the model is nonzero at T = 0 and that the amplitude of fluctuations about the average position of the interface diverges as the system size (i.e., the distance between the boundaries) goes to infinity. Similar arguments hold for all faces other than (100). A comparison of interface energies for various crystal models is given in Fig.7.2.

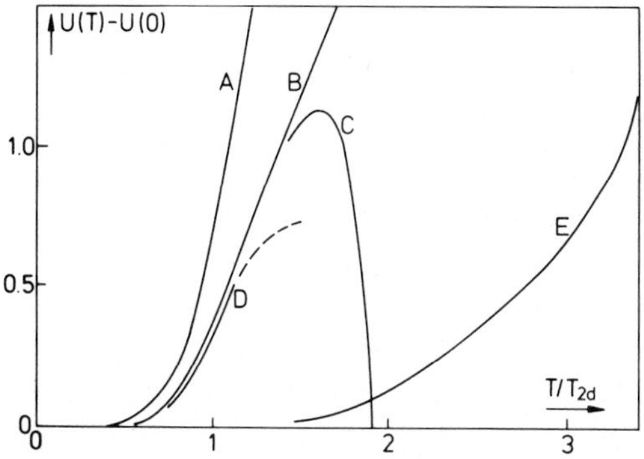

Fig.7.2 Energy *versus* temperature curves for various models for a crystal surface. *A*: Orthorhombic SOS model (anisotropic interaction energies within the surface layer); *B*: Standard sc-100 face of an SOS model; *C*: Interface in a three-dimensional spin-½ Ising model; *D*: fcc-100 face; *E*: fcc-111 face.
The energy u(T) is normalized to one nearest-neighbor bond per particle, the temperature scale is given in units of the critical temperature T_{2d} of the two-dimensional Ising square lattice [7.3]

Since these large amplitude fluctuations involve an enormous number of highly correlated lattice points, the simulation of faces other than (100) in quasi-equilibrium is extremely costly. Even at temperatures very close to zero the decay of long wavelength modes is of very slow random-walk-type because there is no energy difference that could locally prescribe a direction for an exponential relaxation. Hence the system will remember the long wavelength structure of the initial condition for a very long time [7.36] (on the time scale of an elementary adsorption-desorption process). While the energy of such a surface does not see these fluctuations it is, therefore, not easy to obtain unbiased simulation results for the geometric structures of such non-(100) faces).

In a real crystal this may not be a problem since there are always other than nearest-neighbor interactions present, which may change the situation significantly. Closely related is the simulation of lattices with different symmetry and different coordination number [7.4,19,23,29,37] , such as fcc-lattices. Here the (111) face has the minimum energy. From the total coordination number of 12 there are only 3 missing for the atoms in the top layer. At an (001) face there are 4 neighbors missing for the top layer atoms. At an (011) face there are 5 neighbors missing for an atom in the top layer and one more neighbor for an atom in the layer below. The coordination numbers within the top layer are 6 (111), 4 (001) and 2 (011), the latter structure forming straight rows in (001) direction.

The simulations of the fcc-system give results which agree with the qualitative coordination number arguments. The (111) plane has the lowest energy and stays flat up to temperatures about 6/4 above the temperatures obtained for the sc-(001) face. The (001) face in fcc has similar properties as the one in the sc-lattice. Again it should be noted that the (011) face has now similar problems for the simulation concerning the ground-state degeneracy as the (011) face in the sc-lattice.

In agreement with the computer simulations one finds experimentally [7.76] that crystals tend to form facets in the planes of highest coordination numbers.

7.2.2 Surface Steps

Surface steps on crystals are created either by cutting a crystal in an (h,k,ℓ) plane, where one index is much larger than the others, e.g., sc-(0,1,10), or occur during growth by two-dimensional nucleation or stem from screw dislocations. At low enough temperatures (well below the roughening temperature [7.73]) steps are well defined and may be treated as one-dimensional interfaces in a two-dimensional lattice which has the structure of the surface layer of the respective crystal between the steps. An $(0,1,\ell)$ face ($\ell \gg 1$) of a sc-crystal in the neighborhood of a step then is expected to have equilibrium properties like the interface in a two-dimensional spin-½ Ising model [7.77]. This is substantiated by Monte Carlo calculations [7.26,46]. Very close to $T = 0$ one even may neglect overhanging structures and thus approximate the step by a one-dimensional SOS model [7.78-80]. At higher temperatures (approaching the roughening point) fluctuations like holes in the layer below the step and adatoms clusters in the layer above interact with the step [7.14,46], which is not taken into account in the Ising model analogue. The importance of overhangs within the planes adjacent to the step is reflected in the fact, that the energy per step length becomes independent of the mean step orientation as the roughening temperature is approached [7.44].

Again it should be noted that the results of simulations of stepped surfaces depend sensitively upon the boundary conditions. A stepped surface is not a minimum energy surface. Its average direction has to be fixed by the boundary conditions. Prescribing the step height and step density the minimum energy then is associated with steps along certain low-indexed directions, for example (10) direction for a step on a sc-(100) face. At zero temperature this step would run in a straight line. An (11) step, being tilted by 45° against the (10) step, has a higher ground-state energy. Its average direction therefore has to be fixed by the boundary conditions. But this step now does not have a well defined structure due to the same type of ground-state degeneracy as discussed for non-(100) faces in Sect.7.2.1b; there are many different ways to make a zig-zag path from one point on a (100) face to another one in an average (11) direction [7.36]. The analysis of simulated quasi-equilibrium structure therefore has to be done with caution, since the width of such a cross-running step diverges even at $T = 0$ with the step length going to infinity.

Steps on crystals other than with sc-symmetry have not attracted much attention sofar, even though one might interpret the fcc-(011) face as a densely stepped face. Experiments on steps can be performed with various methods. Geometric structures can be observed using decoration techniques [7.81], cooperative effects on parallel step trains are tractable with LEED [7.82] and the step energy can be studied in electrocrystallization experiments [7.83].

7.2.3 Roughening Transition

Low-index faces such as sc-(100) or fcc-(111) were predicted by BURTON et al. [7.72] to undergo a roughening transition from a smooth plane at low temperatures to a rough, stepped structure at high temperatures. Their argument was apparently based on the assumption, that a correct treatment of the odd-level models (corresponding to spin-N Ising models) in two dimensions would exhibit a phase transition similar to the even-level models (simplest case: spin-½ Ising model).

This is obviously not the case, since the first class of models has a uniquely defined ground state while the second group possesses the broken symmetry of the low temperature state which is necessary for such a phase transition [7.73].

The infinite-level models (like the SOS model), being realistic for a crystal-vapor interface, have an infinite number of equivalent discrete ground-state positions of the interface. They are, therefore, expected to show a phase transition at some temperature T_R from a state where the interface is localized ($T < T_R$) to a state where the interface fluctuates freely ($T > T_R$).

Presently available theories [7.73,84] are not completely satisfactory regarding the roughening singularity in the most frequently used models and, therefore, Monte Carlo methods proved to be helpful in the quantitative understanding of this transition.

The first indication of a singularity in the SOS model (except for a series analysis mentioned in [7.85]) came from Monte Carlo simulations of a stepped interface [7.26], which indicated that the excess energy of a step vanishes at a temperature somewhat above T_{2d} (Ising). A later investigation [7.46] which also made an extrapolation of the system size confirmed this result, although the temperature T_R appeared to be closer to T_{2d} (this difference is due to a bias in the data analysis [7.26]). This indicates that the step energy vanishes at T_R similar to the one-dimensional interface in the two-dimensional Ising model, in accordance with the exactly solved "BCSOS" model [7.84] (see Fig.7.3).

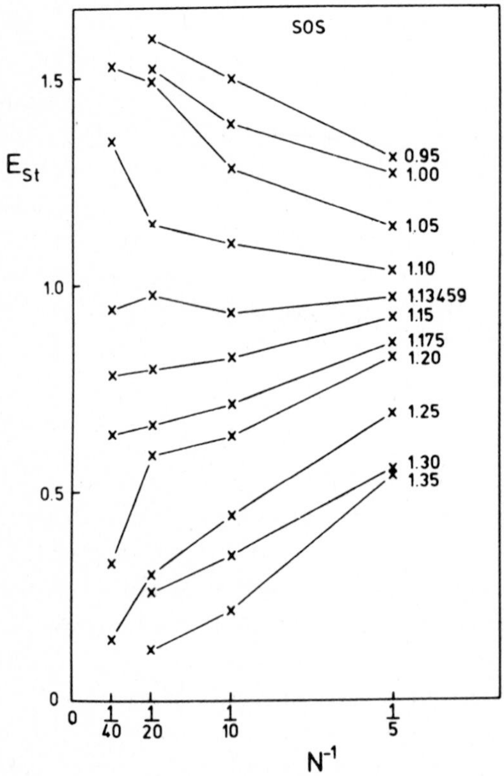

Fig.7.3 Excess energy of a step E_{St} on the SOS model (sc-100 face), plotted versus the inverse side-length N^{-1} of the lattice. Parameter of the curves is the temperature ($T_{2d} \equiv 1.13459$). For lower temperatures $T<T_{2d}$ the step energy increases with the system size, for higher temperatures the energy decreases. [7.46]

The gradient of atomic concentration across the interface was predicted to vanish with some power law $(T_R - T)^{-\vartheta_g}$, which seems to be supported by Monte Carlo calculations [7.4]. The correct value of the exponent is unknown sofar. If T_R would be exactly T_{2d}, one must have [7.2] $\vartheta_g \leqq \beta = 1/8$, the critical exponent of the order parameter in the two-dimensional Ising model at T_{2d}. Mean field-type theories [7.73] give $\vartheta_g = 1/2$, while series expansions indicate larger values [7.85,32]. These latter values would be possible if $T_R > T_{2d}$, which again is not known exactly.

The higher order moments $\langle h_i^{2n} \rangle$ are predicted [7.73] to diverge as $\langle h^{2n} \rangle \sim (T_R - T)^{-n}$ which is difficult to show numerically, since the divergence of these moments above T_R with the system size L (L: edge length of the simulated two-dimensional interface area in the lattice units) is expected to be only logarithmic $\langle h^{2n} \rangle \sim \ln L$. The latter prediction is consistent with the corresponding Monte Carlo simulations [7.46]. The variation of the moments with temperature is shown in Fig.7.4.

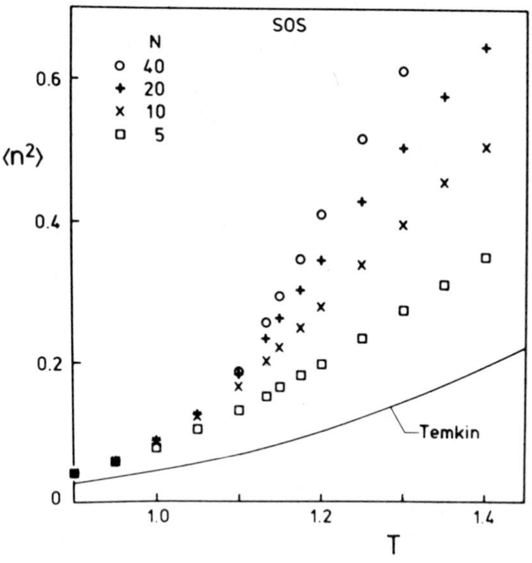

Fig.7.4 Second interface moment $\langle n^2 \rangle \equiv \langle h^2 \rangle$ plotted versus temperature T of the SOS lattice. Parameter of the curves is the sidelength N. For the temperature above T=1.15 the moment increases continuously with increasing N($N \leqq 40$), while for the values below $T=1.15$ the moment apparently has converged to a constant value. [7.46]

The present picture of the static properties of a (100) crystal surface around the roughening temperature therefore is the following. Below T_R isolated clusters of adatoms and holes appear on the surface. Surface steps are clearly distinguishable. Approaching T_R the clusters and - if present - surface steps interact strongly. Above T_R surface steps of infinite extent are formed by thermal fluctuations, causing very long wavelength fluctuations for the local position of the interface. This does not necessarily mean that the "local roughness" (being proportional to the interface energy or the number of broken bonds within the interface) has an observable singularity. The predicted singularities [7.73] are extremely weak and it may thus not be possible to locate the roughening temperature experimentally within a few percent. The overall effect, however, would be that below T_R the crystal is clearly faceted while above T_R all structures become rounded but may still reflect the crystallographic anisotropies.

The quantitative details furthermore depend markedly upon the crystallographic symmetries and range of interactions between atoms. A sc-(110) face or a sc-(111) face is always rough ($T_R = 0$) if only nearest-neighbor interactions are present. But already very weak next-nearest neighbor interactions suppress the otherwise possible fluctuations of the non-(100) faces at low temperatures and a nonzero T_R should exist. This is observed in the exactly soluble BCSOS model [7.84], which can be mapped onto a six-vertex model. An fcc-lattice with nearest-neighbor interaction [7.4,29] as a further example may have two different roughening temperatures $T_R > 0$ for the (111) and the (100) face and $T_R = 0$ for the (110) face. The latter effect is due to the arrangements of atoms in straight rows with only two bonds per atom parallel to the surface.

These various equilibrium structures are of fundamental importance for the growth laws as it is discussed in Sect.7.3. The experimental location of the roughening point will be very difficult because of the expected weak singularities. There is, however, some evidence for the existence of two temperature regimes: one with sharp faceted structures at low temperatures and one where surface structures are smooth and rounded [7.86,87]. New experimental techniques [7.74,75,88] may provide the tools to obtain more quantitative information about this phenomenon.

7.3 Growth Kinetics

7.3.1 General Aspects of Kinetic Simulations in Crystal Growth

The most frequently used lattice model for a crystal surface (SOS model) is best suited for the simulation of crystal-vapor or crystal-solution interfaces. In these cases one may assume that the atoms in the noncrystalline phase are not interacting with each other and that they have a random impingement frequency onto the crystal surface. For this case one can interpret the kinetics as a stochastic process which is well described by a master equation.

The change in probability $P(\{h_i\},t)$ with time of a given configuration $\{h_i\}$ of the local height variables is then expressed as a single-particle process

$$\begin{aligned}\frac{\partial}{\partial t} P(\{h_i\},t) = &- \sum_j P(h_1, \ldots, h_j, \ldots, h_N,t)\{W(h_j \to h_j + 1)+W(h_j \to h_j - 1)\} \\ &+ \sum_j \{P(h_1, \ldots, h_j + 1, \ldots, h_N,t)W(h_j + 1 \to h_j) \\ &+P(h_1, \ldots, h_j - 1, \ldots, h_N,t)W(h_j - 1 \to h_j)\}\end{aligned} \tag{7.3}$$

where $W(h_j \rightarrow h_j \pm 1)$ represents the conditional probability for adsorption or desorption at site j in the two-dimensional height-matrix.

Surface diffusion [7.52,15] may be introduced by additional two-site processes of the form

$$- \sum_j \sum_{q(j)} P(h_1, \ldots, h_j, h_{q(j)}) W(h_j, h_{q(j)} \rightarrow h_j \pm 1, h_{q(j)} \mp 1) + \ldots \qquad (7.4)$$

on the right hand side of (7.3). The index q(j) runs over all nearest neighbors of h_j and the conditional probabilites $W(h_j, h_q \rightarrow h_j \pm 1, h_q \mp 1)$ denote a one-atom displacement from h_j to h_q or vice versa.

The conditional probabilities have to fulfill a detailed balance condition, see (1.4), but otherwise can be freely chosen to simulate the physico-chemical interface processes.

The most frequently used probabilities [7.14,15] are based on the assumption, that the adsorption frequency corresponds to the impingement frequency and hence is independent of the particular adsorption site, while the desorption depends upon the local neighborhood of the adsorbed particle.

The use of other types of transition probabilities (such as, e.g., the symmetrized function [7.4] used by GLAUBER [7.89]) does not lead to qualitative differences, as long as the kinetics can be described by some linear response theory. This is the case for small driving forces across interfaces which are above their roughening temperature (Sect.7.3.4). The effect of different transition probabilities and inclusion of surface diffusion is most pronounced in the regime of plane interfaces, where strong nonlinear effects such as two-dimensional nucleation control the growth mechanism [7.67].

The realization of the master equation (7.3) on a computer may be carried out in several ways, as already mentioned in Sect.1.3.1. In crystal growth simulation two methods have particular advantages. The direct method [7.15] first chooses a particular lattice point at random, then selects one of the allowed processes, adsorption, desorption or diffusion-jump to one of the neighboring positions, and finally either accepts or discards this selected microprocess. The probability for this last decision is the conditional probability W, (7.3,4), which is realized with an additional random number. The advantage of this direct method is the simple programming and the fair performance in the sufficiently high temperature regime. Its disadvantage at low temperatures comes from the small transition probabilites W in that regime due to reduced thermal activation. Since the time scale is given by the fastest rate (e.g., desorption of an isolated atom) most attempts for a transition are discarded at low temperatures.

At the cost of considerable programming effort this problem can be circumvented by inverting the selection procedure [7.40]. For a given configuration of the model system all possible transition processes are stored in tables (i.e., arrays in the computer program) such that a particular table contains all processes which occur with the same probability. At first a table is selected with the corresponding transition probability (of course, normalized with respect to the relative number of elements in the table). Then one of the processes is selected at random from the table and the corresponding change in state of the system is performed. Finally the tables are updated.

This whole procedure of course has more steps than the direct method. But since even at low temperatures every iteration step here performs a transition, a net gain in speed of up to a factor of 1000 is obtained over the direct method within the interesting temperature region.

A number of further modifications (e.g., complete neglect of obviously irrelevant processes) of the Monte Carlo method was used which will not be discussed in detail [7.7,19,28,40,56].

Somewhat related to the above described methods are numerical methods, which make use of stochastically generated initial conditions or of random noise [7.58, 62]. Their main difference with the above procedures is the missing decision whether to accept or discard a process according to a given probability. They will, therefore, not be discussed explicitly.

7.3.2 Low-Index Faces

a) Two-Dimensional Ising Model

At low enough temperatures a low-index crystal surface is essentially flat with only a few adatoms and holes. Growth then proceeds by two-dimensional nucleation of the next surface layer. It therefore appears to be reasonable to study first the formation of a single new layer with time.

A good model for this single layer formation is the two-dimensional spin-½ kinetic Ising model. The advance of the interface by one layer in the crystal-vapor system then corresponds to the spin-reversal in the Ising model. The model has been studied in two versions in considerable detail. The first version is the kinetic one-spin-flip Ising model [7.35] where the order parameter is not conserved and only single-site processes are considered. This version is closely related to the crystal-growth problem. The second version keeps the number of particles in the layer constant (conserved order parameter), but allows for diffusional jumps between neighboring sites [7.55]. This version is appropriate to study surface diffusion and phase separation at surfaces. For crystal growth most appropriate would be a combination of both versions, where single- and two-site processes were considered. Since this has not been treated sofar we will concentrate on the first version [7.2,35].

The condensation of an additional layer is a first-order phase transition, which is usually described by nucleation theory [7.90,91]. When at time $t = 0$ a small chemical potential difference as a driving force is "switched on", the adatoms tend to incorporate additional adatoms into larger clusters. The adsorption rate is slightly increased above the desorption rate until a new quasi-stationary metastable state [7.2] is reached. As long as the clusters are below a critical size they will not grow in the average. But when due to thermal fluctuations a cluster larger than the critical size is generated it will grow in the average, until it covers the whole system.

In the one-spin-flip Ising model only the sites along the perimeter of the cluster contribute to the cluster growth. If surface diffusion (i.e., two-site processes) would come into play, all the sites within a certain range $\lambda = (D\tau)^{\frac{1}{2}}$ away from the cluster would contribute to its growth (D is the diffusion contant, τ is the average time during which a single reevaporating atom has been adsorbed at the surface). Interactions between clusters therefore occur in the first version only if the distance between the clusters is within two lattice constants, while in the second version the range of dynamic interactions is 2λ. In the latter siutation a supercritical cluster therefore grows much faster and at the cost of smaller clusters. The density of supercritical growing clusters in the system accordingly can be considerably smaller than in the first situation.

Metastable states in the first version of the model [7.35] may exist over long times during which supercritical clusters form and grow. The concentration of adatoms changes very slowly with time. When the supercritical clusters have reached such a large concentration that their interaction leads to coagulation, the metastable state terminates and the coverage of the surface layer with adatoms increases rapidly. This is marked by the change in slope of the relative coverage versus time plotted in Fig.7.5. This changeover region is not predicted by nucleation theory and in fact represents the characteristic time needed for the formation of a new layer.

In the case where surface diffusion between the clusters is important, the time limitation for metastable states may be reached at a much lower density of supercritical clusters since their large diffusion fields may lead to much higher spreading rates. But this process sofar has not been investigated in detail.

Mean-field theories predict the existence of metastable states with infinite lifetime [7.92], which terminate only if the chemical potential difference $\Delta\mu$ between the two phases (solid-vapor) exceeds a critical "spinodal" value. This and the existence of a spinodal limit is not supported by Monte Carlo simulations [7.35], but the metastable states only become less well defined as $\Delta\mu$ increases.

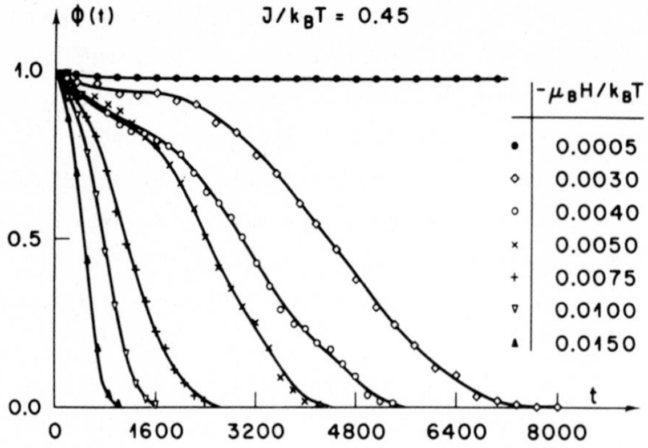

Fig.7.5 Nonequilibrium relaxation function $\phi(t)=\langle c(t)-c(\infty)\rangle / \langle c(0)-c(\infty)\rangle$ for the coverage c(t) of a monolayer with adatoms (= two-dimensional Ising model). At time t = 0 a chemical potential difference $\Delta\mu(\sim -\mu_B H)$ is switched on which causes a first-order phase transition from an uncovered to a covered layer. For small enough driving forces $\Delta\mu$ very long-lived metastable states exist. (The time t is given in Monte Carlo steps per site). [7.35]

b) Kossel Model (100) Face

The Kossel model (or SOS model) (7.1) does not restrict the movement of the interface to a finite number of layers and is therefore well suited to study growth kinetics.

From the different equilibrium structures (Sect.7.2) one expects at least two temperature regimes, above and below the roughening transition, with different dynamic behavior. Further qualitative changes then may come from the special choice of transition probabilities (7.3,4) and in different ranges of the driving force $\Delta\mu$.

Let us first discuss the regime of low temperatures and low supersaturations $|\Delta\mu| \ll \varepsilon$ (7.1). The existence of a metastable behavior similar to the two-dimensional Ising model (Fig.7.5) is clearly seen in Fig.7.6. As in the two-dimensional Ising model a nucleation barrier has to be overcome for the translation of the interface by one layer, and accordingly one expects the growth rate v to behave as

$$v \sim \exp(-c/\Delta\mu) \tag{7.5}$$

where c is some constant proportional to the edge free energy squared of a critical cluster. For the two-dimensional cluster this factor $c \equiv c_I$ is obtained from [Ref.7.36, eqs. (23,24)], not too close to the critical point. For the Kossel crystal this constant would be smaller $c_K = (1/3)c_I$, since in this case nucleation on top of other nuclei partly allows for simultaneous nucleation in several layers [7.93]. Various modifications of these predictions have been tested numerically [7.11, 14-17, 24, 31-34, 36, 38, 47], the quantitatively

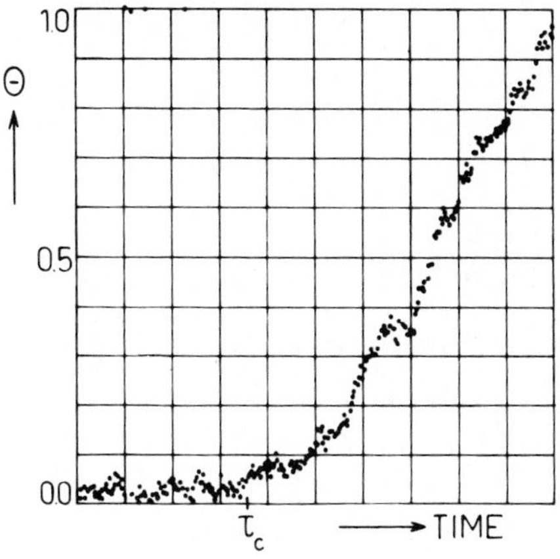

Fig.7.6 Metastable behavior of a Kossel crystal (SOS sc-100 face) after switching on a chemical potential difference $\Delta\mu$ at time t = 0. Plotted is the degree of coverage θ of a monolayer versus time. Starting from an equilibrium surface it takes a characteristic time τ_c of nucleation and coagulation before the interface can advance. Compare this behavior with Fig.7.5. [7.33]

most reliable Monte Carlo data probably being given in [7.24]. The existence of the nucleation barrier with the associated exponential behavior (7.5) for small $\Delta\mu$ is exhibited in Fig.7.7, where growth rate as a function of supersaturation is plotted. The nonlinear region vanishes above some temperature T_R (see Sect.7.3.4, "roughening"), where a linear law $v \sim \Delta\mu$ holds. A problem here is, however, the determination of the prefactor which multiplies the exponential function in (7.5). Since one cannot go to arbitrarily small values of $\Delta\mu$ in Monte Carlo simulations due to time limitations, it is hard to decide whether one is already in the $\Delta\mu$ -regime where the exponential is the dominating factor. The growth rate, of course, does not just depend upon the nucleation rate but also on the spreading rate [7.36] of the supercritical clusters. This is not important as long as the radial spreading rate depends linearly upon $\Delta\mu$, i.e., in the limit $\Delta\mu \to 0$. But in experiments as well as in computer simulations the critical clusters may be so small, that the shape fluctuations of the clusters are no longer small on the length-scale of their diameter. Hence one cannot expect a linear law $v_s \sim \Delta\mu$ to hold for the radial spreading rate v_s in these cases.

The low-temperature, low-supersaturation regime accordingly is characterized by the appearance of metastable states as growth proceeds from layer to layer (Fig.7.6). Mean-field theories [7.66b,67] and experiments [7.94] reflect this series of metastable states dynamically by showing a periodic variation in growth rate. Monte Carlo calculations [7.36] sofar were not conclusive on this point, presumably due to insufficient system size and simulation time. If nucleation on top of nuclei obscures this periodic variation surface diffusion should tend to make it more pronounced.

The main effect of an increasing chemical potential difference is to reduce the size of the critical two-dimensional nucleus. When the critical size is of the order of one atom, every deposited adatom serves as a condensation seed and the crystallization takes place all over the surface. The surface is no more flat but

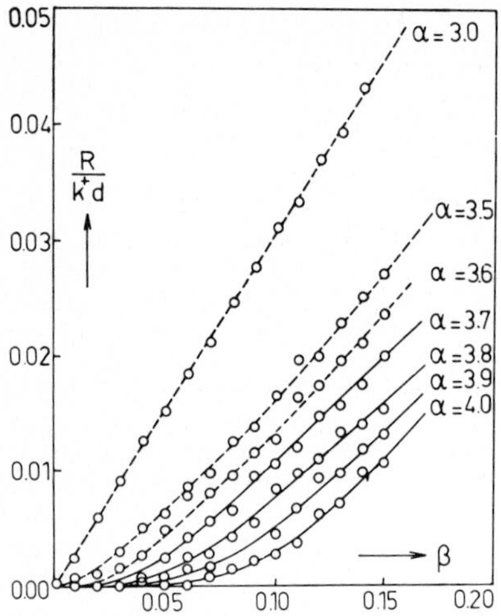

Fig.7.7 Growth rate R(=v) depending upon supersaturation $\Delta\mu = \beta k_B T$. The model is a sc-100 face (SOS model), simulated with a special-purpose computer [7.24]. The growth rate R is normalized by impingement frequency k^+ and lattice constant d. Parameter of the curves is the inverted temperature $\alpha = n\varepsilon/k_B T$. ($\alpha_{2d} = n\varepsilon/k_B T_{2d} = 3.53$). It is clearly seen that for $\alpha = 3.0$ ($T > T_{2d}$) the growth law $R \sim \Delta\mu$ is linear, while at low temperatures ($\alpha > 3.6$) a nucleation-type behavior $R \sim \exp(-c/\Delta\mu)$, [7.5] exists

becomes diffuse ("kinetic roughening"), as has been predicted by mean-field calculations [7.67]. Correlations between the adsorption sites are unimportant and the growth rate is finally controlled by the impingement frequency.

In the regime safely above the roughening transition (see Sects.7.2.3,4) the nucleation barrier for two-dimensional nucleation is missing. Surface steps (Sect.7.3.3) exist due to thermal excitation. In the regime of small $|\Delta\mu|$ the steps are moved over the surface with a velocity proportional to $\Delta\mu$. Coagulation between surface clusters takes place and new steps are generated by thermal fluctuations. When $\Delta\mu$ is increased the critical nucleus size again is reduced and, as discussed above, finally "normal" growth [7.3] is obtained as a single-particle attachment process. Growth in the high temperature regime [7.95] therefore is characterized by a smooth growth rate v versus $\Delta\mu$ curve, with a linear relation $v \sim \Delta\mu$ for $\Delta\mu \to 0$ [7.24].

c) Orientational Dependence of the Growth Rate (sc-, fcc-Lattices)

The main influence of other lattice symmetries and other low-index faces (110,111) on growth kinetics, namely anisotropy [7.96] of the growth rates, can be attributed to the different number of interactions involved in attachment-evaporation processes. In the low temperature regime, where two-dimensional nucleation theory may be applied, one can separate the lattice influence on interface kinetics in two parts: a static part, which is based on the quasi-equilibrium derivation of the size and shape of the critical nucleus, and a dynamic part, which defines the single-particle attachment kinetics [7.36]. (But not the importance of coagulation for the time-dependence of the coverage [see Ref.7.35, Sect.C].

In the sc-lattice with nearest-neighbor interaction all non-(100) faces are "rough" even at very low temperatures. No nucleation barrier exists therefore and a linear growth law holds. In an fcc lattice [7.4,29] there are alternative possibilities. The (111) face is tightly packed within the surface layer. The formation of a two-dimensional nucleus costs more energy and hence the nucleation barrier is higher than in the sc-(100) face. The fcc-(100) face is very similar to the sc-(100) face, since it contains the same number of nearest-neighbor bonds within a surface layer. Its growth rate (at given $\Delta\mu$) is considerably higher than the (111) face and compares with the rate of the sc-(100) face [7.29]. The fcc-(110) face should be comparable to the sc-(110) face since in both cases there are only two nearest neighbors along a row within the surface layer. Two-dimensional nucleation therefore is not necessary and normal growth ($v \sim \Delta\mu$) exists at all temperatures. The orientational variation of the growth rate for various fcc-faces is plotted in Fig.7.8.

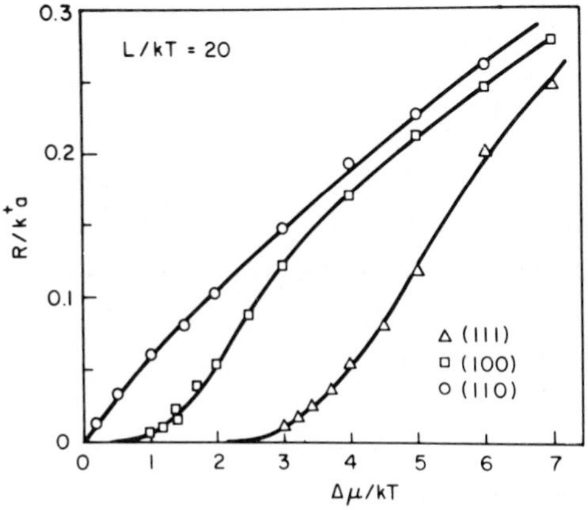

Fig.7.8 Growth rate R versus $\Delta\mu$ for different faces of an fcc-lattice at the same low temperature (normalization: $L/k_BT_{2d} = 4.9$). The (110) face obeys a linear law $R \sim \Delta\mu$ while both the (110)- and (111) face show nucleation barriers for small $\Delta\mu$, the (111) face growing with lowest rate. [7.4]

An additional difference in growth rates comes via the kinetics of single-particle processes. There is first a normalization problem concerning the impingement frequency [7.4]. The packing density within the surface layer depends strongly upon the orientation of the faces. An additional complication is introduced by the existence of half-layers in the fcc-lattice for (100) and (110) faces and finally there is the problem of the fcc-hcp degeneracy if only nearest-neighbor interaction exists. The latter two problems are treated numerically by using two sets of columns with appropriate interactions and SOS restrictions. The impingement frequency is most adequately normalized with respect to some unit area which is kept independent of orientation, and not with respect to the density of adsorption sites per layer or half-layer. The growth rate accordingly should be measured in unit distances which also are independent of orientation.

Surface structures in the fcc and other lattices accordingly may show considerably more details which depend sensitively upon additional parameters. The fcc-hcp degeneracy, e.g., can be removed by an additional longer-ranged interaction or by orientational preference of inter-atomic potentials. Next-nearest neighbor interaction could introduce new bonds within the surface layer and stabilize, e.g., the sc-(111) or fcc-(110) face. Surface diffusion finally (Sect.7.3.2e) may involve the simulation of local minima, e.g., hcp positions on an fcc-(111) face as important intermediate steps.

As a final point we include some remarks about interaction anisotropy [7.25]. An fcc-(110) face with weak next-nearest neighbor attraction is namely similar to a sc-(100) face where the interaction in one direction within the surface is much weaker than in the orthogonal direction. In the nucleation regime (low temperatures, small supersaturations) the growth rate is higher than in the case where both interaction energies would have the same high value. In the region of large $\Delta\mu$ the growth rate is lower because of the higher evaporation rate.

d) Adsorption on a Substrate

In all models discussed so far the interaction energies between two atoms were assumed to be constant, independent of their position relative to the interface. But it seems very plausible that the interaction within the layers closest to the interface differs from the layers in the bulk. The consequences of this effect so far have only been studied on fixed surfaces of magnetic model systems [7.5].

In crystal growth somewhat related problems occur during absorption on a foreign substrate or heterogeneous two-dimensional nucleation. The substrate is assumed to be microscopically flat and immovable. The material to be adsorbed has a different interaction strength among its atoms than between an atom and the substrate. If the substrate has a long-ranged attractive potential ($\sim r^{-3}$) towards the adsorbed material, mean-field theory predicts a series of first-order phase transitions as material is adsorbed [7.65,97].

In computer experiments only short-ranged (nearest-neighbor) potentials have been considered sofar. The first simulation assumed, that all material was condensed in a monolayer, and only the rearrangement of atoms was studied [7.52]. Several investigations along similar lines followed [7.9,11,18,28,53-56,59].

Extensions to thicker adsorption films with more layers exhibited two characteristic growth forms: island and layer growth [7.54]. Layer growth occurs, if the interaction energy between the substrate and the adsorbed atoms is larger than the interaction among the adatoms. Island growth occurs in the opposite case (at small enough supersaturations) when nucleation on top of nuclei is easier than directly on the substrate. Then towers of adatoms will be built up which only coalesce after they have reached a certain height. As supersaturation increases, this height is reduced and relatively sharp transition from island to layer growth occurs. Inclusion of surface diffusion and longer-ranged potentials has not been performed but may easily be included to give a good representation of experiments [7.98].

e) Surface Diffusion

Surface diffusion of adsorbed atoms may lead to a considerable increase in growth rates. Atoms which are adsorbed at site isolated from other adatoms would usually reevaporate before they could be incorporated by an already existing larger cluster. If surface diffusion is allowed these atoms diffuse over the surface and thus have a better chance of meeting other adatoms and clusters before they reevaporate [7.15].

Equilibrium structures, which are completely determined by the configurational free energy, of course are not influenced. For small supersaturations where kinetic theories are obtained by linearizing around equilibrium structures sometimes only the kinetic prefactors (or the time scales) have to be adjusted to the enhanced mobility due to surface diffusion. Nucleation theory (even though it is not a linear response approach) represents a special case. The characteristic quantity here is the edge free energy for a two-dimensional nucleus on a plane surface which is equivalent to the equilibrium free energy of a surface step in the limit of large radii of the nucleus.

Even in the linear nonequilibrium system ($\Delta\mu \to 0$) surface diffusion may lead to qualitative changes in the dynamic behavior since it opens additional paths in the configuration space of the system, over which it may lose its excess free energy faster than without surface diffusion.

Monte Carlo simulations have proven to be a useful tool for the investigation of these effects. Considerable increase in growth rates both above and below the roughening temperature are observed [7.15]. In the low temperature regime, where two-dimensional nucleation and spiral growth (Sect.7.3.3) are dominant, the surface steps act as almost perfect sinks for the diffusion adatoms. A peculiar effect in spiral growth is the increase in interstep distance [7.4] near the spiral center because there the adatom concentration is reduced due to competition between adjacent spiral arms. Anisotropy of the surface diffusion coefficients finally may lead to changes in shape of surface structures [7.98], which are caused by the different capture areas of surface steps moving in different directions.

As a special application the diffusion and self-diffusion of adatoms, adsorbed in clusters on a substrate, was studied recently [7.56]. A strong decrease of the cluster-size dependent diffusion constant D_n with the number n of particles in the cluster was observed. There is clear evidence for the experimental importance [7.98] of these effects both in growth from the vapor and from solution.

7.3.3 Surface Steps

a) Straight Steps

The existence of steps on an otherwise flat crystal surface leads to a qualitative change in the growth law at low temperatures [7.72]. (An example for computer-generated steps on a Kossel model for a crystal surface is shown in Fig.7.9). A step is an essentially one-dimensional system (with assumed short-ranged interaction between neighboring atoms) which does not show a phase transition [7.99]. The divergence [7.78] of the step width $d \sim L^{\frac{1}{2}}$ with the step length L shows that the step is not localized and therefore an infinitesimal driving force moves the step over the surface without a hindering nucleation barrier. A mean-field approximation (as in the one-dimensional discrete gaussian model [7.71]) is expected to be qualitatively correct and gives a linear growth law $v \sim \Delta\mu$. This is also easily understood from the kink-adsorption mechanism [7.72]. The kinks are shifted by infinitesimal driving forces with a velocity $v_k \sim \Delta\mu$ and lead to the observed growth law [7.20] as long as the kink density in a step is maintained by thermal fluctuations. If several steps are present on the surface then every step will make an equal contribution to the growth rate [7.14], which is then proportional to step density and step velocity. (The change in growth law for surfaces with and without steps is shown in Fig.7.10). When surface diffusion comes into play this proportionality ceases to hold if the interstep-distance is less than the effective diffusion range [7.72].

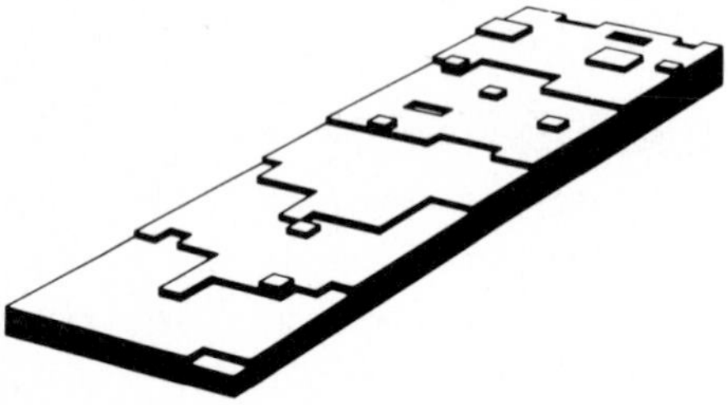

Fig.7.9 Computer simulation of steps on a crystal surface (40 × 10 lattice units). There are four steps and several clusters of adatoms and holes present. [7.27]

The growth rates due to step movement of course depend upon the orientation of the steps. If they run in a low-index direction [e.g., (10) on a sc-(100) face] they do not have many kinks at low temperatures and thus the step velocity at low supersaturations will be small. In a (111) direction on the other hand there are always many kinks present and the step velocity will be much larger. This leads to a kinetic faceting of two-dimensional nuclei [7.36] and spirals [7.40,44] as is explained in Sect.7.3.3b.

Experimental observation often reveals macrosteps of more than unit height. From the models in [7.14,15,4] this phenomenon is not easily obtained. Without surface diffusion two steps do not interact as long as they are more than one

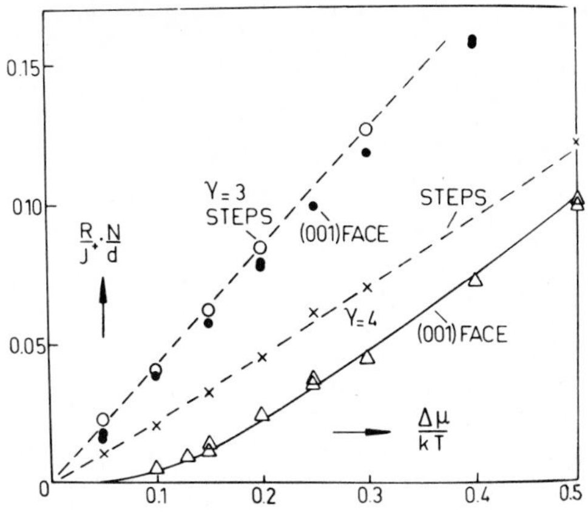

Fig.7.10 Normalized growth rate R versus chemical potential difference Δμ for two temperatures (γ ≡ α as in Fig.7.7). At the lower temperature γ = 4 the growth law with steps is linear, while without there is a nucleation barrier. At the high temperature γ = 3 there is almost no difference in growth rate with and without steps. [7.14]

lattice unit apart. In that case there is an infinite repulsion effective because the SOS restriction excludes overlaps. Inclusion of surface diffusion now may have a stabilizing or destabilizing effect on the equidistancy of step trains [7.100]. If the incorporation of diffusing adatoms into the step is easier from the lower side of the step than from the higher side the equidistant distribution is stable. In the reversed case it is unstable. This "Schwoebel" effect [7.101] was investigated quantitatively by Monte Carlo simulations [7.27]. Once two neighboring steps have collapsed the velocity of the double step may be smaller by an order of magnitude.

Experimentally it is not quite clear at present, whether this effect is responsible for the formation of macrosteps. It is conceivable that longer-ranged interatomic potentials, van der Waals forces between steps or elastic strain fields provide the cooperative effects which lead to bunching. The existence of the Schwoebel effect is supported by measurements [7.102,103] of adatom concentrations in step trains, which show an inhomogeneous distribution between the steps.

b) Spiral Growth and Spreading of Two-Dimensional Nuclei

If a surface step is generated by a screw dislocation it will wind up into a spiral as the crystal grows, the one end of the step being pinned down at the surface [7.72].

Experimental observations showed many growth forms of such spirals, round shaped, polygonized, cooperating, rotating in the same sense and counterrotating [7.104]. Computer simulations of spirals, which allow the interpretation of experiments, require considerable programming effort. The characteristic quantity describing the size of the spiral is the radius r_c of the critical nucleus. The interstep distance between successive spiral turns was predicted to be 8-19 times the critical radius r_c [7.44]. In order to simulate only a few turns on a finite lattice, one needs to simulate enormous lattices if one wants the critical nucleus to consist of some 10-100 lattice sites. This was achieved by using the inverted Monte Carlo method [7.40] (with tables of the active sites, see Sect.7.3.1) and neglecting the most improbable single-particle processes. The latter restriction is allowed in the region where spiral growth dominates over nucleation growth, i.e., below the roughening temperature T_R (Sect.7.3.4) at sufficiently small $\Delta\mu$. A series of Monte Carlo generated spiral pictures is shown in Fig.7.11. Plotted is an evaporation spiral, which is most frequently observed in experiments [7.81]. The next plot, Fig.7.12, shows a pair of counterrotating spirals.

Relatively close to T_R the spirals are round shaped. As the temperature decreases the spirals become more and more polygonized, reflecting the underlying lattice symmetry [7.4], in agreement with theory [7.44]. The anisotropy of the spirals at low temperatures accordingly is due to the anisotropic edge free energy of a step and its anisotropic mobility.

The continuum theory [7.72,44] gives a quadratic growth law for a crystal with a spiral dislocation at small $\Delta\mu$: $v \sim \Delta\mu^2$. This of course, is only expected to be valid if the lattice constant is small compared to all other lengths in the system. It was a surprise to find [7.40] that this law and the interstep distance $d \sim r_c \sim \Delta\mu^{-1}$ follow the predictions up to values of $\Delta\mu$, where the critical radius would be only a single lattice spacing (see Fig.7.13).

The continuum theory ceases to be valid at low temperatures, when the average distance between kinks in a step is smaller than the diameter of the critical nucleus. In that case [7.105] the mobility of a step is no longer constant but increases linearly with its length until more than one thermally generated kink is always present.

From the equilibrium properties of surface steps (Sect.7.2.2) one expects an increase of the fluctuations normal to the steps with increasing step length. This effect is neither observed in experiments [7.81,103] nor clearly indicated by the computer simulations [7.4,44]. The reason for that may be that the time needed to develop a large fluctuation [7.36] is longer than the time during which a step moves from the spiral center outward to the edge of the crystal.

A very interesting phenomenon is the cooperation of several spirals [7.72,41]. If there are two spirals operating in opposite direction there is a critical distance between the spiral centers (namely the diameter of the critical nucleus), above which the spirals rotate freely and below which a nucleation barrier has to be overcome. This nucleation distance as obtained from Monte Carlo calculations [7.36,40,41] is in good agreement with the diameter of the critical nucleus as calculated from the interstep distance d at high temperatures [7.72,44]: $r_c = d/19$, and also agrees with direct estimations of the critical nucleus size [7.33]. At this stage, therefore, the spiral theory and nucleation theory have been fully confirmed by Monte Carlo simulations of growth spirals.

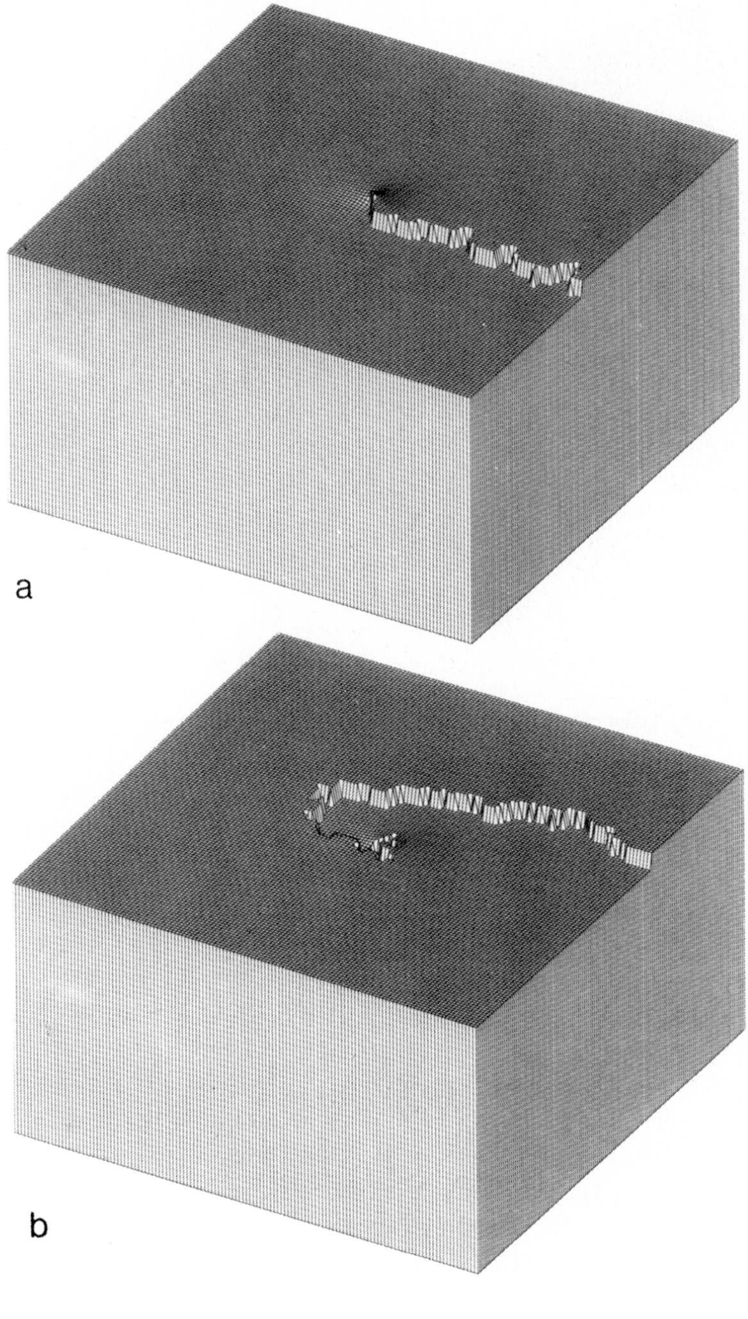

Fig.7.11a-c Development of an evaporation spiral with time from an initially straight step, which is pinned down at a screw dislocation. (Computer simulation as in [7.40])

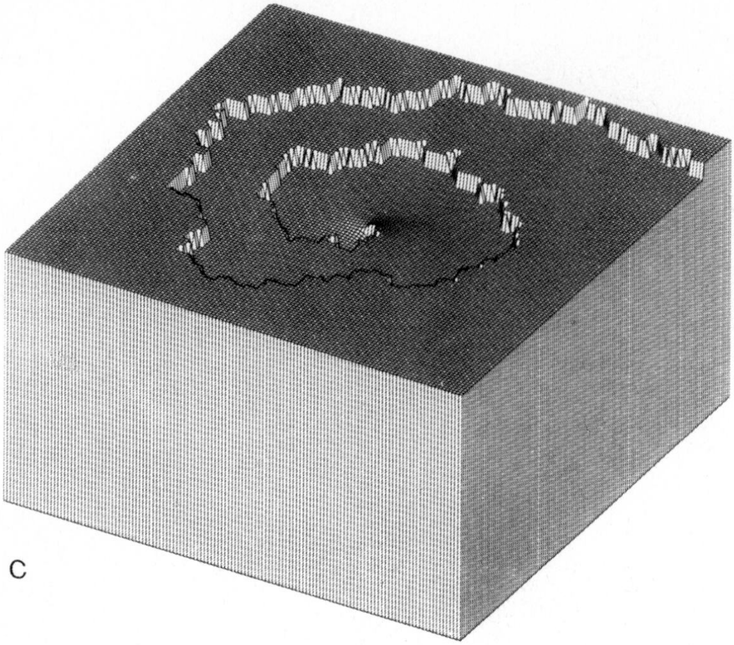

C

Close to the roughening temperature or at sufficiently large supersaturation the nucleation of clusters on the surface becomes a competing mechanism. The relative influence of spiral growth on the growth rate, including two-dimensional nucleation, was studied by direct Monte Carlo simulations [7.4] and was compared to a combined nucleation-spiral-growth theory. For given driving force $\Delta\mu$ it was found that the temperature regime, where the spiral mechanism dominates, agrees with the theory within ≈ 20% [Ref.7.4, Fig.13].

Surface diffusion has two main effects in spiral growth. First, it increases the growth rate, as discussed in Sect.7.3.2e. Second, it may change the spiral shape. At low temperatures where the spiral becomes polygonized partly due to the low advance velocity of a step in (10 direction, diffusion along the step may increase the kink velocity considerably and thus increase the advance velocity of the (10) step. The spiral then becomes rounder. This effect is easily incorporated theoretically [7.44,72]. The nonlocality of the step mobility due to the coupling between different portions of the step via surface diffusion in the plane is more difficult to treat [7.72]. The influence on the spiral shape has only partly been investigated [7.4].

Macrospirals [7.106,108] may be formed by screw dislocations with Burgers vectors of more than one lattice unit. In the simplest case without surface diffusion, without "Schwoebel" effect (Sect.7.3.2e), without elastic strain fields taken into account, however, the original double steps (or higher) separate and lead to equidistant single-step spirals [7.4].

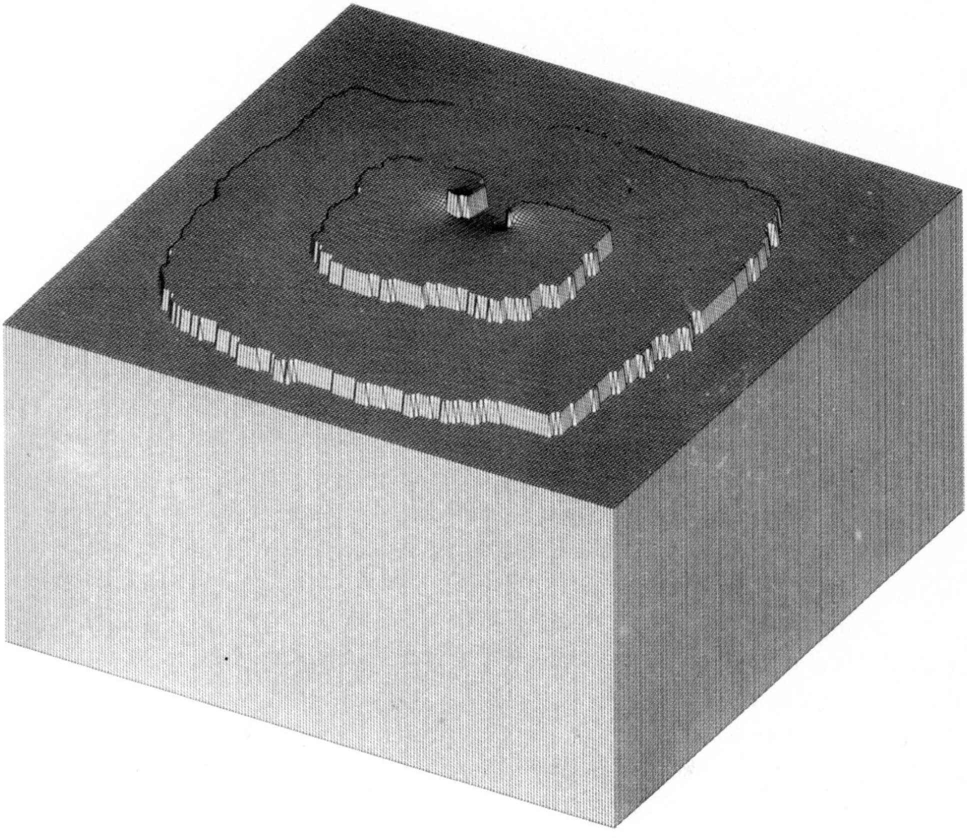

Fig.7.12 A pair of counterrotating spirals on an sc-100 face as in Fig.7.11, but now for growth and at lower temperature. The spiral arms touch after every turn and produce one new island on top of the other. These growth islands increase continuously in size and finally form a growth pyramid with the symmetry of the underlying lattice [7.40]

Large sets of spirals finally have been predicted [7.72,100] to give different growth laws, but computer simulations using the inverted method have only recently been started.

Experiments have clearly confirmed the predictions of the spiral theory [7.72, 44] concerning the growth velocity $v \sim \Delta\mu^2$ of the crystal [7.109] and the interstep distance $d \sim \Delta\mu^{-1}$ [7.110].

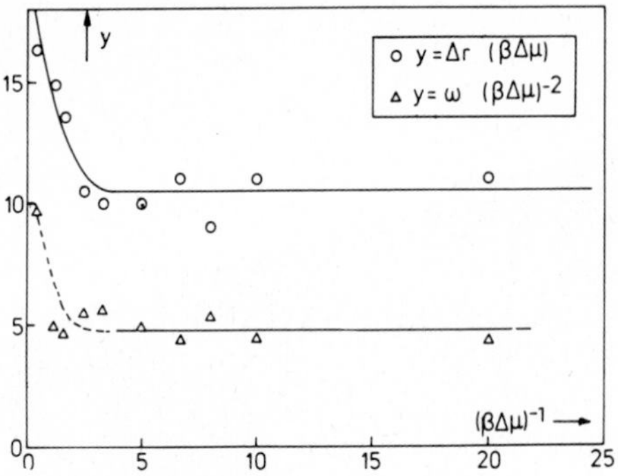

Fig.7.13 Monte Carlo test of the growth laws [7.72,44] for spirals; the rotation frequency $\omega \sim \Delta\mu^2$, the interstep distance $\Delta r \sim \Delta\mu^{-1}$ ($\beta = 1/k_BT$). The theory is confirmed down to critical nuclei of only a few particles. [7.40]

7.3.4 Roughening Transition

The definition of surface roughness [7.73] in the literature is not unique and sometimes leads to confusion when the influence of roughness on the growth rate is discussed. We therefore first have to clarify the terminology.

The intuitive and historically first definition of roughness [7.72] is given by the deviation of the actual surface from a smooth plane in terms of laterally broken bonds (parallel to the surface). The so defined roughness is proportional to the excess energy of the interface (surface) and is a completely locally defined quantity in systems with short-ranged interactions between atoms. This quantity will therefore be called "local roughness".

The more recent definition [7.73,111] of roughness is based on the moments [7.85] of the interface position $\langle(h_i - \langle h_i\rangle)^{2n}\rangle$ where h_i is the local normal deviation of the interface from the average position $\langle h_i\rangle$, the brackets denoting sample averages at a given time. The divergence of these moments at the roughening temperature is a much stronger singularity than expected for the interface energy and is based on long-wavelength fluctuations of the interface. In this sense one accordingly may speak about "global roughness" of the interface.

Three aspects of roughness of a growing crystal are considered here, at first the influence of the roughening transition on the kinetics of growth, then the influence of the nonequilibrium condition on the roughness of the interface and finally time correlation functions in thermal equilibrium as the basis of dynamic linear response theory.

The most significant effect of surface roughening is the enormous increase in growth rate at small chemical potential differences $|\Delta\mu| \ll kT$ as the roughening temperature is approached. This was investigated in detail with a special-purpose computer [7.24] in simulations of a Kossel sc-(100) face. The nucleation barrier which hinders lateral growth at low temperatures obviously decreases as $T_R > T_{2d}$ (Ising) is approached and a growth law $v \sim \Delta\mu$ appears to become valid even at very small $\Delta\mu$, instead of the nucleation behavior $v \sim \exp(-c/\Delta\mu)$ (see Fig.7.7). This change in growth laws can be understood in terms of the decreasing excess free energy of a step [7.84] as T_R is approached. Above T_R steps causing the long-wavelength fluctuations are always present and can be moved over the surface with a step velocity $v_s \sim \Delta\mu$(see Sect.7.3.3). According to this picture the global roughness is responsible for the linear growth law above T_R.

The local roughness or interface energy has been investigated [7.17,29] and it was found that it increases with increasing driving force $\Delta\mu$. At temperatures below the roughening transition, one defines a local roughness averaged over the time needed to add one layer to the crystal. There is a strong increase in roughness R (laterally broken bonds) around some value $\Delta\mu$ of the chemical potential difference [7.29], while for larger and smaller values of $\Delta\mu$ there is only a slow increase of R with $\Delta\mu$, see Fig.7.14. Sufficiently far below T_R this may be understood in terms of two-dimensional nucleation. When the radius of the critical nucleus $r_c \sim \Delta\mu^{-1}$ is of the order of the correlation length ξ (in two dimensions) thermal fluctuations lead to a rapid breakdown of the metastable state of the advancing interface.

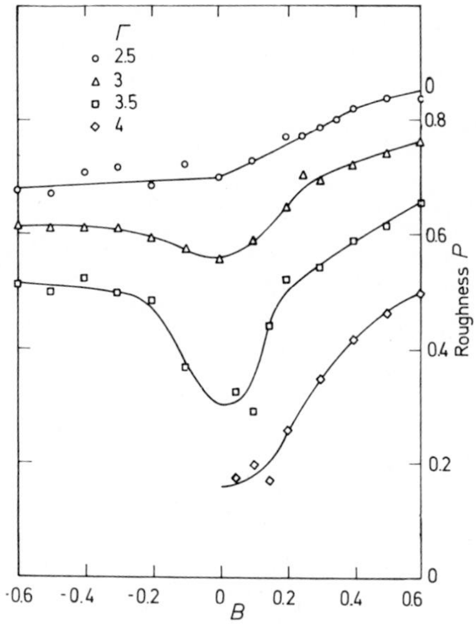

Fig.7.14 Surface energy (local roughness) of an fcc-(111) face under growth conditions ($Bk_BT = -\Delta\mu \neq 0$), [7.29]. Parameter of the curves is the inverse temperature ($\Gamma = \alpha$ as in Fig.7.7). Below the roughening temperature ($\Gamma = 2.2$ [7.72]) there is a characteristic value of $\Delta\mu$ (for each temperature). Around which the roughness increases significantly in qualitative agreement with [7.67]

The global roughness is more difficult to investigate, since it requires very large model-systems and very long simulation times [7.46,71]. In particular it is not clear whether it is significantly different in equilibrium and under growth conditions.

The different dynamic regimes of growth are expected to be reflected in dynamic quantities at equilibrium: below T_R fluctuations should decay exponentially with time, while above T_R the long wavelength modes should show a diffusional behavior

[7.72] for long times $t \to \infty$. This prediction is consistent with recent Monte Carlo simulations of autocorrelation functions for the local and global roughness [7.45, 46].

The global roughness (autocorrelation of interface moments) has extremely long relaxation times above T_R which are very sensitive to the system size. In addition, both the global and the local roughness (or interface energy) have strong peaks in the initial relaxation time τ_I at T_R [7.45] (see Fig.7.15), which is not easily understood in terms of long-wavelength fluctuations, but may depend explicitly on the type of short-ranged interaction between lattice sites (Ising-like, SOS or DG model).

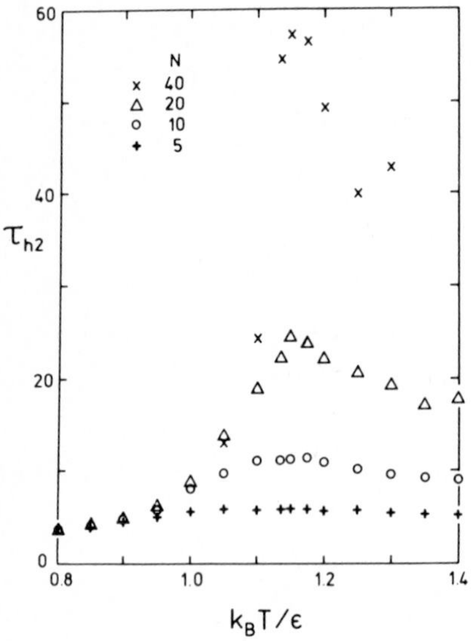

Fig.7.15 Initial relaxation time $\tau_{h2} = \tau_I$ for the autocorrelation of the second interface moment $\langle h^2 \rangle$ as a function of temperature. At $T_R(\approx T_{2d})$ there exists a pronounced peak which sharpens as the system size N (side-length) is increased to 40 lattice units. [7.45]

7.3.5 More Component Crystals and Segregation of Impurities

The simulations of growing N-component crystals [7.4,8,48-51] are still in a very preliminary state, even though historically two-component crystals were among the first systems being simulated [7.67]. One reason is the considerably larger parameter space already in a two-component (plus "fluid" phase) model. The simplest model is the spin-1-Ising model [7.112] with nearest-neighbor interaction. The three local spin states (1,0,-1) then correspond to particles of type A, B or F ("fluid" for simulation of growth from solution, "empty" for simulation of growth from the vapor). Complete phase diagrams require the investigation of a five-dimensional parameter space, the kinetics in addition introduce two basic kinetic coefficients k_A^+, k_B^+ at least. The simulations therefore have to be restricted to the physically most relevant subspaces spanned by the interaction parameters.

It is usually assumed [7.3,6,7,48,49] that there is a stronger attractive interaction between A-A, A-B, B-B and F-F atoms than between A-F and B-F atoms, which causes a separation into an {A,B}-rich phase and a F-rich phase. If the interactions between A-B atoms are stronger than between A-A and B-B, the system tends to form an ABAB... superlattice [7.7] at sufficiently low temperatures. (This corresponds to antiferromagnetic interaction in spin language). More complicated superlattice structures [7.113] of course may exist if next-nearest and longer-ranged interactions are present.

All the presently treated models [7.4,6,7,48,49,51] have some kind of a SOS restriction built in. The solution ahead of the interface therefore consists of mere fluid ("F") particles. The SOS restriction accordingly only requires no A or B atom to be on top of F atoms, while A and B may be arbitrarily on top of each other. It would therefore be desirable to treat a full three-dimensional system without SOS restriction in order to treat the concentrations of particles (A,B) both in the solid and in the liquid part simultaneously. With the SOS restriction one has to relate the concentrations in the liquid to the impingement frequencies k_A^+, k_B^- in order to define the liquidus line [7.48]. The solidus line then is defined by the different concentrations on the solid side of the interface. Depending upon the different interaction energies between atoms of different kinds one may obtain different phase diagrams in the temperature versus concentration plots, e.g., two-phase regions with or without azeotropic concentrations [7.3]. The presently available data, however, are not systematic enough to draw general conclusions. An example for a phase-diagram with a single miscibility gap in the temperature-concentration diagram is given in Fig.7.16.

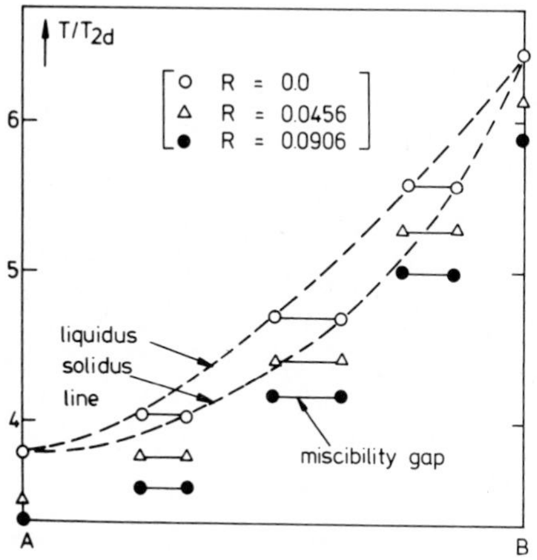

Fig.7.16 Phase diagram (temperature versus concentration) for a compound crystal. The model used [7.48] allows an occupation of each lattice site with A or B atoms or with a "fluid" (F) atom. The SOS restriction excludes F above A or B. For increasing growth rate R it is found that the miscibility gap between solid and "liquid" phase is shifted to lower temperatures

The definitions of driving forces for growth also depend upon further model assumptions. For the kinetic coefficients k_A^+, k_B^+ one may assume proportionality with the concentrations in the liquid. But the chemical potential differences $\Delta\mu_A$, $\Delta\mu_B$ between a particle in the solid or in the liquid phase may be different for the two different species. Increasing the concentration of one species (A) in the (dilute) solution means changing only $\Delta\mu_A$, while reducing the temperature may increase both $\Delta\mu_A$ and $\Delta\mu_B$ by (different) amounts. One should, therefore, carry out the simulations in close analogy to specific experimental conditions in order to make effective use of the many degrees of freedom in these systems.

Analogous to the simulations of one component crystals (Sect.7.3.2) an increase of the driving forces $\Delta\mu$ leads to a reduction of the correlated deposition of atoms at the interface. It was therefore observed [7.7,48] that the ordered superlattice structure, which is formed at small supersaturations, is destroyed at higher growth rates (see Fig.7.17). This was termed kinetic transition [7.7], and the growth rate versus chemical potential difference curve shows a cusp [7.3] at the characteristic $\Delta\mu$ value [Ref.7.3, see Fig.10].

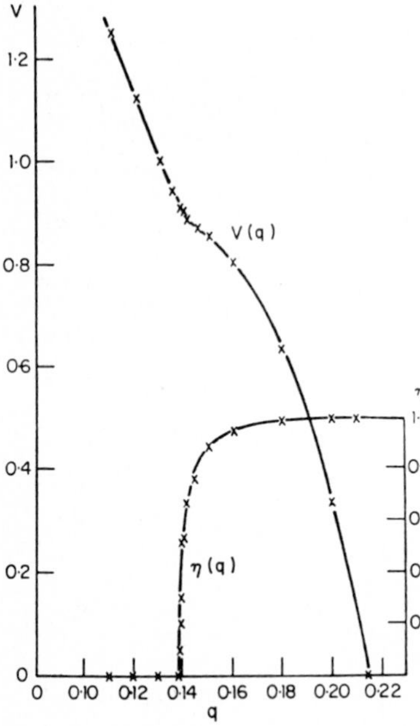

Fig.7.17 Long-range order $\eta(0 \leq \eta \leq 1)$ versus temperature parameter $q = \exp(-\varepsilon_{AA}/k_BT)$ (ε_{AA} is the interaction energy between one type of atoms). Also plotted is the advance velocity V of a kink in a surface step. The flow of particles onto the kink was constant. Above a particular temperature the long-range order breaks down $\eta = 0$ and the advance velocity V increases significantly. For a higher flow of particles (not shown here) the transition temperature is lower [7.7]

A particular aspect of two-component crystal growth is the segregation of impurities at the solid-vapor or solid-liquid interface. In the limit of low impurity concentrations where impurity-impurity interactions can be neglected this has been investigated in some detail [7.4]. Under the assumption that the impurities have a weaker attraction to the crystal than the crystal atoms only a fraction K of the concentration in the liquid is incorporated. The distribution coefficient K [7.114] at low temperatures and low supersaturations decreases strongly with decreasing attractive energy between crystal and impurity. At sufficiently high supersaturations the incorporation of impurities becomes independent of the adsorption sites and all impinging atoms are incorporated into the crystal. These computer simulations are well represented by a simple kink-adsorption theory [7.4]. The segregation and capture of impurities is such an important and widely studied phenomenon that we only refer to [7.1,115] for the documentation of experimental results.

7.4 Outlook

The presently available Monte Carlo simulation results on simple lattices and one-component crystals were of considerable use for the qualitative understanding and illustration of the growth mechanism. In particular, the method as a tool to test theoretical models directly, has been well exploited.

Monte Carlo simulation as a tool to explain experiments, however, is not used too frequently. This is not due to an intrinsic limitation of the method but due to some technical problems. The application to vapor growth on simple cubic lattices is straightforward. The experimentally more interesting monatomic lattices fcc and bcc have also come to attention. But the application of the interface models to solution growth already causes certain problems.

Most of the models, e.g., (7.1), only give a representation of the interface, not of the bulk properties in the solid and nonsolid phases. The effective bond energies ε within the interface therefore are differences in the bond strengths between solid-phase and fluid-phase atoms. The models are only good for sufficiently low concentrations of solute in the solution. If namely the concentration approaches unity the model corresponds to a solid-melt system which is not well represented by these lattice models.

At intermediate concentrations correspondingly cooperative effects of solute atoms in the solution become important. This means at least a breakdown of the SOS assumptions while the lattice model may still be used to study interface properties in solid solutions.

Even at low concentrations these cooperative effects (correlations between solute atoms) may be important if polarization effects or clustering of solvent atoms around solute comes into play. One therefore has to decide for each experimental case whether the properties of the system are sufficiently well represented by a simple SOS type model. Noncritical adaption of JACKSON's rule [7.116,100] for solid vapor systems onto solid-liquid (solution) systems, namely assuming the effective bond energy within the interface to be a fraction of the latent heat L of dissolution, may lead to inconsistencies, as L may be negative [7.3,87]. We have, therefore a clear need for the extension of the present models to account for more complicated systems. The following selection of possibilities may sketch the future development in this field.

The first generalization, which already was started [7.3], is the simulation of n-component crystals, still in a simple (sc, fcc, bcc) lattice. The additional degrees of freedom contained in the larger number of interaction parameters require to restrict oneself to experimentally most important problems: segregation, structural phase transformations, and phase diagrams. Inclusion of further than next-nearest neighbor interactions allows for a large number of phases with complicated superlattices and transitions with multicritical points characteristic for competing interactions [7.117].

A more fundamental adaption to experimental conditions [7.118] would be to include elastic strains near the interface. This requires major changes in the simulation procedure. The interaction between atoms has to be continuous in space. The local distance between neighbors has to be a result of the computation. The local minima or adsorption sites have to be newly determined after every single adsorption process. Obviously this will be an extremely costly procedure and hence will be restricted to only a few, experimentally most relevant, situations.

Realistic simulation of solution growth requires the study of clustering and transport ahead of the interface. Since many single diffusion jumps are necessary for the advancement of the interface by a single layer, transport simulation also requires long computing times.

Table 7.1 Cross listing: Subjects discussed in the References.

Categories of subjects

1 Equilibrium structure of surface (Kossel crystal)
2 Growth (Kossel crystal) and evaporation
3 Ising models (2-dimensional, or 3-dimensional with interface)
4 Other lattices such as fcc and more then nn-interaction
5 Anisotropy of interaction energy
6 Surface steps: equilibrium structure
7 Surface steps: growth
8 Screw dislocation spirals
9 Metastability and nucleation
10 Two-dimensional clusters and thin films
11 Roughening transition
12 Surface diffusion
13 More-component crystals and impurities
14 Grain boundaries and dislocations
15 Microcrystals and three-dimensional nucleation

Subjects discussed in References

Subject No.:	1	2	3	4	5	6	7	8	9	10	11	12	13	14	15
Ref.															
7.6	-	-	-	-	-	-	+	-	-	-	-	-	+	-	-
7.7	-	-	-	-	-	-	+	-	-	-	-	-	+	-	-
7.8	-	-	-	-	-	-	+	-	-	-	-	-	+	-	-
7.9	-	-	-	-	-	-	-	-	+	+	-	-	-	-	-
7.10	-	-	-	-	-	-	-	-	+	-	-	-	-	-	-
7.11	-	+	-	-	-	-	+	-	+	+	-	-	-	-	+
7.12	-	-	-	-	-	-	-	-	+	-	-	-	-	-	-
7.13	+	-	-	-	-	-	-	-	-	-	-	-	-	-	-
7.14	+	+	-	-	-	-	+	-	-	-	-	-	-	-	-
7.15	-	+	-	-	-	-	+	-	-	-	-	+	-	-	-
7.16	+	+	-	-	-	-	-	-	-	-	-	-	-	-	-
7.17	+	+	-	-	-	-	-	-	-	-	-	-	-	-	-
7.18	-	-	-	-	-	-	-	-	-	+	-	-	-	-	-
7.19	-	-	-	+	-	-	-	-	-	-	-	+	-	-	-
7.20	-	+	-	-	-	-	+	-	+	-	-	+	-	-	-
7.21	-	-	+	-	-	-	-	-	-	-	-	-	-	-	-
7.22	+	-	-	+	-	-	-	-	-	-	-	-	-	-	-
7.23	-	-	-	+	-	-	-	-	-	-	-	-	-	-	-
7.24	-	+	-	-	-	-	-	-	+	-	-	-	-	-	-
7.25	+	-	-	-	+	-	-	-	-	-	-	-	-	-	-
7.26	-	-	-	-	-	+	-	-	-	-	+	-	-	-	-
7.27	-	-	-	-	+	-	+	-	-	-	-	-	-	-	-
7.28	-	-	-	-	-	-	+	-	+	+	-	-	-	-	-
7.29	-	+	-	+	-	-	-	-	-	-	-	-	-	-	-
7.30	-	-	-	-	-	+	-	-	-	-	-	-	-	-	-
7.31	-	+	-	-	-	-	-	-	-	-	-	-	-	-	-
7.32	+	+	-	-	-	-	-	-	+	-	+	-	-	-	-
7.33	-	+	-	-	-	-	+	-	+	-	-	-	-	-	-
7.34	-	+	-	-	-	-	-	-	-	-	-	-	-	-	-
7.35	-	-	+	-	-	-	-	-	+	+	-	-	-	-	-
7.36	-	+	-	-	-	-	+	+	+	+	-	-	-	-	-
7.37	-	-	-	+	-	-	-	-	-	-	-	+	-	-	-
7.38	-	+	-	-	-	-	-	-	-	-	-	-	-	-	-
7.39	-	-	-	-	-	-	-	+	-	-	-	-	-	-	-
7.40	-	-	-	-	-	-	-	+	-	-	-	-	-	-	-
7.41	-	-	-	-	-	-	-	+	-	-	-	-	-	-	-
7.42	-	-	-	+	-	-	-	+	+	-	+	+	-	-	-
7.43	-	-	-	-	-	-	-	+	+	-	-	+	-	-	-
7.44	-	-	-	-	-	+	-	-	-	-	-	-	-	-	-
7.45	-	-	-	-	-	-	-	-	-	-	+	-	-	-	-
7.46	+	-	-	-	-	+	-	-	-	-	+	-	-	-	-
7.47	-	+	-	-	+	-	-	-	+	-	-	-	-	-	-
7.48	-	-	-	-	-	-	-	-	-	-	-	-	+	-	-
7.49	-	-	-	-	-	-	-	-	-	-	-	-	+	-	-
7.50	-	-	-	-	-	-	-	-	-	-	-	-	+	-	-
7.51	-	-	-	-	-	-	-	-	-	-	-	-	+	-	-
7.52	-	-	-	-	-	-	-	-	-	+	-	+	-	-	-
7.53	-	-	-	-	-	-	-	-	-	+	-	-	-	-	-
7.54	-	-	-	-	-	-	-	-	+	+	-	-	-	-	-
7.55	-	-	+	-	-	-	-	-	-	+	-	+	-	-	-
7.56	-	-	-	-	-	-	-	-	+	+	-	+	-	-	-
7.57	-	-	-	-	-	-	-	-	-	-	-	-	-	+	+
7.58	-	-	-	-	-	-	-	-	-	-	-	-	-	+	-
7.59	-	-	-	-	-	-	-	-	+	+	-	+	-	-	-
7.60	-	-	-	-	-	-	-	-	-	-	-	-	-	-	+
7.61	-	-	-	-	-	-	-	-	-	+	-	-	-	-	-
7.62	-	-	-	+	-	-	-	-	+	+	-	+	-	-	+
7.63	-	+	-	+	-	-	-	-	-	-	-	+	-	-	-

A very interesting problem is the propagation of lattice dislocations. Screw dislocations are usually assumed to be pinned down at some position within the surface. A series of papers on the propagation of dislocations is coming up in the literature [7.64], but no particular method seems to be the best suited for most purposes, because again the influence of elastic deformations becomes important. As a final task there is still the problem of solid-melt interface simulations. The problem here is the simulation of two phases of continuously moving atoms in the same box and in particular the formulation of appropriate boundary conditions. A few attempts exist [7.2] but the simulations are far from allowing the study of crystal-growth phenomena.

This summary here is certainly not complete in any respect. Various special topics like investigation of cooperating spirals or growth of small clusters will remain active in the future. But the selection clearly shows, that the complicated programs required for several groups of problems in this list will force us to an even stronger incorporation of experimental facts into the structure of future models.

References

7.1 As a general introduction into problems of crystal growth, the reader is referred to
D. Elwell, H.J. Scheel: *Crystal Growth from High Temperature Solutions* (Academic Press, London 1975)

7.2 The theoretical aspects of crystal surface dynamics have been reviewed by H. Müller-Krumbhaar: In *Current Topics in Materials Science*, ed. by E. Kaldis, Vol. I (North-Holland, Amsterdam 1977) Chap.1

7.3 J.P. v.d. Eerden, P. Bennema, T.A. Cherepanova: In *Progress in Crystal Growth and Assessment*, ed. by B.R. Pamplin, No. 3 (Pergamon Press, Oxford 1977) (Review)

7.4 G.H. Gilmer, K.A. Jackson: In *Proc. of the 1st European Conference on Crystal Growth, Current Topics in Materials Science*, Vol. II, ed. by E. Kaldis, H.J. Scheel (North-Holland, Amsterdam 1977) (Review)

7.5 K. Binder, P.C. Hohenberg: Phys. Rev. B *6*, 3461 (1972); B *9*, 2194 (1974) concerning the difference between "surface" and "interface"

7.6 A.A. Chernov: International Conference on Crystal Growth, ed. by H. Steffen (Peiser, Oxford 1967) p. 25

7.7 A.A. Chernov, J. Lewis: J. Phys. Chem. Sol. *28*, 2185 (1967)

7.8 A.A. Chernov: Sov. Phys.-Usp. *13*, 101 (1970)

7.9 F.L. Binsbergen: Kolloid Z. *237*, 289 (1970)

7.10 F.L. Binsbergen: Kolloid Z. *238*, 389 (1970)

7.11 F.L. Binsbergen: J. Cryst. Growth *16*, 249 (1972)

7.12 F.L. Binsbergen: J. Cryst. Growth *13/14*, 44 (1972)

7.13 H.J. Leamy, K.A. Jackson: J. Appl. Phys. *42*, 2121 (1971)

7.14 G.H. Gilmer, P. Bennema: J. Cryst. Growth *13/14*, 148 (1972)

7.15 G.H. Gilmer, P. Bennema: J. Appl. Phys. *43*, 1347 (1972)

7.16 V.V. Solov'ev, V.T. Borisov: Sov. Phys.-Dokl. *17*, 8 (1972)

7.17 V.V. Solov'ev, V.T. Borisov: Sov. Phys.-Cryst. *17*, 814 (1973)
I. Horsak, S.J. Skrivanek: Ber. Bunsenges. *79*, 433 (1975)

7.18 A.C. Adams, K.A. Jackson: J. Cryst. Growth *13/14*, 144 (1972)
7.19 C.S. Kohli, M.B. Ives: J. Cryst. Growth *16*, 123 (1972)
7.20 P. Bennema, J. Boon, C. van Leeuwen, G.H. Gilmer: Krist. U. Techn. *8*, 659 (1973)
7.21 H.J. Leamy, G.H. Gilmer, K.A. Jackson, P. Bennema: Phys. Rev. Lett. *30*, 601 (1973)
7.22 C. van Leeuwen: J. Cryst. Growth *19*, 133 (1973)
7.23 V.O. Esin, V.J. Daniluk, J.M. Plishkin, G.L. Podchinenova: Sov. Phys.-Cryst. *18*, 578 (1974)
7.24 S.W.H. de Haan, V.J.A. Meeussen, B.P.Th. Veltman, P. Bennema, C. van Leeuwen, G.H. Gilmer: J. Cryst. Growth *24/25*, 491 (1974)
7.25 D.J. van Dijk, C. van Leeuwen, P. Bennema: J. Cryst. Growth *23*, 81 (1974)
7.26 H.J. Leamy, G.H. Gilmer: J. Cryst. Growth *24/25*, 499 (1974)
7.27 C. van Leeuwen, R. van Rosmalen, P. Bennema: Surf. Sci. *44*, 213 (1974)
7.28 U. Bertocci: Surf. Sci. *15*, 286 (1969)
7.29 U. Bertocci: J. Cryst. Growth *26*, 219 (1974)
7.30 C. van Leeuwen, F.H. Mischgofsky: J. Appl. Phys. *46*, 1056 (1975)
7.31 G.H. Gilmer, H.J. Leamy, K.A. Jackson: J. Cryst. Growth *24/25*, 495 (1974)
7.32 H.J. Leamy, G.H. Gilmer, K.A. Jackson: In *Surface Physics*, ed. by J.H. Blakeley (Academic Press, New York 1976)
7.33 C. van Leeuwen, P. Bennema: Surf. Sci. *51*, 109 (1975)
7.34 J.D. Weeks, G. Gilmer, K.A. Jackson: J. Chem. Phys. *65*, 712 (1976)
7.35 K. Binder, H. Müller-Krumbhaar: Phys. Rev. B *9*, 2328 (1974)
7.36 C. van Leeuwen, J.P. van der Eerden: Surf. Sci. *64*, 237 (1977)
7.37 A.I. Michaels, M.B. Ives: Ref. 7.64, p. 133
7.38 J.P. van der Eerden, R. Kalf, C. van Leeuwen: J. Cryst. Growth *35*, 241 (1976)
7.39 H. Müller-Krumbhaar: In *Fluctuations, Instabilities and Phase Transitions*, ed. by T. Riste (Plenum Press, New York 1975)
7.40 R.H. Swendsen, P.J. Kortman, D.P. Landau, H. Müller-Krumbhaar: J. Cryst. Growth *35*, 73 (1976)
7.41 D.P. Landau, R.H. Swendsen: Bull. Am. Soc. Phys. *22*, 392 (1977)
7.42 G.H. Gilmer: Ref. 7.64, p. 964
7.43 G.H. Gilmer: J. Cryst. Growth *35*, 15 (1976)
7.44 H. Müller-Krumbhaar, T. Burkhardt, D. Kroll: J. Cryst. Growth *38*, 13 (1977)
7.45 R.H. Swendsen: Phys. Rev. Lett. *37*, 1478 (1976)
7.46 R.H. Swendsen: Phys. Rev. B *15*, 5421 (1977)
7.47 J.P. van der Eerden, C. van Leeuwen, P. Bennema, W.L. van der Kruk, B.P.Th. Veltman: J. Appl. Phys. *48*, 2124 (1977)
7.48 T.A. Cherepanova, A.V. Shirin, V.T. Borisov: Scientific Letters of Latvian State University, Riga *237*, 40 (1975) (in Russian)
7.49 T.A. Cherepanova, A.V. Shirin, V.T. Borisov: In *Industrial Crystallization*, ed. by J.W. Mullin (Plenum Press, New York 1976)
7.50 D. de Fontaine, O. Buck: Phil. Mag. *27*, 967 (1973)
7.51 T.A. Cherepanova, J.P. v.d. Eerden, P. Bennema: J. Crystal Growth *44*, 537 (1978); Prog. Cryst. Growth and Charact. *1*, 219 (1979)
7.52 A.J.W. Moore: J. Australian Inst. Metals *11*, 220 (1966)
7.53 I. Hors'ak, S.J. Skrivanek: Ber. Bunsenges. Physik Chem. *77*, 336 (1973)
7.54 D. Kashchiev, J.P. v.d. Eerden, C. v. Leeuwen: J. Cryst. Growth *40*, 47 (1977)
7.55 M. Rao, M.H. Kalos, J.L. Lebowitz, J. Marro: Ref. 7.64, p. 180
7.56 J.P. van der Eerden, D. Kashchiev, P. Bennema: J. Cryst. Growth *42* (1977)
7.57 G.D. Quinn, G.H. Bishop, R.J. Harrison: Ref. 7.64, p. 1215
7.58 S.I. Zaitsev, E.M. Nadgornyi: Ref. 7.64, p. 707
7.59 F.F. Abraham, G.H. White: J. Appl. Phys. *41*, 1841 (1970)
7.60 F.F. Abraham: J. Chem. Phys. *61*, 1221 (1974)
7.61 J.J. Couts, B. Hopewell: Thin Solid Films *9*, 37 (1972)
7.62 V.S. Sundaram, P. Wynblatt: Ref. 7.64, p. 143
7.63 K.W. Mahin, K. Hanson, J.W. Morris, Jr.: Ref. 7.64, p. 39
7.64 R.J. Arsenault, J.R. Beeler, Jr., J.A. Simmons (eds.): *Nuclear Metallurgy*, Vol. 20, 1976. Proc. Intern. Conf. on Computer Simulation for Materials Application April 19-21, 1976 (Nat. Bur. Stand., Gaithersburg, Md)
7.65 T.L. Hill: J. Chem. Phys. *15*, 761 (1947)

7.66 a) D.E. Temkin: In *Crystal Growth* (Akademia Nauk, Moscow 1965) p. 89 (in Russian)
b) D.E. Temkin: Sov. Phys. Cryst. *14*, 344 (1969)
7.67 H. Müller-Krumbhaar: Phys. Rev. B *10*, 1308 (1974)
7.68 See e.g., C. Domb: Ref. 7.69, Vol. 3, p. 357
7.69 C. Domb, M.S. Green (eds).: *Phase Transitions and Critical Phenomena* (Academic Press, New York 1972)
7.70 H.J. Leamy, K.A. Jackson: J. Cryst. Growth *13/14*, 140 (1972); but note that the particular surface considered (110-fcc) is globally rough at all temperatures
7.71 H. Müller-Krumbhaar: Z. Phys. B *25*, 287 (1976)
7.72 W.K. Burton, N. Cabrera, F.C. Frank: Phil. Trans. R. Soc. A *243*, 299 (1951)
7.73 H. Müller-Krumbhaar: In Proc. of the 1st European Conference on Crystal Growth, *Current Topics in Materials Science*, ed. by E. Kaldis, H.J. Scheel, Vol. 2 (North-Holland, Amsterdam 1977) p. 115
7.74 L.J. Cunningham, A.J. Braundmeier, Jr.: Phys. Rev. B *14*, 479 (1976)
7.75 A.A. Maradudin, W. Zierau: Phys. Rev. B *14*, 484 (1976)
7.76 K.A. Jackson: In *Crystal Growth* (Pergamon Press, New York 1967) p. 17
7.77 G. Gallavotti: Nuovo Cimento *2*, 133 (1972) and references therein
7.78 H.N. Temperley: Proc. Camb. Phil. Soc. *48*, 683 (1952)
7.79 D.E. Temkin, V.V. Shevelev: Sov. Phys.-Cryst. *21*, 588 (1976)
7.80 R.H. Swendsen: J. Cryst. Growth *36*, 11 (1976)
7.81 H. Bethge, K. Keller: J. Cryst. Growth *23*, 105 (1974)
7.82 K.H. Besocke, H. Wagner: Surf. Sci. *52*, 653 (1975)
7.83 E. Budevski, W. Bostanoff, T. Vitanoff, Z. Stoinoff, A. Kotzewa, R. Kaishew: Phys. Stat. Sol. *13*, 577 (1966)
7.84 H. v. Beijeren: Phys. Rev. Lett. *38*, 993 (1977)
7.85 J.D. Weeks, G. Gilmer, H. Leamy: Phys. Rev. Lett. *31*, 549 (1973)
7.86 D.E. Ovsienko, G.A. Alfintsev, E.V. Maslov: J. Cryst. Growth *26*, 233 (1974)
7.87 J.R. Bourne, R.J. Davey: J. Cryst. Growth *36*, 278 (1976)
7.88 J. Venables: In Proc. of the 1st European Conference on Crystal Growth, *Current Topics in Materials Science*, ed. by E. Kaldis, H.J. Scheel, Vol. 2 (North-Holland, Amsterdam 1977)
7.89 R.J. Glauber: J. Math. Phys. *4*, 294 (1963)
7.90 A.C. Zettlemoyer (ed.): *Nucleation* (Marcel Dekker, New York 1969); *Nucleation Phenomena* (Elsevier, New York 1977)
7.91 A.E. Nielsen: *Kinetics of Precipitation* (Pergamon Press, Oxford 1964)
7.92 K. Binder: Phys. Rev. B *8*, 3423 (1973)
7.93 W.B. Hillig: In *Proc. Int. Conf. Crystal Growth*, Boston 1966, ed. by H. Steffen (Peiser, Oxford 1967) p. 779
7.94 R. Kaishew, E. Budevski: Contemp. Phys. *8*, 489 (1967)
7.95 "Normal growth" ($v \sim \Delta\mu$) is often found in organic materials such as succinonitrite or cyclohexanol[7.87]
7.96 For a general discussion see F.C. Frank: In *Growth and Perfection of Crystals* (Wiley and Sons, New York 1958) p. 411
7.97 M.J.de Oliveira, D. Furman, R.B. Griffiths: Phys. Rev. Lett. *40*, 977 (1978)
7.98 P. Bennema: J. Cryst. Growth *5*, 29 (1969)
7.99 L.D. Landau, E.M. Lifshitz: *Statistische Physik* (Akademie Verlag, Berlin 1970)
7.100 P. Bennema, G.H. Gilmer: In *Crystal Growth: An Introduction*, ed. by P. Hartman (North-Holland, Amsterdam 1973) p. 282
7.101 R.L. Schwoebel: J. Appl. Phys. *40*, 614 (1969)
7.102 M. Klauda: Rost Kristallov *11*, 65 (1975) (in Russian)
7.103 K.W. Keller: In *Crystal Growth and Characterization*, ed. by R. Ueda, J.B. Mullin (North-Holland, Amsterdam 1975) p. 361
7.104 There is a vast literature on spiral observation. A number of characteristic forms are published in[7.81,94,103]
7.105 V.V. Voronkov: Sov. Phys.-Cryst. *18*, 19 (1973)
7.106 R.J. Davey, J.W. Mullin: J. Cryst. Growth *26*, 45 (1974)
7.107 S. Amelinckx, W. Bontinck, W. Dekeyser: Phil. Mag. *2*, 1264 (1957)

7.108 A.A. Chernov: Sov. Phys-Usp. *4*, 116 (1961)
7.109 Reviews are given by P. Bennema: In *Crystal Growth* (Supplement to J. Phys. Chem. Solids) (Pergamon Press, Oxford 1967), and[7.20]
7.110 R. Kaishew, E. Budevski: Nova Acta Leopoldina (Suppl.) *34*, 105 (1968)
7.111 S. Chui, J.D. Weeks: Phys. Rev. B *14*, 4978 (1976)
7.112 D. Furman, S. Dattagupta, R.B. Griffiths: Phys. Rev. B *15*, 441 (1977) and references therein
7.113 U. Poppe, A. Hüller: J. Phys. C *11*, 245 (1978)
7.114 A.A. Chernov: Sov. Phys.-Usp. *13*, 101 (1970)
7.115 D.J.T. Hurle: In *Crystal Growth: An Introduction*, ed. by P. Hartman (North-Holland, Amsterdam 1973) p. 120
7.116 K.A. Jackson: Trans. Quart., Am. Soc. Metals *50*, 174 (1958)
7.117 M.E. Fisher: AIP Conf. Proc. *24*, 273 (1975)
E.K. Riedel: AIP Conf. Proc. *18*, 865 (1973)
7.118 R. Kern: In Proc. of the 1st European Conference on Crystal Growth, *Current Topics in Materials Science*, ed. by E. Kaldis, H.J. Scheel, Vol. 2 (North-Holland, Amsterdam 1977) (Review)

8. Monte Carlo Studies of Systems with Disorder

K. Binder and D. Stauffer

With 17 Figures

The main emphasis in this chapter is on two types of magnetic systems: on the *percolation* problem which corresponds to a zero-temperature dilute quenched ferromagnet; and on the *spin glass* problem, where the interactions between nearest-neighbor Ising spins can be both ferromagnetic and antiferromagnetic. In both problems, a square or simple cubic lattice structure is assumed in most studies, and the distribution of ferromagnetic ions and exchange energies, respectively, is taken as random. Near the percolation threshold, the percolation cluster numbers seem well described by scaling assumptions analogous to usual critical phenomena. While for spin glasses the Monte Carlo work does not reliably answer the question whether or not a phase transition exists in true equilibrium, results for specific heat, susceptibility, remanent magnetization, etc., favorably compare with experimental data.

Further topics briefly discussed are then the effects of isolated nonmagnetic impurities in ferromagnets and the structural properties of glasses and other amorphous systems.

Monte Carlo simulations of systems containing some disorder, in general, consist of two steps. Firstly, a (random) disorder is introduced in the system by some "Monte Carlo method" [8.1], e.g., lattice sites are chosen at random where one replaces the atoms of the considered solid by impurities. Secondly, thermal fluctuations in this disordered system are simulated by Monte Carlo techniques just as described in the other chapters of this book. Obviously, the simplest calculations refer to zero temperature problems where only the first step has to be taken (like the percolation problem, Sect.8.2), and no thermal fluctuations need to be taken into account.

In this chapter, mainly magnetic systems are considered. Section 8.1 briefly discusses effects due to isolated nonmagnetic impurities, while Sect.8.2 discusses the case where the impurity concentration is so high that ferromagnetism breaks down. Section 8.3 is devoted to simulations of spin glasses at both zero and nonzero temperatures; there also the dynamic properties are discussed. Heteropolymers are mentioned in Sect.8.4 and noncrystalline solids are mentioned in Sect.8.5. Clearly, this selection of material is somewhat arbitrary; for example, one could regard all liquids (as opposed to crystals) as "disordered" (for simulations of liquids see Chap.2), and we do not discuss order-disorder phenomena in alloys (see Chap.3). Some of the many other subjects omitted here are computer simulations on excluded volume-problems, polymer chains in solution and related problems (random and self-avoiding walks on a lattice, etc.), since a recent review [8.2] exists in that field.

8.1 Dilute Impurities in Magnets

Since an ideal crystal, i.e., a crystal with perfect lattice and periodic boundary conditions, does not exist in nature, the nonideal behavior due to disorder [surfaces (see Chap.9), impurities, dislocations, etc.] deserves consideration. Impurities in ferromagnets may have two kinds of effects, (I) effects on global properties (Sect.8.2.1) and (II) effects on local properties. Monte Carlo methods can be used to study both static [8.3,4] and dynamic [8.5] local properties close to defects. Here we concentrate on static effects.

The simplest model assumes nearest-neighbor exchange interaction J between the magnetic atoms, while there is no interaction between a magnetic atom and the nonmagnetic impurity. A classical Heisenberg magnet with one such defect situated at $\underline{R} = 0$ was studied [8.3], and the deviation $\lambda_{\underline{R}}$ of the local magnetization $m_{\underline{R}}$ from the bulk value m_b was recorded for various $\underline{R}$, $\lambda_{\underline{R}} = m_b - m_{\underline{R}}$. A sample of N = 512 spins with "self-consistent" boundary conditions was used (see Chap.1), and m_b was determined by making runs under identical conditions but without the impurity. Figure 8.1 shows that the temperature dependence of $\lambda_{\underline{R}}$ is quite similar to simple mean-field predictions, except very close to T_c, where mean-field theory fails to reproduce the correct exponents, and at low temperatures, where mean-field theory fails to reproduce the correct spin-wave behavior.

Obviously such results are quite relevant also for practical purposes, if one tries to measure the magnetization with local measurements, like Mössbauer effect, γ - γ angular correlations, EPR, NMR, where one dilutes nonmagnetic atoms more suitable for the resonance technique in the magnet. Clearly, these atoms will disturb their environment in some way or another, and the above model calculation where λ_{100} reaches 10% in its maximum shows that the deviation from the ideal bulk behavior need not be negligible. For more extended defects even more drastic deviations occur [8.4]. These effects are particularly cumbersome, if one wants to determine critical exponents [8.4,6], Fig.8.2. While many arguments exist [8.3,4, 6,7] that the "local" critical exponent β close to an impurity is the same as the bulk one, it will appear only extremely close to the critical temperature T_c while further away from T_c a larger "effective" exponent is seen (Fig.6.2). These simulations possibly explain nicely the experimental findings in nickel, where bulk

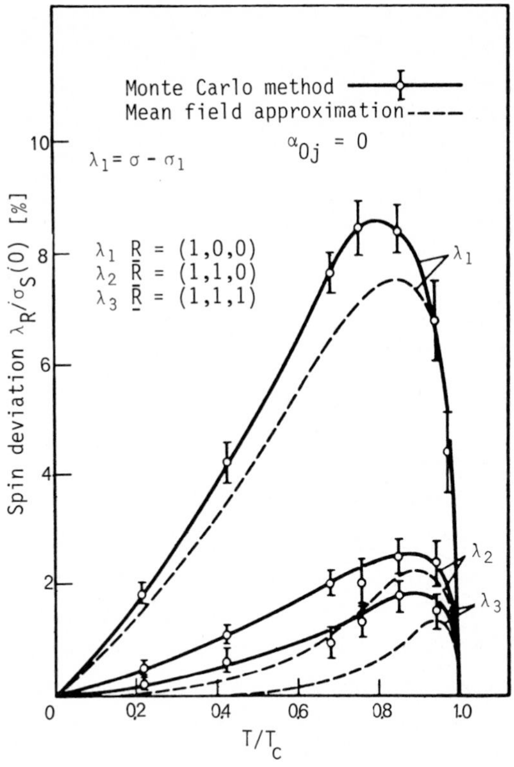

Fig.8.1 Magnetization deviations λ_R (in percent of the saturation magnetization) plotted versus the reduced temperature T/T_c for the case of a simple nonmagnetic impurity at $\underline{R} = 0$. The error bars at the points denote the standard deviations. Mean-field calculations are shown as dashed curves. [8.3]

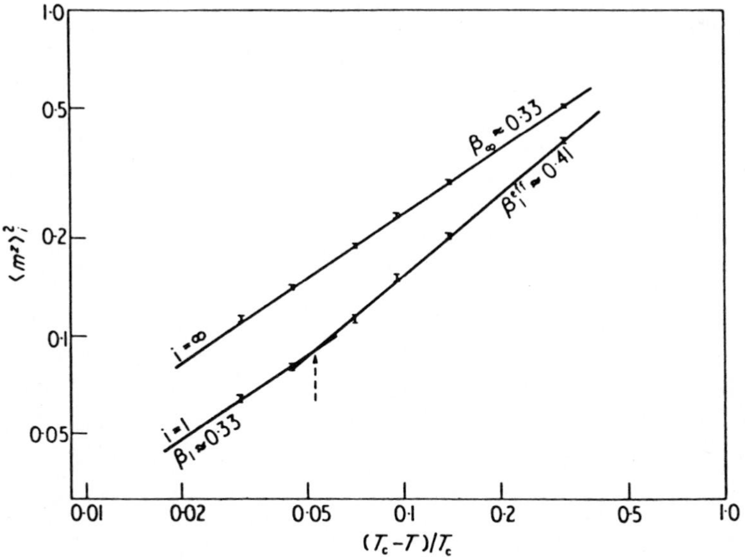

Fig.8.2 Log-lot plot of the magnetization of the unperturbed system (curve labelled by i=∞) and of the local magnetization around a cluster of 2^3 lattice sites (curve labelled by i=1). Arrow shows temperature where the "effective exponent" has a crossover from β_1^{eff} 0.41 to its asymptotic value $\beta_1 \simeq \beta_\infty \simeq 0.33$. [8.4]

measurements [8.8] gave $\beta \approx 0.35$, while $\beta \approx 0.42$ with ^{66}Cu in Ni or ^{67}Zr in Ni [8.9], or $\beta \approx 0.38$ with Fe^{57} in Ni [8.10], or $\beta \approx 0.41$ for ^{181}Ta in Ni [8.11], etc.

Of course, it would be valuable to have similar Monte Carlo investigations for other systems too, e.g., ferroelectrics, or systems undergoing structural phase transitions like $SrTiO_3$, or molecular crystals like KCN, etc. There also the dynamic coupling of the defect to the order parameter is of interest, since it might explain the narrow "central peak" in the dynamic structure function [8.12].

8.2 Dilute Ferromagnets and the Percolation Problem

8.2.1 Thermodynamic Properties at Nonzero Temperatures

Pure ferromagnets have been studied extensively by Monte Carlo methods (for reviews, see Chap.3 and [8.13,14]). Much less work has been done for dilute magnets, where a fraction (1-p) of the magnetic atoms is replaced at random by nonmagnetic ones. These nonmagnetic atoms may be either *quenched* (i.e., fixed in their position) or *annealed* (i.e., mobile due to thermal fluctuations). One expects a quite different behavior in these two cases. Monte Carlo work was made [8.3,15-19] on the quenched case only to which we restrict ourselves here.

Since for pure magnets the Curie temperature T_c is proportional to the magnetic exchange interaction, one expects T_c to decrease with increasing (1-p). In fact, a spontaneous mangetization ("long-range order") is possible only if an infinite network exists of spins mutually connected by exchange interactions, e.g., by nearest-neighbor bonds. If the concentration p of the spins is very low and if the interactions have a finite range, then most of the spins will have no neighbors to interact with, and no such infinite network of interacting spins is possible. Then there is neither a spontaneous magnetization nor a Curie temperature T_c. Thus $T_c(p)$ decreases with decreasing p until it reaches zero at some critical value p_c, at which (presumably [8.20]) also this infinite network is destroyed. The quantity p_c is called the *percolation treshold* [8.20]. Thus two limits are of particular interest: $p \to p_c$ and $p \to 1$. For the latter, the slope of $T_c(p)$ was obtained both for the three-dimensional Heisenberg model [8.3] $[T_c^{-1}(dT_c/dp)_{p=1} = 1.3 \pm 0.2]$, the two-dimensional Ising model [8.15-17] $[T_c^{-1}(dT_c/dp)_{p=1} \approx 1.5]$, and the three-dimensional Ising model [8.18,19]$[T_c^{-1}(dT_c/dp)_{p=1} \approx 1.1]$. At the phase transition $T = T_c(p)$ one expects critical singularities characterized by critical exponents, which might be different from that of the pure case [8.21]. However, it was not possible to find from the Monte Carlo work [8.3,15-19]reliable differences in these critical exponents, as compared to the pure case [8.13]. Reference [8.17] warns that finite-size effects (Chap.1) in dilute systems are stronger than in the corresponding pure system. Lattice sizes used range from 40 × 40 [8.15] to 110 × 110 [8.17] or 20 × 20 × 20 [8.18], respectively.

All these Monte Carlo calculations became difficult to interpret for p close to p_c (from series expansions, p_c = 0.59 for the square lattice while p_c = 0.31 for the simple cubic lattice with nearest-neighbor interactions [8.22]). In particular, no reliable information on the behavior at the phase transition could be obtained. Experimentally some recent results concern the neutron scattering [8.23] from $Rb_2Mn_p Mg_{1-p} F_4$, which is a dilute two-dimensional antiferromagnet. Figure 8.3 shows, as a challenge for future Monte Carlo work, the experimental data for the inverse correlation length (more references to experimental work can be found in [8.24]). The Monte Carlo work [8.15-19]is quite recent and more progress should be expected in the near future.

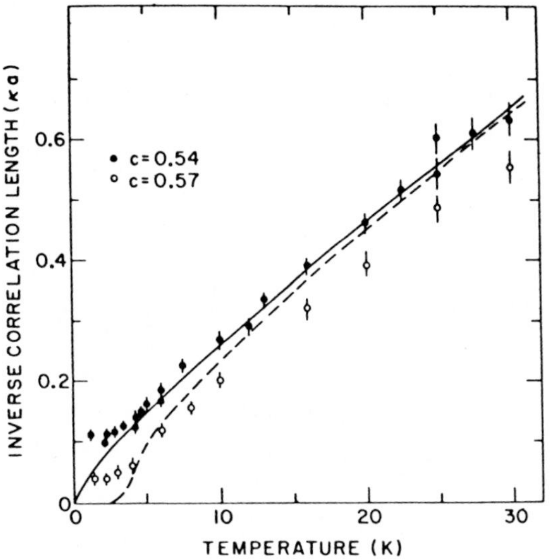

Fig.8.3 Inverse correlation length in $Rb_2Mn_cMg_{1-c}F_4$ in reduced units (a is the lattice spacing) plotted versus temperature at two values of concentration. [8.23]

8.2.2 Percolation Cluster Numbers at Zero Temperature

If a dilute ferromagnet is cooled down to extremely low temperatures, thermal fluctuations are frozen out, and quantum fluctuations are often negligible too. Thus two spins connected by a ferromagnetic exchange interaction (and not connected to any other spins) will always be parallel. This pair then just acts as a single large "superspin", and in an Ising or anisotropic Heisenberg model this "superspin" points either up or down. Then, for temperature T and field H both going to zero at constant ratio H/T, the contribution of this pair to the equilibrium magnetization is $2\mu \tanh(2\mu H/k_BT)$, if μ is the magnetic moment of a single spin. Similarly we can define a cluster of n spins (called a *n-cluster*) as a set of n spins connected with each other (directly or indirectly) by exchange interactions and not connected with any spins outside this cluster. Then each cluster acts as a "superspin" as described above, just as in the theory of superparamagnetism [8.25], and contributes $n\mu \tanh(n\mu H/k_BT)$ to the magnetization. Hence in this $T \to 0$ limit the total magnetization M in units of its saturation value is [8.26-29]

$$M = \sum_{n=1}^{\infty}{}' \; nc_n \tanh(n\mu H/k_BT) \pm M_0 \quad , \tag{8.1a}$$

$$M_0 = 1 - \sum_{n=1}^{\infty}{}' \; nc_n \quad . \tag{8.1b}$$

Here c_n denotes the average number of n-clusters per spin, M_0 is the spontaneous magnetization which arises from the infinite network of mutually connected spins ($M_0 \neq 0$ for $p > p_c$ only). Since all spins must belong either to a finite cluster (including $n = 1$ for isolated spins) or to the infinite network, the number of spins in the infinite network equals the total number of spins N minus the number of spins in the finite clusters, $N\sum' nc_n$, and hence follows (8.1b), (the prime on the sum means that the sum does not include the infinite network $n = \infty$). For $p < p_c$ no infinite network exists, $M_0 = 0$, and hence

$$\sum_{n=1}^{\infty}{}' \; nc_n = 1 \quad . \tag{8.1c}$$

For $p > p_c$, on the other hand, all spins in the infinite network are assumed to be parallel to each other in equilibrium; they point either all up or all down and thus give rise to the spontaneous magnetization M_0, (8.1b).

Thus the magnetic equation of state for $T \to 0$ is known from (8.1) once the cluster numbers c_n are known, there is no need to make any Monte Carlo simulations as a function of the applied field H/T, for example. For a given lattice and range of interaction only the spin concentration p remains as a variable. The problem remains to calculate the $c_n(p)$ and to determine p_c.

This "percolation problem" [8.20,30-32] can be defined much more mathematically without recourse to the picture of low temperature ferromagnets, and has many other applications [8.32,33]. We simply ask: how many n-clusters exist on the average at a lattice where each site is occupied randomly with probability p and empty with probability (1-p), and where an n-cluster is defined as a set of occupied sites connected by distances not exceeding some specified range. Then for $p > p_c$ "percolation" occurs, i.e., an infinite network of occupied sites is formed, just as water can percolate through humid sand if enough pores are filled with water. Exact results for the $c_n(p)$ are restricted to the Bethe lattice (Cayley tree) of coordination number q, where $p_c = 1/(q - 1)$ [8.34], and to small cluster sizes up to n = 10 to 20 for the common two- and three-dimensional lattices with nearest-neighbor bonds [8.22].

Numerous Monte Carlo simulations of percolation problems have been made [8.34-47]. Unfortunately, not always were the results adequately compared with the knowledge accumulated in earlier work, and thus progress was rather slow. Details of how to count clusters, etc., are given in Chap.5, and in [8.36,45a,46]. Reference [8.45a] is based on the largest lattice size (64×10^6 sites) known to us as used for Monte Carlo studies. Thus we refer the reader to that paper for more numerical information on cluster numbers and give here only some impressions. (Even better statistics, but on a much smaller lattice, were obtained in [8.45b], with the results confirming some conclusions of [8.45a]).

The cluster numbers $c_n(p)$ are the smaller (and, unfortunately, more inaccurate) the larger the cluster size n is. For fixed n, $c_n(p)$ reaches a maximum at some $p_{max}(n) < p_c$, with apparently $p_{max} \to p_c$ as $n \to \infty$ (see [8.45a] for a more detailed discussion of this maximum). Since the $c_n(p)$ are known exactly for $n \leq 10$ from other methods [8.22,26,48], the Monte Carlo method is rather useless for p far away from p_c, since there are only too few large clusters. Only for p rather close to p_c, many large clusters appear and the Monte Carlo evaluation of the $c_n(p)$ is really useful. Thus, we concentrate on the *critical behavior* $n \to \infty$, $p \to p_c$ in the following. In our previous language appropriate for a dilute ferromagnet at T = 0, we consider the phase transition from a paramagnet ($p < p_c$, no infinite network) to a ferromagnet ($p > p_c$, spontaneous magnetization due to infinite network). Right at p_c, the c_n seem to decay with a simple power law $c_n \propto n^{-\tau}$ [8.26,41,45a], as shown in Fig.8.4 for the triangular lattice where $p_c = 1/2$ exactly [8.30]. A similar behavior was found for other lattices in both two and three dimensions as well [8.26,41]. For p close to p_c a "scaling" assumption has been proposed for large clusters [8.26,49,50]

$$c_n(p) = q_0 n^{-\tau} f(q_1 \varepsilon n^{\sigma}) \quad , \quad \varepsilon = p - p_c \quad , \tag{8.2}$$

where q_0, q_1 are constants, τ and σ are "critical exponents", and f is a "scaling function". This scaling assumption, (8.2) is made in analogy with cluster models for other phase transitions [8.51,52]. Figure 8.4 shows that τ is slightly larger than 2, in agreement with series expansion results [8.22] and theoretical expectations [8.26].

As a further test of (8.2), we plot in Fig.8.5 (similar to [8.45a]) the ratio $c_n(p)/c_n(p_c)$ as a function of εn^{σ} with $\sigma = 0.39$ for p slightly above p_c. According to (8.2) this ratio equals $f(q_1 \varepsilon n^{\sigma})$ and thus is a single function of the product εn^{σ} and does not depend on ε,n separately. Figure 8.5 shows that the Monte Carlo data for different p indeed follow the same curve within their rather small sta-

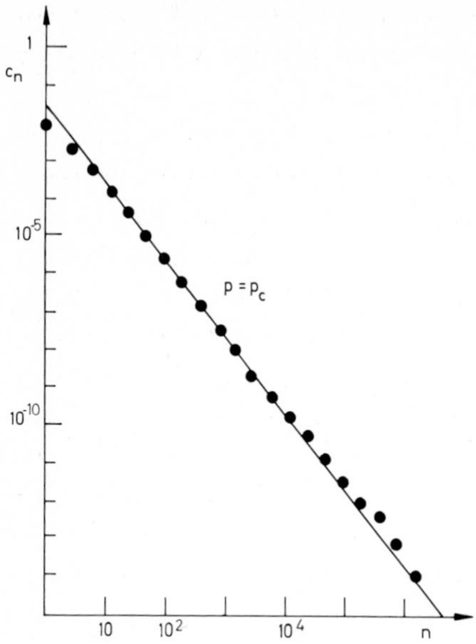

Fig.8.4 Log-log plot of cluster numbers $c_n(p)$ versus n in the triangular lattice at the percolation threshold $p=p_c=1/2$. The dots are exact results from [8.22], the crosses are Monte Carlo data from [8.37] ($n \leqq 10^3$) and [8.45a] ($n \geqq 10^3$)

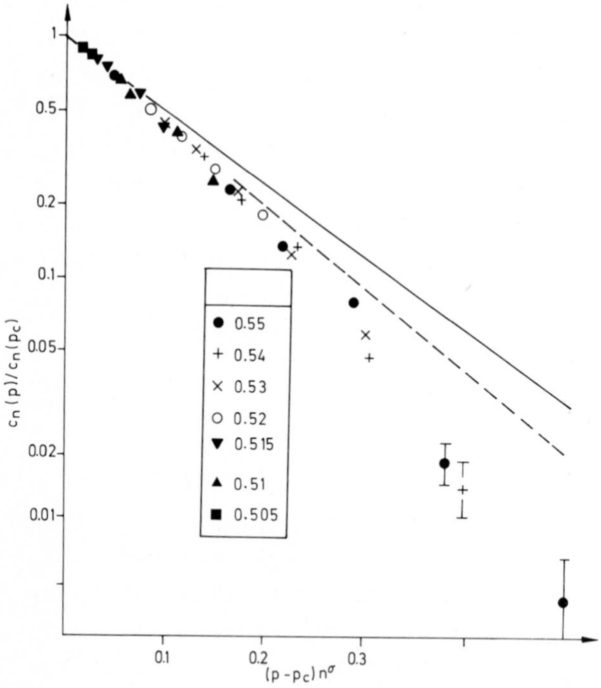

Fig.8.5 Test of scaling behavior, (8.2,3): The ratio $c_n(p)/c_n(p_c)$ is plotted logarithmically versus $(p-p_c)n^{0.39}$ for the triangular site percolation problem, using $p_c = 1/2$. [8.45a]

tistical errors. The solid line in that figure is the tangent to the origin, based on additional data [8.45a]. It shows that the data do not simply follow a straight line; the simple droplet model approximation, known from other phase transitions [8.51,52]

$$f(x) = e^{-x} \quad , \qquad x > 0 \tag{8.3}$$

would predict such a simple line, as shown by the dashed line in Fig.8.5. Thus this droplet model is only a rough approximation.

If the cluster size distribution is known its various moments can be calculated. Of particular interest are: the total number F of clusters,

$$F = \sum_{n=1}^{\infty}{}' \; c_n \tag{8.4a}$$

the fraction M_0 of occupied sites belonging to the infinite network, as defined by (8.1b), (this quantity is often called the percolation probability P [8.30]), the mean cluster size χ

$$\chi = \sum_{n=1}^{\infty} n^2 c_n \quad , \tag{8.4b}$$

and the function

$$M'(\lambda,p) = 1 - \sum_{n=1}^{\infty} nc_n \lambda^n \quad , \tag{8.4c}$$

where λ is some constant between 0 and 1. As an example, Fig.8.6 shows M_0 in three dimensions over the whole range of concentrations.

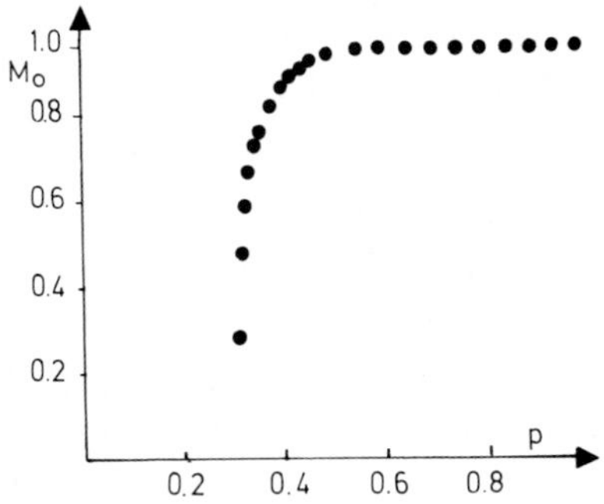

Fig.8.6 Fraction M_0 of occupied sites belonging to the infinite ("percolating") network plotted versus concentration p, using Monte Carlo data for a 30 × 30 × 30 simple cubic lattice. [8.24]

Now critical exponents $\alpha,\beta,\gamma,\delta$ can be defined for p close to p_c and λ close to unity

$$F - F_{reg} \propto |p - p_c|^{2-\alpha} \qquad (8.5a)$$

$$M_0 \propto (p - p_c)^{\beta} \quad , \qquad (8.5b)$$

$$\chi \propto |p - p_c|^{-\gamma} \quad , \qquad (8.5c)$$

$$M'(\lambda,p_c) \propto (1 - \lambda)^{1/\delta} \quad . \qquad (8.5d)$$

Here F_{reg} is an analytic "background" function of p. From the scaling assumption (8.2) and (8.1b,3,4) it is easy to derive the famous "scaling laws" [8.50]

$$2-\alpha = (\tau-1)/\sigma, \quad \beta = (\tau-2)/\sigma, \quad \gamma = (3-\tau)/\sigma, \quad 1/\delta = \tau-2 \qquad (8.6a)$$

or

$$2-\alpha = \gamma+2\beta = \beta(\delta+1) \quad . \qquad (8.6b)$$

These scaling laws are well known from other phase transitions [8.53], in particular from droplet model assumptions [8.51,52], where our F, M_0, χ and $M'(\lambda,p_c)$ here correspond to free energy, spontaneous magnetization, susceptibility and magnetization on the critical isotherm. The main difference to usual phase transitions is the use of $p - p_c$ instead of $T_c - T$ as a variable.

Monte Carlo estimates [8.44,45a,54] for the exponents β and γ from which the other exponents can be calculated via (8.6b) are shown in Fig.8.7 for lattice dimensionality d between 2 and 6; they agree reasonably well with the (presumably more accurate) results from series expansions [8.22,55] also given there. Taking $\beta = 0.14$, $\gamma = 2.43$ for d = 2 we find $\sigma = 1/(\beta + \gamma) = 0.39$ and $\tau = 2 + 1/\delta = 2.054$, compatible with Figs.8.4,5; taking $\beta = 0.4$, $\gamma = 1.66$ for d = 3 yields $\sigma = 0.48$ and $\tau = 2.19$. As expected [8.56], for d = 6 the critical exponents approach the Bethe lattice values [8.34,50,57,58] $\gamma = \beta = 1$, $\sigma = 1/2$, $\tau = 5/2$. (For usual phase transitions the exponents become "classical" in $d_c = 4$ dimensions [8.59]; but there $\beta = 1/2$ in classical Landau theory whereas for percolation on the Bethe lattice Flory [8.57] already showed $\beta = 1$. This difference shifts the "marginal dimensionality" d_c from the usual $d_c = 4$ to $d_c = 6$ for percolation [8.56,60]). At present for no other phase transition has a variation of critical exponents with dimensionality been observed as nicely as in Fig.8.7.

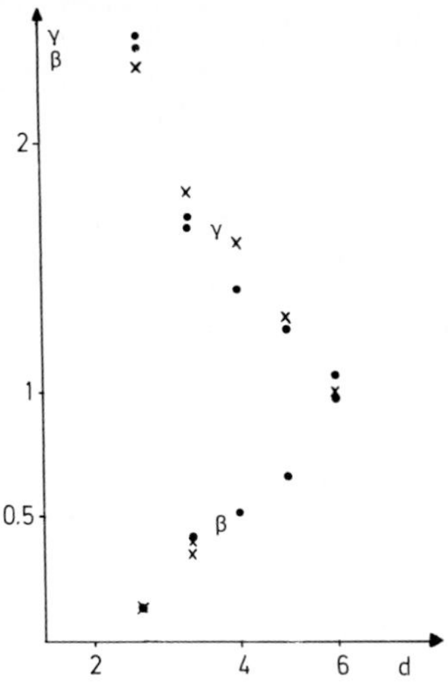

Fig.8.7 Variation of critical exponents β and γ with dimensionality d for d = 2,3,4,5, and 6. Crosses are Monte Carlo data from [8.44,45a,54], dots are series expansion results from [8.22,27, 55]. The "classical" FLORY [8.57] - STOCKMAYER [8.58] -theory for Bethe lattices predicts $\beta = \gamma = 1$, as valid for other lattices also for $d > 6$ dimensions

Having then established the behavior of $c_n(p)$, one can calculate quantities of physical interest, e.g., the equation of state $M = M(p,H/T)$ of a dilute ferromagnet at low temperatures from (8.1), for numerical results see [8.61]. However, we are not aware that experimental data yet exist with which these results could be compared.

An additional exponent ζ can be defined [8.62] both close to p_c and far away from p_c by looking at the asymptotic decay of the c_n for $n \to \infty$

$$\log[c_n(p)] \propto -n^{\zeta} , \quad n \to \infty , \quad p \neq p_c \quad . \tag{8.7}$$

Figure 8.5 suggests ζ to be somewhat larger than $\sigma \simeq 0.4$ because of the curvature. Indeed the estimate $\zeta = 1/2$ fits quite well [8.45a,b] (for a different interpretation of Monte Carlo results see [8.42,44]). On the other hand, below p_c the estimate $\zeta \simeq 1$ was found from Monte Carlo results [8.62,46,45a] and from exact cluster numbers for small sizes, $n \sim 10$. Thus we have perhaps in two dimensions $\zeta(p < p_c) = 1$ and $\zeta(p > p_c) = 1/2$, or more generally in d dimensions

$$\zeta = 1 \ (p < p_c) \tag{8.8a}$$

$$\zeta = 1 - 1/d \quad (p > p_c) \quad . \tag{8.8b}$$

This result is also suggested by analogy with other phase transitions [8.64]. Moreover, rigorous inequalities [8.49] have shown, about simultaneously with these numerical results [8.62,46,63,48], that indeed (8.8) must be valid for sufficiently small and large p, respectively. They also show the connection of cluster numbers with the external surface, as suggested by this behavior $c_n \sim \exp(-\text{const}\cdot n^{1-1/d})$. Thus in the next section we look, among other things, at various "surfaces".

Future work could attack various problems with Monte Carlo methods; for example, the exponents τ,σ and the function f are supposed to be "universal", i.e., for fixed d to be independent of lattice structure and range of "interaction". Only the proportionality constants q_0, q_1 in (8.2) are supposed to depend on such "details" [8.26,89]. Although at least roughly correct [8.26], this universality hypothesis (which is again analogous to that at ordinary phase transitions [8.59]) has not yet been checked in detail by Monte Carlo calculations (but see [8.42a,45a]).

8.2.3 Cluster Surfaces and Correlations

Cluster surfaces usually are defined in terms of "broken bonds" of the cluster. However, if one wants to define a surface of clusters in the usual sense of the word "surface", one should restrict this definition to the outer surface of the cluster and not also count all interior broken bonds as "surface" [8.63]. Only the outer surface of a large spherical cluster varies as $(\text{volume})^{2/3}$ in three dimensions; if broken bonds in the interior are also counted, then for large clusters this "surface" is proportional to the cluster volume n. For example, near T_c of a pure spin-½ nearest-neighbor ferromagnet on a triangular lattice, each spin is surrounded by 6 nearest neighbors; in the average 5 of them are parallel to the central spin but one is antiparallel. If the surface of any large ferromagnetic domain would be defined through the total number of antiparallel neighbor pairs, then this surface would measure basically the internal energy of the domain and be proportional to the domain volume and not to a domain surface in the usual sense.

For percolation clusters, one can look at the "perimeter", i.e., the number of lattice sites which do not belong to the cluster but which are separated from cluster sites only by one bond length or less. Alternatively, one can look at the "energy", i.e., the number of occupied-empty bonds for the cluster sites. In both cases one can distinguish between the external energy or perimeter (which both might serve as a measure of the "surface") and the total energy (or the related "cyclomatic number" [8.65]) or perimeter which include all "internal surfaces". The latter quantities thus say more about the internal structure of a large cluster than about its surface or shape.

The total perimeter was estimated by Monte Carlo methods for d = 2 [8.42a,b]. Compatible with the theoretical prediction near p_c it was found for large n that

$$\text{average perimeter} = n(1 - p_c)/p_c \quad . \tag{8.9a}$$

Analysis of the external perimeter is in progress [8.42a]. The total energy was studied both for d = 2 [8.65,66] and d = 3 [8.44], with the result

$$\text{average energy} \propto n \quad . \tag{8.9b}$$

No analysis of external energies for percolation clusters is known to us; such a calculation would be analogous to the old external energy studies for Ising model clusters in pure magnets [8.67], (see also Chap.5).

Future work should concentrate on the perimeter, which we think is more fundamental for percolation problems than the energy: the average number of clusters with size n and perimeter s varies with concentration p as $p^{n-1}(1-p)^s$; no such formula holds for the energy. Of particular interest would be the behavior of the average external perimeter s_{ext} for large n, $s_{ext} \propto n^{\zeta'}$, to determine the exponent ζ' (which might - or might not - be identical with ζ) on both sides of p_c [8.49].

This discussion of external and internal "surfaces" already indicated that the clusters have "holes" inside. The topology of large clusters was studied in greater detail for the bcc lattice [8.41] when the topological genus g (number of different holes in the cluster interior) was given for $n < 9000$, with an average g_n of

$$g_n \cong n/28 \quad (p = p_c = 0.175) \quad . \tag{8.10}$$

Other quantities of interest are the average cluster volume V_n and radius R_n. The volume V_n is not just n times the volume of a unit cell since again the internal holes make an additional contribution to the volume. For the square lattice, a mean square radius of gyration R_n^2 was obtained as [8.42a]

$$R_n^2 \propto n^{1.22} \; (p = 0.5) \; , \quad R_n^2 \propto n^{1.13} \; (p = 0.55) \quad . \tag{8.11}$$

Since $d = 2$ this quantity should be proportional to V_n. The theoretical prediction [8.63] is $V_n \propto n^{\tau-1} = n^{1.05}$ for percolation clusters of intermediate size and $p \geqq p_c$ and $d = 2$. For $d = 3$ this prediction was confirmed at p_c [8.41].

A related quantity is the correlation function $G(\underline{r})$ measuring the probability that the origin and site $\underline{r}$ belong to the same cluster. One expects [8.63,27-29] that $G(\underline{r})$ decays with $\underline{r}$ on the scale of the correlation length ξ, with

$$\xi \propto |p - p_c|^{-\nu} \quad . \tag{8.12}$$

This quantity was estimated from Monte Carlo work applying the concept of finite size scaling [8.68] (see Chap.1 and 5), which was successfully used for ordinary phase transitions too [8.69,70]. In short, one expects drastic finite size effects when ξ equals L, the linear dimension of the simulated finite system. For instance, $p_c(L)$ should be shifted away from the true p_c [$= p_c(\infty)$] by an amount

$$p_c(L) - p_c(\infty) \propto L^{-1/\nu} \quad , \tag{8.13a}$$

and $M_0(\varepsilon,L)$ should be given by a scaling representation [8.69]

$$M_0(\varepsilon,L) = L^{-\beta/\nu}\tilde{M}_0(\varepsilon L^{1/\nu}) \quad , \tag{8.13b}$$

where $\tilde{M}_0(x)$ is the corresponding scaling function. It turns out that this concept works well for the percolation transition both for d = 2 [8.71] and d = 3 [8.44,71]. The resulting estimates

$$\nu = 1.3 \; (d = 2) \quad , \qquad \nu = 0.8 \; (d = 3) \tag{8.14}$$

are compatible with the scaling law $d\nu = \gamma + 2\beta$, as known from other phase transitions [8.53]. Thus although $G(\underline{r})$ itself has not yet been estimated from Monte Carlo work, ξ has been estimated indirectly by these finite size effects which now seem quite well understood for the percolation transition.

8.2.4 Conductivity and Spin Waves

For cluster size distributions and correlation lengths the Monte Carlo method is one of several successful methods to find out, for instance, the desired critical behavior. Now we discuss another percolation problem, where the Monte Carlo method is mainly the most successful method to date: the *conductivity of random resistor networks*. Imagine that in our previous percolation problem each occupied site is occupied by an electrically conducting material, whereas each empty site is insulating. Only neighboring occupied sites are electrically connected (on the atomic scale this would mean a "hopping conductivity" of electrons). More simply, in the *bond percolation problem* [8.20,30] each nearest-neighbor bond is either conducting or insulating with probabilities p and 1 - p, respectively. Then the lattice as a whole again separates into finite clusters and (for p above p_c) one infinite network. An electric current can flow from one end of the sample to the other only through the infinite network, but not through the many finite clusters. Thus the bulk conductivity $\sigma(p)$ is nonzero only for $p > p_c$ and again one can define a critical exponent μ by

$$\sigma(p) \propto (p - p_c)^{\mu} \quad , \qquad p \to p_c \quad . \tag{8.15}$$

Analogously one can look at the flow of water through a porous medium, a process which has given percolation its name [8.40]. Earlier Monte Carlo studies of $\sigma(p)$ were reviewed in [8.20]; besides on the computer such studies can be done by randomly punching holes into graphite paper (for both d = 2 [8.72] and d = 3 [8.73]) or by randomly cutting wires in a steel-wire mesh [8.74]. Such experimental techniques are out of consideration here. Fig.8.8 shows a computer experiment for d = 3 [8.24,33]. It was concluded [8.24] that

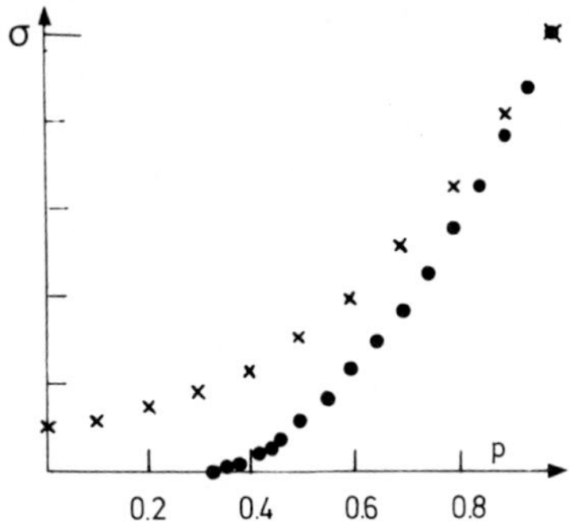

Fig.8.8 Conductivity $\sigma(p)$ in arbitrary units of a random resistor network on a 24 × 24 × 24 simple cubic lattice (dots [8.24]) plotted versus p. Data referring to mixtures of poor and bad conductors (with a conductivity ratio of 10) are included (crosses [8.33])

$$\mu \cong 1.1 \ (d = 2) \ , \quad \mu \cong 1.6 \ (d = 3) \quad . \tag{8.16}$$

Other work [8.40,73-75a,b], gives similar or somewhat higher values. These estimates are roughly consistent with De GENNES' [8.76] suggestion that $\mu = 1 + \nu\,(d - 2)$.

For a comparison with real experiments one can, for example, randomly mix conducting and nonconducting spheres and measure the bulk conductivity of that composite material. We analyzed a recent experiment of that type [8.77] and found roughly $\mu \approx 2$ for $d = 3$. Less artificial systems include alkali-tungsten bronzes (like Na_xWO_3) [8.78,79], metal-ammonia solutions (like Na_xNH_3) [8.79] and electrons in amorphous semiconductors [8.80]. While in the latter case it is doubtful to what extent a hopping conductivity-picture is appropriate [8.81], in the former case the distribution of the metal atoms is not entirely random but shows short-range order effects [8.79]; thus a quantitative comparison with percolation theory may be difficult.

Moreover, it has been suggested [8.57,58,32,76] that the gelation process of branching polymers in (concentrated) solutions is described by percolation theory and its nonclassical critical exponents; in particular [8.76] the shear modulus of the gel phase ($p > p_c$) corresponds to the conductivity in percolation theory. In fact, elastic data close to p_c look similar to Fig.8.8 although the quantitative value for μ was not determined [8.82]. (Besides the conductivity of mixtures of insulators and conductors, one can study random mixtures of good and bad conductors [8.33]. In that case, however, $\sigma(p)$ remains finite on both sides of p_c, as shown in Fig.8.8).

The conductivity of random networks is closely related [8.20,83] to the "hydrodynamic" (i.e., long-wavelength) spin-wave frequency spectrum $\omega(k)$ in dilute isotropic Heisenberg ferromagnets at T = 0. Identifying the conducting and isolating sites of the percolation conductivity problem with magnetic atoms (with isotropic interaction) and nonmagnetic atoms, one finds that $\omega(k) = Dk^2$ with [8.20,24,83]

$$D \propto \sigma(p)/M_0(p) \quad . \tag{8.17a}$$

Thus near p_c we have

$$D \propto (p - p_c)^{\mu-\beta} \quad . \qquad (8.17b)$$

For $p < p_c$ no connected network of spins and thus no magnetic long-range order exists, and long-wavelength spin excitations are impossible. References [8.24,84] discuss $\omega(k)$ in detail, also for antiferromagnets, on the basis of Monte Carlo and other methods. We refer the reader to these extensive papers and note only that once the T = 0 spin wave spectrum is known one can calculate some properties of dilute ferromagnets at nonzero temperature (Sect.8.2.1), if the interaction between different spinwaves is neglected [8.24,84,85].

8.2.5 Miscellaneous Topics

(I) *Bond percolation*: As already mentioned above, there exists a "bond percolation problem" besides the "site percolation problem" on which this review concentrates. Whereas in the latter problem every lattice site is either occupied or empty with probability p and 1 - p, respectively, in the former problem each bond between (neighboring) lattice sites is either closed or open with probability p and 1 - p, respectively. So far we know of no evidence for any qualitative difference (e.g., different exponents etc.) between bond and site percolation.

(II) *Dependence on Interaction Range*: Bond percolation theory has been suggested to describe the gel to sol-transition of branching polymers in concentrated solutions [8.49,76], as mentioned above. In that connection only one Monte Carlo study [8.86] is known to us, which we interpret as percolation in the limit of infinitely long interaction range [8.49]. Besides that study little is known from Monte Carlo on the dependence of percolation on the range of interaction, since usually only nearest-neighbor bonds are taken into account. No significant difference (besides a shift of p_c from 0.59 to 0.41) was detected for a square lattice in a comparison [8.37] of nearest-neighbor and next-nearest neighbor percolation.

(III) *Nonrandom Occupancy*: So far the occupation of lattice sites was assumed to be random. Nonrandom correlations between the occupation probabilities at different sites were discussed in the context of clusters of reversed spins in the pure Ising model [8.87-89]. This is outside the scope of the present chapter (but see Chap.5).

(IV) *Percolation in a Continuum* instead of a lattice is much more difficult to simulate [8.43,79] and is also not discussed here.

(V) *Dynamic Properties of Low-Temperature Dilute Ferromagnets*: In a simple picture [8.26] every n-cluster switches randomly its orientation with a rate r_n. Then the magnetization relaxes with time t as [8.26] $M = \sum_{n=1}^{\infty} nc_n\mu \tanh(n\mu H/k_BT) \exp(-r_nt)$, if for $p < p_c$ a small magnetic field H is switched off at t = 0 (cf.

8.1a). This sum can be evaluated directly [8.26] without the need of dynamic Monte Carlo calculations (cf. Chaps.1 and 6). On the other hand, the rates r_n for low temperatures come from quantum fluctuations (tunneling [8.90]) instead of thermal fluctuations, and thus cannot be calculated by the usual Monte Carlo methods. Thus the Monte Carlo method does not seem to be useful to study the dynamics in dilute ferromagnets in the $T \to 0$ limit. At nonzero T, of course, it would be useful but there already the study of static phenomena is quite difficult (Sect.8.3.1).

8.3 Spin Glasses

Literature expansion in the field of spin glasses is even more rapid than for the percolation problem. Unfortunately, we are far from a simple quantitative theoretical understanding. Recent reviews are available both on experimental [8.91-94] and theoretical [8.95-100] aspects, including computer simulations [8.100], therefore we keep this section rather short.

8.3.1 Physical Properties of Spin Glass Systems

The "classical" spin glass systems are iron-gold alloys (with a few % of iron only) which show a freeze-in transition (?) of the spins at a temperature T_f ($T_f \approx$ 5-15 K depending on concentration). The Fe spins are distributed randomly among the nonmagnetic Au ions, and interact with each other by indirect RKKY exchange interactions, which behave for large distances r as $J(r) \propto r^{-3} \cos(2k_F r)$ where k_F is the Fermi wave number of the Au conduction electrons. The Fe concentration is high enough to make this interaction not negligibly small (in contrast to Kondo systems), but small enough that direct exchange (if two Fe atoms are at nearest-neighbor lattice sites) is negligible [8.91-94], in contrast to the dilute ferromagnets of Sect.8.2. At present experimental observation leads to the following properties as rather reliably established [8.91-94]:

(I) no long-range ferro- or antiferromagnetic order, but magnetic short-range order.

(II) A sharp cusp in the zero-field susceptibility at T_f, which is rounded by the magnetic field (Fig.8.9a [8.101]).

(III) No sharp cusp or singularity in the specific heat [8.102] (Fig.8.9b).

(IV) Hysteresis effects below T_f with relaxation lower than exponential (Fig. 8.9c) [8.103].

Using the above interaction J(r) which extends to infinity would be very time consuming for Monte Carlo simulations. Simulations are hence very difficult but one study [8.146] of a model for CuMn (0.9 at.%) indeed roughly reproduces the

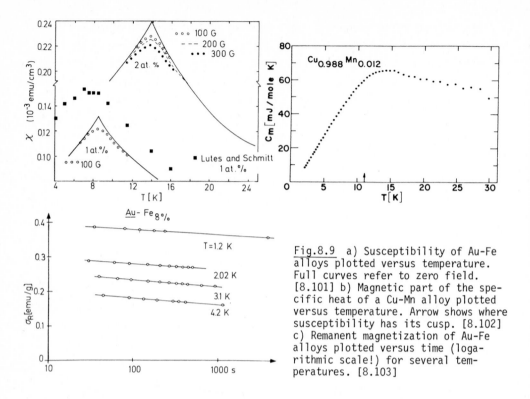

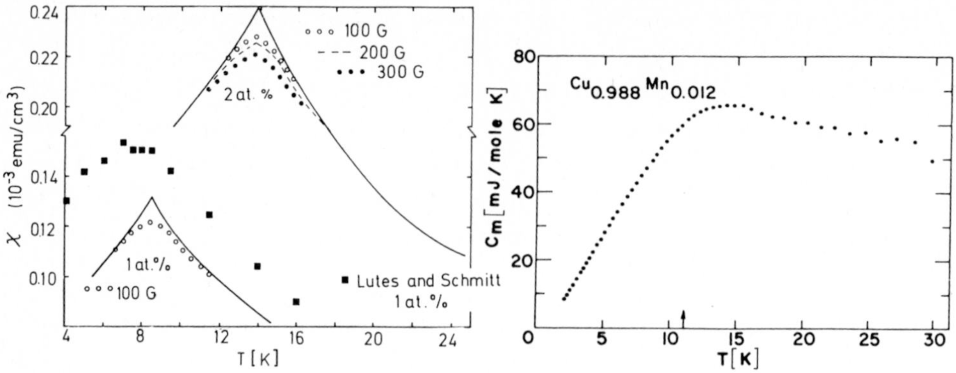

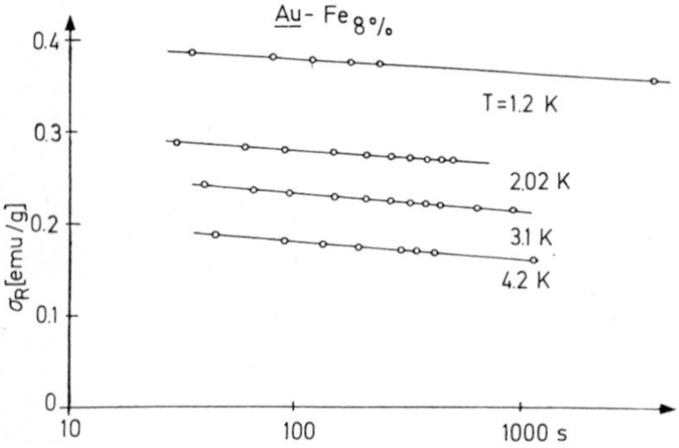

Fig.8.9 a) Susceptibility of Au-Fe alloys plotted versus temperature. Full curves refer to zero field. [8.101] b) Magnetic part of the specific heat of a Cu-Mn alloy plotted versus temperature. Arrow shows where susceptibility has its cusp. [8.102] c) Remanent magnetization of Au-Fe alloys plotted versus time (logarithmic scale!) for several temperatures. [8.103]

location of the susceptibility maximum in agreement with experiment. Fortunately, a simpler model was proposed by EDWARDS and ANDERSON [8.104]; there the spins form a (regular) lattice and are no longer mixed with nonmagnetic ions, but their exchange interaction is chosen randomly according to some probability $P(J)$, e.g.,

$$P(J) \propto \exp\left[-\frac{1}{2}(J/\Delta J)^2\right] \quad , \tag{8.18a}$$

or

$$P(J) = (1 - X)\delta(J - J_0) + X\delta(J + J_0) \;, \quad 0 < X < 1 \quad . \tag{8.18b}$$

Both in the original RKKY problem and in (8.18) the essential feature is the competition between ferromagnetic and antiferromagnetic bonds, i.e., contrary to dilute ferromagnets (Sect.8.2) where the exchange was either J or zero now the exchange may be either positive or negative.

Monte Carlo studies of a model with (8.18) may have several aims: (I) comparison with the more realistic RKKY description (II) test of approximate theories based also on (8.18) (III) comparison with experiment to check if (8.18) is a sensible model. Both static and dynamic properties can be investigated and compared. Clearly, the Monte Carlo method should be particularly suitable for this purpose, and hence numerous studies have been made [8.100,145-149,105-116,95]. Nevertheless there exist many unsolved problems.

8.3.2 Distribution of Interactions and Effective Fields

Whereas in the RKKY model the interactions are fixed for a given distance but the positions of the spins (and hence distances between them) are random, (8.18) has fixed spin positions and random exchange energies. Thus the first question is: how well is the distribution P(J) which follows from the RKKY model approximated by a gaussian, (8.18a). This question, in which neither time nor temperature enters, can be answered by "simple" Monte Carlo sampling (cf. Chap.1). It was discussed in [8.105] both for dimensionality d = 3 and d = 2, distributing 10^3 Ising spins randomly in a cube (square) of length $50/k_F$ and calculating P(J) from the RKKY formula. It was found [8.105] that P(J) is roughly symmetric about J = 0 with a singular maximum $[P(J \to 0) \to \infty]$, and further smaller side-maxima occuring at nonzero J, making the monotonic gaussian (8.18a) a bad approximation, although it is better than a sum of two delta functions, (8.18b). It is hoped (but not proven) that mean free path effects etc. smooth out these undesired smaller maxima as well as the singularity at J = 0.

For any given configuration of spins the influence of other spins on a fixed spin via the exchange interaction J can be expressed as an effective field H_{eff}. Mean-field theory [8.117] draws attention to the distribution of effective fields, $P(H_{eff})$. $P(H_{eff})$ has been obtained for the case of (8.18a) and nearest-neighbor interaction both for d = 2 Ising and d = 3 Heisenberg spin glasses [8.107]. For $H_{eff} \to 0$, $P(H_{eff}) \propto H_{eff}^2$ in the Heisenberg case, while for large H_{eff} the predicted $\log P(H_{eff}) \propto -H_{eff}^2$ behavior [8.117] was found. But, in contrast to [8.117], $P(H_{eff})$ was only very slightly temperature-dependent. Thus one expects that mean-field predictions for the thermodynamics of the Edwards-Anderson model (8.18) are quite unreliable. This expectation is borne out in the next Section.

8.3.3 Susceptibility and Specific Heat

Ferromagnetic susceptibility χ and specific heat were calculated for d = 2 [8.105-107] and d = 3 Ising [8.106] models as well as the d = 3 Heisenberg model [8.107, 111] using (8.18a), and for the d = 3 Ising model [8.113-115] using (8.18b). For the d = 2 Ising calculations typical system sizes were 80×80 and observation times of t = 2000 Monte Carlo steps spin [8.105,115] (runs [8.113,114] with both smaller size and shorter time yield only too inaccurate results), while 50×50 systems were followed up to t = 16000 [8.108] and $8 \times 8 \times 8$ systems were followed up to t = 64000 [8.106]. Fig.8.10 gives a typical result for the case of (8.18a) [8.105]. While χ shows a rather sharp peak at $T_f \approx 1.0\ J/k_B$, the specific heat has a much broader peak at a temperature somewhat higher than T_f. Obviously these re-

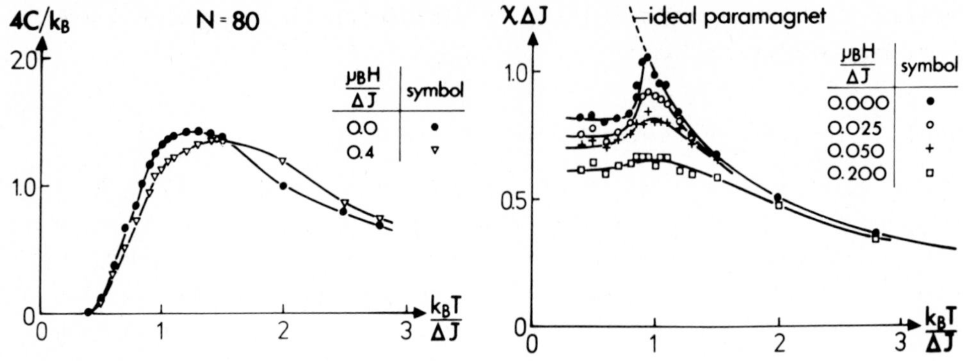

Fig.8.10 Specific heat (left part) and susceptibility (right part) of an 80 × 80 Ising spin glass plotted versus temperature at various fields. Data points were obtained from numerical differentiation ($\chi = \partial M/\partial H$, $C = \partial \langle \mathcal{H} \rangle / \partial T$). [8.105]

sults are qualitatively consistent with experiment (Fig.8.9) in contrast to mean-field theory [8.104,117] which yields sharp cusps both for χ and for C. Using the discrete distribution (8.18a) KIRKPATRICK [8.115] obtained also a very smooth peak for C but centered at T_f, - presumably the formation of small clusters of strongly coupled spins is enhanced by the continuous distribution, (8.18a), and hence one gets a larger contribution to the entropy for $T > T_f$.

Similar results were obtained for the d = 3 Ising model [8.106]. In particular, no evidence was found to suggest that the "phase transition" at $T = T_f$ is less sharp or less reliable in two than in three dimensions (theoretically the existence of a phase transition to a spin glass state is much more controversial for d = 2 than for d = 3 [8.100]). Still, since some relaxation phenomena extend over extremely large time scales [8.106,108] the Monte Carlo method cannot rule out the possibility that spin-glass freezing is a nonequilibrium effect rather than a thermodynamic phase transition [8.100]. In fact, BRAY et al. [8.146-148] argue that the spin glass transition is an artifact due to finite observation time, since they find the susceptibility below T_f to depend distinctly on the observation time. A similar effect is found in a 1d spin glass where it is clear that no phase transition occurs, while observation time effects in simulations of a MATTIS spin glass [8.150] are found to be much smaller [8.147]. On the other hand, these observation time effects could be a consequence of the high degeneracy of the ordered state and the existence of many metastable states with energies being only closely higher [8.108]. In fact, similar observation time effects are found in simulations of a spin glass with infinite range of interaction [8.151], which model is known to have a transition from analytic treatments (see [8.149] for further references). Thus the question if spin glasses have a true phase transition for d = 2 and d = 3 cannot be answered by Monte Carlo methods. Although good reasons exist that there might be no transition, we still keep the notation of a phase transition in the following. Of course, this question is less important as far as comparison with experiment is concerned, since there precisely the same difficulty occurs. The situation is even more unclear in the d = 3 Heisenberg case, Fig.8.11: the apparent peak in the susceptibility at $T \approx 0.3J/k_B$ is no reliable indication of a transition. since the system did not reach thermal equilibrium yet (note the inconsistency in the specific heat data!). Thus while probably $T_f \approx 1.5J/k_B$ for a d = 3 Ising spin glass we might have $T_f = 0$ for a d = 3 Heisenberg spin glass. As mentioned above, SHERRINGTON and KIRKPATRICK [8.149] simulated the Edwards-Anderson model with infinite range of interaction, and obtained impressive agreement between the simulation results and available exact solutions. In this case both susceptibility and specific heat exhibit cusps.

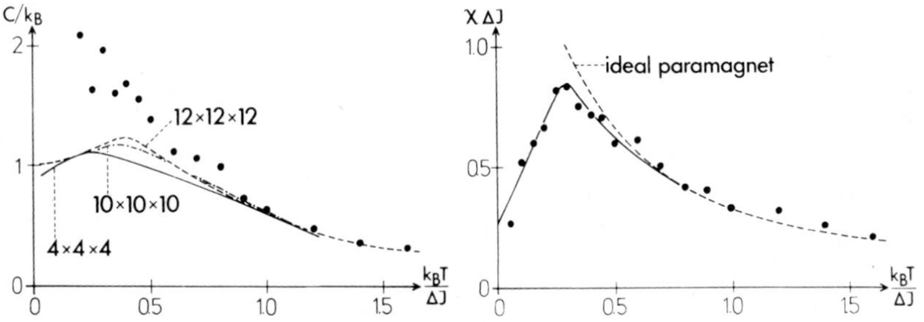

Fig.8.11 Specific heat (left part) and susceptibility (right part) of a 12 × 12 × 12 classical Heisenberg spin glass at zero field. Data points [8.107] were obtained from magnetization or energy fluctuations, respectively. The curves for C were obtained from numerical differentiation ($C = \partial\langle\mathcal{H}\rangle/\partial T$). Data for smaller lattices are from [8.111]

8.3.4 Magnetization and Order Parameters

If a Monte Carlo simulation of a ferromagnet starts with all spins pointing up, the magnetization $M(t) = \langle\sigma_i(t)\rangle$ starts with $M(0) = 1$ and diminishes until equilibrium is established with eventually the spontaneous magnetization. Applying the same procedure to spin glasses, one finds for $T < T_f$ a remanent magnetization [8.105,106] which decreases as T increases and tends to zero for T close to T_f. Closer examination shows, however, that this remanent magnetization very slowly decays to zero, Fig.8.12, as was found in the experiment (Fig.8.9c). In contrast to usual phase transitions, this decay is nonexponential

$$M(t) \propto t^{-a} \ , \quad a \approx k_B T/\Delta J \quad . \tag{8.19}$$

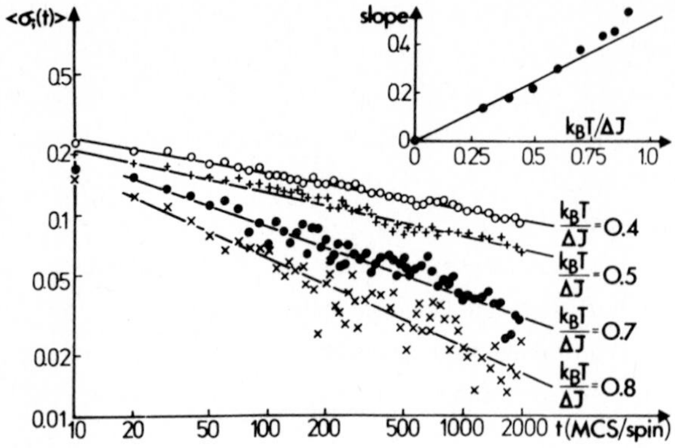

Fig.8.12 Log-lot plot of the magnetization versus time; for 80 × 80 systems. The inset shows the temperature variation of the exponent a in (8.19). [8.105]

Similar results were obtained [8.106] for d = 3, while use of (8.18b) results in a quicker exponential relaxation [8.115]. Clearly, the continuous distribution of J's and hence energy barriers for the relaxation leads to a much broader spectrum of relaxation times than does the discrete one, (8.18b). For T small the exponent a is also small and (8.19) becomes $M(t) \propto \exp[-a\cdot\ln(t)] \simeq 1 - a\cdot\ln(t)$, in agreement with experiments [8.103].

Obviously, (8.19) implies that M is not an appropriate order parameter in the sense that it is zero above T_f and nonzero below T_f in thermal equilibrium. Another suggestion for the order parameter is the quantity [8.104]

$$q = \overline{\langle\sigma_i\rangle^2} \quad , \tag{8.20a}$$

where <...> denotes the thermal average for one configuration of exchange interactions, and the bar symbolizes the average over various distributions of exchange constants according to (8.18). Since in the Monte Carlo method thermal averages are calculated as time averages according to the corresponding master equation (cf. Chaps.1.6), (8.20a) is then represented by

$$q(t) = \frac{1}{N}\sum_{i=1}^{N}\left[\int_0^t \sigma_i(t')dt'/t\right]^2 \quad . \tag{8.20b}$$

Here we explicitly include the dependence on the total time t of the averaging, since q(t) is very slowly relaxing: for $T > T_f$ $q = 0$, (8.20a), but $q(t) \to 2\tau_A/t$ for $t \to \infty$, where $\tau_A = \int_0^\infty dt'\langle\sigma_i(0)\sigma_i(t')\rangle$ [8.107]. Thus Monte Carlo data on q which necessarily use a finite t can exhibit at T_f a rather smooth transition only (Fig.8.13). Apart from these finite-time rounding effects the results agree quite well with corresponding experimental data on $Gd_{0.37}Al_{0.63}$ [8.118].

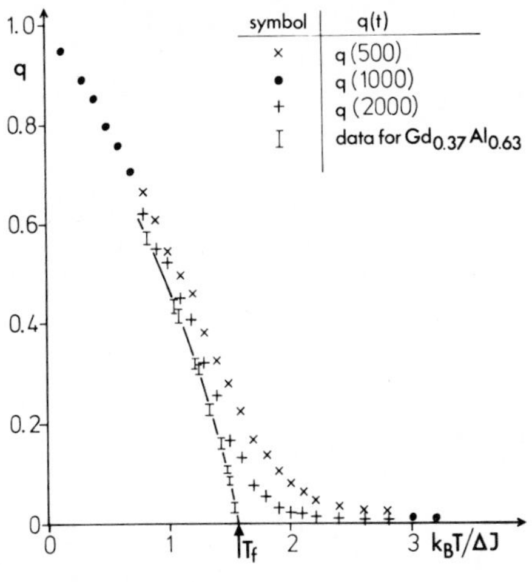

Fig.8.13 Order parameter q of a 16 × 16 × 16 Ising spin glass plotted versus temperature [8.106], as compared to experimental data on $Gd_{0.63}$ [8.118]. [8.100]

However, this quantity q has no thermal fluctuations diverging at T_f, as one usually expects for an appropriate order parameter. Thus another choice for the order parameter was suggested [8.105,107] in analogy with the staggered magnetization used as order parameter for antiferromagnets. Imagine that we know all spin directions in the ℓth ground state: $\sigma_i^{(\ell)}$ (T = 0) = $\phi_i^{(\ell)} = \pm 1$, i = 1,2, ..., N. Then, if in any particular configuration all spins σ_i happen to be parallel to the appropriate ground state ϕ_i, the order parameter is "saturated" and set equal to unity. If in another configuration half of the spins are parallel, half antiparallel, the order parameter is zero. Finally, if all σ_i are antiparallel to the $\phi_i^{(\ell)}$ the order parameter equals minus unity. Thus we can define a local order parameter ψ (which can fluctuate) and a bulk equilibrium order parameter $\langle\psi\rangle$ by

$$\langle\psi\rangle = \frac{1}{N}\sum_{i=1}^{N} \langle\psi_i\rangle \ , \qquad \psi_i = \sigma_i\phi_i^{(\ell)} \quad . \tag{8.21}$$

Actually, since the ground state may be highly degenerate it is nontrivial that this definition works. In the Monte Carlo simulations [8.107,108] one starts letting the system run for a while at $T \approx T_f$ and then cooling down to T = 0 to find some (pseudo-) ground state by minimizing the energy with a few MCS/spin. This state then defines the $\phi_i^{(\ell)}$ and is used as an initial condition for the Monte Carlo simulation at finite temperatures T. For each T several runs are made, using different ground states (ℓ) and thus different sets $\{\phi_i^{(\ell)}\}$. In this way the degeneracy of the ground state does not affect the calculation, and it is rather straightforward to obtain ψ.

Figure 8.14 gives the results for q(t) and $\langle\psi\rangle$ in the d = 2 Ising case. Both q and ψ seem to vanish at the same temperature T_f at which the susceptibility χ had its cusp (Fig.8.11). Having thus established that $\langle\psi\rangle$, (8.21), is a promising candidate for the order parameter, we look at the order parameter susceptibility χ_ψ defined by the second derivative of the free energy $F(\langle\psi\rangle,T)$,

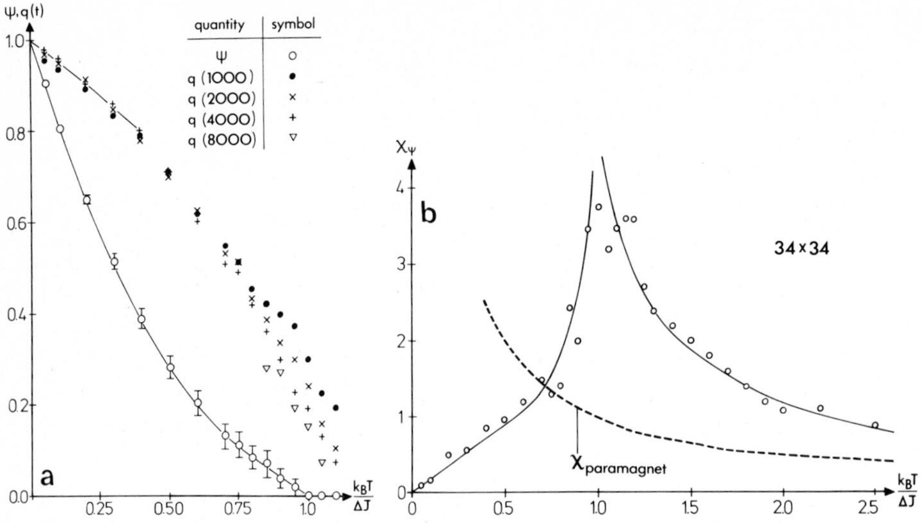

Fig.8.14 a) Order parameters $q(t),\psi$ of 34 × 34 Ising spin glasses plotted versus temperature.
b) Order parameter susceptibility χ_ψ plotted versus temperature. [8.107]

$$\chi_\psi \equiv [\partial^2F/\partial\psi^2]^{-1} = [\langle\psi^2\rangle - \langle\psi\rangle^2]/k_BT \quad , \tag{8.22a}$$

just as the usual susceptibility is given by

$$\chi \equiv [\partial^2F/\partial M^2]^{-1} = [\langle M^2\rangle - \langle M\rangle^2]/k_BT \quad . \tag{8.22b}$$

Figure 8.14b shows that χ_ψ probably diverges at the same T_f where χ has its cusp and ψ and q vanish. Moreover, if one attempts to estimate critical exponents β_ψ,γ_ψ defined as $\langle\psi\rangle \propto (1 - T/T_f)^{\beta_\psi}$, $\chi_\psi \propto |1 - T/T_f|^{-\gamma_\psi}$, one finds [8.107] that the specific heat exponent $\alpha(C \propto |T - T_f|^{-\alpha})$ as determined from the scaling law $2 - \alpha = \gamma_\psi + 2\beta_\psi$ turns out to be smaller than minus unity; this means the slope dC/dT is continuous at T_f, only higher derivatives would clearly show a singular behavior. Such a result would of course be consistent with both experiment (Fig.8.9b) and computer experiment (Fig.8.10), remembering the restricted accuracy of these data. The conclusion then would be that the spin glass transition is an ordinary second-order phase transition, its only peculiarities being a very complicated order parameter ψ and a strongly negative α. Clearly, this conclusion [8.107] is highly speculative, since the accuracy of the data (Fig.8.14) does not allow us to obtain exponent estimates with meaningful accuracy. But the same conclusion was reached by renormalization group calculations too, although their accuracy is again doubtful [8.119]. (Recently up to 10^6 MCS/spin were used [8.108] to study q.)

8.3.5 Kinetic Phenomena

We have already mentioned the strange slow nonexponential relaxation of the remanent magnetization for $T < T_f$, Fig.8.12. Thus it was interesting to investigate the time dependence of other quantities as well. The self-correlation function f(t) $= \langle\sigma_i(0)\sigma_i(t)\rangle$ was obtained both for $T < T_f$ and $T > T_f$ [8.105]. It was found that $f(t) \to q(t)$ for large t and $T < T_f$, while $f(t) \to 0$ for $T \geqq T_f$. Strong deviations from a simple exponential relaxation were found for $T \geqq T_f$. But at the same time it was found that $\psi(t)$ relaxes as $\psi(t) \propto \exp(-t/\tau_\psi)$ for not too small times. Both τ_ψ and $\tau_A = \int_0^\infty f(t)dt$ (which is most reliably determined in terms of q(t), as noted above [8.107]) increased very strongly as T approaches T_f, i.e., the exponents describing the critical slowing down are quite large. Thus it is clear that a very broad spectrum of relaxation times exists in the system, which fact may explain the seemingly nonexponential relaxation of both f(t) and $\langle\sigma_i(t)\rangle$. Of course, these results are far from giving a complete picture, and more complete and more accurate studies would be desirable.

8.3.6 Ground-State Properties

For calculations at T = 0 a Monte Carlo simulation merely requires to search for a minimum of the energy; no thermal fluctuations have to be taken into account. Thus for a given amount of computer time one can look at spin glasses which are more complicated than those on which the $T \neq 0$ simulations were based.

The RKKY model mentioned in Sect.8.3.1 was simulated [8.109] using a $30 \times 30 \times 30$ fcc-lattice with 324 spins distributed in it at random (i.e., the concentration was 0.3%). These spins turned out to be locally aligned but without ferro- or antiferromagnetic long-range order. The spin wave spectrum was obtained and the experimental specific heat [8.102] at low temperatures accounted for. It was found that the lower frequencies corresponded to delocalized modes while the higher frequencies were due to localized modes. Similar results were obtained for a nearest-neighbor Edwards-Anderson Heisenberg spin glass [8.120] of a $4 \times 4 \times 4$ sc-lattice, but there all modes are more or less delocalized. In the first ground-state investigation of that type [8.110] 10^3 Heisenberg spins with RKKY interaction were randomly distributed in a continuum, and the ground-state energy was found as well as the effective field distribution, which agreed with the prediction [8.121] $P(H_{eff}) \propto H_{eff}^2/(\text{const} + H_{eff}^2)^2$. Their study of the spin-spin correlation function revealed some short-range order consistent with both the above spin-wave analysis [8.109] and with neutron small-angle scattering results [8.122].

The studies of KIRKPATRICK [8.115] are devoted to elucidate the nature of the ground state and its degeneracy, questions which will be basic for any theoretical understanding of spin glasses. His work is based on the concept of "frustrated squares" [8.123],i.e., squares of neighboring spins for which no spin configuration exists which minimizes the energy of all 4 nearest-neighbor pairs of the square. A basic quantity for this approach is the average "string" length L (a string connects the broken bonds of two frustrated squares). Since one can show [8.115] for the model of (8.18b) that the ground-state energy of the square lattice is $E_0/J_0 = -2 + x_f L$, the number of frustrated squares x_f being $x_f = 4x(1 - x)[x^2+(1-x)^2]$ where x is the fraction of antiferromagnetic bonds, it is easy to obtain L(x) from the ground-state energy. Fig.8.15 shows that L(x) probably has a cusp for $x_c \approx 0.16$ (d = 2) and $x_c \approx 0.24$ (d = 3, sc-lattice). This behavior is interpreted [8.115] in terms of a first-order phase transition from a ferromagnetic phase for $x < x_c$ to a spin glass phase at $x > x_c$. In addition, the ground-state degeneracy was estimated for x = 0.5 to be of the order $2^{0.14N}$ (d = 2) or $2^{0.09N}$ (d = 3), respectively.[[8.108] discussed ground-state properties of (8.18a)].

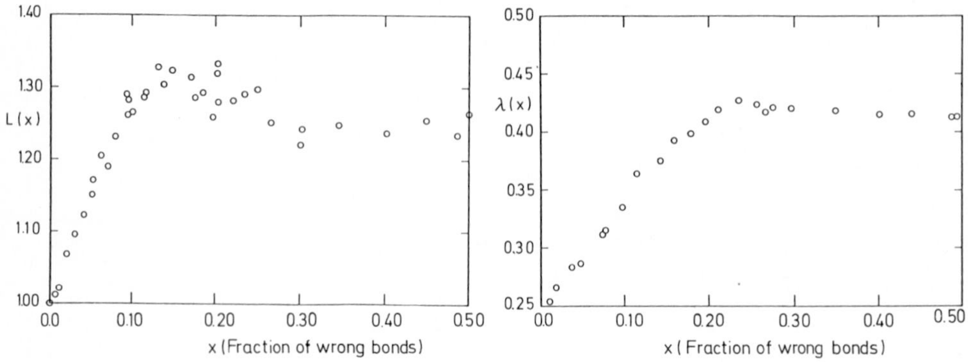

Fig.8.15 Average ground-state string length L(x) plotted versus concentration of antiferromagnetic bonds x in 80 × 80 systems (left part) and ratio of defect surface to perimeter for three-dimensional 16 × 16 × 16 systems (right part). [8.115]

8.4 Disordered Heteropolymers and Their Helix-Coil Transition

Here we consider the helix-coil transition of a biopolymer [8.124]. A heteropolymer consists of a sequence of N peptid units or nucleic acids; each unit may be one of two species (denoted symbolically by A or B). While synthetic heteropolymers with ordered sequence ABABAB ... exist, the sequence of the two species in natural heteropolymers is not ordered. In fact, in the case of nucleic acids the precise information contained in the sequence ABAABABBBA ... corresponds to the genetic information. Form the viewpoint of statistical physics this genetic information is unknown, of course, and one may approximate the sequence as random [8.124]. For one particular biopolymer, the chemical sequence is of course fixed.

The heteropolymer is held together by various chemical bondings: one has a strong covalent bonding along the "backbone" of the chain molecule, and in addition there are much weaker hydrogen bonds between sidegroups of the chain. Specifically, the hydrogen of a base-NH in one unit may form a hydrogen bond with a base -C = 0 in a unit four units away from the first one. These hydrogen bonds lead to a spiral arrangement of the polymer (the α-helix [8.124]). If a unit does not take part in this hydrogen binding, then the structural arrangement (with respect to the neighboring units) is not specified. Therefore the polymer may also lie in a random structure (the coil [8.124]). It turns out that one comes gradually from one state to the other if one varies either the temperature or the pH-value of the (aqueous) solution in which the biopolymer is kept. Denoting the states of the hydrogen bonds by h if it is closed (giving rise to the h̲elix state) or by c if it is open (giving rise to the c̲oil state in that unit), the configurational state of the chain is described by the sequence cchhhchhcccch ... .Introducing a structure variable ("pseudo-spin") $\mu_K(t) = +1$ if the Kth bond is in state h and $\mu_K(t) = -1$ if it is in state c, it is clear that the helix-coil transition can be discussed in terms of an Ising model, in which the sites constitute a one-dimensional chain. In the hamiltonian of the system one often takes into account only interaction energies between neighboring units. The situation is similar to the case of homopolymers but now we must take into account that three types of nearest-neighbor pairs AA,AB,BB occur randomly along the chain, and the interaction parameters will depend on the type of the pair. It hence turns out that the model can be cast in the following form [8.124] (cf. Chap.6)

$$\mathcal{H} = -\sum_{j=1}^{N-1} J_j \mu_j \mu_{j+1} - \sum_{j=1}^{N} H_j \mu_j \quad , \qquad (8.23)$$

where now $J_j = J_0 \pm \Delta J$ and $H_j = H \pm h$ if the jth unit is an A [or B, respectively]. In this representation, the system is thus equivalent to a one-dimensional "spin glass" (cf. Sect.8.3). While only rather preliminary Monte Carlo studies of the full model (8.23) are available [8.125], a more detailed study [8.126] exists on the special case $J_0 = 0$, $h = 0$. Fig.8.16 shows, as an example, the magnetization process at the reduced temperature $T/\Delta J = 0.4$. It is seen that the magnetization process of this random bond Ising chain is quite different from that of a purely ferromagnetic chain (dash-dotted curve) or antiferromagnetic chain (broken curve), and it is also only rather poorly approximated by a perturbation expansion [8.126] (dotted curves).

No comparison of these results with experimental "melting curves" of heteropolymers has been attempted so far. Clearly, the available studies [8.125,126] are rather preliminary first steps only, but further work would be quite promising.

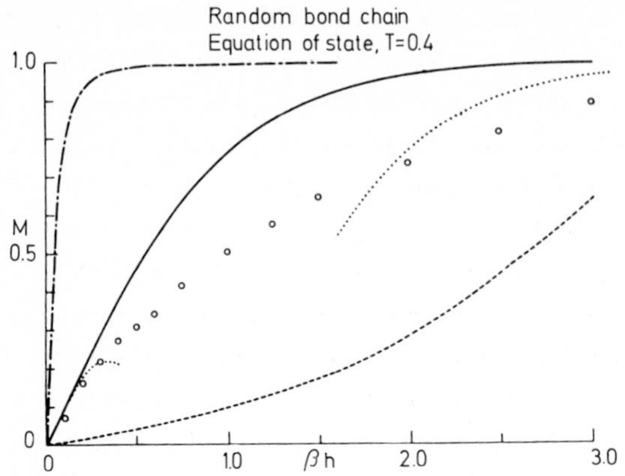

Fig.8.16 Magnetization of the random-bond Ising model plotted versus magnetic field H/k_BT. Dots are Monte Carlo points, curves represent analytic results (cf. text). [8.116]

8.5 Structurally Disordered Solids (Glasses etc.)

Solids may be structurally disordered for several reasons.

(I) Far above the critical temperature of a binary alloy, one has a random mixture of the two components A,B where each atom still occupies a lattice site. The configurational aspects of random mixing ("clusters" of B atoms in the A phase etc.) lead again to the percolation problem (Sect.8.2). However, in these systems one is rather interested in the case where the distribution of the two species is annealed rather than quenched, and not completely random but exhibiting "short-range order" (see Chap.3 for Monte Carlo simulations on that problem).

(II) Far above the critical temperature of a system undergoing a structural transition, an ion (or molecule group) randomly occupies one (or two or more equivalent) position in the unit cell defined by the (nonrandom) positions of the other ions (or molecule groups) of the system. No Monte Carlo simulations of such systems are known to us.

(III) The positions of all the atoms are no longer restricted to lattice sites but the atoms still have "equilibrium" positions, in contrast to the liquid state (Chap.2). This happens in glasses, on which we will concentrate in the following. There one is interested in investigating both their structure and their vibrational properties. We give here only a very brief discussion since many recent reviews [8.127-132] exist, and refer in particular to [8.132] for computer simulations.

The similarity between some of these calculations and the work on spin glasses (Sect.8.3) is particularly evident from a paper of DEAN [8.133], restricted to one dimension. To account for the difference between a disordered chain and a periodic (ordered) one, in principle one would have to use the actual positions in the chain for calculating the distance-dependent interaction potentials. This procedure is difficult just as the RKKY spin glass model is complicated. But just as the Edwards-

Anderson model [8.104], (8.18) gives a simpler approximation by varying randomly the spin exchange forces and using periodic lattice sites for the spin, [8.133] uses randomly varying interaction forces between the atoms on periodic chain sites in order to approximate a truly disordered chain. Even a gaussian approximation for the probability distributions of these forces has been used [8.134] just as (8.18a) for spin glasses.

However, in these papers [8.133,134] as in many others the only Monte Carlo method used consists of using random numbers to fix the positions and/or force constants of the atoms. The main work then may consist of calculating the dynamic matrix of vibrations and determining the eigenvectors and eigenvalues (phonon frequencies) of that matrix. Thus random numbers play only a marginal role in this type of work, and thus we refer to existing reviews [8.127-132].

If only the structure and not the vibrational properties are calculated, then the random construction of a disordered solid is the main part of the work. Such calculations have been done for instance for dense random packing of hard spheres, which can be done both for a pure disordered solid (one sort of spheres) and a mixed glass (e.g., two sorts of spheres mixed randomly). But such computer simulations still have little in common with the Monte Carlo calculations in the rest of this book, where thermal fluctuations at finite temperature are simulated by Monte Carlo, or when at zero temperature a complicated ground state is found by minimizing the energy in a Monte Carlo procedure. Thus, we disregard this dense random packing here, referring to a recent paper [8.135] for further literature, etc.

The analogy with the Monte Carlo methods in other chapters is closest in the work of AVERBACH's group [8.136-140] on the structure of vitreous selenium and As-Se glasses. However, a crucial difference remains: usually the Monte Carlo calculation assumes the energy of every configuration (i.e., the hamiltonian) to be known, and then proceeds (in a $T = 0$ calculation) to minimize that energy. For these glasses, on the other hand, this hamiltonian is not known exactly, although some information on average binding energies is available. Instead of the energy, accurate information exists on the shape of the radial distribution function $g(r)$ through x-ray diffraction or transmission electron diffraction, - i.e., one knows the average density surrounding one atom as a function of the distance r from this atom (cf. also Chap.2). Thus these papers [8.136-140] do not minimize the energy of the computer model but rather minimize through Monte Carlo steps the difference between the $g(r)$ of the computer model and the experimental $g(r)$. The aim of such work is to get a detailed structure of the microscopic configurations from the experimental $g(r)$, which would not follow directly from $g(r)$.

Each Monte Carlo step thus consists of changing slightly the structure of a given configuration. Then g(r) of that configuration is computed as well as the mean-square deviation from the experimental g(r); if the agreement is improved the Monte Carlo step is accepted and otherwise rejected. Then another step is tried, etc., until further repetitions give no significant improvement of the fit. Typically one uses 100-200 atoms, corresponding to a sphere of about 20 Å diameter in these glasses, and 10^4 - 10^5 steps are necessary to achieve "equilibrium" (i.e., no further improvement of the fit). As initial condition, [8.136] used various lattices; but a more realistic approach was taken in the later papers [8.137-140] where the initial arrangement was taken as "quasirandom". Quasirandom here means that certain restrictions were used in building up the initial configuration, taking into account the known chemical binding energies and bond angles. Thermal Boltzmann probability factors $\exp(-E/k_BT)$ were taken into account [8.138] in the build-up of the initial configuration, where E is the energy change due to the addition of an atom and is estimated from the binding energies. A bond was defined simply as a neighbor distance of less than 3 Å, and the number of (energetically unfavorable) As-As bonds was taken as small as possible for the given composition of the As-Se glass. Calculations were always done in three dimensions, but the experimental g(r) to compare the model with was taken from both bulk data [8.136,138,139] and from sputtered films of about 10^2 Å thickness [8.137,139,140].

In pure bulk vitreous Se the dominant "equilibrium" configuration for the Se atoms turned out to be slightly distorted Se_8 rings. In the mixed As-Se system, clusters of connected atoms appeared; Fig.8.17a shows [8.139] that the average cluster size (counting only clusters with more than 3 atoms) has a maximum as a function of the mixing ratio. A similar effect is known from percolation theory [8.30,39]. The dominant feature is the tendency to form As_2Se_3-like structures consisting of Se chains (2 bonds/atom) which can branch at those points where an As atom (3 bonds/atom) is located (Fig.8.17b) [8.140].

8.6 Conclusions and Outlook

In this chapter it was demonstrated that Monte Carlo calculations can make meaningful contributions to the physics of systems containing disorder, and clarify questions relevant for both theory and experiment. We first considered the local distortion of the order parameter of a magnet around an impurity (Sect.8.1), and showed that these results are helpful for the interpretation of measurements with local probes (Mössbauer effect, NMR, EPR, etc.). Clearly, similar investigations could be valuable for other systems too (distortions around defects in anharmonic crystals, effects of defects on the order parameter of molecular crystals, ferro-

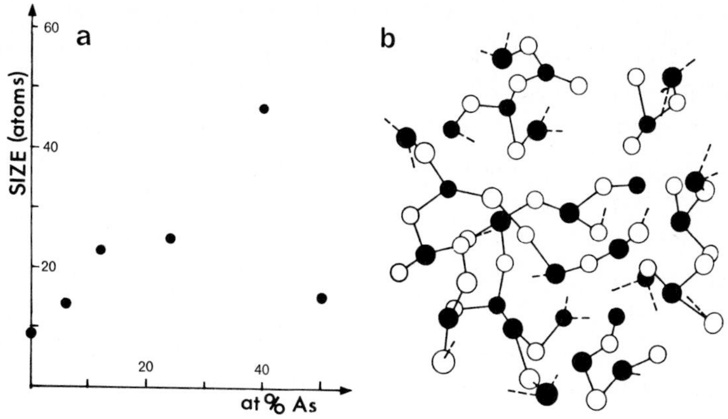

Fig.8.17 a) Average cluster size as a function of composition in As-Se films [Ref.8.139, Table I].b) Cross section of atomic arrangement for $As_{40}Se_{60}$ (dark circles: As; open circles: Se). Dashed lines correspond to bonds to atoms above (or below) the cross section plane. [8.140]

electrics, etc., undergoing structural transitions, etc.). Also, dynamic effects should be treated.

Then the thermal properties of dilute magnets could be treated (Sect.8.2.1). Such systems present a challenge to the Monte Carlo method, since the temperatures of interest where the phase transition occurs are much lower than the interaction energy J between two spins; hence slow relaxation and metastability effects inevitably hamper the approach to equilibrium (cf. Chap.1). Thus additional efforts are necessary to investigate these systems by Monte Carlo methods in order to compare with the available experiments. But, on the other hand, there are large classes of disordered magnets which also deserve Monte Carlo studies: (I) amorphous magnets on which a variety of interesting experimental Monte Carlo studies [8.141] have been made; (II) random magnetic alloys like $Rb_2Mn_{0.5}Ni_{0.5}F_4$, which have also been studied experimentally in detail [8.142]. There both random constituents A, B are magnetic while the exchange constants J_{AA}, J_{AB} and J_{BB} differ. (III) Systems containing random anisotropies or random magnetic fields, where one expects a completely different critical behavior [8.143] and only preliminary Monte Carlo studies are underway [8.144] (they indicate [8.42b] a smaller value of the order parameter exponent, as well as the occurrence of a tricritical point at the critical field curve); etc. And, of course, related problems exist for molecular crystals, ferroelectrics, and systems undergoing any other structural transitions.

While in these areas most work still has to be done, the state of the art with regard to the percolation problem (Sect.8.2.2-5) is much more advanced. The Monte Carlo studies, taken together with series expansion work, gave a rather simple although still tentative picture for the phase transition in percolation theory;

this transition seems to be qualitatively similar to albeit quantitatively different from usual phase transitions and their description by cluster models [8.51], one only has to replace $T-T_c$ by p_c-p. For cluster numbers and perimeters, further refinement of existing data and more accurate tests of scaling and universality are desirable; for cluster surfaces, radii, and density profiles, still much is unknown. The same is true for the structure of the percolating infinite network. A reliable understanding of the percolation conductivity still seems to be lacking, and it is doubtful whether more accurate simulations will help in that understanding at the present stage.

Due to lack of knowledge on the ground state and hence very slow approach to equilibrium, Monte Carlo simulations on spin glasses (Sect.8.3) are also hard to perform. In contrast to the percolation problem, the detailed critical behavior is not yet the central question, but whether, if at all, a phase transition in the usual sense exists. Monte Carlo data nicely agree with available experiments but share their disadvantage that it is hard to distinguish a nonequilibrium freeze-in from a thermodynamic transition. It appears so far that both two-and three-dimensional Ising-Edwards-Anderson spin glasses do have a transition, while little can be said on the Heisenberg case. Also the investigations on how the choice of interaction distribution $P(J)$ influences physical properties are still at the beginning. The same is true for the simulations of one-dimensional spin glass systems which at the same time are models of heteropolymer melting (Sect.8.4).

Finally, simulations of amorphous solids (glasses) have been considered (Sect. 8.5). Apparently very little has been done to study their thermal properties by the usual Monte Carlo simulation of thermal fluctuations. Available work either tries to construct dynamical matrices with suitable randomness built-in which then are diagonalized in order to yield the phonon spectrum in harmonic approximation, or it aims at constructing the random atomic structure in agreement with experimental radial distribution functions. Obviously, also in this field one has reached only a rather crude understanding of the physics of these systems, and the Monte Carlo simulation can be expected to yield valuable contributions in the future.

Acknowledgements. We are indebted to a large number of colleagues working in this area who have sent us preprints of their work.

References

8.1 J.M. Hammersley, D.C. Handscomb: *Monte Carlo Methods* (Methuen, London 1964)
8.2 D.S. McKenzie: Phys. Rpts. *27*, 95 (1976)
8.3 V. Wildpaner, H. Rauch, K. Binder: J. Phys. Chem. Sol. *34*, 925 (1973)
8.4 H. Müller-Krumbhaar: J. Phys. C *9*, 345 (1976)
8.5 E. Stoll, T. Schneider: in *Statistical Physics-Statphys. 13, PtII, Haifa Israel, 24-30 Aug. 1977* (Bristol, England: Adam Hilger Ltd.) p. 431
8.6 A. Holz, A. Sorgen, K.H. Bennemann: Physica *87* A, 145 (1977)
8.7 A.J. Bray, M.A. Moore: J. Phys. A *10*, 1927 (1977)
8.8 E.E. Anderson, S. Arajs, A.A. Stelmach, B.L. Tehan, Y.D. Yas: Phys. Lett. A *36*, 173 (1971)
8.9 W.D. Zeitz: Dr. Dissertation Thesis, Free University of Berlin (1974)
8.10 D. Gumprecht, P. Steiner, G. Crecelius, S. Hüfner: Phys. Lett. A *34*, 79 (1971)
H.C. Benski, R.C. Reno, C. Hohenemser, R. Lyons, C. Abeledo: Phys. Rev. B *6*, 4266 (1972)
8.11 L.J. Oddou, J. Berthier, P. Peretto: Phys. Lett. A *45*, 445 (1973)
8.12 B.I. Halperin, C.M. Varma: Phys. Rev. B *14*, 4030 (1976)
8.13 K. Binder: Advan. Phys. *23*, 917 (1974); and in *Phase Transitions and Critical Phenomena*, Vol. 5b, ed. by C. Domb, M.S. Green (Academic Press, New York 1976)
8.14 D.P. Landau: AIP Conf. Proc. *18*, 819 (1974)
8.15 W.Y. Ching, D.L. Huber: Phys. Rev. B *13*, 2962 (1976)
8.16 R. Fisch, A.B. Harris: AIP Conf. Proc. *29*, 488 (1976)
8.17 E. Stoll, T. Schneider: J. Phys. A *9*, L67 (1976)
8.18 D.P. Landau: Physica B *86-88*, 731 (1977)
8.19 D.P. Landau: AIP Conf. Proc. (1978) to be published
8.20 S. Kirkpatrick: Rev. Mod. Phys. *45*, 574 (1973)
8.21 D.E. Khmelnitskii: Sov. Phys.-JETP *41*, 981 (1975)
T.C. Lubensky: Phys. Rev. B *11*, 3573 (1975)
U. Krey: Z. Phys. B *11*, 355 (1977); B *26*, 325 (1977)
8.22 M.F. Sykes, D.S. Gaunt: J. Phys. A *9*, 87, 97, 1109, 1705 (1976)
8.23 R.J. Birgeneau, R.A. Cowley, G. Shirane, H.J. Guggenheim: Phys. Rev. Lett. *37*, 940 (1976)
8.24 A.B. Harris, S. Kirkpatrick: Phys. Rev. B *15*, 542 (1977)
8.25 E. Kneller: "Theorie der Magnetisierungskurve kleiner Kristalle", in *Ferromagnetismus*, Hrsg. H.P.J. Wijn, Handbuch der Physik, Bd. XVIII/2 (Springer, Berlin, Heidelberg, New York 1966) p. 438
8.26 D. Stauffer: Phys. Rev. Lett. *35*, 394 (1975)
8.27 A.G. Dunn, J.W. Essam, J.M. Loveluck: J. Phys. C *8*, 743 (1975)
8.28 A.G. Dunn, J.W. Essam, D.S. Ritchie: J. Phys. C *8*, 4219 (1975)
8.29 J.W. Essam, K.M. Gwilym, J.M. Loveluck: J. Phys. C *9*, 365 (1976)
8.30 J.W. Essam: In *Phase Transitions and Critical Phenomena*, Vol. II, ed. by C. Domb, M.S. Green (Academic Press, New York 1972)
8.31 V.K.S. Shante, S. Kirkpatrick: Advan. Phys. *20*, 325 (1971)
8.32 P.G. deGennes: La Recherche *7*, 919 (1976)
8.33 Y. Yuge: J. Statist. Phys. *16*, 339 (1977)
J.P. Straley: AIP Conf. Proc. *40*, 118 (1978)
8.34 M.E. Fisher, J.W. Essam: J. Math. Phys. *2*, 609 (1961)
8.35 V.A. Vyssotksky, S.B. Gordon, H.L. Frisch, J.M. Hammersley: Phys. Rev. *123*, 1566 (1961)
8.36 P. Dean: Proc. Cambridge Phil. Soc. *59*, 397 (1963)
8.37 P. Dean, N.F. Bird: Natl. Physical Laboratory Rpt. Ma61, Teddington, England (1966)
8.38 J.W.J. Coey: Phys. Rev. B *6*, 3260 (1972)
8.39 K.J. Duff, V. Canella: In *Amorphous Magnetism*, ed. by H.O. Hooper, A.M. de Graaf (Plenum Press, New York 1973) p. 207
8.40 K. Onizuka: J. Phys. Soc. Jpn. *39*, 527 (1975)
8.41 G.P. Quinn, G.H. Bishop, R.J. Harrison: J. Phys. A *9*, L9 (1976) and in *Computer Simulation for Materials Applications*, ed. by R.J. Arsenault, J.R. Beeler, Jr., J.A. Simmons (Natl. Bur. Stand., Gaithersburg 1976) and J. Statist. Phys. *19*, 53 (1978)
8.42a P. Leath: Phys. Rev. Lett. *36*, 921 (1976); Phys. Rev. B *14*, 5046 (1976)
P. Leath, G.R. Reich: J. Phys. C *11*, 4017 (1978)

8.42b D.P. Landau, H.H. Lee, W. Kao: J. Appl. Phys. *49*, 1356 (1978)
8.43 D.H. Fremlin: J. Phys. (Paris) *37*, 813 (1976)
8.44 A. Sur, J.L. Lebovitz, J. Marro, M.H. Kalos, S. Kirkpatrick: J. Statist. Phys. *15*, 345 (1977)
8.45a J. Hoshen, R. Kopelman: Phys. Rev. B *14*, 3638 (1976)
J. Hoshen, D. Stauffer, G.H. Bishop, R.J. Harrison, G.D. Quinn: J. Phys. A*12*, 1285 (1979)
8.45b E. Stoll, C. Domb: J. Phys. A *11*, L57 (1978)
8.46 H. Müller-Krumbhaar, E.P. Stoll: J. Chem. Phys. *65*, 4294 (1976)
8.47 A. Sur: Private communication
8.48 A. Flammang: Z. Phys. B *28*, 47 (1977)
8.49 H. Kunz, B. Souillard: J. Statist. Phys. *19*, 77 (1978); Phys. Rev. Lett. *40*, 132 (1978)
8.50 J.W. Essam, K.M. Gwilym: J. Phys. C *4*, L228 (1971)
8.51 M.E. Fisher: Physics *3*, 255 (1967)
8.52 K. Binder: Ann. Phys. *98*, 390 (1976)
8.53 H.E. Stanley: *An Introduction to Phase Transitions and Critical Phenomena* (Oxford University Press, Oxford 1971)
8.54 S. Kirkpatrick: Phys. Rev. Lett. *36*, 69 (1976)
8.55 D.S. Gaunt, M.F. Sykes, H. Ruskin: J. Phys. A *9*, 1899 (1976)
8.56 A.B. Harris, T.C. Lubensky, W.K. Holcomb, D. Dasgupta: Phys. Rev. Lett. *35*, 327 (1975)
8.57 P.J. Flory: J. Am. Chem. Soc. *63*, 3083, 3091, 3096 (1941)
8.58 W.H. Stockmayer: J. Chem. Phys. *11*, 45 (1943)
8.59 S.-K. Ma: *Modern Theory of Critical Phenomena* (Benjamin Reading, Mass. 1976)
8.60 G. Toulouse: Nuovo Cimento B *23*, 234 (1974)
8.61 D. Stauffer: In *Amorphous Magnetism II,* ed. by R.A. Levy, R. Hasegawa (Plenum Press, New York 1977) p. 17, and unpublished
8.62 M.M. Bakri, D. Stauffer: Phys. Rev. B *14*, 4215 (1976); see also 8.46
8.63 D. Stauffer: Z. Phys. B *25*, 391 (1976)
8.64 R. Kretschmer, K. Binder, D. Stauffer: J. Statist. Phys. *15*, 267 (1976)
8.65 C. Domb, E. Stoll: J. Phys. A *10*, 1141 (1977)
8.66 C. Domb, T. Schneider, E. Stoll: J. Phys. A *8*, L90 (1975)
8.67 K. Binder, D. Stauffer: J. Statist. Phys. *6*, 49 (1972)
8.68 M.E. Fisher: In *Critical Phenomena,* ed. by M.S. Green (Academic Press, New York 1971)
8.69 K. Binder: Thin Solid Films *20*, 367 (1974)
8.70 D.P. Landau: Phys. Rev. B *13*, 2997 (1976); B *14*, 255 (1976)
8.71 M.E. Levinshtein, B.I. Shklovskii, M.S. Shur, A.L. Efros: JETP *42*, 197 (1976)
8.72 D.J. Thouless, B.J. Last: Phys. Rev. Lett. *27*, 1719 (1971)
M. Levinshtein: J. Phys. C *10*, 1895 (1977)
8.73 M.E. Levinshtein, M.S. Shur, A.L. Efros: Sov. Phys.-JETP *42*, 2203 (1976)
8.74 B.P. Watson, P.L. Leath: Phys. Rev. B *9*, 4893 (1974)
8.75a A.S. Skal, B.I. Shklovskii: Sov. Phys. Semicond. *8*, 1029 (1975)
8.75b I. Webman, J. Jortner, M.H. Cohen: Phys. Rev. B *16*, 2593 (1977)
8.76 P.G. de Gennes: J. Phys. (Paris) *37*, L1 (1976)
8.77 J. Clerq, G. Giraud, J. Roussenq: C.R. Acad. Sci. Paris *281*, B227 (1975); see also
B. Abeles, H.L. Punch, J.I. Gittleman: Phys. Rev. Lett. *35*, 247 (1975)
N.T. Liang, Y. Shan, S.Y. Wang: Phys. Rev. Lett. *37*, 526 (1976)
8.78 P.A. Lightsey: Phys. Rev. B *8*, 3586 (1973)
8.79 I. Webmann, J.Jortner, M.H. Cohen: Phys. Rev. B *11*, 2885 (1975); B *13*, 713 (1976); B *15*, 1936 (1977) and references contained therein; see also
G.E. Pike, C.H. Seager: Phys. Rev. B *10*, 1421 (1974)
8.80 A.L. Efros, B.I. Shkovskii: Phys. Stat. Sol. B *76*, 475 (1976); Sov. Phys.-Uspekhi *18*, 845 (1975)
H. Böttger, V.V. Bryksin: Phys. Stat. Sol. B *78*, 9, 415 (1976); *81*, 433 (1977)
8.81 D.J. Thouless: Phys. Rpts. C *13*, 93 (1974)
8.82 C.A.L. Peniche-Covas, S.B. Dev, M. Gordon, M. Judd, K. Kajivara: Faraday Discuss. Chem. Soc. *57*, 165 (1974)
8.83 A.B. Harris, S. Kirkpatrick: AIP Conf. Proc. *24*, 99 (1975)
8.84 S. Kirkpatrick, A.B. Harris: Phys. Rev. B *12*, 4980 (1975)

8.85 E.F. Shender, B.I. Shklovskii: Phys. Lett. A *55*, 77 (1975)
E.F. Shender: J. Phys. C *9*, 1309 (1976); Phys. Lett. A *62*, 161 (1977)
8.86 M. Falk, R.E. Thomas: Can. J. Chem. *52*, 3286 (1974); see also
H.J. Wintle, T.P.T. Williams: Can. J. Phys. *55*, 635 (1977)
8.87 H. Müller-Krumbhaar: Phys. Lett. A *50*, 27 (1974)
8.88 T. Odagaki, N. Ogita, H. Matsuda: J. Phys. Soc. Jpn. *39*, 618 (1975)
8.89 J. Marro: Phys. Lett. A *59*, 180 (1976)
8.90 D. Stauffer: Solid State Commun. *18*, 533 (1976)
8.91 V. Canella, J.A. Mydosh: AIP Conf. Proc. *18*, 651 (1974)
8.92 J.A. Mydosh: In *Amorphous Magnetism II*, ed. by R.A. Levy, R. Hasagawa (Plenum Press, New York 1977), p. 73
8.93 J.A. Mydosh: In *Proc. of the Internal Conference on Magnetic Alloys and Oxides, Haifa 1977*; J. Magn. Mag. Mater. *7*, 237 (1978)
8.94 G. Heber: J. Magn. Mag. Mater. *2*, 47 (1975)
8.95 D. Sherrington: AIP Conf. Proc. *29*, 224 (1976)
8.96 G. Heber: Appl. Phys. *10*, 101 (1976)
8.97 P.W. Anderson: In *Amorphous Magnetism II*, ed. by R.A. Levy, R. Hasegawa (Plenum Press, New York 1977) p. 1
8.98 K.H. Fischer: Physica B *86-88*, 813 (1977)
8.99 U. Krey: J. Magn. Mag. Mater. *7*, 150 (1978)
8.100 K. Binder: In *Advances in Solid State Physics*, Vol. 17, ed. by J. Treusch Vieweg, Braunschweig 1977) p. 55
8.101 V. Canella, J.A. Mydosh: Phys. Rev. B *6*, 4220 (1972)
8.102 L.E. Wenger, P.H. Keesom: Phys. Rev. B *13*, 4053 (1976)
8.103 F. Holtzberg, J.L. Tholence, R. Tournier: In *Amorphous Magnetism II*, ed. by R.A. Levy, R. Hasegawa (Plenum Press, New York 1977) p. 155
8.104 S.F. Edwards, P.W. Anderson: J. Phys. F *5*, 965 (1975)
8.105 K. Binder, K. Schröder: Solid State Commun. *18*, 1361 (1976); Phys. Rev. B *14*, 2142 (1976)
8.106 K. Binder, D. Stauffer: Phys. Lett. A *57*, 177 (1976)
8.107 K. Binder: Z. Phys. B *26*, 339 (1977); Physica B *86-88*, 871 (1977)
8.108 D. Stauffer, K. Binder: Z. Phys. B *30*, 313 (1978)
8.109 L.R. Walker, R.E. Walstedt: Phys. Rev. Lett. *38*, 514 (1977)
8.110 F.A. de Rozario, D.A. Smith, C.H.J. Johnson: Physica B *86-88*, 861 (1977)
8.111 W.Y. Ching, D.L. Huber: Phys. Lett. A *59*, 383 (1977); AIP Conf. Proc. *34*, 370 (1977)
8.112 B.W. Southern, A.P. Young: J. Phys. C *10*, 2179 (1977)
8.113 I. Ono: J. Phys. Soc. Jpn. *41*, 345 (1976)
I. Ono Y. Matsuoka: J. Phys. Soc. Jpn. *41*, 1425, 1427 (1976)
8.114 M. Sabata, F. Matsubara, Y. Abe, S. Katsura: J. Phys. C *10*, 2887 (1977)
8.115 S. Kirkpatrick: Phys. Rev. B *16*, 4630 (1977)
8.116 G.J. Nieuwenhuys, J.A. Mydosh: Physica B *86-88*, 880 (1970)
8.117 M.W. Klein: Phys. Rev. *173*, 552 (1968); B *14*, 5018 (1976)
8.118 T. Mizoguchi, T.R. McGuire, S. Kirkpatrick, R.J. Gambino: Phys. Rev. Lett. *38*, 89 (1977)
8.119 C. Jayaprabash, J. Chalupa, M. Wortis: Phys. Rev. B *15*, 1495 (1977)
8.120 W.Y. Ching, K.M. Leung, D.L. Huber: Phys. Rev. Lett. *39*, 729 (1977)
8.121 C. Held, M.W. Klein: Phys. Rev. Lett. *35*, 1783 (1975)
8.122 A.P. Murani: Phys. Rev. Lett. *37*, 450 (1976)
8.123 G. Toulouse: Commun. Phys. *2*, 115 (1977)
8.124 D. Poland, H.A. Scheraga: *Theory of Helix-Coil Transitions in Biopolymers* (Academic Press, New York 1970)
8.125 I. Morgenstern, K. Binder, A. Baumgärtner: J. Chem. Phys. *69*, 253 (1978)
8.126 D.P. Landau, M. Blume: Phys. Rev. B *13*, 187 (1976)
8.127 P. Dean: Rev. Mod. Phys. *44*, 127 (1972)
8.128 R. Bell: Rept. Progr. Phys. *35*, 1315 (1972)
8.129 H. Böttger: Phys. Stat. Sol. B *62*, 9 (1974)
8.130 D. Weaire: Contemp. Phys. *17*, 173 (1976)
8.131 C.S. Cargill, III, S. Kirkpatrick: AIP Conf. Proc. *31*, 339 (1977) and other articles therein

8.132 W.M. Visscher, J.E. Gubernatis: "Computer Experiments and Disordered Solids", in *The Dynamical Properties of Solids*, ed. by G.K. Horton, A.A. Maradudin, Vol. 3 (North-Holland Amsterdam, to be published)
8.133 P. Dean: Proc. Phys. Soc. *84*, 727 (1964)
8.134 I. Ishida, T. Matsubara: J. Phys. Soc. Jpn. *38*, 534 (1975)
8.135 D.S. Baudreaux, J.M. Gregor: J. Appl. Phys. *48*, 152 (1977)
8.136 R. Kaplow, T.A. Rowe, B.L. Averbach: Phys. Rev. *168*, 1068 (1968)
8.137 M.D. Rechtin, B.L. Averbach: Solid State Commun. *13*, 491 (1974)
8.138 M.D. Rechtin, A.L. Renninger, B.L. Averbach: J. Noncryst. Sol. *15*, 74 (1974)
8.139 A.L. Renninger, M.D. Rechtin: J. Noncryst. Sol. *16*, 1 (1974)
8.140 M.D. Rechtin, B.L. Averbach: Phys. Stat. Sol. A *28*, 283 (1975)
8.141 S. Kobe: Phys. Stat. Sol. *41*, K13 (1970)
8.142 J. Als-Nielsen, R.J. Birgeneau, H.J. Guggenheim, G. Shirane: Phys. Rev. B *12*, 4963 (1975); J. Phys. C *9*, L121 (1976)
8.143 Y. Imry, S.-K. Ma: Phys. Rev. Lett. *35*, 1399 (1975)
8.144 A. Aharony: Private communication
D.P. Landau: Private communication
8.145 W.Y. Ching, D.L. Huber: Phys. Lett. A *59*, 383 (1977); AIP Conf. Proc.*34*, 370 (1977); J. Phys. F *8*, L63 (1978)
8.146 A.J. Bray, M.A. Moore: J. Phys. F *7*, L333 (1977)
8.147 A.J. Bray, M.A. Moore, P. Reed: J. Phys. C *11*, 1187 (1978)
8.148 P. Reed, M.A. Moore, A.J. Bray: J. Phys. C *11*, L139 (1978)
8.149 D. Sherrington, S. Kirkpatrick: Phys. Rev. B *17*, 4384 (1978)
8.150 D.C. Mattis: Phys. Lett. A *56*, 421 (1976)
8.151 S. Kirkpatrick: Private communication

9. Applications in Surface Physics

D. P. Landau

With 11 Figures

Recent progress in the development of techniques for the study of solid surfaces has presented an ever increasing amount of information which can be effectively interpreted with the aid of Monte Carlo studies. Simulation results are presented which explain the possible surface critical behavior in magnets with modified surface exchange. Critical behavior in binary alloy surfaces is examined and the effects of surface enrichment are studied. Surface adatom deposition has been examined and Monte Carlo data are described for order-disorder transitions and kinetic phenomenon in adsorbed monolayers.

9.1 Introductory Remarks

Surface physics is an active field in which new developments are continuously adding to our basic understanding of surface behavior. In recent years refined experimental techniques have made possible the study and unambiguous classification of a variety of static phenomena on solid surfaces, e.g., crystallographic "reconstruction", surface segregation in alloys, ordering of adsorbates, as well as dynamic behavior, e.g., surface modes and adsorption kinetics. Until recently, these surface phenomena had not been investigated nearly as thoroughly as the corresponding bulk behavior.

Phase transitions and ordered structures have been extensively investigated in both a wide range of magnetic materials as well as in binary alloys for quite some time [9.1]. A variety of theoretical approaches have been used to treat simple models believed to be appropriate for these systems, but in virtually all cases surface effects have been ignored (usually by simply making the model infinite in extent). The usual justification for neglecting surface effects is that in a d-dimensional lattice of N sites typically only on the order of $N^{(d-1)/d}$ of these sites lie on the surface. If the model is truly intended to describe a real physical system, N should be so large as to render surface effects negligible and the approximation of an infinite lattice should be satisfactory. More recently, however, it has become clear that this picture is oversimplified and that accurate, detailed knowledge of surface behavior is needed for two reasons: (I) real physical systems are generally composed of relatively small grains and surface behavior may be significantly different from bulk behavior. Thus measurements (e.g., specific heat, magnetization, susceptibility) which do not differentiate between bulk and surface contributions may, in fact, measure an average value which differs from the true bulk value (particularly near the critical temperature of a second-order phase transition); (II) improved resolution of various experimental methods and improved techniques for producing new types of materials are making "direct" measurements of surface behavior possible so that theoretical predictions can now be tested. A better understanding of the properties of surfaces may provide a key for important applications such as in catalysis, etc.

The Monte Carlo method is particularly well suited to the study of two types of "surface" problems. Several simple magnetic models as well as the binary alloy (Ising) lattice models form a class of systems in which the bulk and surface constituents are essentially the same and interact strongly and directly. Yet a different class of surface problems includes absorbed layers of atoms on an essentially inert substrate. The underlying surface which causes the atoms to adhere defines a periodic potential and thus a "lattice" of possible occupation sites. In this latter case the characteristics of the adsorbed layer on an exposed surface of a three-dimensional solid should correspond to that of a two-dimensional system. In this chapter we shall consider the entire range of surface problems to which the Monte Carlo method has been applied. In Sect.9.2 we shall review the behavior of magnetic particles and films and in Sect.9.3 examine binary alloys. Sect.9.4 shall be devoted to order-disorder transitions in adsorbed gas layers on surfaces and Sect.9.5 shall consider kinetic phenomena at surfaces.

9.2 Critical Behavior of Magnetic Systems with Surfaces

In real systems the presence of surfaces has two effects: (I) by limiting the size of the system a limit is also placed on the maximum value which the correlation length may have leading to the finite size effects which have been described in Chap.5; (II) the presence of the surface leads to an inhomogeneous distribution of the magnetization and other thermodynamic properties. A general scaling theory which includes the effects of surfaces has been developed [9.2]; we shall sketch the scaling treatment since we want to compare the simulations to this theory. The singular part of the free energy of a d-dimensional system of n (d - 1) dimensional layers is

$$F(n,h,h_1,T) = n^{\psi} f(n^{-\phi_1} h_1, n^{\theta} t) \cdot \tag{9.1}$$

where $t = (1 - T/T_c)$, h is a magnetic field acting uniformly on *all* spins and h_1 is a magnetic field which acts on the surface spins *only*. (h_1 is introduced for computational convenience and must be set equal to zero in relating the final results to properties of physical surfaces). The free energy may then be separated into a bulk term F_b and a surface term F_s where

$$F(n_1 h,h_1,T) = F_b(h,t) + 2n^{-1} F_s(h,h_1,t) + \ldots \tag{9.2}$$

with

$$F_b(h,t) = t^{2-\alpha} f_b(h/t^{\Delta}) \tag{9.3a}$$

$$F_s(h,h_1,t) = t^{2-\alpha_s} f_s(h/t^{\Delta}, h_1/t^{\Delta_1}) \quad . \tag{9.3b}$$

The surface term contains new exponents α_s, Δ_1 which are absent in the bulk expression. Differentiating F_s with respect to h yields a new quantity, the "surface" magnetization $m_s = \lim_{n\to\infty} [\sum_{i=1}^{i=n} (m_b - m_i)]$ where m_b is the bulk magnetization and m_i is the value in the ith layer only. Similarly, a "layer" magnetization m_1 can be defined by $m_1 = \partial F_s/\partial h_1$. m_1 describes the local magnetization at the surface and is experimentally accessible using LEED (low energy electron diffraction). Whereas the exponent relation $m_1 \propto t^{\beta_1}$ describes the critical behavior of the surface layer only as T_c is approached, the variation of m_s describes the total difference between the bulk magnetization and the average over the system. For a fully finite system the finite-size scaling relation then contains a surface correction term and total magnetization m of this system is given by

$$mn^{\beta/\nu} = f(x) = Bx^{\beta} + B_s x^{\beta_s} \tag{9.4}$$

where $x = tn^{1/\nu}$ with t small and $n \to \infty$. Extensive Monte Carlo data are available for a variety of fully finite Ising systems [9.3-6] and Heisenberg models [9.7-9], thin Ising films [9.2,8,10-12]; and "semi-infinite" magnets [9.2,13]. The general form for the Ising model hamiltonian is

$$\mathscr{H} = - J \sum_{\text{bulk}} s_{iz}s_{jz} - J_s \sum_{\text{surf}} s_{iz}s_{jz} - h \sum_{\text{bulk}} s_{iz} - h_1 \sum_{\text{surf}} s_{iz} \tag{9.5}$$

and the Heisenberg models differ only in that the spin-spin interactions are isotropic. Rather accurate magnetization profiles have been determined for both Ising films and the Heisenberg half-space. The results, shown in Fig.9.1, indicate that at distances z from the surface, which are large compared to the bulk correlation length ξ_b, the magnetization essentially assumes the bulk value. Near to the surface the magnetization begins to fall off exponentially towards the value which it has in the surface layer. Note that when the film becomes too thin (i.e., $L \lesssim \xi_b$) the magnetization in the interior never reaches the bulk value. In the Heisenberg half-space the behavior is qualitatively similar, however, the decrease which occurs as the surface is approached is not exponential but varies approximately as z^{-1} instead. The critical behavior of the layer magnetization in Ising films is compared with that of the bulk in Fig.9.2. In the case where the surface exchange is identical to that of the bulk the data suggest $\beta_1 \sim 2/3$ whereas for $J_s = J/2$ the value $\beta_1 = 0.86$ appears to be the best estimate. BRAY and MOORE [9.14] have used an ε-expansion approach to study a surface perturbation hamiltonian which they believe will yield the same result as for the exposed surface. Their estimate of $\beta_1 = 1/2 + \nu(d - 2)/2 = 0.82$ is somewhat larger than the unenhanced Monte Carlo value. The best estimates [9.15,16] from series expansions yield $\beta_1 \sim 1.0$ and $\beta_1 = 0.72 \pm 0.03$. These various estimates may be compared with experimental LEED data [9.17,18] which suggest that $\beta_1 \sim 0.9$. (LEED data on NiO are shown in Fig.9.2 for comparison). All treatments [9.2,14,19] suggest that with sufficiently enhanced surface exchange J_s^c a new multicritical point should appear where distinct surface and bulk transitions become simultaneously critical. Since β_1 is predicted [9.14, 19] to be quite small at the multicritical point, it would be expected that for $J_s < J_s^c$ the magnetization should exhibit crossover from multicritical to critical behavior as $T \to T_c$. This means that over a limited temperature range the "effective" value of β_1 should lie between the critical and multicritical values. Far from J_s^c, however, essentially the critical behavior will dominate. The Monte Carlo data in Fig.9.2 are then consistent with the interpretation that for weakened surface exchange the true critical behavior is observed whereas for $J_s = J$ the crossover region includes the temperatures for which data are available and only for $t \lesssim 10^{-1}$ will the behavior be purely critical. The behavior of the surface magnetization m_s was also determined for their films. For $t \lesssim 0.2$ the data are consistent with the scaling prediction $\beta_s = -0.33$.

Estimates for the location of the multicritical point differ significantly. Mean-field theory predicts $J_s^c/J = 1.25$ whereas series expansion analyses suggest $J_s^c/J \approx 1.6$ (the series are only slowly convergent and in view of the expected crossover behavior should be treated with caution). Estimates from Oguchi and constant-coupling methods [9.20] fall in between these values. Clearly careful Monte Carlo calculations probably present the best way to locate the multicritical point.

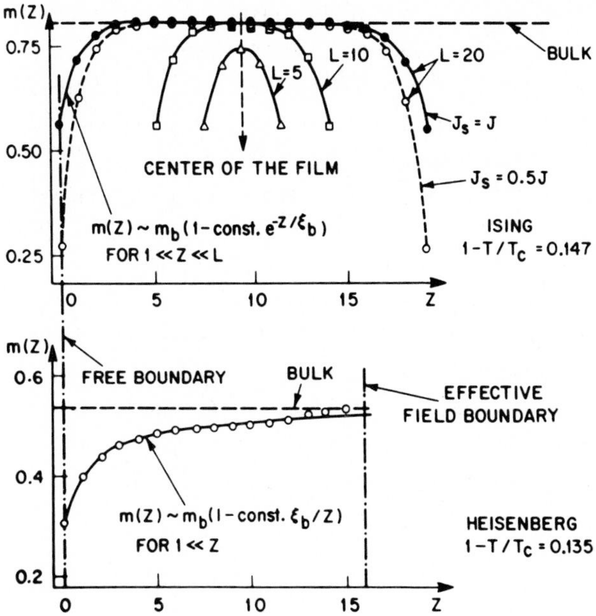

Fig.9.1 Upper portion: magnetization profiles for Ising films of thickness L where z is the distance (in lattice constants); lower portion: magnetization profile for a Heisenberg "half space". [9.11]

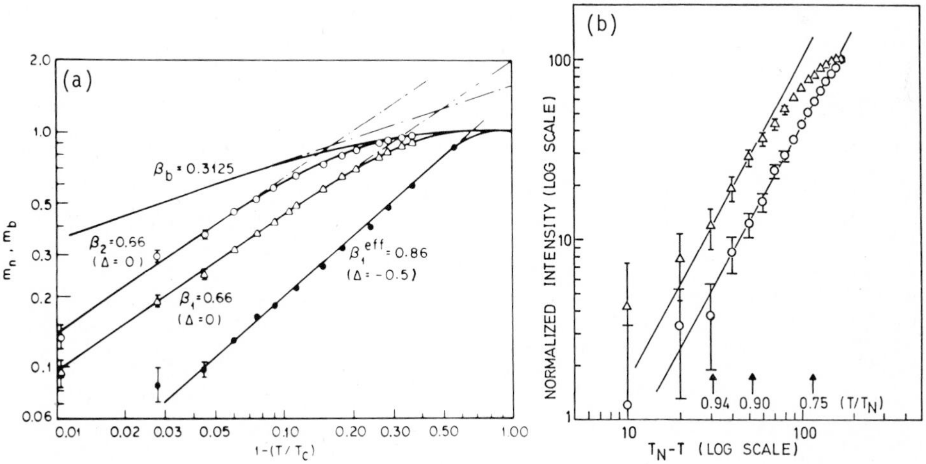

Fig.9.2 (a) Critical behavior of the layer magnetization for an Ising film with modified surface exchange. Curves labelled β_1 show surface layer data, curves labelled β_2 indicate results for the 2nd layer. The curve labelled β_b shows the critical behavior of the bulk magnetization [9.2].

(b) Experimental data (LEED) on NiO. (Note the intensity is proportional to the *square* of the surface magnetization). [9.18]

Detailed Monte Carlo data are also available for fully finite two- and three-dimensional Ising ferromagnets with free edges [9.3-6]. The thermodynamic properties of these systems are significantly different from those of the same size lattices with periodic boundary conditions (and thus having no surfaces). Examples are shown in Figs.9.3,4. By examining the difference between the measured average quantity and the bulk quantity (e.g., $\Delta m = m - Bt^{\beta}$, $\Delta\chi = \chi - Ct^{-\gamma}$) the scaling behavior of the surface contribution may be extracted, see (9.4). The data for the simple cubic Ising model are shown in Fig.9.5. Scaling appears to work quite well and the resultant exponents are clearly consistent with the predicted values $\beta_s = \beta - \nu$, $\gamma_s = \gamma + \nu$. These data have important implications for the interpretation of experimental data near critical points. The "rounding" of the phase transition due to the exposed surface is apparently much greater than that due to the finite size alone. The magnitude of the rounding is not inconsistent with what the present data suggest should occur due to grain boundaries in real systems.

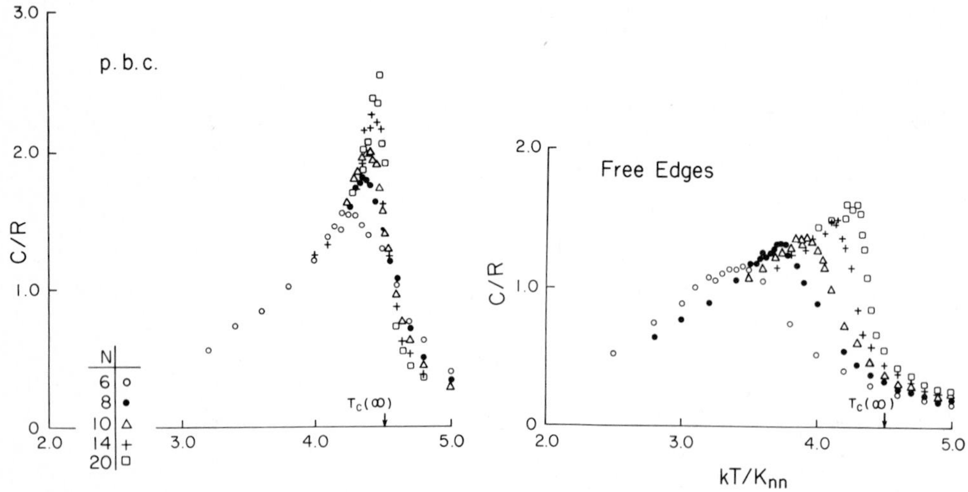

Fig.9.3 Comparison of the specific heat behavior for finite Ising simple cubic lattices with exposed surfaces (free edges) and with no surfaces (periodic boundary conditions). For comparison the infinite lattice critical temperature $T_c(\infty)$ is marked. [9.6]

Heisenberg-model studies have been carried out [9.2] using an effective-field boundary condition opposite the exposed surface and periodic boundary conditions on the sides. The resultant data yield the estimate $\beta_1 = 0.75 \pm 0.10$. Using the same method MÜLLER-KRUMBHAAR [9.13] measured the layer magnetization as a function of field along the critical isotherm and from the relation

$$m_1 = m_1^{(h)} h^{1/\delta_1} \qquad (9.6)$$

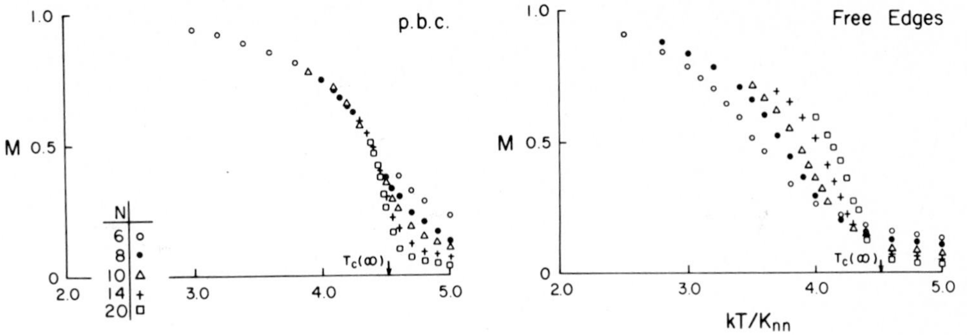

Fig.9.4 Comparison of the order parameters for finite Ising simple cubic lattices with exposed surfaces (free edges) and with no surfaces (periodic boundary conditions). For comparison the finite lattice critical temperature $T_c(\infty)$ is marked. [9.6]

extracted the estimate $\delta_1 = 2.3 \pm 0.1$. Experimental data for Nickel [9.21] yield $\delta_1 \approx 2.9$, but they are probably affected by crossover to bulk behavior since they actually refer to a mean depth of about 20 layers near the surface rather than the topmost layer alone.

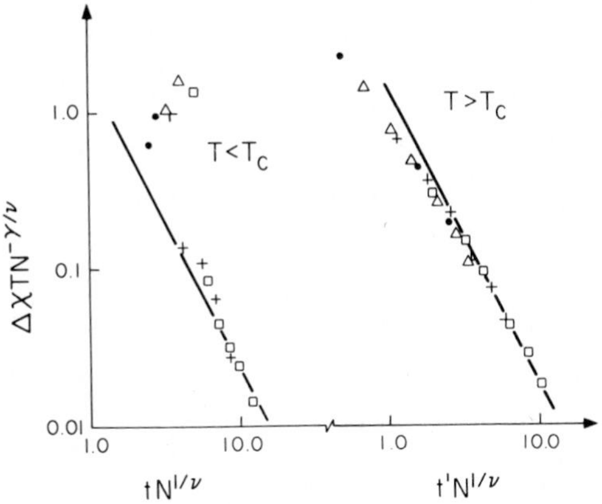

Fig.9.5 Finite-size scaling plot for the surface correction to bulk behavior of the magnetic susceptibility of a simple cubic Ising lattice with free surface. The solid lines are best fits to the date using the predicted form $C_s x^{-\gamma_s}$ where $x = tN^{1/\nu}$ and where $\gamma_s = \gamma + \nu$. [9.6]

9.3 Surface Effects in Binary Alloys

9.3.1 Surface Enrichment

The variation in composition near the surface of a binary alloy has been observed in numerous systems. The surface "enrichment" is a phenomenon which is well suited to Monte Carlo study in that a simple lattice (Ising) model is appropriate. Neglecting elastic interactions due to the lattice "misfit" of the constituents the hamiltonian of a binary alloy can be expressed in terms of local site occupation variables c_i^A, c_i^B (where $c_i^A = 1$, $c_i^B = 0$ if the ith site is occupied by an A atom and $c_i^A = 0$, $c_i^B = 1$ if it is occupied by a B atom)

$$\mathcal{H} = \sum_{i,j} [\phi_{AA}(\bar{r}_i,\bar{r}_j)c_i^A c_j^A + 2\phi_{AB}(\bar{r}_i,\bar{r}_j)c_i^A c_j^B + \phi_{BB}(\bar{r}_i,\bar{r}_j)c_i^B c_j^B] - \sum_i [c_i^A \mu_A(\bar{r}_i) + c_i^B \mu_B(\bar{r}_j)] + \mathcal{H}_0 \tag{9.7}$$

where the ϕ's are pair potentials, μ_A,μ_B are chemical potentials and $\mathcal{H}_0$ represents other degrees of freedom. (In most cases the interactions are restricted to a few near neighbors). A transformation to spin variables $\sigma_i = \pm 1$ is easily carried out

$$c_i^A = (1 + \sigma_i)/2 \tag{9.8a}$$

$$c_i^B = (1 - \sigma_i)/2 \quad . \tag{9.8b}$$

The hamiltonian may then be rewritten in the same form as that of the Ising model in a magnetic field

$$\mathcal{H} = \sum_{(i,j)} J\sigma_i\sigma_j - H\sum_i \sigma_i - H_1 \sum_j \sigma_j + \mathcal{H}_0 \tag{9.9}$$

where $\sigma_i,\sigma_j = \pm 1$ and J, H, and H_1 are suitable combinations of the pair potentials and chemical potentials of (9.7). In this form the problem is clearly amenable to computer simulation using the same technique as applied to problems in magnetism. An alternative approach is to fix the concentration, i.e., $\langle\sigma_i\rangle$, and use a pairwise exchange method which conserves $\langle\sigma_i\rangle$ at each step. Although the pairwise exchange method describes the dynamics more realistically, it also suffers from long relaxation time effects as described in Chap.1. Both approaches have been used to study the static behavior of alloys.

Monte Carlo calculations were carried out [9.22] on such a lattice model for Cu_3Au with particular emphasis on the surface behavior so that comparison could be made with the experimental results from a LEED investigation. The calculations were carried out on a cubic section of an fcc lattice with periodic boundary conditions in the horizontal direction and the bottom two layers fixed so as to have the bulk disorder [9.23]. In these studies the chemical potentials were adjusted so as to maintain the correct bulk composition. Since the experimental results showed no surface segregation the surface field was set H_1 - 2.5J, the resultant behavior of the surface order parameter is compared with experimental results in Fig.9.6.

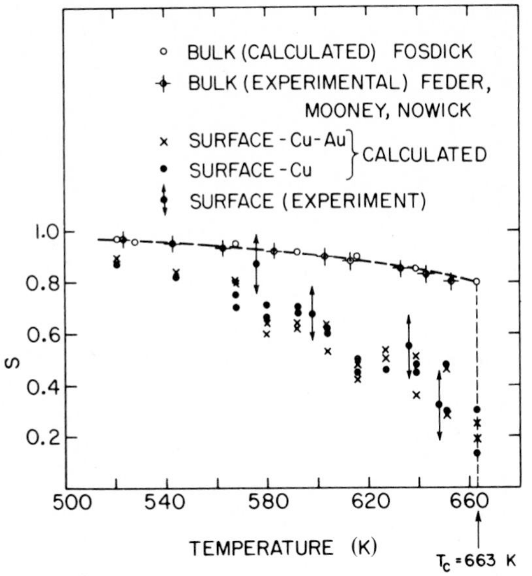

Fig.9.6 Measured and calculated order parameters for the {100} surface of Cu_3Au. Closed circles without arrows correspond to model boundary conditions extrapolating to a Cu surface; crosses to a Cu-Au surface. Open circles give calculated values for the temperature dependence of the order parameter in the bulk. The experimental results for the surface order parameter are shown by the closed circles with error bars (arrows). [9.22]

Monte Carlo calculations have also been used to study the problem of surface segregation directly [9.24]. In this study a pairwise interchange of atoms was used instead of the "spin-flip" method. Calculations were made for semi-infinite solids, thin films and small particles (cubo-octahedral solids). The results show that the alloys with the lowest surface energy tend to segregate to the first few surface layers. In alloys where $J < 0$ like atoms tend to group together ("clustering" alloys) the segregation is more pronounced than in $J > 0$ ("ordering" alloys). Studies also show that in {110} surfaces the effect is more pronounced than in {100} surfaces. Both grains and thin films showed similar behavior on a smaller scale (due to mass conservation effects).

In Fig.9.7 we show a comparison of Monte Carlo data [9.24] with Bragg-Williams calculations and experimental data [9.25] for surface segregation in Au-Ni alloys. The general agreement between experiment and theory in alloys which tend to order is not good. The Monte Carlo results, however, show a strong tendency to composition modulation near the surface and it is therefore possible that experimental probes such as AES (Auger electron spectroscopy) may not be able to detect the true surface composition.

9.3.2 Surface Critical Phenomena

Monte Carlo studies near phase transitions have also been carried out [9.26]. New surface exponents can be defined in such a way as to maintain the analogy with the Ising model. Data obtained on $30 \times 30 \times 10$ thin films yield the estimate $\gamma_1 \simeq 0.93$. These latter studies allow surface irregularities by allowing for the presence of vacancies. Using the same formalism as in (9.6) we write the new hamiltonian

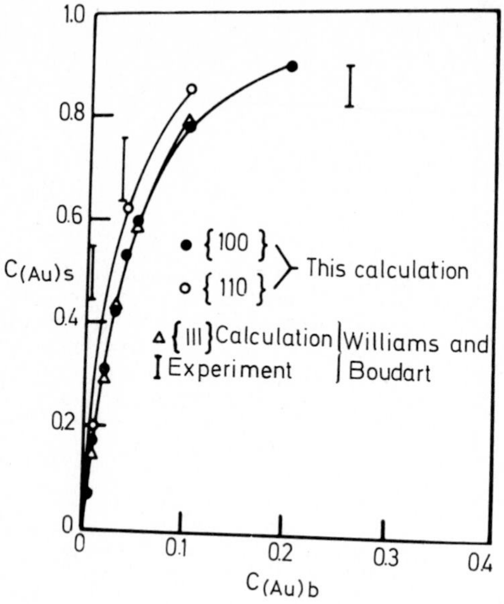

Fig.9.7 Calculated surface layer composition vs. bulk composition for a lattice gas model for AuNi at T = 1000 K with ΔH=15 kcal, V_1= 200 cal/mole and V_2/V_1 = 0.5 [9.24]

$$\begin{aligned}\mathscr{H} = \sum_{(i,j)} \Big[& c_i^A c_j^A \phi_{AA}(\bar{r}_i,\bar{r}_j) + 2c_i^A c_j^B \phi_{AB}(\bar{r}_i,\bar{r}_j) + 2c_i^A c_j^V \phi_{AV}(\bar{r}_i,\bar{r}_j) \\ & + c_i^B c_j^B \phi_{BB}(\bar{r}_i,\bar{r}_j) + 2c_i^B c_j^V \phi_{BV}(\bar{r}_i,\bar{r}_j) + c_i^V c_j^V \phi_{VV}(\bar{r}_i,\bar{r}_j)\Big] \\ & + \sum \Big[c_i^A \mu_A(\bar{r}_i) + c_i^B \mu_B(\bar{r}_i) + c_i^V \mu_V(\bar{r}_i)\Big] + \mathscr{H}_0 \quad .\end{aligned} \tag{9.10}$$

This hamiltonian can be easily transformed into a spin-1 representation with

$$c_i^A = s_i(s_i + 1)/2 \tag{9.11a}$$

$$c_i^B = s_i(s_i - 1)/2 \tag{9.11b}$$

$$c_i^V = 1 - s_i^2 \quad . \tag{9.11c}$$

The resultant hamiltonian is quite complex but a simplification of it gives the BLUME-CAPEL model [9.27]

$$\mathscr{H} = J \sum s_i s_j + \Delta \sum s_i^2 - H \sum s_i \quad . \tag{9.12}$$

BINDER et al. [9.26] have studied spherical particles of a simple cubic Blume-Capel model for several different values of Δ. Results for the concentration and density profiles are shown in Fig.9.8. For $\Delta = -\infty$ the energy for producing vacancies is infinite but as Δ becomes less negative vacancies begin to occur with increasing probability. The profiles shown in Fig.9.8 show that the presence of vacancies tends to enhance surface enrichment.

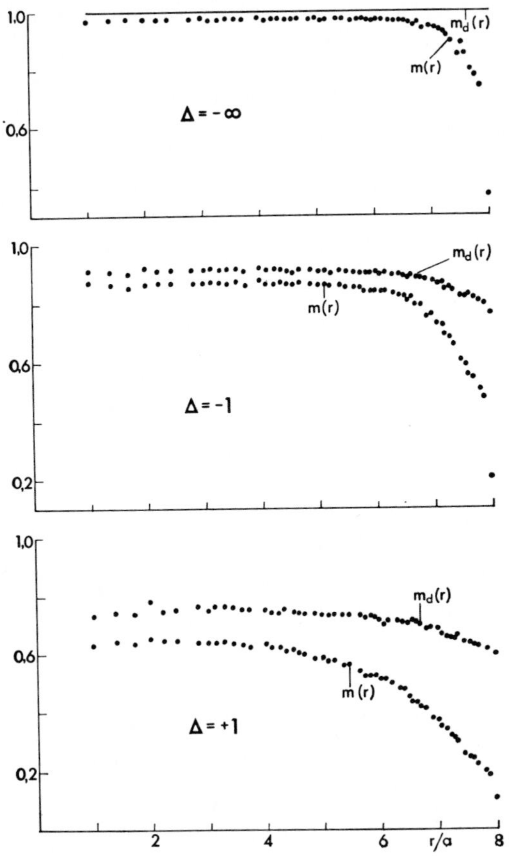

Fig.9.8 Calculated profiles of the order parameter $m(r)$ and density $m_d(r)$ of a pseudospherical particle of radius $R = 8$ lattice spacing. Data are from Monte Carlo calculations on a simple cubic Blume-Capel model with nearest-neighbor interactions at $T_{cb}(\Delta=-\infty)/T = 1.75$ and three values of Δ. [9.26]

9.4 Phase Transitions in Adsorbed Surface Layers

9.4.1 Lattice Gas Models

One simple approach to the study of adatom deposition on a substrate has been to consider a model in which atoms of the adsorbed species may occupy only positions specified by the bottom of the potential wells of the periodic potential provided by the substrate materials. All of the possible occupation sites thus form a two-dimensional periodic lattice where each site is labelled with index i. We may then introduce a local occupation variable c_i where $c_i = +1$ if the site is occupied by an adatom and $c_i = 0$ if the site is empty. The coverage of the surface θ is then given by $\theta = \langle c_i \rangle$. At each lattice site i there will be a binding energy ε_i between the adatom and the substrate. In addition there will be interactions $\phi(\bar{r}_i - \bar{r}_j)$ between adatoms on different sites. The hamiltonian for this model is

$$\mathscr{H} = \sum_{(i,j)} c_i c_j \phi(\bar{r}_i - \bar{r}_j) + \sum_i \varepsilon_i c_i + \mathscr{H}_0 \tag{9.13}$$

where $\mathscr{H}_0$ includes all other contributions due to other degrees of freedom. If we now carry out the transformation

$$c_i = (1 + s_i)/2 \tag{9.14}$$

where $s_i = \pm 1/2$ the hamiltonian is transferred into that appropriate for an Ising model

$$\mathscr{H} = -\sum_{(ij)} J_{ij} s_i s_j - \sum_i H_i s_i + \mathscr{H}_0 \tag{9.15}$$

with

$$J_{ij} = -\frac{1}{4}\phi(\bar{r}_i - \bar{r}_j) \tag{9.16a}$$

$$H_i = -\frac{1}{2}\left[\varepsilon_i + \sum_{j \neq 1} \phi(\bar{r}_i - \bar{r}_j)\right] \quad . \tag{9.16b}$$

The very earliest simulations [9.28] of lattice gas model adsorption were carried out in hopes of explaining experimental data for tungsten adatoms being adsorbed on a low temperature tungsten substrate [9.29]. Atoms were allowed first to rain down at randomly chosen lattice sites. If the site was occupied the atom was "deflected" to a nearest-neighbor site. If sufficient first-layer sites were occupied, the atom would be adsorbed in the second layer. In addition to this simple "no-jump" calculation, a second calculation was made in which each adatom was allowed to make two random jumps following adsorption. The no-jump calculation indicated that after 440 atoms had been adsorbed about 28% were in the 2nd or 3rd layer whereas only 4% were missing from a similar two-jump simulation. A comparison with the experimental data showed the no-jump results to be in better agreement. This study did, of course, neglect any thermally induced motion of adatoms on the surface and would therefore not be suitable for predicting the surface atom arrangement after thermal equilibrium had been reached following rearrangement.

A more detailed study (including thermal rearrangement) was carried out by GORDON [9.30] for monolayer adsorption onto a 10×10 hexagonal close-packed planar lattice with nearest-neighbor interactions. Both adsorption and desorption were allowed for and the fraction of occupied sites at equilibrium was computed as a function of chemical potential μ. The results showed a clear and distinct increase in the fractional coverage as μ/kT was increased but both fluctuations as well as the small surface size make it impossible to interpret the results quantitatively. There are now a number of experimental studies of order-disorder transitions in adsorbed surface layers. (Extensive references to investigations in chemisorbed systems are given in [9.31]).

The actual arrangements of the adatoms on the substrate were first studied by the Monte Carlo method [9.32] using the pairwise exchange method. Equilibrium configurations for given coverages were determined and the LEED scattering intensity from such a configuration was then determined in k-space. Appropriate parameters were chosen to represent the adsorption of O_2 on Pd(110). For small values of θ the resultant pattern shows maxima at $k = 1/3$, $2/3$ corresponding to a 1×3 ordered structure. For $1/3 < \theta < 1/2$ the scattering peaks move continuously together and for $\theta = 1/2$ only one peak is seen at $k = 1/2$, typical of a 1×2 structure. Smoothed calculated scattering patterns are shown in Fig.9.9 for several coverages. These results agree with experiment [9.33] even to the increasing coverage. The values of the interaction constants were adjusted so as to provide the best agreement with the experimental data.

More recently order-disorder transitions in adsorbed layers have been studied by both "pairwise exchange" [9.34] and by "spin-flip" [9.35-38] methods. The "pairwise exchange" results [9.34] were for a $\theta = 1/2$ square lattice gas model only; however, more recent "spin-flip" studies [9.35] have shown how the model may be extended to all θ values by using nonzero field, see (9.15). These studies have included next-nearest neighbor interactions and show that this may produce a wide variety of pure and mixed ordered phases. For nearest-neighbor coupling alone on the square lattice only a single pure ordered phase is possible. If attractive nnn-(next-nearest neighbor) interactions are present, coexistence regions involving the ordered state and the liquid-gas phase appear. If the nnn-coupling is repulsive two new pure phases appear along with multiple coexistence regions, as shown in Fig.9.10 [9.39]. In fact, molecular orbital calculations [9.40] of the indirect interaction between adatom pairs on a (100) surface of a simple-cubic tight-binding solid suggest that the nearest-neighbor coupling is generally repulsive with smaller, attractive coupling between more distant neighbors.

Similar studies have been carried out for a triangular lattice which corresponds to the surface of graphite for which many experimental studies have been made [9.41, 42]. Because of the large "effective area" of the gas atoms, occupation of nnn-sites is very unfavorable but an ordered state in which nnn-sites are filled has been observed experimentally. An appropriate lattice gas model is thus one with repulsive nn-interactions and attractive nnn-coupling. The simulations were carried out for

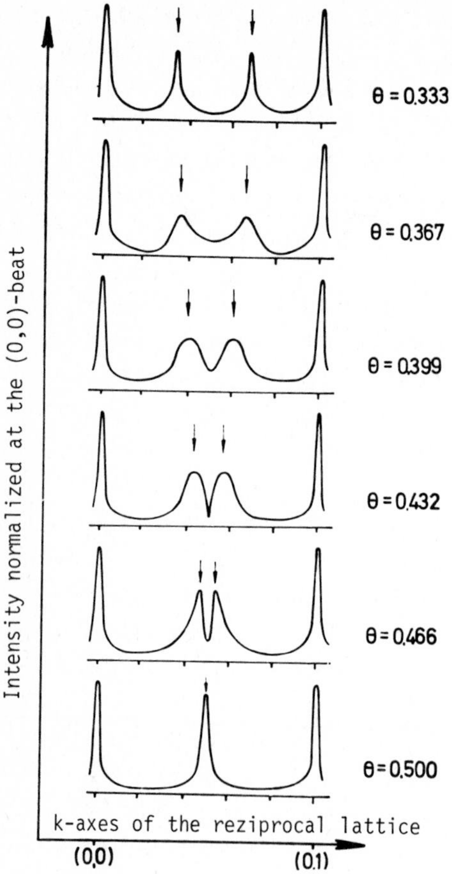

Fig.9.9 Calculated LEED intensity pattern [angular distribution between the (0,0) and (0,1) reflections] for a square lattice surface. Patterns are determined from the equilibrium distribution of adsorbed atoms (as determined from Monte Carlo calculations for various values of the fractional coverage θ. [9.32]

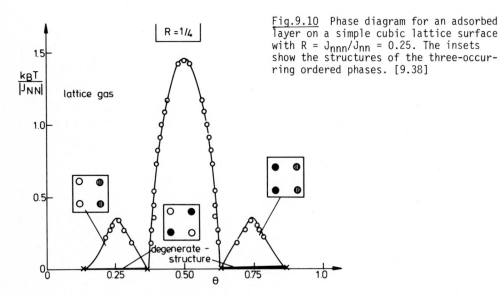

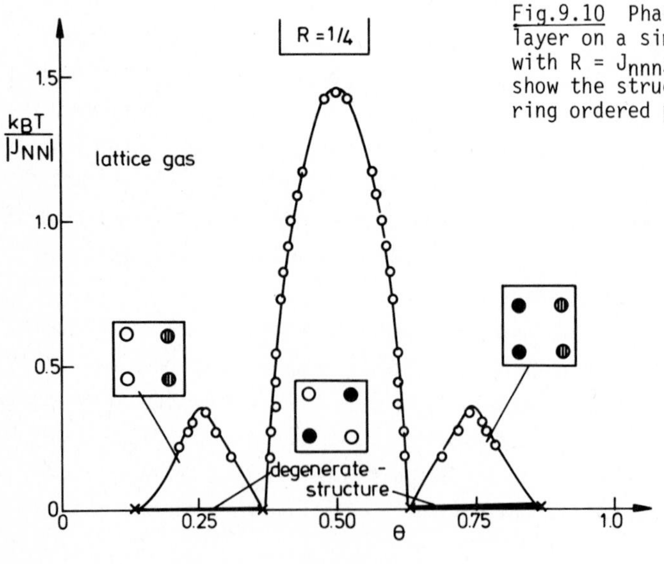

Fig.9.10 Phase diagram for an adsorbed layer on a simple cubic lattice surface with $R = J_{nnn}/J_{nn} = 0.25$. The insets show the structures of the three-occurring ordered phases. [9.38]

$J_{NN}/J_{NNN} = -1$ on rhombohedral shaped lattices as large as 60 × 60 with periodic boundary conditions. The results differ *qualitatively* from those found earlier by the Bethe-Peierls-Weiss method [9.43]. As a function of chemical potential two different ordered phases occur, one with 1/3 the sites occupied and one with 2/3 the sites filled. The phase diagram in coverage temperature space is shown in Fig.9.11. Portions of this diagram have apparently been observed experimentally [9.42] but for the simple systems studied so far, e.g., He or Ar on graphite, the ordered adsorbed layer goes out of registry with the substrate before a coverage of 1/2 is reached. Recent theoretical work [9.43-45] has predicted that phase transitions in a variety of different universality classes should be experimentally observable in common adsorbed systems.

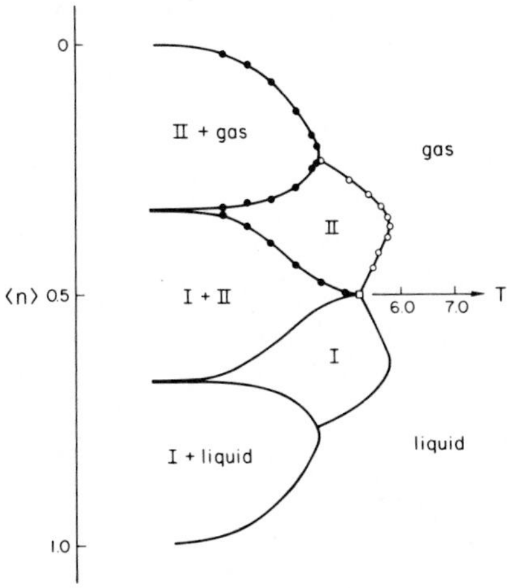

Fig.9.11 Phase diagram in density (coverage) - temperature space for an adsorbed surface layer on a triangular lattice. Data points are shown for the upper portion of the diagram only (second-order points are open, first-order points are filled). [9.36]

Very recent studies [9.37] have been made on a square lattice gas model for dissociative adsorption of CO on Mo(100). A spin-1 system was used to allow for the possibility of a site being empty, occupied by a C atom, or occupied by an O atom. With nn-coupling repulsive (between any atoms) and nnn-coupling attractive for unlike atoms an ordered state results which is identical to that observed experimentally. A higher density state also appears and the phase diagram appears to have both a bicritical as well as a tricritical point. Similar two-species ordered structures may occur in other coadsorbed systems, e.g., H_2 and CO on Pd(110) [9.46, 47].

9.4.2 Continuum Models

A valuable alternative approach to the adsorbed layer problem has been taken by ROWLEY et al. [9.48,49]. They have simulated the behavior of a collection of particles (adatoms) which may move continuously within a given volume above a substrate. The particles interact with each other via a Lennard-Jones 12-6 potential and are attracted to the (continuous) substrate by a 9-3 potential which depends only on the distance of a particle from the surface. Because of the rather large volume available to the particles the system was treated within the grand canonic ensemble and particles could be created and destroyed as well as moved. Typically between 35 and 130 particles were present. The resultant data showed a definite tendency for well defined layers to deposit themselves, one upon the other with essentially constant separation on the substrate. Data for the pair distribution functions and fractional occupancies suggest a solid-like state in the lowest adsorbed layers at a low temperature and a more liquid-like configuration at a higher temperature. Since the substrate potential shows no periodicity, these data cannot provide explicit information regarding the behavior of real adsorbed layers on a specific substrate.

9.5 Kinetic Phenomena at Surfaces

Several different kinetic processes which occur in surface layers are well suited for study by computer simulation. Heterogeneous nucleation on solid surfaces has been studied by a number of authors [9.50-63], largely in connection with crystal growth. Our interest here is limited to the condensation, evaporation, and migration which occur in a single surface layer. (The broader consideration of surface roughening and crystal growth will be discussed in detail in Chap.7). Because of the complex nature of this problem, most studies have been carried out using simple Kossel models in which atoms are constrained to lie on a simple lattice with no vacancies or over-hangs allowed (the "solid-on-solid" approximation). Adatoms are bound to the surface with uniform energy difference $\Delta\mu$ between the solid as opposed to "liquid" or "vapor" states. In addition, the adatoms interact with each other via near-neighbor pairwise interactions $\phi(\bar{r}_i,\bar{r}_j)$ where the range of the interaction $r = \bar{r}_i - \bar{r}_j$ is generally truncated at the nearest-or next-nearest neighbor spacing. (For realistic interactions this is quite a reasonable approximation and certainly makes the computations more tractable). Atoms are allowed to deposit on the surface at a constant rate and evaporation and migration rates are determined by the relative energies of the different possible states. A detailed study of heterogeneous nucleation under conditions of high supersaturation ($\Delta\mu/kT$ large) was carried out by MICHAELS et al. [9.53] on square lattice surfaces as large as 20×20 with periodic boundary conditions. The adsorption rate was arbitrarily chosen to produce nucleation at a relatively small cluster size and in a reasonable amount of computing time. The simulation data are in good agreement with the cluster size distribution predicted by the atomistic theory of Walton-Rhodin for heterogeneous nucleation. Similar studies by ADAMS and JACKSON [9.52] have examined the time dependence of cluster concentration for a range of (small) cluster sizes. Ideally these simulations need to be repeated on much larger lattices and with models chosen to be more appropriate to specific real, physical systems.

Surface diffusion of adatoms is now amenable to direct study using a field ion microscope. In the standard representation the diffusion coefficient is

$$D = D_0 \exp(-V/kT) \tag{9.17}$$

where V is the activation energy. The low temperature variation of D has been calculated using a zeroth-order (Bragg-Williams) approximation [9.54], Monte Carlo data should provide substantially more realistic information over a much wider temperature range. Existing Monte Carlo studies however, concentrate on the effects of diffusion on nucleation and not on the nature of the diffusion itself.

9.6 Conclusions

Sufficient results of Monte Carlo calculations now exist to demonstrate the utility of the method to studies of surface behavior. We do wish to emphasize, however, that Monte Carlo studies have only dealt with a small fraction of the types of surface behavior to which they are applicable so that future work in this area is quite promising. Analytical approaches to these problems are often cumbersome due to the anisotropy which the surface introduces. Moreover, for 2-dimensional systems approximate techniques are often less reliable than in 3-dimensional. It would be quite straightforward to include a broader range of interactions in the Monte Carlo studies including new features, such as surface roughness, which may be very important to the description of behavior of real systems. Additional simulation results could stimulate further experimental work using new, sophisticated techniques (LEED, AES, neutron scattering, etc.) and the interplay between real experiments and Monte Carlo "experiments" may well accelerate our progress in understanding a broad range of surface phenomena.

References

9.1 Extensive details and references to the application of the Monte Carlo method to the study of bulk phase transitions are given in Chap.3

9.2 See K. Binder, P.C. Hohenberg: Phys. Rev. B *6*, 3461 (1972); B *9*, 2194 (1974) and references contained therein

9.3 K. Binder: Physica *C2*, 508 (1972)

9.4 D.P. Landau: AIP Conf. Proc. *24*, 304 (1975)

9.5 D.P. Landau: Phys. Rev. B *13*, 2997 (1976)

9.6 D.P. Landau: Phys. Rev. B *14*, 255 (1976)

9.7 K. Binder, H. Rauch, V. Wildplaner:J. Phys. Chem. Sol. *31*, 391 (1970)

9.8 K. Binder, H. Rauch, V. Wildpaner: In *Proc. Intern. Conf. of Magnetism VI* (Nawka, Moscow 1974)

9.9 V. Wildpaner: Z. Phys. *270*, 215 (1974)

9.10 K. Binder: Thin Solid Films *20*, 367 (1974)

9.11 P.C. Hohenberg, K. Binder: AIP Conf. Proc. *24*, 300 (1975)

9.12 K. Binder: Proc. 7th Intern. Coll. on Magnetic Films (Regensburg 1975)

9.13 H. Müller-Krumbhaar: J. Phys. C *9*, 345 (1976)

9.14 A.J. Bray, M.A. Moore: J. Phys. A *10*, 1927 (1977)

9.15 M.N. Barber: Phys. Rev. B *8*, 407 (1973)

9.16 M.N. Barber: J. Phys. C *6*, L262 (1973)

9.17 T. Wolfram, R.E. DeWames, W.F. Hall, P.W. Palmberg: Surface Sci. *28*, 45 (1971)

9.18 K. Namikawa: Tech. Rpt. of ISSP #A805 (1977)

9.19 T.W. Burkhardt, E. Eisenriegler: Phys. Rev. B *16*, 3213 (1977)

9.20 H.H. Chen: J. Phys. C *9*, 2395 (1976)

9.21 G. Schwarzl: Thesis, University of Regensburg (1976)

9.22 V.S. Sundaram, B. Farrell, R.S. Alben, W.D. Robertson: Phys. Rev. Lett. *31*, 1136 (1973)

9.23 R. Feder, M. Mooney, A.S. Nowick: Acta Met. *6*, 266 (1958)

9.24 V.S. Sundaram, P. Wynblatt: Surface Sci. *52*, 569 (1975); In *Proc. 1976 Intern. Conf. on Computer Simulations for Materials Applications, Nuclear Metallurgy 20*, ed. by R.J. Arsenault, J.R. Beeler, Jr., J.H. Simmons (1976) p. 143

9.25 F.L. Williams, M. Boudart: J. Catalysis *30*, 438 (1973)

9.26 K. Binder, D. Stauffer, V. Wildpaner: Acta Met. *23*, 1191 (1975)

9.27 M. Blume: Phys. Rev. *141*, 517 (1966)
H.W. Capel: Physica *32*, 966 (1966)

9.28 R.D. Young, D.C. Schubert: J. Chem. Phys. *42*, 3943 (1965)

9.29 T. Gurney, Jr., F. Hutchinson, R.D. Young: J. Chem. Phys. *42*, 3939 (1965)

9.30 R. Gordon: J. Chem. Phys. *48*, 1408 (1968)

9.31 M.G. Lagally, G.C. Wang, T.M. Lu: CRC Rev. Solid State Sci. (to be published)

9.32 G. Ertl, J. Küppers: Surface Sci. *21*, 61 (1970)

9.33 G. Ertl, P. Rau: Surface Sci. *15*, 443 (1969)

9.34 G. Doyen, G. Ertl, M. Plancher: J. Chem. Phys. *62*, 2957 (1975)

9.35 K. Binder, D.P. Landau: Surface Sci. *61*, 577 (1976)

9.36 B. Mihura, D.P. Landau: Phys. Rev. Lett. *38*, 977 (1977)

9.37 H.H. Lee, D.P. Landau: Bull. Am. Phys. Soc. *22*, 498 (1977)

9.38 K. Binder, D.P. Landau: In *Proc. 7th Intern. Vac. Congr. and 3rd Intern. Conf. Solid Surfaces* (Vienna, 1977), Vol. 1 (Berger and Sons, Vienna 1977) p. 811; Phys. Rev. B*21*, 1941 (1980)

9.39 Earlier versions of this diagram (see 9.35,38) were marred by an extrapolation of the Monte Carlo data to incorrect T = 0 limits

9.40 T.L. Einstein, J.R. Schrieffer: Phys. Rev. B *7*, 3629 (1973)

9.41 See, for example, M. Bretz, J.G. Dash, D.C. Hickernell, E.O. McLean, O.E. Vilches: Phys. Rev. A *8*, 1589 (1973)
J.K. Kjems, L. Passell, H. Taub, J.G. Dash, A.D. Novaco: Phys. Rev. B *13*, 1446 (1976)
J.G. Dash: *Films on Solid Surfaces* (Academic Press, New York 1975)

9.42 D. Butler, J. Litzinger, G.A. Stewart: Unpublished
F.A. Putnam, T. Fort, Jr.: J. Phys. Chem. *79*, 549 (1975)
A. Thomy, X. Duval: J. Chem. Phys. *67*, 1101 (1970)
9.43 C.E. Campbell, M. Schick: Phys. Rev. A *5*, 1919 (1972)
9.44 E. Domany, M. Schick, J.S. Walker: Phys. Rev. Lett. *38*, 1148 (1977);
E. Domany, M. Schick, J.S. Walker, R.B. Griffiths: Phys. Rev. B*18*, 2209 (1978)
9.45 E. Domany, E.K. Riedel: Phys. Rev. Lett. *40*, 561 (1978)
9.46 H. Conrad, G. Ertl, E.E. Latta: J. Catalysis *35*, 363 (1974)
9.47 T. Felter, P.J. Estrup: Surface Sci. *54*, 179 (1976)
9.48 L.A. Rowley, D. Nicholson, N.G. Parsonage: Mol. Phys. *31*, 365 (1976)
9.49 L.A. Rowley, D. Nicholson, N.G. Parsonage: Mol. Phys. *31*, 389 (1976)
9.50 F.F. Abraham, G.M. White: J. Appl. Phys. *11*, 1841 (1970)
9.51 G.H. Gilmer, P. Bennema: J. Appl. Phys. *43*, 1347 (1972)
9.52 A.C. Adams, K.A. Jackson: J. Cryst. Growth *13/14*, 144 (1972)
9.53 A.I. Michaels, G.M. Pound, F.F. Abraham: J. Appl. Phys. *45*, 9 (1974)
9.54 H.J. Leamy, K.A. Jackson: J. Cryst. Growth *13/14*, 140 (1972)
9.55 A.C. Adams, K.A. Jackson: J. Cryst. Growth *13/14*, 144 (1972)
9.56 C. van Leeuwen, J.P. van der Eerden: Surface Sci. *64*, 237 (1977)
9.57 F.L. Binsbergen: J. Cryst. Growth *13/14*, 44 (1972)
9.58 J.P. van der Eerden, C. van Leeuwen, P. Bennema, W.L. van der Kruk, B.P.Th. Veltman: J. Appl. Phys. *48*, 2124 (1977)
9.59 K. Binder: Advan. Colloid Interface Sci. *7*, 279 (1977)
9.60 U. Bertocci: Surface Sci. *15*, 286 (1969)
9.61 A.A. Chernov, J. Lewis: J. Phys. Chem. Sol. *28*, 2185 (1967)
9.62 G. Ayrault, G. Ehrlich: J. Chem. Phys. *60*, 281 (1974)
9.63 J.P. van der Eerden, R.L. Kalf, C. van Leeuwen: J. Cryst. Growth *35*, 241 (1976)

Addendum

Additional work [9.64] has now been completed for the simulation of thin film deposition on a Kossel model substrate. In addition WILLIAMS et al. [9.65] have simulated the order-disorder transition of oxygen on a W(110) surface using pairwise interactions. CHING et al. [9.66] have extended this study to larger coverages and have included three-particle interactions in order to produce the experimentally observed asymmetry about 50% coverage.

9.64 D. Kashchiev, J.P. Van der Eerden, C. Van Leeuwen: J. Cryst. Growth *40*, 47 (1977)
9.65 E.D. Williams, S.L. Cunningham, W.H. Weinberg: J. Vac. Sci. Technol. *15*, 417 (1978); J. Chem. Phys. *68*, 4688 (1978)
9.66 W.Y. Ching, D.L. Huber, M.G. Lagally, G.-C. Wang: Surf. Sci. *77*, 550 (1978)

10. Recent Trends in the Development and Application of the Monte Carlo Method

K. Binder

With 6 Figures

This chapter tries to summarize some of the general progress which has been made during the last decade in the understanding of the Monte Carlo method in statistical physics and in its efficient implementation; but it will also point out some problems which still remain to be solved. In fact, there has been an enormously growing interest in the use of this method, and even a brief review of its recent applications fills a whole book [10.1]; thus, the present chapter will not attempt to cover these applications in complete detail, but rather attention will be focused on a few highlights which serve to illustrate some general points.

There are two general lines of development:

i) The Monte Carlo method is used as a successful tool for more and more branches of physics and chemistry; there are now areas of application where its results are well-established and accepted but were simply not in existence a few years ago, such as applications to lattice gauge theory [10.2].

ii) Many problems of interest could only be treated qualitatively due to the restricted accuracy of the Monte Carlo "data" available a couple of years ago. For several of these problems a quantitative and reliable analysis is possible now. This progress is due to both an enormous progress with respect to the efficiency of computers in general, and the development of implementations of Monte Carlo programs which execute much faster than previously. The good statistical quality of such Monte Carlo results allows, in turn, to use techniques of analysis which have not been accessible earlier.

In Sect.10.1, we shall start with a discussion of what has now been achieved in increasing the speed of Monte Carlo programs. Section 10.2 contains some comments on our present understanding of finite-size phenomena and finite-size scaling; although this point has been made many times by the present author [10.3-6], it is said again that a proper consideration of finite-size effects is a prerequisite of any credible simulation analysis. In fact, the theory of finite-size effects on phase transitions is still an active area of research [10.7-14]. In Sect.10.3 we turn to a very brief discussion of new fields now very actively studied by Monte Carlo techniques such as quantum chromodynamics [10.2], irreversible kinetic growth and aggregation phenomena [10.15], the Monte-Carlo renormalization group, etc. Then Sect.10.4 will contain some comments on Monte Carlo simulations of quantum mechanical problems, which is a particular active area of research (recent reviews can be found in [10.16-18]). Finally, Sect.10.5 will contain a few conclusions, as well as speculations on the future directions of research using Monte Carlo methods.

10.1 Performance of Monte Carlo Programs

10.1.1 General Comments

Of course, sampling of thermal fluctuations in a many-body system by the standard Metropolis [10.19] Monte Carlo technique [10.1-5] implies that the same basic computational step, the updating of a selected single degree of freedom of the system according to some transition probability, is repeated again and again, an enormous number of times. Thus, Monte Carlo programs are rather time-consuming, and, in fact, this is the main drawback of the method under consideration. This problem can get less severe due to several developments which we list in order of their importance.

i) The general increase of speed of general purpose computers. In fact, there has been an increase of about one order of magnitude of most scientific large general-purpose machines produced by the main computer manufacturers during the last decade. Also the access to these machines has become much easier—a terminal on the desk of the researcher no longer is an exception.

ii) The increase of speed due to the use of vector processors or array processors. At such machines, considerable speed up is gained because the above basic computational steps can be executed in an overlapping manner (pipeline architecture of the vector processor) or in a truly parallel fashion.

However, it is nontrivial to actually achieve this gain—mere use of a program which previously ran on a general-purpose machine with marginal changes (use of another pseudorandom number generator, etc.) will not work; there is need to 'vectorize' the program in order to make full use of the computational power of these machines.

It depends very much on the type of physical problem whether this can be done with little effort or only with large effort. There is also no general formula to compute the gain factor in speed in comparison with a non-vectorizable program on the same machine; typically, the gain factor is of the order of 10.

iii) The availability of machines where CPU time is orders of magnitude cheaper than on large general-purpose computers. For problems with minor storage requirements such as the simulation of an Ising model, one even may use microcomputers, at least for rough explorative studies [10.20,21]. Here, the advantage is that—using a personal computer for such a purpose—hundreds of hours of CPU time needs not be a problem, the computer can run right in the office of the researcher, etc. A related point is the convenience and ease obtained for developing and testing programs.

Of course, the personal computer is not a solution for problems which need either large storage already for the basic Monte Carlo step, or an excessive number of Monte Carlo steps per degree of freedom to achieve sufficient statistical accuracy, or both. But here an interesting alternative is the construction of special-purpose Monte Carlo processors [10.22-25] (note that related special purpose machines for molecular dynamics simulations do exist [10.26] and have produced useful results on the problem of two-dimensional melting [10.27]).

At this point, we must emphasize that much of what has been said to point (iii) is credit for the future: so far, this author is not aware of many important problems of actual physics research, which have been solved by doing Monte Carlo calculations on microcomputers – papers such as [10.21] are rather feasibility studies.

10.1.2 Monte Carlo Calculations on Special-Purpose Computers

So far, the experience with special-purpose computers is somewhat mixed. The first construction of an Ising-model machine was started at Delft [10.22]. In 1983 when the machine started to become available for applications, its hardware was already somewhat dated, however. Thus, it is not too surprising that its speed (1.5×10^6 Monte Carlo steps/second) is only marginally larger than speeds reported for standard general-purpose computers, if multi-spin coding [10.28] is used: e.g., KALLE and WINKELMANN [10.29] reported 1×10^6 Monte Carlo steps per second already in 1982 – and with vector processors substantially larger speeds are possible, as will be explained below. Nevertheless, the Delft Ising special-purpose computer should be a rather useful machine, superior at least to such studies of Ising problems which are run on general-purpose machines and, for the sake of simplicity, do not even use multispin-coding, thus with updating rates of, e.g., 1.2×10^5 Monte Carlo steps/second only (including, however, full analysis of the generated spin configurations). Such applications have yielded results of great physical interest recently [10.30]. Although the Delft machine can be used for studying a large space of different interaction parameters, it so far has mainly been used to study the ANNNI model ('axial next nearest neighbour Ising model' [10.31,32]) in two space dimensions [10.33]. The results appear to be promising and interesting [10.33], but since two years after the publication describing the machine [10.22] no paper on the results obtained is as yet available, a final answer regarding the usefulness of this machine cannot yet be given.

The second Ising machine constructed at Santa Barbara [10.23] is restricted in its use to the three-dimensional nearest-neighbour ferromagnetic Ising model. This machine updated 25×10^6 spins on a $64 \times 64 \times 64$ lattice which in 1983 was a world record in speed for a Monte Carlo program. However, in spite of its speed this machine has become an example for the fact that one must be most careful in designing and applying such special purpose processors: first of all, it was intended to check the validity of the so-called hyperscaling relation between critical exponents [10.34] by means of a new finite-size scaling analysis [10.35,36]. However, it was pointed out later [10.37] that this particular extension of finite-size scaling was wrong, and hence with the data obtained from the calculation [10.35,36] there was no way to check the hyperscaling relation whatsoever: from the data on bulk magnetization, susceptibility and higher-order moments one could, in principle, deduce just two exponents [10.37]. To get a third independent one a study of the spin pair correlation function would be desirable—which is not at all straight-forward to implement on that machine.

But what is even worse than this failure is the fact that the finite-size behavior of some data [10.35] did exhibit a rather strange cusp at linear dimension $L = 24$ lattice spacings. Calculations on the DELFT machine [10.38] did not reproduce this feature and attributed it to an artefact due to insufficient pseudorandom numbers implemented. In fact, for very precise Monte Carlo calculations the availability of sufficiently 'random' pseudorandom numbers is a problem that deserves careful consideration. Particular choices of linear-feedback shift-register random-number generators are normally used in these applications, proposed by TAUSWORTHE [10.39]. Although some versions of this random-number generator have been studied in detail [10.40-42,22] it is still possible that a Monte-Carlo simulation pushed to extreme limits may reveal deficiencies of a particular choice of a random-number generator, which were not known before. So, the only firmly established result of [10.35,36] is a very accurate estimate for the critical temperature of the Ising model, $J/k_BT_c = 0.221650 \pm 0.000005$.

Hence, so far the most successful example of a special-purpose Monte-Carlo processor seems the Bell AT & T Ising spin-glass computer [10.24]: runs which took about one half year of CPU time at this machine, updating at nearly the same rate as the Santa Barbara machine, have yielded interesting results [10.43,44]. These results furthermore seem to be corroborated by other work performed independently on the ICL-DAP machine [10.45].

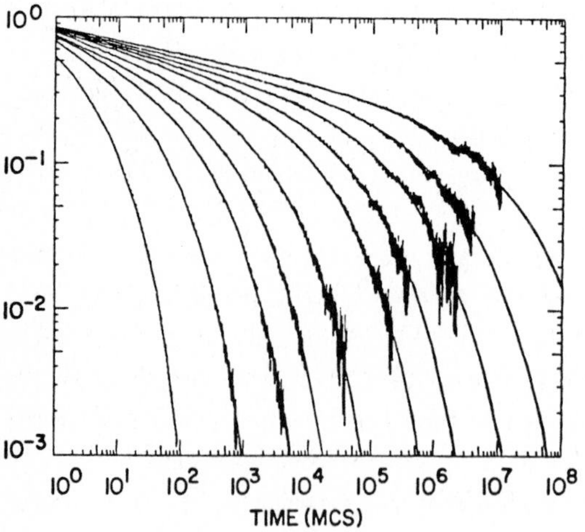

Fig. 10.1 Log-log plot of the spin autocorrelation function of the three-dimensional nearest-neighbour Ising spin glass (symmetrical $\pm$ J model) plotted vs. time; data points are shown together with error bars. From left to right, the temperatures are $T/J = 2.50$, 2.0, 1.8, 1.7, 1.6, 1.5, 1.45, 1.4, 1.35, and 1.4. Lattice size 64^3. Solid curves represent best fit to the function Ct^{-x} $\exp[-(t/\tau)^y]$, with C, r, y, τ being four adjustable parameters. From OGIELSKI (1985) [10.44]

Typical data obtained by this investigation [10.44] are shown in Fig.10.1 where the time dependence of the spin auto-correlation function of the ± J spin-glass model is shown, together with a phenomenological function which fits these data. This figure illustrates the difficulty of such an investigation: the decay at the lowest temperature shown spans more than seven decades of time. Moreover, there are considerable statistical fluctuations, particularly at late times. Note that these data are not the results of single Monte-Carlo runs, but typically averages of 170 trajectories (starting from different initial spin configurations) are taken. It is this enormous effort which clearly is impractical at a general-purpose computer. In fact, in the first Monte-Carlo work on spin glasses [10.46] already an attempt was undertaken to study the spin auto-correlation function of spin glasses – but in this work, nine years earlier on a TR 440 computer, only two decades in time were spanned and only about half a dozen trajectories for a lattice size 24^2 were taken. Since this first pioneering work, the effort in simulations of spin glasses has steadily increased – see [10.47-50] and Chap.8 for reviews –, and the work of OGIELSKI and collaborators [10.24,43,44] certainly is a world's record in Monte Carlo effort spent on a single problem. Still this should not be considered as the final stage – the behaviour of the ±J model at temperatures $T/J < 1.2$ still remains somewhat uncertain. The question as to whether in this region there occurs a nonzero order parameter needs further clarification.

All the work on special-purpose processors described so far refers to Ising-type problems, i.e., problems relating to traditional statistical mechanics of phase transitions and magnetism. There are lots of other problems in condensed-matter physics where Monte Carlo special purpose processors could be useful potentially – percolation, Potts models, aggregation kinetics, fluids, etc. are just a few examples of such fields. Of course, the field where the future computational needs seem to be most demanding is lattice-gauge theory [10.2]; various proposals to construct special-purpose computers for lattice-gauge Monte Carlo calculations have been made [10.25]. After this outlook for current and future activities with special purpose Monte Carlo computers we return to Item (ii) and briefly review the state of the art of Monte Carlo programs on recent vector processors.

10.1.3 Speeding-up of Monte Carlo Programs on Vector Processors

There are classes of problems where the gain due to vectorization is immediately obvious: usually for the updating of a chosen degree of freedom it is necessary to compute the energy change of those neighbours, with which the considered degree of freedom interacts. For problems with long-range interactions, this amounts – on a conventional serial computer – to running through a DO loop the size of which equals the total number of degrees of freedom in the system. Even for lattice systems, the available largest lattice linear dimensions (L) have thus been very restricted (e.g., Ising systems with dipolar interactions [10.51], where $L \leq 12$).

However, availability of vector processors speeds up such programs typically by factors of 30 to 40. There are numerous examples of problems which now have become accessible to study on machines such as CRAY-1, Cyber 205 or CRAY-XMP: models for Cu-Mn spin glasses, the interaction being the long-range RUDERMAN-KITTEL interaction [10.52]; the infinite range SHERRINGTON-KIRK PATRICK [10.53] spin glass model [10.54]; spin glasses in one dimension with random inverse power law interaction [10.55]; models of polyelectrolytes [10.56]; etc.

Somewhat more tricky are lattice problems of the nearest-neighbour interaction type; but now there are examples of highly speeded-up programs, making large use of vectorization of the innermost DO loops with vector processors, for problems of lattice-gauge theory [10.57,58], Potts models [10.12], and Ising systems [10.59-63,45]. For lattice-gauge problems where the degree of freedom has matrix character, the time for one Monte Carlo update depends very much on the chosen gauge group and type of action; therefore we shall not give any details on the actual performance of such programs here, but rather restrict attention to the Ising model which now is considered [10.63] "as one of the yardsticks by which the suitability of different computers (and more generally computer architectures) for scientific calculations is measured". Combining multi-spin coding techniques (for descriptions, see [10.1,28, 29] with the well-known device [10.64] of dividing the lattice into a red and a black sublattice, WANSLEBEN et al [10.60] obtained a speed of 10×10^6 Monte Carlo steps per second on a CDC Cyber 205; improving this code KALLE [10.61] reached the speed of 22×10^6 updates per second, and thus nearly matches the speed of the Santa Barbara machine. Applying the same technique at the ICL-DAP PAWLEY et al. [10.59] made a Monte Carlo renormalization-group study of the three-dimensional Ising model, with an updating rate of 2.7×10^6 updates/second; using for random-number generation a shift-register generator written in assembler languages, the program could even be speeded up to 6×10^6 updates/second: this example shows that one has reached the point where the speed of random-number generation limits the speed of the Monte Carlo program. Therefore, microcanonical Monte Carlo simulations (the demon Algorithm [10.65]) may become of great interest in the future because there one no longer needs random numbers. However, already the standard [10.1] Monte Carlo simulation in the canonical ensemble which needs random numbers could be brought to a speed of 218×10^6 updates/second on the ICL-DAP, in a special assembler implementation for a $128 \times 128 \times 144$ lattice [10.63]. This new world record again requires a new very-high-performance pseudo-random number generation [10.66].

Now it is quite clear that this program makes particular use of the specific DAP architecture with its 64×64 parallel processors, and already the choice of other linear dimensions which are no simple multiples or fractions of 64 may slow down the code. Therefore, one should not conclude from this discussion that that computer is better suited for Monte Carlo simulation than any of the other products recently available. In fact, an important feature is also the ease by which one can write such a fast simulation program. Therefore, it is worth mentioning that programs such as that of WANSLEBEN et al. [10.60] and that of KIKUCHI and OKABE [10.62] are still written in FORTRAN. The latter researchers used a HITAC S810-20 vector computer and reach a speed of 11.5×10^6 updates/second for a three-dimensional Ising system with fully periodic boundary conditions, and a speed of 9.6×10^6 updates/second for a geometry with two free surfaces (system size 32^3).

What new physics has then emerged from the availability of these very fast programs? In this case it is easy to prove that all that effort indeed has been worthwhile: (1) In their Monte Carlo renormalization-group study, PAWLEY et al. [10.59], for the first time, were able to obtain critical exponent estimates $\nu = 0.629 \pm 0.004$, $\eta = 0.031 \pm 0.005$ *competitive in accuracy, and compatible with,* the best field-theoretic renormalization-group estimates [10.67]. Previous attempts of treating this problem [10.68], as well as related work using phenomenological renormalization [10.69] were unable to reach this desired accuracy, due to much less efficient programs. (2) In his study of critical relaxation of the Ising model, KALLE [10.61], for the first time, was able to obtain the dynamic exponent z [10.70] with high accuracy and again this resolves a long-standing problem. Already the first papers utilizing the possibility of applying Monte Carlo methods to simulating stochastic dynamical models [10.71] addressed the same problem [10.72] and since then there have been numberous other attempts (e.g., [10.73]), or the dynamic Monte Carlo renormalization group work [10.74], which all gave somewhat unsatisfactory results. There are a number of related problems (critical dynamics of kinetic Potts models [10.75] or of Ising models with two-component order parameters [10.76] where a related speed-up of the programs is conceivable and also likely to yield a breakthrough with results. (3) In their study of the 10-state Potts model in two dimensions, CHALLA et al. [10.12], for the first time, were able to quantitatively test the phenomenological theory of finite-size scaling at first-order transitions [10.7,8,12] for a thermally driven transition. (4) In their study of surface critical phenomena, KIKUCHI and OKABE [10.62] succeeded in obtaining the scaling function for the surface equation of state, as well as the scaling function for the correlation length, and several critical amplitude ratios. The scaling function for the correlation length compares well with a renormalization group prediction [10.77], while most other quantities calculated have not been obtained by other methods before. The same researchers [10.78] presented also the first Monte Carlo study of critical relaxation near a surface.

In conclusion of this section, we emphasize that the Oed's method [10.64] of introducing sublattices such that degrees of freedom on the same sublattice do not interact, and hence all degrees of freedom on one sublattice can be updated simultaneously, should be of general usefulness for all array processors and vector processors. In fact, this work [10.64] was the basis for the first study of the critical behaviour of the surface tension of Ising models with Monte Carlo methods [10.79] on a FPS-AP 190 processor, on which also speed-up factors of about one order of magnitude were obtained [10.64]. Of course, if the interaction is of medium range, one would need a large number of sublattices. A variant of this method, where one updates simultaneously an array of sites for a fcc lattice with five different pair interaction parameters, was introduced by SCHWEIKA [10.80]. He constructed this array such that the first site of the array is randomly selected, while the n'th site is a fixed constant distance away from the (n-1)'th site, using helical boundary conditions. This program, used for modelling the phase diagram and short-range order for Ni-Cr-alloys, updates 1×10^6 sites at a CRAY-XMP.

10.2 Some Comments on Finite-Size Effects

10.2.1 Statement of the Problem

Recent reviews on finite-size effects at phase transitions have been given in [10.5, 6]. We shall not repeat this discussion here, but recall only the main points.

At a second-order phase transition, in the thermodynamic limit ($N = L^d \to \infty$, in d dimensions) there is an order-parameter correlation length ξ, an order-parameter susceptibility χ_T and a relaxation time τ, which diverge when the critical temperature is approached [10.34],

$$\xi \propto |1 - T/T_c|^{-\nu} \quad , \quad \chi_T \propto |1 - T/T_c|^{-\gamma} \quad , \quad \tau \propto |1 - T/T_c|^{-\nu z} \quad , \tag{10.1}$$

while the order parameter m vanishes continuously

$$m \propto (1 - T/T_c)^{\beta} \quad . \tag{10.2}$$

In (10.1,2) ν, γ, z, β are the respective critical exponents, and we consider an approach towards criticality by variation of temperature with zero conjugate field. Now, in a finite system, the transition is rounded: $\chi(T,L)$ and $\tau(T,L)$ reach maxima of finite height $\chi_{max} = \chi(T_c(L), L)$, $\tau_{max} = \tau(T_c(L), L)$ only, and significant deviations of $\chi(T,L)$, from $\chi_T \equiv \chi(T,\infty)$ occur in a region $\pm \Delta T$ around $T_c(L)$, with

$$\Delta T \propto L^{-\Theta} \quad , \tag{10.3}$$

Θ being the rounding exponent. In addition, the transition is shifted,

$$T_c(L) - T_c(\infty) \propto L^{-\lambda} \quad , \tag{10.4}$$

λ being the shift exponent. Of course, (10.3,4) hold in the asymptotic region, $L \to \infty$ only, *which is not always entered by a simulation*. The same remark holds with respect to the power laws for χ_{max} and τ_{max},

$$\chi_{max} \propto L^{\gamma_m} \quad , \quad \tau_{max} \propto L^{z_m} \quad , \tag{10.5}$$

where γ_m and z_m are the respective exponents describing the divergence of χ_{max} and τ_{max}, as $L\to\infty$.

10.2.2 Standard Finite-Size Scaling Theory

Now the following statements can be made regarding the various exponents that have been introduced: If the hyperscaling relation [10.34] $d\nu = \gamma + 2\beta$ among the critical exponents is valid, standard finite-size scaling theory holds which implies that in the asymptotic regime $T\to T_c(\infty)$, $L\to\infty$ susceptibility and relaxation time behave as [10.6,34,81,82].

$$\chi(T,L) = L^{\gamma/\nu}\,\tilde{\chi}[(1 - T/T_c)L^{1/\nu}] \quad , \tag{10.6}$$

$$\tau(T,L) = L^{z}\,\tilde{\tau}[(1 - T/T_c)L^{1/\nu}] \quad , \tag{10.7}$$

where $\tilde{\chi}$ and $\tilde{\tau}$ are scaling functions and their argument $(1-T/T_c)L^{1/\nu} \propto (L/\xi)^{1/\nu}$ simply expresses the fact that "L scales with the correlation length ξ". Equations (10.6,7) yield the following relations for the exponents Θ, λ, γ_m, z_m defined in (10.3-5):

$$\Theta = \lambda = 1/\nu \quad ; \quad \gamma_m = \gamma/\nu \quad ; \quad z_m = z \quad . \tag{10.8}$$

As is well known [10.1,5,6], (10.6-8) form the basis of two standard methods for studying second-order phase transitions with the Monte Carlo technique, aiming at estimating T_c, the critical exponents, and other critical parameters:

i) Adjusting simultaneously T_c and the exponents γ/ν, $1/\nu$, one should be able to collapse the family of curves $\chi(T,L)$ on a single curve, representing the scaling function $\tilde{\chi}$, by plotting $\chi(T,L)L^{-\gamma/\nu}$ versus $(1-T/T_c)L^{1/\nu}$. Similarly, the family of curves $\tau(T,L)$ collapses on a single curve, representing $\tilde{\tau}$, by plotting $\tau(T,L)L^{-z}$ versus $(1-T/T_c)L^{1/\nu}$. For the order-parameter susceptibility (and the order parameter itself) this procedure is quite standard, successful examples are found in [10.5,6, 8,45,83-85], for instance, while for the relaxation time this method has been demonstrated only once [10.86]. It must be admitted, however, that none of these examples reach a very high precision—the estimated error of the exponents quoted typically is a few percent. A high precision analysis of this type was presented in [10.35], but there also questions due to possible systematic errors because of a bad choice of the pseudorandom number generator can be raised [10.38].

ii) One simply tries to estimate the maxima χ_{max}, τ_{max}, as well as the position $T_c(L)$ where these maxima occur and estimates the exponents λ, γ_m, z_m from the slopes of the curves on double-logarithmic plots [assuming (10.4,5)]. This method has also

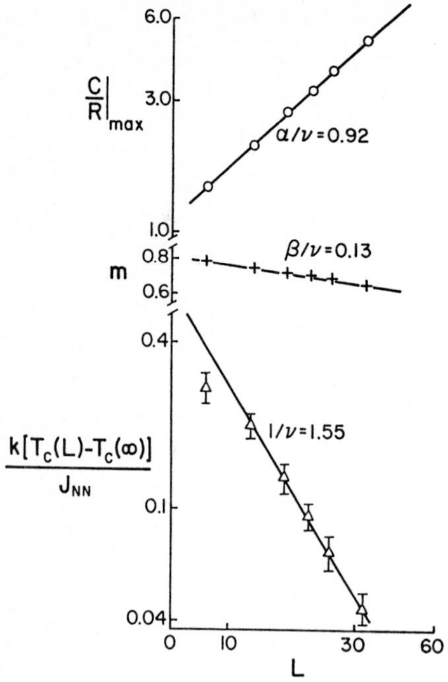

Fig. 10.2 Finite-size dependence of the specific heat maximum (upper part), order parameter (middle part) and shift of T_c (lower part), for an Ising square lattice with nearest-neighbour interaction $J_{NN} < 0$ and next-nearest neighbour interaction $J_{NNN} = RJ_{NN}$, $R = 0.65$. Numbers at the straight lines of these log-log plots show the critical exponents which are deduced from this analysis. From LANDAU and BINDER (1985) [10.87]

been tried occasionally [10.35,36,84,86,87], examples are shown in Fig.10.2. This method is more difficult than the first one, because it emphasizes those data for which the statistical error has its maximum {recall from [10.1,71] that the sampling time needed, in units of Monte Carlo steps per site, must by far exceed $\tau(T,L)$ in order that meaningful data are obtained, and $\tau(T,L)$ also controls this statistical error.}

Now it is seen from Fig.10.2 that in this example the range in L spanned is only about half a decade – therefore it is hard to judge whether one has actually reached the asymptotic regime where (10.3-7) hold, or whether there is still some effect due to *corrections to finite-size scaling*. In Fig.10.2, such corrections to finite-size scaling would show up as a mild curvature of the plot for small L which gradually disappears, as L becomes larger and larger. An indication of such an effect is only seen for the data on the shift of T_c in Fig.10.2.

In the data collapsing method, such corrections to scaling then show up in a systematic displacement of curves for different values of L from its asymptotic limit representing the function $\tilde{\chi}$. However, since T_c and the exponents are usually not known beforehand, one may find an optimum choice for the collapse of $\chi(L,T)L^{-\gamma_m}$ on $\tilde{\chi}$ with a choice γ_m^{eff}, T_c^{eff}, $1/\nu^{eff}$, where the critical temperature T_c^{eff} and the exponents γ_m^{eff}, ν^{eff} are effective values, systematically differing from the true ones. Of course, the collapse in such a case will never be perfect, but this fact may be easily obscured due to a statistical scatter of the data. Such a problem typically occurs if there is crossover from one type of critical behaviour to another. An example where the naive application of finite-size scaling would give misleading results can be found in a recent study of the so-called ϕ^4-model on the square lattice [10.88]: there one crosses over from critical behaviour in the universality class of the two-dimensional Ising model to mean-field critical behaviour (which is only reached at the displacive limit); such crossover behaviour is rather difficult to distinguish from the case of true variation of critical exponents with a parameter, as studied in [10.84,87].

10.2.3 Phenomenological Renormalization of Monte Carlo Data

An alternative method which is closely related to finite-size scaling ideas, however, is the phenomenological renormalization of Monte Carlo data [10.69,89]: It is based on working with moments $\langle s^k \rangle_L$ of the order-parameter probability distribution. Of course, the lowest moments are trivially related to the quantities considered before,

$$m = \langle |s| \rangle_L \quad , \quad s = \frac{1}{L^d} \sum_i m_i \quad , \tag{10.9}$$

m_i being the local value of the order parameter at the i'th lattice site and

$$k_\beta T \chi_T = L^d (\langle s^2 \rangle_L - \langle |s| \rangle_L^2) \quad . \tag{10.10}$$

However, of particular additional relevance is the fourth-order reduced cumulant U_L of the distribution {sometimes $g_L = -3U_L$ is called the renormalized coupling constant [10.35-37]}

$$U_L = 1 - \langle s^4 \rangle_L / (3 \langle s^2 \rangle_L^2) \quad . \tag{10.11}$$

Now one can show [10.69] that $U_L \to 0$ for $T > T_c$ as $L \to \infty$, while $U_L \to 2/3$ for $T < T_c$, and $U_L \to U^*$ (in between 0, 2/3) for $T = T_c$. As a function of T, U_L furthermore decreases monotonically. Consequently a pair of curves $U_L, U_{L'}$ with $L = bL'$ plotted versus T should intersect at $T = 0$, $T \to \infty$, and $T = T_c$. Thus, looking for such intersection points one is able to locate T_c, independent of any prejudice about the critical exponents. Furthermore, if corrections to finite-size scaling are negligible, all pairs (L',bL') must intersect in a unique point; however, if they are not neglible, there is a systematic scatter of intersection points (Fig.10.3). In this case it is necessary [10.69] to extrapolate the data on T_c and the data on the critical exponents as a function of the variable $1/(\ln b) = 1/[\ln(L/L')]$, Fig.10.4. The exponent estimates $1/\nu$, $2\beta/\nu$ follow from the moments as [10.69]

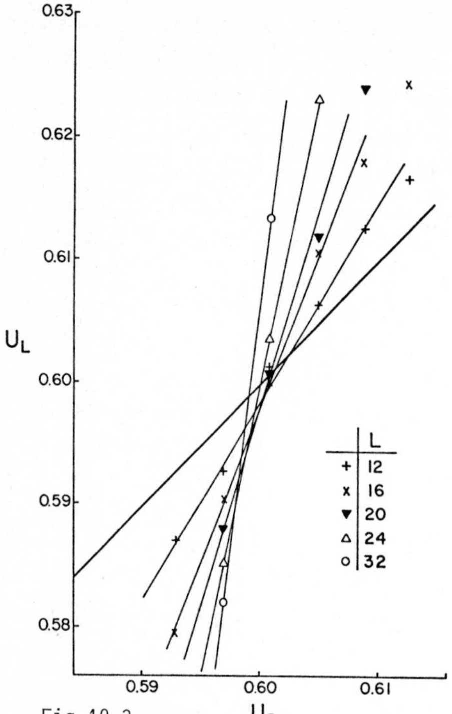

Fig.10.3

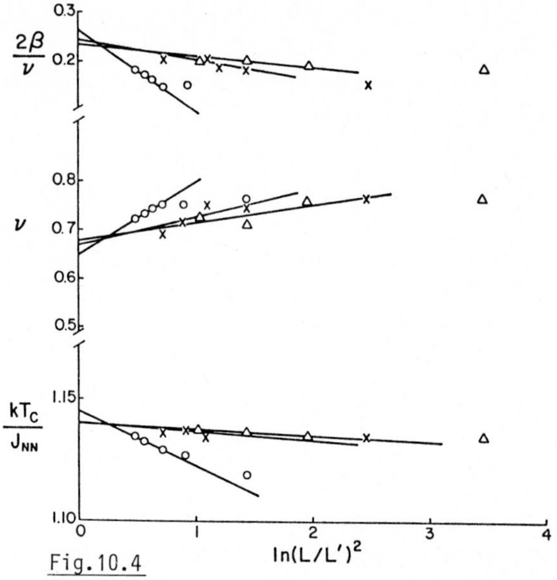

Fig.10.4

Fig. 10.3 Plot of cumulant U_L versus U_8 for the same model as in Fig. 10.2, and temperatures near T_c. From LANDAU and BINDER (1985) [10.87]

Fig. 10.4 Variation of critical parameters with $\ln b = \ln(L/L')$ for the model of Fig. 10.2. Data are for L' = 4 (circles), L' = 8 (crosses) and L' = 12 (triangles). From LANDAU and BINDER (1985) [10.87]

$$\frac{1}{\nu} = \frac{\ln(\partial U_{bL'} / \partial U_{L'})_{U^*}}{\ln b} \quad , \quad \frac{2\beta}{\nu} = \frac{\ln(\langle s^2 \rangle_{bL'} / \langle s^2 \rangle_{L'})_{T_c}}{\ln b} \quad , \tag{10.12}$$

and one can show [10.69] that corrections to scaling show up as corrections of order $1/(\ln b)$ in (10.12). Figure 10.4 shows that data for $L' = 4$ are clearly outside the scaling regime, while data for $L' = 8$ and $L' = 12$ extrapolate linearly to about the same value. These estimates are consistent, in accuracy, with those of Fig.10.2 and confirm the expectation that in this case for $L' \gtrsim 10$ corrections to finite-size scaling are very small.

Figure 10.5 now shows the results so obtained in comparison with other methods: Monte Carlo data collapsing [10.84], Monte Carlo renormalization group (MCRG) [10.90], high-temperature series extrapolation [10.91], phenomenological renormalization of transfer-matrix calculations for finite strips [10.92] and standard real-space renormalization group methods [10.93]. The conclusion is the following: in this case the Monte Carlo study was superior in accuracy to standard analytical methods such

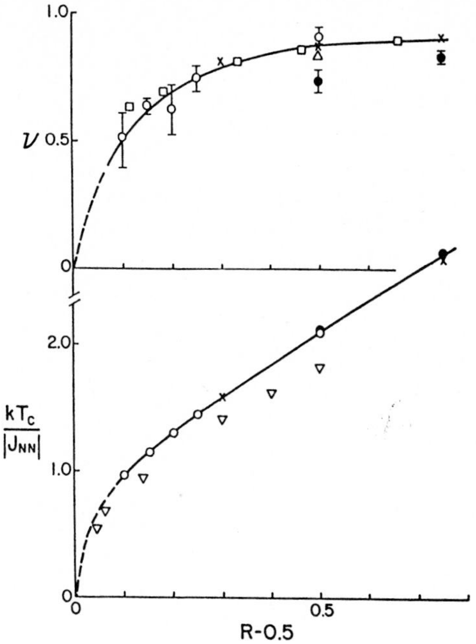

Fig. 10.5 Variation of the correlation-length exponent (upper part) and critical temperature T_c (lower part) with R. Results of phenomenological renormalization, as described in Fig. 10.4, open circles; Monte-Carlo results of [10.84], open triangle in upper part; MCRG results of [10.90], crosses; series expansion results of [10.91], full dots; transfer-matrix renormalization [10.92], squares; real-space renormalization group, open triangles (lower part) [10.93]. From LANDAU and BINDER (1985) [10.87]

as high-temperature series extrapolation or real-space renormalization; the accuracy of the MCRG calculation [10.90] is probably somewhat superior, while the transfer matrix calculation [10.92] has comparable accuracy. On the other hand, the transfer-matrix method is probably no longer competitive for the case where a third-nearest neighbour interaction is added [10.87] in which case ordered structures with large unit cells [(4×4) and (4×2), respectively] occur [10.87]. The first-order transitions between those structures and the disordered phase are still easily dealt with by the Monte Carlo analysis, as described here. Also the Monte Carlo method still works similarly well for continuum problems (e.g., XY model with 4-fold anisotropy [10.95]) and higher dimensionalities [10.69] – cases where other methods such as the transfer matrix renormalization are inapplicable, and methods such as MCRG are no longer superior in accuracy, or only with enormous effort [10.59].

For the benefit of those readers who feel discouraged by our discussion of what can be done to speed up Monte Carlo programs on supercomputers (Sect.10.1) and the fact that such an effort is indeed needed in some cases we emphasize that the recent example described here in detail (Figs.10.2-5) is based on calculations performed on standard computers such as IBM 3081, with a standard Monte Carlo program (a very similar version thereof has already been used as early as 1971 [10.96]). So this example proves that it still is possible to obtain results of great current interest with relatively modest effort. Of course, for keeping this effort modest, it was essential *not* to use too large linear dimensions (in [10.87] $L \leq 60$ was used): Since the maximum value of the statistical error of the magnetization, $\overline{\delta m^2_{max}}$, increases with L as [10.1,5]

$$\overline{\delta m^2_{max}} \propto \chi_{max}\, \tau_{max} \propto L^{\gamma/\nu+z} \quad , \tag{10.13}$$

and on serial computer the time for one Monte Carlo step/site increases as L^d, we conclude that the computational effort needed to obtain data for m at $T_c(L)$ with constant given accuracy increases as $L^{d+\gamma/\nu+z}$ with L. The exponent $d+\gamma/\nu+z$ in the present case is $\{z \approx 2.4 \pm 0.2$ [10.76], $\gamma/\nu = 1.75$ [10.90]$\}$ about 6; thus an increase by a factor of 2 in the linear dimension implies a factor of 64 more CPU time! This is why for such large linear dimensions $L \geqslant 64$ in three dimensions, where the exponent $d+\gamma/\nu+z$ is about 7, vector or array processors (or specialized Monte Carlo computers, respectively) are so valuable.

We now leave this specific example taken from [10.87] and resume our general discussion of size effects at second-order phase transitions. There are two cases where the analysis presented above needs be modified:

(i) At a KOSTERLITZ-THOULESS transition [10.97] such as it occurs for the $d=2$ XY model, instead of (10.2) the orderparameter m vanishes identically for $T>0$, and the divergence of ξ, χ_T, and τ is stronger than any power law

$$\xi \propto \exp\left[\text{const}\left(\frac{T}{T_c} - 1\right)^{-1/2}\right] \quad , \quad \chi_T \propto \xi^{2-\eta} \quad , \quad \tau \propto \xi^z \quad , T > T_c \quad ; \tag{10.14}$$

for $T<T_c$ the correlation function decays algebraically which implies that ξ, χ_T and τ remain infinite for $T \leqslant T_c$. Now finite-size scaling written in a form slightly more general than (10.6), (10.7) remains valid,

$$\chi(T,L) = L^{\gamma/\nu}\, \tilde{\chi}'(L/\xi) \quad , \quad \tau(T,L) = L^z\, \tilde{\tau}'(L/\xi) \quad , \tag{10.15}$$

where $\tilde{\chi}'$ and $\tilde{\tau}'$ are again scaling functions. Equation (10.15) imply that (10.5,8) are still valid; the formal result $\lambda = 0$ for the shift should be understood as

$$T_c(L) - T_c(\infty) \propto (\ln L)^{-2} \quad , \tag{10.16}$$

and the same holds for the width of the region where rounding occurs. Equations (10.14,15) have been used successfully for several models where KOSTERLITZ-THOULESS transitions occur, such as the $d=2$ 6-state clock model [10.98]. With respect to the phenomenological renormalization based on (10.12) we note that at T_c a curve of $U_{bL'}$ versus $U_{L'}$ must merge at the diagonal with slope of unity, and stay on the diagonal for $T<T_c$ since the low-temperature phase is a line of critical points. This behaviour has, in fact, been seen for the XY model [10.95], while for the 6-state clock model the curves $U_{bL'}$, versus $U_{L'}$ intersect at T_c, with $U_{bL'} > U_{L'}$, for $T<T_c$, parti-

cularly for small L' [10.98,99]. A similar behaviour was found in a phenomenological renormalization study of the ANNNI model [10.89], where plots of the curves $\ln[\chi(T,bL')/\chi(T,L')]$ and $\ln[\chi(T,bL'')/\chi(T,L'')]$, which should merge at T_c with a value γ/ν and superimpose for $T<T_c$, were found to intersect instead. However, since there is a systematic tendency of the difference $U_{bL'}-U_{L'}$ to decrease for $T<T_c$ with increasing L' [10.98], we interpret this behaviour as an effect due to corrections to finite-size scaling. Thus, one has to be rather careful with the interpretation of such finite-size scaling analyses.

(ii) If the hyperscaling relation $d\nu=\gamma+2\beta$ among the critical exponents fails (as happens, e.g., for the d = 5 Ising model, where $\nu=1/2$, $\gamma=1$, $\beta=1/2$), a simple behaviour emerges only when one uses fully periodic boundary conditions [10.9]. In this case the correlation length ξ does not enter the finite-size scaling description: rather the linear dimension L scales with a thermodynamic length ℓ

$$\ell \propto |1-T/T_c|^{-(\gamma+2\beta)/d} \quad . \tag{10.17}$$

Of course, $\ell\propto\xi$ if hyperscaling were valid. Thus, (10.17) means that in (10.5-8) we simply should replace the exponent $1/\nu$ by the exponent $d/(\gamma+2\beta)$, i.e.,

$$\chi(T,L) = L^{\gamma d/(\gamma+2\beta)}\,\tilde{\chi}''[(1-T/T_c)L^{d/(\gamma+2\beta)}] \quad , \tag{10.18}$$

$$\tau(T,L) = L^{dz\nu/(\gamma+2\beta)}\,\tilde{\tau}''[(1-T/T_c)L^{d/(\gamma+2\beta)}] \quad , \tag{10.19}$$

where $\tilde{\chi}''$, $\tilde{\tau}''$ are again some scaling functions, and

$$\Theta = \lambda = d/(\gamma+2\beta) \quad , \quad \gamma_m = \gamma d/(\gamma+2\beta) \quad , \quad z_m = dz\nu/(\gamma+2\beta) \quad . \tag{10.20}$$

For mean-field exponents, as they apply to Ising model with $d\geq 4$, we thus have $\Theta=\lambda=\gamma_m=z_m=d/2$. Equation (10.13) becomes replaced by $\overline{\delta m^2}_{max}\propto L^{d(\gamma+\nu z)/(\gamma+2\beta)}$ or L^d in the mean-field case, respectively, and the computational effort at T_c increases hence as L^{2d} for $d\geq 4$. Therefore, it is no surprise that work in d = 5 so far was restricted to $L\leq 7$ [10.9,100] — doubling L would require a factor of 10^3 more effort!

If one uses free instead of periodic boundary conditions, the scaling behaviour would be much more complicated: both ratios L/ℓ, L/ξ enter the description [10.14], it is the latter ratio which controls the shift [(10.14) with $\lambda=1/\nu$ (= 2 in the meanfield case) remains valid], while very close to $T_c(L)$ $\Theta=d/(\gamma+2\beta)$ controls the rounding.

10.2.4 Finite-Size Effects at First-Order Phase Transitions

As a final point of this section, we briefly turn to first-order phase transitions. At a first-order phase transition, there is no diverging length, of course. There finite-size rounding and shifting becomes important on a parameter scale controlled by the square root of the *volume* of the system [10.1,5-8,12]. This case can be understood trivially in terms of elementary thermodynamic fluctuation theory, expanding an argument due to IMRY [10.101], and one can make quantitative predictions not only for the exponents describing shift and rounding but also for amplitude prefactors and scaling functions [10.8,12]. As the most interesting case, we consider a first-order phase transition driven by temperature; at T_c the specific heat then for $V\to\infty$ has a delta-function singularity, representing the latent heat, the jump in internal energy from E_- to E_+ as the temperature is increased through T_c. In addition, the specific heat C(T) also has a jump singularity, $\lim C(T) = C_-$, $\lim C(T) = C_+$. In a finite system, the specific heat singularity is a rounded peak of finite height which also occurs at a temperature shifted away from T_c,

$$C_{max} \propto L^{\alpha m} \quad , \tag{10.21}$$

while the relaxation time increases stronger than any power law if the interfacial free energy f_{int} between the phases coexisting at T_c in the thermodynamic limit does not vanish,

$$\ln \tau_{max} \propto L^{d-1} \quad . \tag{10.22}$$

Equation (10.22) is understood from the fact that in order to go from one pure-phase state in a finite system to the other, one first has to nucleate a state containing interfaces of size L^{d-1}, which hence means one has to overcome a barrier of the order of $f_{int}\, L^{d-1}$ [10.69,79]. Now the finite-size behaviour of the specific heat can simply be understood from the observation [10.12] that the energy-distribution $P_L(E)$ is a superposition of two Gaussians ($\Delta T = T - T_c$)

$$P_L(E) \propto g_+ \exp -\left(\frac{(E - E_+ - C_+\Delta T)^2 L^d}{k_B T^2 C_+}\right) + g_- \exp -\left(\frac{(E - E_- - C_-\Delta T)^2 L^d}{k_B T^2 C_-}\right) \tag{10.23}$$

where (assuming that the low-temperature phase has degeneracy q_+ and the high-temperature phase q_-) the weights g_+, g_- are, for $T\to T_c$

$$g_+ = q_+ \exp\left(\frac{(T - T_c)(E_+ - E_-)}{2k_B T T_c} L^d\right) \quad , \quad g_- = q_- \exp\left(- \frac{(T - T_c)(E_+ - E_-)}{2k_B T T_c} L^d\right). \tag{10.24}$$

From (10.23,24) it is straightforward to obtain the specific heat peak in the finite system as

$$C_L = \frac{g_+C_+^{3/2} + g_-C_-^{3/2}}{g_+\sqrt{C_+} + g_-\sqrt{C_-}} + \frac{q_+q_-\sqrt{C_+C_-}}{k_BT^2} L^d \left(\frac{(E_+ - E_-) + (C_+ - C_-)(T - T_c)}{\sqrt{C_+}g_+ + \sqrt{C_-}g_-}\right)^2 \quad , \quad (10.25)$$

using $\langle E\rangle = \int dEP_L(E)E$, $C_L = d\langle E\rangle / dT$. From (10.24,25) it is seen that a maximum of height [10.12]

$$C_L^{max} \approx \frac{(E_+ - E_-)^2}{4k_BT_c^2} L^d + \frac{C_+ + C_-}{2} \quad (10.26)$$

occurs at a temperature $T_c(L)$, where the *weights* of the two Gaussians are equal, $\sqrt{C_+}\, g_+ = \sqrt{C_-}\, g_-$, which yields [10.12]

$$\frac{T_c(L) - T_c}{T_c} = \frac{k_BT_c}{E_+ - E_-} \ln\left(\frac{q_-}{q_+}\sqrt{\frac{C_-}{C_+}}\right) \frac{1}{L^d} \quad . \quad (10.27)$$

Since the width of the rounding can be defined by requiring that the argument of the exponentials in (10.24) is ± 1, we find

$$\frac{\Delta T}{T_c} = \frac{2k_BT_c}{E_+ - E_-} \cdot \frac{1}{L^d} \quad . \quad (10.28)$$

Comparing now with the definitions (10.3,4) we recognize the exponents

$$\Theta = \lambda = \alpha_m = d \quad . \quad (10.29)$$

Moreover, in (10.26-28) the prefactors are simply determined in terms of thermodynamic quantities, too. If the first term on the right-hand side of (10.25) is omitted, since it is a small correction in comparison with the leading term of order C_L^{max} only, it is seen that $C_L\, L^d$ indeed is a function of the scaled argument $(T - T_c)L^d$ only [in the region of interest the correction $(C_+ - C_-)(T - T_c)$ in (10.25) is a small correction to scaling only].

While at the transition temperature T_c in the thermodynamic limit the free energy densities f of all q_- low-temperature phases have to be equal to the q_+ high-temperature phases, which then implies (10.23) because $P_L(E) \propto \exp[-L^df(E) / k_BT]$, i.e., all coexisting phases are described by *peaks of equal heights*, the effective transition temperature $T_c(L)$ in the finite system (defined via the specific-heat maximum) occurs where the *peaks have equal weights*. This simple principle is nicely born out by explicit calculations for the 10-state Potts model in two dimensions [10.12].

Thus, first-order transitions of this type are easy to understand. However, due to the enormous increase of the relaxation time with system size, (10.22), the equilibrium behaviour described here can be observed only rarely in practice, and needs a very large computational effort. If the observation time of the Monte Carlo run does not exceed τ_{max}, one will observe hysteresis.

It may also happen that a characteristic length (and associate susceptibilities) diverge when a first-order transition is approached (example: phase transition in the Heisenberg ferromagnet for $T < T_c$ at $H = 0$ as a function of the field). For a discussion of the more complex behaviour expected in this case see [10.11].

10.3 New Directions for the Application of the Monte Carlo Method

10.3.1 Monte Carlo Renormalization Group (MCRG)

The combination of Monte Carlo computer simulation with renormalization-group techniques first suggested by MA [10.102] in 1976, and since 1979 pioneered by SWENDSEN and coworkers [10.103-107,59,68,90], is by now a well established and powerful tool for the study of critical phenomena. It has significantly contributed to our knowledge of critical exponents. Since several brilliant reviews are available [10.105-107], we shall present here a rough outline of this method only.

What is done in any real-space renormalization [10.108] is a mapping from the set of original degrees of freedom (e.g., Ising spins $S_i = \pm 1, i = 1, \ldots, N$, say) to degrees of freedom on a larger length scale (block spins $S_i' = \pm 1, i' = 1, \ldots, N/b^d$, where b is the scale factor of the transformation). This coarsegraining of the configuration of the system can be formally expressed in terms of a projection operator $P(\{S_i\}, \{S_{i'}'\})$

$$\exp[-\mathcal{H}'(\{S_{i'}'\})/k_BT] = \sum_{\{S_i=\pm 1\}} P(\{S_i\}, \{S_{i'}'\}) \exp[-\mathcal{H}(\{S_i\})/k_BT] \qquad (10.30)$$

with

$$\sum_{\{S_{i'}'=\pm 1\}} P(\{S_i\}, \{S_{i'}'\}) = 1 \quad , \qquad (10.31)$$

to ensure that the partition function remains invariant under the transformation, $Z' = \sum \exp(-\mathcal{H}'/k_BT) = Z$. A typical example for P is the "majority rule" [10.108]: suppose we consider $b = 3$, then $S_i' = \mathrm{sign}(\sum_i S_i)$ where the sum extends over those sites which belong to the block labeled by i'. In this way, the Hamilton $\mathcal{H}$ is mapped on a renormalized Hamiltonian $\mathcal{H}'$.

For a more detailed description of this mapping, it is convenient to write $\mathcal{H}$ as follows

$$\mathcal{H}/k_BT = \sum_\alpha K_\alpha S^\alpha \quad , \quad S^1 \equiv \{S_i\} \quad , \quad S^2 \equiv \{S_iS_j\} \quad , \quad S^3 \equiv \{S_iS_jS_k\}, \ \ldots \quad ,(10.32)$$

which contains all types of couplings K_α possible to construct with Ising spins. What (10.30) hence amounts to, is a mapping in the infinite-dimensional space of all coupling parameters onto itself, $\{K_\alpha\} \to \{K'_\alpha\}$.

Since such a renormalization-group transformation is local it leaves the correlation length unchanged when measured in units of the original lattice spacing. That implies that in units of the new lattice spacing, the correlation length decreases by a factor of b. Away from criticality, where the original correlation length is finite, repeated iterations of this mapping will reduce the renormalized correlation length to zero; only right at criticality the renormalized Hamiltonian also has infinite correlation length, and therefore is a critical point. Further iterations continue to produce Hamiltonians at the critical hypersurface, and the sequence of Hamiltonians eventually converges towards a fixed-point Hamiltonian $\mathcal{H}^*$, described by couplings $\{K_\alpha\}$, $\lim_{n\to\infty}\{K_\alpha^{(n)}\} = \{K_\alpha^*\}$.

One can now show [10.108] that the standard critical exponents can be found from the eigenvalues of the matrix $T^*_{\alpha\beta}$,

$$T^*_{\alpha\beta} = \partial K_\alpha^{(n+1)} / \partial K_\beta^{(n)} \Big|_{K_\alpha^{(n+1)} = K_\alpha^*, K_\beta^{(n)} = K_\beta^*} \quad , \tag{10.33}$$

which describes the linearized renormalization-group transformation

$$K_\alpha^{(n+1)} - K_\alpha^* = \sum_\beta T^*_{\alpha\beta}(K_\beta^{(n)} - K_\beta^*) \quad . \tag{10.34}$$

Of course, the mapping described by (10.30) can hardly ever be carried out exactly, and hence all attempts to perform this program analytically involve serious approximations, the accuracy of which often is hard to ascertain—and in many cases, in fact, is not very good (see, e.g., Fig.10.5 for a comparison).

The idea of MCRG now is to study this mapping by Monte Carlo: although at criticality the correlation length is infinite, the *interaction range of the fixed point Hamiltonian typically is short-range* and hence one may be less hampered by the finite-size of the simulated system when one studies (10.34), than one is when one directly considers the thermodynamic functions which become singular in the thermodynamic limit at T_c.

In practice, the matrix $T^*_{\alpha\beta}$ is found numerically by applying the chain rule [10.103]

$$\frac{\partial \langle S^{\gamma(n+1)}\rangle}{\partial K_\beta^{(n)}} = \sum_\alpha \frac{\partial K_\alpha^{(n+1)}}{\partial K_\beta^{(n)}} \frac{\partial \langle S^{\gamma(n+1)}\rangle}{\partial K_\alpha^{(n+1)}} \tag{10.35}$$

and noting that the derivatives in (10.35) can be expressed in terms of correlation functions involving two successive generations of block spins

$$\frac{\partial \langle S^{\gamma(n+1)}\rangle}{\partial K_\beta^{(n)}} = \langle S^{\gamma(n+1)}\, S^{\beta(n)}\rangle - \langle S^{\gamma(n+1)}\rangle\langle S^{\beta(n)}\rangle \quad . \tag{10.36}$$

An implementation of this method on the computer involves two types of truncations: (i) the number N_c of coupling constants $\{K_\alpha\}$ is not infinite but finite; typically $N_c = 7$ [10.59]. One cannot include such couplings which, for the fixed-point Hamiltonian, are relatively very small since then a statistically significant "measurement" of the corresponding correlation function, (10.36), would hardly be possible. Omission of these unimportant couplings, which represent high-order corrections to scaling only, has negligible effect on the leading eigenvalues of $T^*_{\alpha\beta}$. This can be checked explicitly by variation of N_c. (ii) The number of iterations $\{K_\alpha\} \to \{K'_\alpha\}$ which can be performed is limited by the finite linear dimension L of the system because in the n'th iteration one only has an effective linear dimension $L^{(n)} = L/b^n$ for the lattice of block spins of the n'th generation. But $L^{(n)}$ should still be large in comparison to the interaction range of the fixed point Hamiltonian. However, again the accuracy of the procedure can be checked by repeating the analysis of eigenvalues of $T^*_{\alpha\beta}$ for different generations n of blockings, different lattice sizes L, different scale factors b, or different choices of P. This method has been applied to a large variety of models (for a review see [10.106]), and the results have been most encouraging.

Note that by the methods outlined so far the renormalized Hamiltonians are not explicitly calculated, and thus one does not know explicitly the fixed point Hamiltonian $\mathcal{H}^*$. However, knowledge of $\mathcal{H}^*$ would improve the accuracy, since using it as a starting Hamiltonian would allow to obtain accurate exponents already from the first iteration (for which finite-size truncations are least important). In addition, the renormalized couplings are of interest to accurately locate the critical temperature.

Two approaches have been suggested to overcome this problem. The first one is based on a comparison of correlation functions of two lattices which differ in linear dimension by a scale factor b [10.109]. One iteration of the *larger* lattice (L) produces a change in Hamiltonian from $\mathcal{H}^{(0)}$ to $\mathcal{H}^{(1)}$ but now the lattice size is exactly the same as that of the *smaller* lattice (S). If the differences in their correlation functions are not too large, one can expand

$$\langle S^{\alpha(1)}\rangle_L - \langle S^{\alpha(0)}\rangle_S = \sum_\beta \frac{\partial \langle S^{\alpha(1)}\rangle}{\partial K_\beta^{(1)}}\, \delta K_\beta \tag{10.37}$$

where again (10.36) is used to compute the derivatives. If $\mathcal{H}^{(0)}$ were the fixed-point Hamiltonian, we would have $\langle S^{\alpha(1)}\rangle_L = \langle S^{\alpha(0)}\rangle_S$. From the nonzero differences in the correlation functions, we can calculate the deviations in the coupling constants from the fixed point Hamiltonian. By solving the set of linear equations

$$<S^{\alpha(n)}>_L - <S^{\alpha(n-1)}>_S = \sum_{\beta} \left(\frac{\partial <S^{\alpha(n)}>_L}{\partial K_{\beta}^{(0)}} - \frac{\partial <S^{\alpha(n-1)}>_S}{\partial K_{\beta}^{(0)}} \right) \delta K_{\beta}^{(0)} \qquad (10.38)$$

one determines what changes $\partial K_{\beta}^{(0)}$ in the original couplings are necessary to make the correlation functions match. Since (10.38) can treat several coupling constants in a single step, this method is useful to locate multicritical points [10.109]. On the other hand, a practical limitation are the statistical errors involved in finding differences between correlation functions from two independent Monte Carlo simulations [for the large lattice (L) and small lattice (S), respectively]; note that for the application of (10.35,36) the problem of statistical errors seems less severe, as they are highly correlated on successive generations of blockings.

The second approach [10.104] requires only a single Monte Carlo simulation and is based on the idea to formulate two different expressions for the calculation of renormalized correlation functions, one of them in the standard way $<S^{\alpha}> = (1/Z) \sum S^{\alpha} \exp(-\mathcal{H}/k_BT)$, and the other such that it depends on the renormalized coupling explicitly. Differences in the values of the correlation functions from the two expressions can then be used to estimate the differences between the assumed and true values of the renormalized coupling constants; for details, see [10.104].

Note that although we have used throughout a language appropriate to Ising models, the method is useful also for models involving Potts spins or continuum degrees of freedom; the methods involving matching of correlation functions seem to be particularly useful also in lattice gauge models where no fixed point other than the strong coupling limit occurs [10.110] and for critical dynamics [10.111] and kinetics of domain growth [10.112], where dynamic correlations are matched by rescaling of both, space and time.

10.3.2 Applications to Lattice Gauge Theory

Since the pioneering work of CREUTZ et al. [10.113] on the four-dimensional Euclidean lattice gauge theories, with gauge groups Z_N and U(1), Monte Carlo calculations for models relevant to quantum chromodynamics have become a flourishing and active area of research. Here, we shall not attempt to give a review of the many interesting results (see [10.2] for recent detailed reviews), but rather only give a brief formulation of the problem in general terms, to discuss the connection with problems of standard statistical physics.

The starting point of the application of Monte Carlo methods in quantum field theory is the definition of a vacuum expectation value of a quantity Q in terms of a functional integral,

$$\langle Q \rangle = \frac{1}{Z} \int DAD\Psi^* D\Psi Q(A, \Psi^*, \Psi) e^{-S(A,\Psi^*,\Psi)} \qquad \text{with} \tag{10.39}$$

$$Z = \int DAD\Psi^* D\Psi e^{-S(A,\Psi^*,\Psi)} \quad . \tag{10.40}$$

Here Ψ^* and Ψ are matter fields, A denotes the gauge fields, S is the action functional (using units $c = \hbar = 1$), $\int DAD\Psi^* D\Psi$ is a functional integration over all field configurations, and the space-time continuum has been made Euclidean by choosing an imaginary time axis. Now the action depends on coupling constants which we denote symbolically by g, and in the gauge theory of strong interactions one is most interested in the strong coupling regime, where in general no analytic methods to solve (10.39,40) exist.

Of course, a precise mathematical meaning must be given to the functional integrals in these equations, and one way to do this is by a lattice regularization. Then the matter fields $\Psi(\vec{x})$, $\Psi(\vec{x})$ are restricted to the lattice sites i, while the operator involving the gauge-field component $A_\mu^\alpha(\vec{x})$, $\hat{U} = \exp(i\, g\, A_\mu^\alpha\, dx^\mu \tau_\alpha)$, τ_α being the infinitesimal generator of the gauge group in the representation to which Ψ belongs, becomes a dynamical variable U_{ji} associated with the link between neighbouring lattice sites i,j. Of course, the choice of the lattice is arbitrary, and for convenience a hypercubic lattice is used. Of course, the physically relevant properties must not depend on the details of the lattice structure, and so one must again study a sort of continuum limit that emerges from the theory when one would let turn the lattice spacing to zero, which is possible in a meaningful way in the scaling limit associated with the critical points of the model.

In practice, in most work, the matter fields which involve fermionic degrees of freedom and hence require the difficult task of true quantum-mechanical Monte Carlo calculations (Sect.10.4), are omitted and one studies the properties of the pure gauge field itself. This problem is of interest since many relevant features of the interaction between gauge and matter fields are a consequence of the gauge fields alone. Then, (10.39,40) simplify to

$$\langle Q \rangle = \frac{1}{Z} \int \prod_{\{ij\}} dU_{ij}\; Q(\{U_{ij}\}) \exp[-\, S_G(\{U_{ij}\})] \quad , \tag{10.41}$$

and a popular form of the lattice action S_G due to WILSON [10.114] is, for SU(N) groups,

$$S_G = \frac{2}{g^2} \sum_p \mathrm{Tr}\{\hat{I} - \tfrac{1}{2}(\hat{U}_p^+ + \hat{U}_p)\} \quad , \quad U_p = \hat{U}_{i\ell} \hat{U}_{\ell k} \hat{U}_{kj} \hat{U}_{ji} \quad , \tag{10.42}$$

where the sum is extended over all plaquettes p of the hypercubic lattice, and $\hat{I}$ is the identity operator.

From (10.41,42) the analogy to statistical physics is quite evident — the only distinction being that the dynamical variables U_{ij} here are matrices associated with the links of the lattice rather than, say, Ising, Heisenberg spins etc., associated with the sites. The action S_G can be compared with $\mathcal{H}/k_BT$, and hence there is a correspondence between $1/g^2$ and const. / k_BT. As an example, for the group SU(2) an operator $\hat{U}_{ij}$ can be represented as

$$\hat{U}_{ij} = \hat{I} \cos \Theta_{ij} + i\hat{\vec{\sigma}} \cdot \hat{\vec{n}} \sin \Theta_{ij} \tag{10.43}$$

where $\hat{\sigma}_\alpha$ are the Pauli matrices, $\vec{n}$ is a unit vector and Θ_{ij} an angle. An even simpler example occurs for the group $Z_{\tilde{n}}$, where an element U_{ij} can take one of the discrete value $\exp(2\pi\, in/\tilde{n})$, $n = 0,1,\ldots,\tilde{n}-1$, and $U_{ji} = U_{ij}^*$. As $\tilde{n} \to \infty$, the group $Z_{\tilde{n}}$ becomes the group U(1). It is this simple case which was studied in the pioneering work of CREUTZ et al [10.113]. They worked on periodic lattices with a period of 8 lattice spacings in the 3 space-like directions but a period of 20 spacings in the time-like direction; this is an example that for these applications it often is preferable not to have all linear dimensions the same.

Apart from studying various gauge symmetries, it is also interesting to study other action functionals [10.115], which differ from (10.42) although they have the same continuum limit. There are also cases of interest with several coupling constants, and then one studies "phase diagrams" similarly as one does in ordinary statistical mechanics of condensed-matter systems [10.2,116].

For locating phase transitions, the standard concept of defining and sampling an "order parameter" in lattice gauge theory usually is not applicable. A simple way to identify phase transitions then is to locate singularities of the quantity $\langle S \rangle$ corresponding to the internal energy $\langle \mathcal{H} \rangle / k_BT$ in standard statistical mechanics; first-order transitions then show up as jumps and hysteresis [if the observation time is less than τ_{max}, see (10.22)], while second-order transitions are identified from specific-heat singularities as usual. However, in spite of the phase structure found for various models, the numerical results are still consistent with the expectation that the relevant critical point for non-Abelian gauge theories occurs for $g = 0$, the strong coupling limit.

What are the quantities, one then wishes to calculate for $g \to 0$? One quantity of interest is associated with a closed path Γ on the lattice, for which we define

$$\hat{U}_\Gamma = \hat{U}_{i_1 i_N} \cdots\cdots \hat{U}_{i_3 i_2} \hat{U}_{i_2 i_1} \quad , \quad W_\Gamma = \mathrm{Tr}\, \hat{U}_\Gamma \quad . \tag{10.44}$$

If Γ is a rectangular loop extending for m sites along a space-axis and n sites along a time axis, one expects (putting the lattice spacing unity one has $m = r$, $n = t$)

$$\langle W_\Gamma \rangle \underset{t\to\infty}{\to} \exp[-V(r)t] \quad , \quad V(r) \underset{r\to\infty}{\to} \Lambda r \quad , \tag{10.45}$$

where Λ is the so-called "string tension". Estimating this quantity is similarly difficult as obtaining the order-parameter correlation length for lattice models of standard statistical mechanics.

Of course, most interesting is the estimation of masses of elementary particles which can be inferred from the spectra of quantized excitations. Systems containing matter fields coupled to the gauge fields would be of particular interest, especially the SU(3) theory with fermions, which is the problem posed by quantum chromodynamics, the established theory of strong interactions. However, the extension of Monte Carlo methods to such a truly quantum-mechanical problem is difficult (Sect.10.4) and hence it cannot yet be handled on present computers in a satisfactory way. Part of the effort in getting more powerful supercomputers is motivated by this important unsolved problem.

10.3.3 Kinetics of Aggregation and Growth Phenomena

Recently, there has been an enormous interest in "fractal" geometric structures [10.15,117,118]; one typical way how such structures can form in nature is random irreversible growth. Prototype models for this irreversible growth are "diffusion-limited aggregation (DLA)" which leads to the formation of "Witten-Sander aggregates" [10.119], or the "growing self-avoiding walks" (GSAW) [10.120, 10.121], "indefinitely growing self-avoiding walk" (IGSAW) [10.122], "growing self-avoiding trail" (GAT) [10.123], Laplacian random walks [10.124], etc. We shall not describe all these models here (for recent reviews of some of them see also [10.125]), but only mention that all these models are variations of random-walk phenomena, and hence rather straightforward to simulate. In fact, the standard application of Monte Carlo for thermal averaging amounts of constructing a random walk through phase space, using transition probabilities obeying the principle of detailed balance [10.1]. For the irreversible processes considered here, only "forward reactions" and no "backward reactions" occur; thus, there is no detailed balance principle, but nevertheless the Monte Carlo process can easily be used to simulate the object which randomly grows step by step. Of course, for a proper sampling of the statistical properties of these objects, the simulation of the growth has to be repeated again and again, and due to this large statistical effort needed for obtaining meaningful results these problems have become accessible only rather recently.

Now, there are also random-walk problems intended to study diffusion phenomena, and these have been considered by Monte Carlo methods already for a long time (see [10.126,127] for recent reviews), as well as applications of the self-avoiding walk (SAW) problems which may model statics and dynamics of polymers (for reviews, see [10.128,129]). Applications of Monte Carlo methods to such problems have a long tradition, and also many analytical methods and results are available with respect to the statistics of self-avoiding walks, and there is a close connection to the renormalization-group theory of critical phenomena [10.130]. But even in this area there are still interesting questions and model extension, such as the "node-avoiding Lévy flight" [10.131], where Monte Carlo simulations play an illuminating rôle.

In contrast, for the irreversible growth models analytical methods are much less developed; e.g., for DLA the construction of a valid mean-field theory has been a rather controversial topic [10.132]. What one is most interested in, is an understanding of the fractal dimensionality d_f of the object, defined in terms of particle number N and gyration radius R as $N \propto R^{d_f}$ [10.117]. Theoretical methods for relating d_f to d are still under discussion [10.132,133], and the Monte Carlo simulation is clearly the method which guides the field, it is stimulating both for theory and experiment (for an example, see [10.134]). Unlike for critical phenomena, in this area it is not at all clear on which parameters exponents such as d_f can depend, there is even a claim that d_f for DLA clusters on the square lattice differs from those on the triangular lattice [10.135]. Also for other models, such as the GSAW, it has been a matter of considerable debate to clarify what the fractal dimensionality d_f is [10.120] and very extensive and careful Monte Carlo work was, required to show that d_f is the same as that for ordinary SAW's [10.121]. Note that the Monte Carlo algorithm for a GSAW is, in principle, extremely simple: one starts out at a lattice site, fills one of the z neighbouring sites with probability $1/z$, then fills one of the *free* neighbouring sites of that site with probability 1/(number of free sites), etc. Only if all nearest-neighbour sites have been visited before the walk is terminated, while the usual SAW chooses one of the $z-1$ neighbouring sites (other than the previously visited one) with equal probability, and terminates if the chosen site has been visited before. A different fractal dimensionality, however, seems to occur for the IGSAW [10.122] which differs from the GSAW only by the condition that the site the walk can only visit as a next stop must be a "jump site", with probability 1/(number of jump sites). "Jump-sites" are free nearest-neighbour sites which do not necessarily lead to a "cage", where the walk has to stop, during the later growth. Similar to these variations of SAW's, there are several variations of DLA: in standard DLA, one starts with one particle, lets another particle diffuse on the lattice until it gets to a nearest-neighbour site of the particle, where it then sticks; then the next particle is put on the surface of the lattice and diffuses until it sticks, etc., and so one grows the random cluster. But rather than this aggregation of single particles at a cluster, one can also consider cluster-cluster aggregation [10.136], or one can consider diffusion-limited growth of linear [10.124] rather than branched [10.119] objects.

This field is rapidly developing, and we only wish to convey to the reader the flavour of it, rather than giving a thorough review. A related field, intermediate between these growth phenomena, which are irreversible from the outset and equilibrium statistical mechanics, is the problem of growth of ordered domains, which occurs when one brings a system out of equilibrium (in the disordered phase) by a sudden change of thermodynamic parameters such that one crosses the phase boundary of an order-disorder transition. Then the gyration radius of the ordered domains grows with time as $R(t) \propto t^x$, and one wishes to clarify the exponent x. For this problem analytical results also are rare: the question whether the exponent x is universal and, if so, which parameters control the universality classes, are rather controversal (for reviews and references of recent work see [10.137,138]). Again a detailed discussion of this problem, to which Monte Carlo simulation (and recently also MCRG [10.112]) is a valuable tool, is outside the scope of this chapter.

10.4 Quantum Statistical Mechanics on Lattices

In this section, we are concerned with thermal averages for quantum-mechanical Hamiltonians $\hat{\mathcal{H}}$

$$\langle A\rangle_T = \frac{\mathrm{Tr}\hat{A}\,\exp(-\hat{\mathcal{H}}/k_BT)}{\mathrm{Tr}\,\exp(-\hat{\mathcal{H}}/k_BT)} = \frac{\sum_m \langle m|\hat{A}|m\rangle\,\exp(-E_m/k_BT)}{\sum_m \exp(-E_m/k_BT)} \tag{10.46}$$

where we consider the standard situation that eigenstates $|m\rangle$ and eigenvalues E_m of the Hamiltonian $\hat{\mathcal{H}}(\hat{\mathcal{H}}|m\rangle = E_m|m\rangle)$ are not explicitly known (in cases where they are explicitly known, e.g., for Ising model of magnetism, the Monte-Carlo procedure trivially applies exactly as in the case of truly classical statistical mechanics). This problem of quantum statistics still is a challenge for the Monte Carlo method; various approaches have been proposed to solve this problem, some of which we shall briefly review here.

The first method introduced already in 1962 by HANDSCOMB [10.139] is based on performing a high-temperature series expansion of the factor $\exp(-\hat{\mathcal{H}}/k_BT)$ in (10.46) both in the numerator and denominator can then be worked out formally to reduce them to weighting factors which can be estimated by a Monte Carlo procedure.

Let us explain this method for the example of a Heisenberg magnet [10.140] (so far this method has found applications to Heisenberg ferro- and antiferromagnets only [10.141,142], but it is the only method the usefulness of which for three-dimensional systems has been demonstrated for large N). The nearest neighbour Heisenberg Hamiltonian, for spin quantum number $S = 1/2$, is

$$\hat{\mathcal{H}} = -2J\sum_{\langle i,j\rangle}\vec{S}_i\cdot\vec{S}_j - g\mu_BH\sum_i S_i^z \tag{10.47}$$

is rewritten in terms of permutation operators $\hat{E}(i,j) = (1+4\vec{S}_i\cdot\vec{S}_j)/2$ as

$$\hat{\mathcal{H}} = -J\sum_{\langle i,j\rangle}\hat{E}(i,j) - g\mu_BH\sum_{i=1}^{N}S_i^z + \frac{d}{2}JN \equiv \hat{\mathcal{H}}_0 - J\sum_K \hat{E}_K \quad , \tag{10.48}$$

where the index K labels the dN nearest-neighbour bonds $\langle i,j\rangle$. Now the partition function is expanded as $(J>0)$

$$Z = \mathrm{Tr}\exp(-\hat{\mathcal{H}}/k_BT) = \sum_{n=0}^{\infty}\left(\frac{J}{k_BT}\right)^n \frac{1}{n!}\sum_{C_n}\left\{\mathrm{Tr}\,\hat{E}_{i_1}\ldots\hat{E}_{i_n}\,e^{-\frac{\hat{\mathcal{H}}_0}{k_BT}}\right\} \tag{10.49}$$

where $\sum_{C_n}$ denotes the sum overall possible sequences $(i_1 \ldots i_n)$. Defining

$$\Pi(C_n) = \frac{(J/k_BT)^n}{n!}\,\mathrm{Tr}\left\{\hat{E}_{i_1}\ldots\hat{E}_{i_n}\,e^{-\hat{\mathcal{H}}_0/k_BT}\right\} \quad , \tag{10.50}$$

$$\Omega(C_n) = \mathrm{Tr}\left\{\hat{A}\,\hat{E}_{i_1}\ldots\hat{E}_{i_n}\,e^{-\hat{\mathcal{H}}_0/k_BT}\right\} / \,\mathrm{Tr}\left\{\hat{E}_{i_1}\ldots\hat{E}_{i_n}\,e^{-\hat{\mathcal{H}}_0/k_BT}\right\} \tag{10.51}$$

one gets

$$Z = \sum_n\sum_{C_n}\Pi(C_n) \quad , \quad \langle A\rangle_T = \sum_n\sum_{C_n}\Omega(C_n)\,\Pi(C_n) / \sum_n\sum_{C_n}\Pi(C_n) \tag{10.52}$$

By (10.52), the quantum average (10.46), is reduced to a classical expectation value over the distribution $\Pi(C_n)$ — instead of $A_m = \langle m|\hat{A}|m\rangle$ in (10.46) we now have $\Omega(C_n)$, instead of the Boltzmann factor $\exp(-E_m/k_BT)$ the weighting factor $\sum_{C_n}\Pi(C_n)$. Thus it is clear that a Markov process can be defined which performs importance sampling of the average, (10.52), similar to the Markov process of the standard Metropolis method. Of course, the usefulness of this approach rests on the fact that for the decomposition chosen in (10.48) the distribution $\Pi(C_n)$ can readily be evaluated explicitly

$$\Pi(C_n) = \left(\frac{J}{k_BT}\right)^n \frac{1}{n!}\prod_{j=1}^{K} 2\cosh\left(\frac{g\mu_B H a_j}{2k_BT}\right) \tag{10.53}$$

where K is the number of cycles in $\{\hat{E}_{i_1}\ldots\hat{E}_i\}$ and a_j is the length of a cycle. The configuration space to be sampled is the space of all sequences $\{\hat{E}_{i_1}\ldots\hat{E}_i\}$ of permutation operators, for all sequence lengths n. One starts with a random sequence, and adds or removes at a randomly chosen position a permutation operator according to suitable transition probabilities (for details, see [10.140-142]). As an example of the quality of the results which can be obtained, Fig.10.6 shows the temperature variation of the susceptibility of Heisenberg chains containing up to 128 spins [10.140].

An alternative approach which seems to have a much wider range of applicability is based on the use of the Suzuki-Trotter formula [10.143,144]

$$\ell^{\hat{A}_1+\hat{A}_2+\ldots+\hat{A}_p} = \lim_{m\to\infty}\left(e^{\hat{A}_1/m}\,e^{\hat{A}_2/m}\ldots e^{\hat{A}_p/m}\right)^m \quad ; \tag{10.54a}$$

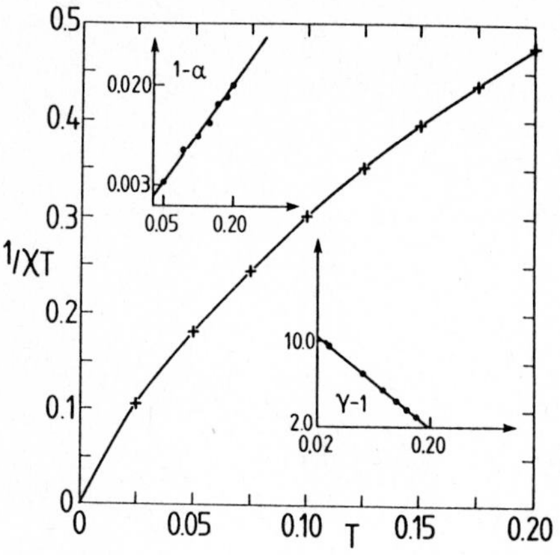

Fig. 10.6 A plot of the Monte-Carlo results for the inverse susceptibility 1/Tχ of the Heisenberg ferromagnetic chain versus temperature. The insert show log-log plots of energy and susceptibility versus temperature, to extract exponents α and γ, which can be considered as "effective exponents" only due to logarithmic corrections (SCHLOTTMANN 1986 to be published). From LYKLEMA (1983) [10.140]

one applies this formula, following the pioneering work of SUZUKI et al. [10.145] to (10.46) suitably decomposing $\hat{\mathcal{H}}$ into a sum of terms and choosing a suitable finite m instead of the limit in (10.54a) to obtain the "m^{th} approximant to Z",

$$Z^{(m)} = \mathrm{Tr}\left(e^{-\hat{\mathcal{H}}_1/mk_BT}\, e^{-\hat{\mathcal{H}}_2/mk_BT} \ldots e^{-\hat{\mathcal{H}}_3/mk_BT}\right)^m \quad . \tag{10.54b}$$

On the basis of this equation, one can relate the d-dimensional quantum-mechanical system to a (d + 1)-dimensional classical system, where m is the linear dimension in the additional "Trotter" direction of this system. For this classical system, Monte Carlo calculations can then be carried out in the usual way.

Of course, this implies that the Hamiltonian $\hat{\mathcal{H}}$ is split into local terms such that traces involving the factors $\exp(-\hat{\mathcal{H}}_i / mk_BT)$ can straightforwardly be evaluated. To be specific, let us consider the anisotropic Heisenberg chain

$$\hat{\mathcal{H}} = - \sum_{i=1}^{N} J_x S_i^x S_{i+1}^x + J_y S_i^y S_{i+1}^y + J_z S_i^z S_{i+1}^z \quad , \tag{10.55}$$

which has recently been carefully studied by CULLEN and LANDAU [10.146]. They used a decomposition suggested by BARMA and SHASTRY [10.147],

$$\hat{\mathcal{H}} = \hat{\mathcal{H}}_0 + \hat{V}_A + \hat{V}_B \quad , \quad \hat{\mathcal{H}}_0 = - \sum_{i=1} J_z S_i^z S_{i+1}^z \quad , \quad \hat{V}_i = -\left(J_x S_i^x S_{i+1}^x + J_y S_i^y S_{i+1}^y\right) \tag{10.56a}$$

$$\hat{V}_A = \sum_{i=1}^{N-1} \hat{V}_i \quad , \text{ i odd;} \quad \hat{V}_B = \sum_{i=2}^{N} \hat{V}_i \quad , \text{ i even.} \tag{10.56b}$$

Then [10.147]

$$Z^{(m)} = \mathrm{Tr}\left\{\exp\left[\frac{J_z}{2k_B T m}\sum_{i=1}^{N}\sum_{r=1}^{2m} S_{i,r}\, S_{i+1,r} + \frac{1}{k_B T}\sum_{i,r} h(i,r)\right]\right\} \qquad (10.57)$$

where h(i,r) is given in terms of the matrix elements

$$\frac{1}{k_B T}\, h(i,r) = \ln \langle S_{i,r} S_{i+1,r}\, | \exp(-\hat{V}_i / m k_B T) |\, S_{i,r+1}\, S_{i+1,r+1}\rangle \quad . \qquad (10.58)$$

Note that the values of the matrix elements in (10.58) depend on the values of the four quantum numbers $S_{i,r}$, $S_{i+1,r}$ $S_{i+1,r}$, $S_{i,r+1}$ around an elementary plaquette of the two-dimensional lattice of the corresponding classical Hamiltonian in (10.57). Now this four-spin interaction has rather special properties, however: there are only 8 states of the 2^4 states of the spins of a plaquette allowed, the other states are forbidden because the matrix element vanishes. This means, however, one cannot apply a straighforward single-spin flip algorithm to (10.57) because this would violate the constraint that one half of the states of a plaquette are forbidden. In fact, each allowed state of a plaquette is obtained from another by either flipping 2 or 4 spins. It was found, however, that one must use both local and nonlocal strings of flipped spins (forming closed loops on the lattice) to equilibrate the system. Thus, the algorithm is rather complicated, and would be even more so if one would try to simulate quantum problems in higher dimension. Another problem, studied in detail by CULLEN and LANDAU [10.146] for the model, (10.55) (and in [10.148] for other models), is the finite-size effect associated with the finiteness of m. But apart from spin models, this approach based on the Trotter formula (10.54), has also been applied to the "small polaron" problem [10.148], to the problem of a two-level system coupled to phonons [10.149], etc. For more details, we refer to [10.17,18].

A closely related formulation of the problem of performing quantum statistics can be obtained from the Feynman path-integral formulation [10.150],

$$Z = \int D\vec{\phi}(\tau)\, \exp\left[-\frac{1}{\hbar}\int_0^{\hbar/k_B T} d\tau\, \hat{\mathcal{H}}(\vec{\phi}(\tau))\right] \quad , \qquad (10.59)$$

where $\vec{\phi}$ is the set of dynamical variables on which the Hamiltonian depends, and τ is an imaginary time variable. Equation (10.59) becomes useful for Monte Carlo simulation, if the "time" interval $\hbar / k_B T$ is discretized into m discrete intervals $\Delta\tau$. Again, the periodicity condition $\vec{\phi}(\tau + \hbar / k_B T) = \phi(\tau)$ leads to an important constraint.

An elegant and rather direct application of (10.59) to the problem of granular superconducting films was presented by JACOBS et al. [10.151], using the model Hamiltonian

$$\hat{\mathcal{H}} = \sum_i \frac{\hbar^2}{2u}\left(\frac{d\phi_i(\tau)}{d\tau}\right)^2 + E_J \sum_{\langle i,j\rangle} [1 - \cos(\phi_j(\tau) - \phi_i(\tau))] \quad , \qquad (10.60)$$

where $\phi_i (0 \leq \phi_i < 2\pi)$ is the phase of the order parameter at site i, E_j the Josephson tunneling amplitude, and the parameter u measures the strength of quantum effects. In order to satisfy the constraint of periodicity in the time direction, JACOBS et al. [10.151] chose a discretization into L_τ time slices, of extent $\varepsilon = \hbar / (k_B T L_\tau)$, and replaced $d\phi / d\tau$ by $\Delta_\tau \phi = 1 - \cos[\phi(\tau+1) - \phi(\tau)]$. It turns out in this problem, however, that for low temperatures one must have $L_\tau \gg u/k_B T$ in order to have results accurately representing the continium limit in the time direction. In practice, systems with $10 \times 10 \times L_\tau$ were studied, with L_τ up to 150.

An alternative formulation based on (10.59) starts by dividing the imaginary time axis into L_τ slices $\varepsilon = \hbar / (k_B T L_\tau)$ as above, and by inserting complete sets of intermediate states $(\sum_i |i><i| = \hat{1})$ between slices [10.152-157]

$$Z = \mathrm{Tr}\, \exp(-\hat{\mathcal{H}}/k_B T) = \sum_{i_1 i_3 \ldots i_{2L_\tau -1}} <i_1|e^{-\varepsilon\hat{\mathcal{H}}}|i_3><i_3|e^{-\varepsilon\hat{\mathcal{H}}}|i_5> \ldots <i_{2L_\tau-1}|e^{-\varepsilon\hat{\mathcal{H}}}|i_1> . \tag{10.61}$$

To compute the matrix elements, for one-dimensional problems one writes $\hat{\mathcal{H}} = \hat{\mathcal{H}}_1 + \hat{\mathcal{H}}_2$ and uses the Trotter formula (10.54), to approximate

$$e^{-\varepsilon\hat{\mathcal{H}}} \simeq e^{-\varepsilon\hat{\mathcal{H}}_1} e^{-\varepsilon\hat{\mathcal{H}}_2} + o(\varepsilon^2[\hat{\mathcal{H}}_1, \hat{\mathcal{H}}_2]) \quad . \tag{10.62}$$

For Hamiltonians describing Fermions or Quantum spins on a lattice, one chooses a real-space break-up so that $\hat{\mathcal{H}}_1$ and $\hat{\mathcal{H}}_2$ are composed of sums of *non-overlapping* cell Hamiltonians. Thus, diagonalizing $\hat{\mathcal{H}}_1$ and $\hat{\mathcal{H}}_2$ involves solving a few sites problem only. We refer to the original literature for details [10.152-157], but wish to emphasize that the methods applied so far, clearly are very useful for a variety of one-dimensional problems, such as solitons in polyacetylene [10.157], for example. For two-dimensional problems such as the 2-d Hubbard problem or spin-polarized Fermion lattice gases [10.156], lattice sizes are restricted to 8×8 or less. Hence the prospects of applying any of the approaches to three-dimensional systems are not so good. However, there are also other approaches which have promising aspects [10.158,159]: lack of space prevents us from discussing them here, and we also do not discuss the very interesting ideas proposed for estimating dynamical properties (i.e., real rather than imaginary time!) by quantum Monte Carlo methods [10.160-162]). But it is quite clear that this field is in rapid progress and a large number of useful applications exist; further methodological developments are likely to occur in the near future.

10.5 Concluding Remarks

In this chapter, we have summarized the state of the art with respect to efficient implementation and analysis of Monte Carlo programs, and then we have briefly reviewed a few areas which have seen enormous recent activity, such as lattice gauge theory, aggregation kinetics, and fields which apply less standard techniques, such as Monte Carlo renormalization-group and quantum Monte Carlo techniques. In no case have we aimed at completeness or at a detailed exposition of these methods; this review can only serve as a guide to the literature.

Of course, there are other fields which see important recent activity as well but have not yet been mentioned at all in this chapter. As examples, we mention static and dynamic properties of dense polymeric systems [10.163], and metallurgical problems such as the study of alloy phase diagrams and short-range order [10.164,80]: this is a field also with interesting new methods, such as the "inverse Monte Carlo method" [10.165] for the estimation of effective interaction parameters from experimental short-range order data, or improved techniques for estimating free energies and entropies [10.166]. Last but not least, there is an enormous effort devoted to the studies of random systems (spin glasses [10.43-50], systems with random fields [10.167], etc.); we have already mentioned that some of the fastest Monte Carlo programs, on particularly well suited machines [10.43-45,167], are developed in this field. Once more we draw attention to Fig.10.1: to the author's knowledge, this work [10.44] is the largest effort ever spent on a problem in computational physics so far (according to OGIELSKI [10.44] the computing time spent on this problem would take "several years" at a CRAY-1). We have not attempted to review the recent work on random systems here — a very recent review [10.50] puts all the work done on spin glasses into perspective. All these areas are by no means exhausted and hence will see much additional effort, as well as the more traditional areas to which Monte Carlo methods have already a longstanding application.

As more and more applications are tackled by this method and many groups are using it, there is still a barrier to overcome: apart from [Ref.10.1,Chap.1] there is hardly a simple pedagogical introduction to the field available for the unexperienced student. A lot of knowledge on practical details is communicated among the practitioners in the field only. This difficulty, however, will be probably eliminated by forthcoming introductory books [10.168,169]. Another useful source of information is the informal newsletter published regularly by the Daresbury Laboratory [10.170].

Since the statistical effort put into certain problems has increased enormously, one must now be more careful about the quality of the pseudorandom numbers than in the traditional work which aimed at a much lower level of accuracy. A rather thorough study of long-range correlations on random-number generators and their influence on Monte Carlo simulations has presently been presented by FILK et al. [10.171], in a study of four-dimensional Ising and Z_2 models. Applying the heat-bath sampling technique and performing regular sweeps through a lattice of linear dimension 2^k, where k is an integer, results were obtained which were systematically wrong. These simulations used 2^{28} sequentially generated random numbers, a number still small in comparison with the period of the random-number generator used (which was 2^{46}). A theoretical analysis of these correlation effects in the random numbers is also given, and thus one gains some insight why it is so dangerous to use lattice linear dimensions which are powers of 2 and hence create a periodicity in the updating process which is also a large power of 2. In any case, ascertaining the randomness of sequences generated by deterministic procedures on a computer is an interesting mathematical problem in its own right [10.172]. We expect that this problem and the practical task of efficient random-number generation for the purpose of Monte Carlo simulations will also see important activity in the future.

References

10.1 K. Binder (ed.): Applications of the Monte Carlo Method in Statistical Physics, Topics Current Phys.,Vol.36 (Springer, Berlin, Heidelberg 1984)
10.2 For recent reviews, see C. Rebbi: In /Ref.10.1, Chap.9/; M. Creutz, L. Jacobs, C. Rebbi: Phys. Repts.95C, 203 (1983); B. Berg: In Proc. Cargèse lectures 1983, to be published
10.3 K. Binder, D. Stauffer: In /Ref.10.1, Chap.1/
10.4 K. Binder: Adv. Phys. 23, 1 (1974); and in Phase Transitions and Critical Phenomena 5b, 1 (Academic, New York 1976)
10.5 K. Binder: J. Comp. Phys. 59, 1 (1985)
10.6 K. Binder: Ferroelectrics, in press
10.7 V. Privman, M.E. Fisher: J. Stat. Phys. 33, 385 (1983)
10.8 K. Binder, D.P. Landau: Phys. Rev. B30, 1477 (1984)
10.9 K. Binder, M. Nauenberg, V. Privman, A.P. Young: Phys. Rev. B31, 1498 (1985)
10.10 E. Brezin, J. Zinn-Justin: Nucl. Phys. B257 /FS14/, 867 (1985)
J. Rudnick, H. Guo, D. Jasnow: J. Stat. Phys. 41, 353 (1985)
10.11 V. Privman, M.E. Fisher: Phys. Rev. B32, 447 (1985)
10.12 M.S.S. Challa, D.P. Landau, K. Binder: preprint
10.13 A. Nemirovsky, K.F. Freed: Phys. Rev. B31, 3161 (1985); J. Phys. A18, L319 (1985); and preprints; E. Eisenriegler: preprints
10.14 J. Rudnick, G. Gaspari, V. Privman: Phys.Rev. B32, 7594 (1985)
10.15 For general introductions to this field see: F. Family, D.P. Landau (eds.): Kinetics of Aggregation and Gelation (Elsevier, New York 1984); R. Pynn, T. Skjeltorg (eds.): Scaling Phenomena in Disordered Systems (Plenum, New York 1985);
R. Jullien, M. Kolb, H.Herrmann, J.Vannimenus: J.Stat.Phys.39,241 (1985)
10.16 K.E. Schmidt, M.H. Kalos: In /Ref.10.1,Chap.4/
10.17 M.H. Kalos (ed.): Monte Carlo Methods in Quantum Problems (Reidel, Dordrecht, Holland 1985)
10.18 H. De Raedt, A. Lagendijk: Phys. Repts. 127, 235 (1985)
10.19 N. Metropolis, A.W. Rosenbluth, M.N. Rosenbluth, A.H. Teller: J. Chem. Phys. 21, 1987 (1953)
10.20 H.E. Müser: Ferroelectrics 39 1099 (1981);
H.D. Maier, H.E. Müser, J. Peterson: Z. Phys. B46, 251 (1982)
10.21 P. Bak: Phys. Today 36, No.12, 25 (1983)
10.22 A. Hooghland, J. Spaa, B. Selman, A. Compagner: J. Comput. Phys. 51, 250 (1983)
10.23 R.B. Pearson, J.L. Richardson, D. Toussaint: J.Comput.Phys.51, 243 (1983)
10.24 J.H. Condon, A.T. Ogielski: Rev. Sci. Instr. 56, 1691 (1985)
10.25 N.H. Christ, A.E. Terrano: Columbia University preprint CU-TP-261
10.26 A.F. Bakker, C. Bruin, F. van Dieren, H.J. Hilhorst: Phys. Lett. A93, 67 (1982)
10.27 A.F. Bakker, C. Bruin, H.J. Hilhorst: Phys. Rev. Lett. 52, 449 (1984)

10.28 R. Zorn, H.J. Herrmann, C. Rebbi: Comp. Phys. Commun. 23, 337 (1981); L. Jacobs, C. Rebbi: J. Comput. Phys. 41, 203 (1981)
10.29 C. Kalle, V. Winkelmann: J. Stat. Phys. 28, 639 (1982)
10.30 K. Binder, D.P. Landau: Phys. Rev. Lett. 52, 318 (1984)
10.31 W. Selke, M.E. Fisher: Z. Phys. B40, 71 (1980); J. Magn. Mag. Mat. 15-18, 403 (1980)
10.32 W. Selke: Z. Phys. B43, 335 (1981)
10.33 H.J. Hilhorst: priv.commun.
10.34 M.E. Fisher: Rev. Mod. Phys. 46, 597 (1974); and in Critical Phenomena, ed. by M.S. Green (Academic, New York 1971) p.1
10.35 M.N. Barber, R.B. Pearson, J.L. Richardson, D. Toussaint: Phys. Rev.B32, 1720 (1985)
10.36 R.B. Pearson: Phys. Repts. 103, 185 (1985)
10.37 K. Binder, M. Nauenberg, V. Privman, A.P. Young:Phys.Rev.B31,1499 (1985)
10.38 A. Hoogland, A. Compagner, H.W.J. Blöte: Physica A132, 593 (1985)
10.39 R.C. Tausworthe: Math. Comp. 19, 291 (1965)
10.40 N. Zierler: Inform. Control. 15, 67 (1969)
10.41 J.P.R. Tootill, W.D. Robinson, A.G. Adams: J.Ass.Computing Machinery 18, 381 (1971)
10.42 S. Kirkpatrick, E.P. Stoll: J. Comput. Phys. 40, 517 (1981)
10.43 A.T. Ogielski, I. Morgenstern: Phys. Rev. Lett. 54, 928 (1985); see also A.T. Ogielski, D.A. Huse: Preprint, for recent work on dilute antiferromagnets in a field
10.44 A.T. Ogielski: Phys. Rev. B32, 7384 (1985)
10.45 R.N. Bhatt, A.P. Young: Phys. Rev. Lett. 54, 924 (1985)
10.46 K. Binder, K. Schröder: Phys. Rev. B14, 2142 (1976); Solid State Commun. 18, 1361 (1976)
10.47 K. Binder: J. Phys. 39, C6-1527 (1978)
10.48 K. Binder, D. Stauffer: In /Ref.10.1, Chap.8/ and Chapter 8 of the present book
10.49 K. Binder, W. Kinzel: In Heidelberg Colloquium on Spin Glasses, ed. by J.L. van Hemmen, I. Morgenstern (Springer, Berlin, Heidelberg 1983) p.279
10.50 K. Binder, A.P. Young: Rev. Mod. Phys., to be published
10.51 R. Kretschmer, K. Binder: Z. Phys. B34, 375 (1979)
10.52 R.W. Walstedt, L.R. Walker: Phys. Rev. Lett. 47, 1624 (1981)
10.53 D. Sherrington, S. Kirkpatrick: Phys. Rev. Lett. 35, 1792 (1975)
10.54 A.P. Young: Phys. Rev. Lett. 51, 1206 (1983); W. Kinzel: preprint
10.55 R.N. Bhatt, A.P. Young: preprint
10.56 A. Baumgärtner: J. Phys. Lett. (Paris) 45, L-515 (1984)
10.57 D. Barkai, M. Creutz, K.J. Moriarty: Nucl. Phys. B225, /FS 9/, 156 (1983)
see also D. Barkai, K.J.M. Moriarty: Comp. Phys. Commun. 25, 57 (1982)
10.58 A.D. Kennedy, J. Kuti, S. Meyer, B.J. Pendleton: Phys. Rev. Lett. 54, 87 (1985)
B. Berg, A. Billoire, S. Meyer, C. Panagiotakopoulos: Phys. Lett. B133, 359 (1983)
10.59 G.S. Pawley, R.H. Swendsen, D.J. Wallace, K.G. Wilson: Phys. Rev. B29, 4030 (1984)
10.60 S. Wansleben, J.G. Zabolitzky, C. Kalle: J. Stat. Phys. 37, 271 (1984)
10.61 C. Kalle: J. Phys. A17, L801 (1984)
10.62 M. Kikuchi, Y. Okabe: Progr. Theor. Phys. 74, 458 (1985)
10.63 S.F. Reddaway, D.M. Scott, K.A. Smith: Comp. Phys. Commun. 37, (1985)
10.64 W. Oed: Appl. Informatics 7, 358 (1982)
10.65 M. Creutz: Phys. Rev. Lett. 50, 1411 (1983)
10.66 K.A. Smith, S.F. Reddaway, D.M. Scott: Comp. Phys. Commun. 37, (1985)
10.67 J.C. Le Guillou, J. Zinn-Justin: Phys. Rev. B21, 3976 (1980)
10.68 H.W.J. Blöte, R.H. Swendsen: Phys. Rev. B20, 2077 (1979); P.S. Sahni, J.R. Banavar: Phys. Lett. A85, 56 (1981)

10.69 K. Binder: Phys. Rev. Lett. 47, 693 (1981); Z. Phys. B43, 119 (1981)
10.70 P.C. Hohenberg, B.I. Halperin: Rev. Mod. Phys. 49, 435 (1977)
10.71 H. Müller-Krumbhaar, K. Binder: J. Stat. Phys. 8, 1 (1973)
10.72 E. Stoll, K. Binder, T. Schneider: Phys. Rev. B8, 3266 (1973)
10.73 B.K. Chakrabarty, H.G. Baumgaertel, D. Stauffer: Z. Phys. B44,333 (1981)
10.74 J. Tobochnik, S. Sarker, R. Cordery: Phys. Rev. Lett. 46, 1417 (1981); M.C. Yalabik, J.D. Gunton: Phys. Rev. B25, 534 (1982); N. Jan, L.L. Moseley, D. Stauffer: J. Stat. Phys. 33, 1 (1983)
10.75 K. Binder: J. Stat. Phys. 24, 69 (1981); J. Tobochnik, A. Jayaprakash: Phys. Rev. B25, 4891 (1972); M. Aydin, M.C. Yalabik: J. Phys. A17, 2531 (1984)
10.76 A. Sadiq, K. Binder: J. Stat. Phys. 35, 517 (1984)
10.77 E. Brezin, J.C. Le Guillou, J. Zinn-Justin: Phys. Lett. A47, 285 (1974)
10.78 M. Kikuchi, Y. Okabe: Phys. Rev. Lett. 55, 1220 (1985)
10.79 K. Binder: Phys. Rev. A25, 1699 (1982)
10.80 W. Schweika: Dissertation, RW TH Aachen (1985) unpublished
10.81 M. Suzuki: Progr. Theor. Phys. 58, 1142 (1977)
10.82 M.N. Barber: In Phase Transitions and Critical Phenomena, Vol.8, ed. by C. Domb, J.L. Lebowitz (Academic, London 1983) Chap. 2
10.83 D.P. Landau: Phys. Rev. B13, 2997 (1976); B14, 255 (1976)
10.84 K. Binder, D.P. Landau: Phys. Rev. B21, 1941 (1980)
10.85 D.P. Landau: Phys. Rev. B21, 1285 (1980)
10.86 S. Miyashita, H. Takano: Progr. Theor. Phys. 73, 1122 (1985)
10.87 D.P. Landau, K. Binder: Phys. Rev. B31, 5946 (1985)
10.88 A. Milchev, D.W. Heermann, K. Binder: preprint
10.89 M.N. Barber, W. Selke: J.Phys. A15, L617 (1982)
10.90 R.H. Swendsen, S. Krinsky: Phys. Rev. Lett. 43, 177 (1979)
10.91 J. Oitmaa: J. Phys. A14, 1159 (1981)
10.92 M.P, Nightingale: Phys. Lett. A59, 486 (1977)
10.93 M. Nauenberg, B. Nienhuis: Phys. Rev.Lett. 33, 944 (1974)
10.94 K. Binder, W. Kinzel, D.P. Landau: Surf. Sci. 117, 232 (1982)
10.95 D.P. Landau: J.Magn. Magn. Mat. 31-34, 11 (1983)
10.96 D.P. Landau: J. Appl. Phys. 42, 1284 (1971)
10.97 J.M. Kosterlitz, D.J. Thouless: J. Phys. C6, 1181 (1973)
10.98 C.S.S. Murty, D.P. Landau: preprint
10.99 C.S.S. Murty, D.P. Landau: J. Appl. Phys. 55, 2429 (1984)
10.100 K. Binder: Z. Phys. B61, 13 (1985)
10.101 Y. Imry: Phys. Rev. B21, 2042 (1980)
10.102 S.-K. Ma: Phys. Rev. Lett. 37, 461 (1976)
10.103 R.H. Swendsen: Phys. Rev. Lett. 42, 859 (1979)
10.104 R.H. Swendsen: Phys. Rev. Lett. 52, 1165 (1984)
10.105 R.H. Swendsen: In Phase Transitions, Cargèse 1980, ed. by M. Levy,J.C. Le Guillou, J. Zinn-Justin (Plenum, New York 1982) p.395
10.106 R.H. Swendsen: In Real-Space Renormalization, ed. by Th.W. Burkhardt, J.M.J. van Leeuwen, Topics Current Phys.,Vol.30 (Springer, Berlin, Heidelberg 1982) p.57
10.107 R.H. Swendsen: J. Stat. Phys. 34, 963 (1984)
10.108 Th. Niemeijer, J.M.J. van Leeuwen: In Phase Transitions and Critical Phenomena, Vol.6, ed. by C. Domb, M.S. Green (Academic, New York 1976) T.W. Burkhardt, J.M.J. Van Leeuwen (eds.): Real Space Renormalization, Topics Current Phys. Vol.30 (Springer, Berlin, Heidelberg 1982)
10.109 D.P. Landau, R.H. Swendsen: Phys. Rev. Lett. 46, 1437 (1981); see also K.G. Wilson: In Physics of Defects, ed.by R. Balian, M.Klēman, J.-P. Poirier (North-Holland, Amsterdam 1982)
10.110 M. Creutz: Phys. Rev. D23, 1815 (1981); see also S. Shenker, J. Tobochnik: Phys. Rev. B22, 4462 (1980) for a related approach for the case of the d=2 Heisenberg ferromagnet
10.111 J. Tobochnik, S. Sarker, R. Cordery: Phys. Rev. Lett. 46, 1417 (1981); J. Tobochnik, C. Jayaprakash: Phys. Rev. B25, 4893 (1982); M. Aydin, M.C. Yalabik: J. Phys. A17, 667, 2531 (1984)
10.112 J. Vinals, M. Grant, M. San Miguel, J.D. Gunton, E.T. Gawlinski: Phys. Rev. Lett. 54, 1264 (1985); J. Vinals, J.D. Gunton: preprint

10.113 M. Creutz, L.Jacobs, C. Rebbi: Phys. Rev. D20, 1915 (1979)
10.114 K. Wilson: Phys. Rev. D10, 2445 (1972)
10.115 B.Berg, S. Meyer, I. Montvay, K. Symanzik: Phys. Rev. Lett. B126, 467 (1983)
10.116 C.B. Lang, C. Rebbi, P. Salomonson, B.S. Shargerstam: Phys. Rev. D21, 1006 (1980)
10.117 B.B. Mandelbrot: Fractals: Form, Chance and Dimension (Freeman, San Francisco 1977)
10.118 L. Pietronero, E. Tosatti (eds.): Fractals in Physics, to be published
10.119 T.A. Witten, Jr., L.M. Sander: Phys. Rev. Lett. 47, 1400 (1981); Phys. Rev. B27, 5686 (1983);
P.A. Meakin: Phys. Rev. B24, 604, 1495 (1983)
10.120 I. Majid, N. Jan, A. Coniglio, H.E. Stanley: In Kinetics of Aggregation and Gelation, ed. by F. Family, D.P. Landay (North Holland, Amsterdam 1984) p.51;
J.W. Lyklema, K. Kremer: ibid. p.241;
I. Majid, N. Jan, A. Coniglio, H.E. Stanley: Phys. Rev. Lett. 52, 1257 (1984);
J.W. Lyklema, K. Kremer: J. Phys. A17, L 691 (1984);
S. Hemmer, P.C. Hemmer: J. Chem. Phys. 81, 584 (1984)
10.121 L. Peliti: J. Phys. (Paris) Lett. 45, 45 (1984);
K. Kremer, J.W. Lyklema: Phys. Rev. Lett. 55, 2091 (1985);
J.W. Lyklema, K. Kremer: J. Phys. A, in print;
J.W. Lyklema, K. Kremer: Phys. Rev. B31, 3182 (1985)
10.122 K. Kremer, J.W. Lyklema: Phys. Rev. Lett. 54, 267 (1985); J. Phys. A18, 1515 (1985)
10.123 J.W. Lyklema: J. Phys. A18 L617 (1985)
10.124 J.W. Lyklema, C. Evertsz: In /10.118/
10.125 J.W. Lyklema: In /10.118/
L. Peliti, L. Pietronero: La Rivista del nuovo Cimento, in press
10.126 K.W. Kehr, K. Binder: In /Ref. 10.1, Chap.6/
10.127 G.E. Murch: Atomic Diffusion Theory in Highly Defective Solids (Trans Tech. House, Adermannsdorf 1980)
10.128 A. Baumgärtner: In /Ref.10.1, Chap.5/
10.129 A. Baumgärtner: Ann. Rev. Phys. Chem. 35, 419 (1984)
10.130 P.G. de Gennes: Scaling Concepts in Polymer Physics (Cornell U. Press, Ithaca, NY 1979)
10.131 J.W. Halley, H. Nakanishi: Phys. Rev. Lett. 55, 551 (1985)
10.132 M. Muthukumar: Phys. Rev. Lett. 50, 893 (1983);
T. Witten, M. Nauenberg: unpublished
10.133 H. Gould, F. Family, H.E. Stanley: Phys. Rev. Lett. 50, 686 (1983);
M. Tokuyama, K. Kawasaki: Phys. Lett. A100, 337 (1984);
H.G.E. Hentschel: Phys. Rev. Lett. 52, 212 (1984)
10.134 D.A. Weitz, M. Oliveria: Phys. Rev. Lett. 52, 1433 (1984)
10.135 L.A. Turkevich, H. Scher: Phys. Rev. Lett. 55, 1026 (1985)
10.136 M. Kolb, R. Botet, R. Jullien: Phys. Rev. Lett. 51, 1123 (1983);
R. Botet, R. Jullien, M. Kolb: J. Phys. A17, 175 (1985)
10.137 A. Sadiq, K. Binder: J. Stat. Phys. 35, 617 (1984)
10.138 K. Binder, D.W. Heermann: In Scaling Phenomena in Disordered Systems, ed. by R. Pynn, T. Skjeltorp (Plenum, New York 1985)
10.139 D.C. Handscomb: Proc. Camb. Phil. Soc. 58, 594 (1962); Proc. Camb. Phil. Soc. 60, 115 (1964)
10.140 J.W. Lyklema: Phys. Rev. B27, 3108 (1983)
10.141 J.W. Lyklema: Phys. Rev. Lett. 49, 88 (1982)
10.142 S. Chakravarty, D.B. Stein: Phys. Rev. Lett. 49, 582 (1982)
10.143 H.F. Trotter: Proc. Am. Math. Soc. 10, 545 (1959)
10.144 M. Suzuki: Progr. Theor. Phys. 56, 1454 (1976); Commun. Math. Phys. 51, 183 (1976)
10.145 M. Suzuki, S. Miyashita, A. Kuroda: Progr. Theor. Phys. 58, 1377 (1977)
10.146 J.J. Cullen, D.P. Landau: Phys. Rev. B27, 297 (1983)
10.147 M. Barma, B.S. Shastry: Phys. Rev. B18, 3351 (1978)

10.148 H. De Raedt, B. De Raedt: Phys. Rev. A28, 3575 (1983);
H. De Raedt, A. Lagendjk: Phys. Rev. Lett. 46, 77 (1981); J.Stat.Phys. 27, 731 (1982); Phys. Rev. B24, 463 (1981); Phys. Rev. Lett. 49, 1552 (1982); Phys. Rev. B27, 921, 6097 (1983); B30, 1671 (1984);
H. De Raedt, B. De Raedt, A. Lagendijk: Z. Phys. B57, 209 (1984)
10.149 B. De Raedt, H. De Raedt: Phys. Rev. Lett. 50, 1926 (1983); Phys. Rev. B29, 5352 (1985)
10.150 R.P. Feynman, A.R. Hibbs: Quantum Mechanics and Path Integrals (McGraw-Hill, New York 1976)
10.151 L. Jacobs, J.V. José, M.A. Novotny: Phys. Rev. Lett. 53, 2177 (1984)
10.152 J.E. Hirsch, D.J. Scalapino, R.L. Sugar, R. Blancenbecler: Phys.Rev. Lett. 47, 1628 (1981); Phys. Rev. B26, 5033 (1982);
see also J.A. Barker: J. Chem. Phys. 70, 2914 (1979);
D. Chandler, P.G. Wolynes: J. Chem. Phys. 74, 4078 (1981);
R.M. Stratt: J. Chem. Phys. 75, 1347 (1982)
10.153 For a review, see J.E. Hirsch: In /10.17/
10.154 M. Takahashi, M. Imada: J. Phys. Soc. Jpn 53, 963 (1984); J. Phys. Soc. Jpn 53, 3765, 3871 (1984)
10.155 J.E. Hirsch, E. Fradkin: Phys. Rev. Lett. 49, 402 (1982); Phys. Rev. B27, 1680, 4302 (1983)
10.156 J.E. Hirsch: Phys. Rev. B28, 4059 (1983);
D.J. Scalapino, R.L. Sugar, W.D. Toussaint: Phys. Rev. B29, 5253 (1984);
J.E. Hirsch: Phys. Rev. Lett. 51, 296 (1983);
J.E. Hirsch, D.J. Scalapino: Phys. Rev. Lett. 50, 1168 (1983); Phys. Rev. B27, 7169 (1983)
10.157 J.E. Hirsch, M. Grabowski: Phys. Rev. Lett. 52, 1713 (1984);
D.K. Campbell, T.A. De Grand, S. Mazumdar: Phys. Rev. Lett. 52, 1717 (1984)
10.158 J. Kuti: Phys. Rev. Lett. 49, 183 (1982)
10.159 D.L. Freeman, J.D. Doll: J. Chem. Phys. 80, 2239, 5709 (1984);
J.D. Doll, R.D. Coalson, D.L. Freeman: Phys. Rev. Lett. 55, 1 (1985);
J.D. Doll, L.E. Myers: J. Chem. Phys. 71, 2880 (1979)
10.160 D. Thirumalai, B.J. Berne: J. Chem. Phys. 79, 5029 (1983);
E.C. Behrman, G.A. Jongeward, P.G. Wolynes: J. Chem. Phys.79, 6227 (1983)
10.161 J.D. Doll: J. Chem. Phys. 81, 3536 (1984)
10.162 H.B. Schüttler, D.J. Scalapino: Phys. Rev. Lett. 55, 1204 (1985);
J.E. Hirsch, J.R. Schrieffer: Phys. Rev. B28, 5353 (1983)
10.163 A. Baumgärtner: In /Ref.10.1, Chap.5/; Ann. Rev. Phys. Chem. 35, 419 (1984); J. Phys. A17, L971 (1984); J. Chem. Phys. 81, 1484 (1984)
10.164 K. Binder: Phys. Rev. Lett. 45, 811 (1980); Z. Phys. B45, 61 (1981);
K. Binder, J.L. Lebowitz, M.K. Phani, M.H. Kalos: Acta Metall. 29, 1655 (1981);
U. Gahn: J. Phys. Chem. Solids 43, 977 (1982);
R.A. Bond, D.K. Ross: J. Phys. F12, 597 (1982);
J.L. Lebowitz, M.K. Phani, D. Styer: J. Stat. Phys. 38, 413 (1985);
K. Binder: preprint
10.165 V. Gerold, J. Kern: preprint
10.166 T.L. Polgreen: Phys. Rev. B29, 1468 (1984);
H. Meirovitch: J. Stat. Phys. 30, 681 (1983); J. Phys. A16, 839 (1983); Phys. Rev. B30, 2866 (1984)
10.167 A.P. Young, M. Nauenberg: Phys. Rev. Lett. 54, 2429 (1985)
10.168 D.W. Heermann: Introduction to Computer Simulation Methods in Theoretical Physics (Springer, Berlin, Heidelberg 1986)
10.169 D.P. Landau, K. Binder: An Introduction to Monte Carlo Simulation Methods in Physics, in preparation
10.170 W. Smith (ed.): Information Quarterly for Computer Simulations of Condensed Phases (Sci. Eng. Res. Council, Daresbury Laboratory, Daresbury, Warrington, WA4 4AD, England; 1st issue: 1981)
10.171 T. Filk, M. Marcu, K. Fredenhagen: DESY preprint 85-098
10.172 S. Wolfram: Phys. Rev. Lett. 55, 449 (1985)

Addendum

Chapter 2

We list, according to the sections of the article, recent references on Monte Carlo computations of classical fluids. References to molecular dynamics simulations are given whenever they are complementary.

Methodological Aspects:

J.C. Owicki, H.A. Scheraga: J. Phys. Chem. *82*, 1257 (1978)
C. Pangoli, M. Rao, B.J. Berne: Chem. Phys. Lett. *55*, 413 (1978)
P.J. Rossky, J.D. Doll, H.L. Friedman: J. Chem. Phys. *69*, 4628 (1978)

2.2.2 Hard Spheres Plus Yukawa Potential:

D. Henderson, E. Waisman, J.L. Lebowitz, L. Blum: Mol. Phys. *35*, 241 (1978)

2.3.2 Lennard-Jones Mixtures:

K. Toukubo, K. Nakanishi: J. Chem. Phys. *65*, 1937 (1976); *67*, 4162 (1977)
K. Nakanishi, K. Toukubo, N. Watanabe: J. Chem. Phys. *68*, 2041 (1978)

2.3.3 Two-dimensional Lennard-Jones System:

D. Henderson: Mol. Phys. *34*, 301 (1977)
Two-dimensional Lennard-Jones system with a three-body interaction:
W. Schommers: Phys. Rev. A *16*, 327 (1977)

2.4.2 Two-dimensional OCP:

R.W. Hockney, T.R. Brown: J. Phys. C *8*, 1813 (1975)
H. Totsuji: Phys. Rev. A *17*, 399 (1978)
R.C. Gann, S. Chakravarty, G.V. Chester: Phys. Rev. B *20*, 329 (1979)

2.4.3 Restrictive Primitive Model:

B. Larsen, S.A. Rodge: J. Chem. Phys. *68*, 1309 (1978)
B. Larsen, S.A. Rodge: J. Chem. Phys. *72*, 2578 (1980)

2.4.4 Molten Potassium Chloride:

D.J. Adams: J. Chem. Soc. Faraday Trans. 2 *72*, 1372 (1976)

Molten Caesium Halides:

M. Dixon, M.J.L. Sangster: J. Phys. C *10*, 3015 (1977)

Liquid Potassium Cyanide:

S. Miller, J.H.R. Clarke: J. Chem. Soc. Faraday Trans. 2 *74*, 160 (1978)

Alkaline Earth Halides:

S.W. de Leeuw: Mol. Phys. *36*, 103 (1978)

Molten Salt Mixtures:

M.L. Saboungi, A. Rahman: J. Chem. Phys. *65*, 2393 (1976)
M.L. Saboungi, A. Rahman: J. Chem. Phys. *66*, 2773 (1977)
D.J. Adams, I.R. McDonald: Mol. Phys. *34*, 287 (1977)

2.4.5 Li:

G. Jacucci, M.L. Klein, R. Taylor: Solid State Commun. *19*, 657 (1976)

Al:

E. Michler, H. Hahn, P. Schofield: J. Phys. F *6*, 319 (1976)

Na, Cs:

T. Lee, J. Bisshop, W. Van der Lugt, W.F. Van Gunsteren: Physica B *93*, 59 (1978)

Rb:

R.D. Mountain: J. Phys. F *8*, 1637 (1978)

Na-K Alloy:

G. Jacucci, I.R. McDonald, R. Taylor: J. Phys. F *8*, L121 (1978)

2.5.1 Hard Spherocylinders:

D.W. Rebertus, K.M. Sando: J. Chem. Phys. *67*, 2585 (1977)
I. Nezbeda, T. Boublik: Czech. J. Phys. B *27*, 953 (1977)
I. Nezbeda, T. Boublik: Czech. J. Phys. B *28*, 353 (1978)
P.A. Monson, M. Rigby: Mol. Phys. *35*, 1337 (1978)
P.A. Monson, M. Rigby: Chem. Phys. Lett. *58*, 122 (1978)

2.5.2 Diatomics

I. Aviram, D.J. Tildesley, W.B. Streett: Mol. Phys. *34*, 881 (1977)
W.B. Streett, D.J. Tildesley: J. Chem. Phys. *68*, 1275 (1978)
R. Agrawal, S.I. Sandler, A.H. Narten: Mol. Phys. *35*, 1087 (1978)

Mixtures of Hard Diatomics:

I. Aviram, D.J. Tildesley: Mol. Phys. *35*, 365 (1978)

Triatomics:

W.B. Streett, D.J. Tildesley: Faraday Discuss. Chem. Soc. *66* (1978)

2.5.3 Dipolar Hard Spheres and Stockmayer Potential:

M. Neumann, F.J. Vesely, O. Steinhauser, P. Schuster: Mol. Phys. *35*, 841 (1978)
A.J.C. Ladd: Mol. Phys. *36*, 463 (1978)
G.N. Patey, D. Levesque, J.J. Weis: Mol. Phys. *38*, 219 (1979)

Dipole in Polarizable Fluid:

E.L. Pollock, B.J. Alder: Phys. Rev. Lett. *39*, 299 (1977)

Polarizable Dipoles:

F.J. Vesely: Chem. Phys. Lett. *56*, 390 (1978)
G.N. Patey, J.P. Valleau: Chem. Phys. Lett. *58*, 157 (1978)

H_2O:

J.C. Owicki, H.A. Scheraga: J. Am. Chem. Soc. *99*, 7403 (1977)
S. Swaminathan, D.L. Beveridge: J. Am. Chem. Soc. *99*, 8392 (1977)
M. Mezei, S. Swaminathan, D.L. Beveridge: J. Am. Chem. Soc. *100*, 3255 (1978)
F.H. Stillinger, A. Rahman: J. Chem. Phys. *68*, 666 (1978)
I.R. McDonald, M.L. Klein: J. Chem. Phys. *68*, 4875 (1978)
W.F. Van Gunsteren, H.J.C. Berendsen, J.A.C. Rullmann: Faraday Discuss. Chem. Soc. *66* (1978)

HF:

M.L. Klein, I.R. McDonald, S.F. O'Shea: J. Chem. Phys. *69*, 63 (1978)
W.L. Jörgensen: J. Am. Chem. Soc. *100* , 7824 (1978)

2.6 Solid-Liquid Interface:

S. Toxvaerd, E. Praestgaard: J. Chem. Phys. *67*, 5291 (1977)
Y. Hiwatari, E. Stoll, T. Schneider: J. Chem. Phys. *68*, 3401 (1978)

Hard Spheres Near a Wall:

I.K. Snook, D. Henderson: J. Chem. Phys. *68*, 2134 (1978)

Lennard-Jones Particles Near a Wall:

F.F. Abraham: J. Chem. Phys. *68*, 3713 (1978)
F.F. Abraham, Y. Singh: J. Chem. Phys. *68*, 4767 (1978)

Gas-Liquid-Solid Coexistence:

A.J.C. Ladd, L.V. Woodcock: Mol. Phys. *36*, 611 (1978)

Liquid-gas interface for molecular systems:

M. Thompson: Faraday Discuss. Chem. Soc. *66* (1978)

Chapter 4

Additional work on optimizing of the energy of a helium system with respect to the pseudopotential has been cariied out by LANTTO, JACKSON, and SIEMENS [4.112] and by CAMPBELL and PINSKI [4.113] who used a Monte Carlo method to generate corrections.

REATTO and collaborators [4.114,115] have been successful in improving the energy and structure of hard-sphere and of Yukawa systems calculated with product trial functions by consideration of the effect of elementary excitations in fluids upon the form of the pseudopotential.

PANDHARIPANDE [4.116] has shed additional light on the discrepancy between the energy of ^{4}He calculated by product potentials and that using GFMC. Using a trial function containing three-body correlations suggested from considerations of back-flow and using HNC, he obtained an energy of -6.72 ± 0.2 K. In related work using trial functions with three-body correlations and integral equations for ^{3}He, SCHMIDT and PANDHARIPANDE [4.117] obtained an equilibrium energy of -2.25 K, a substantial improvement over that obtained with no such correlations.

Further work on liquid helium [4.98] has shown that the "phonon" size correction first obtained in [4.28] is in fact very small numerically.

4.112 L.J. Lantto, A.D. Jackson, P.J. Siemens: Phys. Lett. B *68*, 311 (1977)
4.113 C.E. Campbell, F.J. Pinski: Phys. Lett. B *79*, 23 (1978)
4.114 C. De Michelis, G. Masserini, L. Reatto: Phys. Rev. A *17*, 296 (1978)
4.115 L. Reatto, A. Scotti: Phys. Rev. D *19*, 2304 (1979)
4.116 V.R. Pandharipande: Phys. Rev. B *18*, 218 (1978)
4.117 K.E. Schmidt, V.R. Pandharipande: Phys. Rev.B *19*, 4562 (1979)

Chapter 6

Recent Monte Carlo work concerned the kinetics of first order phase transitions. KAWABATA and KAWASAKI [6.100] studied the time evolution of the structure factor of the two-dimensional single spin-flip kinetic Ising model, performing quenches from temperatures above T_c to $T = 0.2\ T_c$. The results seem to be consistent with a $t^{-1/2}$ decay for the halfwidth of the structure factor. KAWASAKI [6.101] performed a study of spinodal decomposition in a magnetic binary alloy, investigating the influence of the magnetic interaction on the time evolution of the structure factor.

The work on spinodal decomposition described in this chapter has been considerably extended [6.102-105]. KALOS et al. [6.102] performed temperature quenches to states with various concentrations inside but close to the coexistence line of a three-dimensional model system. They found quite well-defined metastable states very close to the coexistence line and established a rather gradual transition from nucleation-like behavior to spinodal decomposition as one moves away from the coexistence line. PENROSE et al. [6.103] showed that the late-time behavior in that concentration regime nicely confirms the LIFSHITZ-SLYOZOV [6.93] theory. On the other hand, data at higher concentrations (MARRO et al. [6.104]) can not be interpreted in terms of unique exponents. The findings on two-dimensional model systems described in this chapter have been extended and confirmed by PUNDARIKA et al. [6.105].

Progress has also been made on several aspects related to this work. An interpretation of the time evolution of the structure factor in terms of the dynamics of clusters was given by BINDER et al. [6.106]. The power law for the cluster diffusivity shown in Fig.6.4 has been explained [6.54] in terms of a crossover between several competing mechanisms.

6.100 C. Kawabata, K. Kawasaki: Phys. Lett A *65*, 137 (1978)
6.101 T. Kawasaki: Progr. Theor. Phys. *59*, 1812 (1978)
6.102 M.H. Kalos, J.L. Lebowitz, O. Penrose, A. Sur: J. Statist. Phys. *18*, 39 (1978)
6.103 O. Penrose, J.L. Lebowitz, J. Marro, M.H. Kalos, A. Sur: J. Statist. Phys. *19*, 243 (1978)
6.104 J. Marro, J.L. Lebowitz, M.H. Kalos: Phys. Rev. Lett. *43*, 282 (1979)
6.105 E.S. Pundarika, K. Hanson, J.W. Morris: Trans. Am. Nucl. Soc. (USA) *27*, 321 (1977)
6.106 K. Binder, C. Billotet, P. Mirold: Z. Phys. B *30*, 183 (1978)

Chapter 8

Another recent review on disordered materials, including Monte Carlo work, was given by KIRKPATRICK [8.152], while percolation clusters are reviewed by STAUFFER [8.153] and spin glasses by BINDER [8.154]. Recent Monte Carlo investigations of scaling properties of percolation are contained in [8.155-8]. HOSHEN et al. [8.159] study the dependence of percolation on the interaction range, and KLEIN et al. [8.160] allow nonrandom occupancy. KINZEL [8.161] studied the remanent magnetization of spin glasses. CHING and HUBER [8.162] and FREUDENHAMMER [8.163] study the thermodynamics of RKKY spin glasses, but using systems of typical size of only N = 96 spins. KIRKPATRICK and SHERRINGTON [8.164] study the infinite-range spin glass and obtain good agreement with the theory of THOULESS et al. [8.165]. RAPAPORT [8.166] studied Ising spin glasses with both symmetric and nonsymmetric distributions of exchange constants, but some of his conclusions may be affected by observation time effects. TATSUMI [8.167] studies random magnetic Ising mixtures in three dimensions, and obtains the magnetic phase diagram. The latter has been studied for the case of a random mixture of two anisotropic magnets with different easy axes by INAWASHIRO et al. [8.168], where a tetracritical point occurs.

8.152 S. Kirkpatrick: in *Ill-condensed Matter* ed.by R. Balian, R. Maynard, and G. Toulouse (North-Holland, Amsterdam 1979)
8.153 D. Stauffer: Phys. Rep. *54*, 1 (1979)
8.154 K. Binder: J. Phys. (Paris) *39*, C6-1527 (1978)
8.155 H. Nakanishi, H.E. Stanley: J. Phys. A *11*, 189 (1978)
8.156 P.J. Reynolds, H.E. Stanley, W. Klein: J. Phys. A *11*, 199 (1978)
8.157 D. Stauffer: Phys. Rev. Lett. *41*, 1333 (1978)
H.J. Herrmann: Z. Phys. B *32*, 335 (1979)
8.158 Y. Yuge, K. Onizuka: J. Phys. C *11*, L763 (1978)
8.159 J. Hoshen, R. Kopelman, E.M. Monberg: J. Stat. Phys. *19*, 219 (1978)
8.160 W. Klein, H.E. Stanley, P.J. Reynolds, A. Coniglio: Phys. Rev. Lett. *41*, 1145 (1978)
8.161 W. Kinzel: J. Phys. (Paris) *39C-6*, 905 (1978)
8.162 W.Y. Ching, D.L. Huber: J. Phys. F *8*, L63 (1978)
8.163 A. Freudenhammer: J. Magn. Mag. Mater. *9*, 46 (1978)
8.164 S. Kirkpatrick, D. Sherrington: Phys. Rev. B *18*, 4384 (1978)
8.165 D.J. Thouless, P.W. Anderson, R.G. Palmer: Phil. Mag. *35*, 593 (1977)
8.166 D.C. Rapaport: J. Phys. C *11*, L111 (1978)
8.167 T. Tatsumi: Progr. Theor. Phys. *59*, 1428, 1437 (1978); *57*, 1799 (1977)
8.168 S. Inawashiro, T. Togawa, R. Kurosaka: J. Phys. *C 12*, 2351 (1979)

Subject Index